Neurochemistry
Cellular, Molecular, and Clinical Aspects

Neurochemistry
Cellular, Molecular, and Clinical Aspects

Edited by

Albert Teelken and
Jaap Korf
University Hospital
Groningen, The Netherlands

Assistant Editors

Jan Albrecht
Stefan Bröer
Piet Eikelenboom
Antonio Guiditta
Jerzy W. Łazarewicz
Konrad Löffelholz
Rona R. Ramsay
Rupert Schmidt
Keith Tipton

Springer Science+Business Media, LLC

Library of Congress Cataloging-in-Publication Data

European Society for Neurochemistry. Meeting (11th : 1996 : Groningen, Netherlands)
 Neurochemistry : cellular, molecular, and clinical aspects / edited by Albert Teelken and Jaap Korf ; assistant editors, Jan Albrecht ... [et al.]
     p. cm.
 "Proceedings of the 11th European Society for Neurochemistry Meeting, held June 15-20, 1996, in Groningen, The Netherlands."
 Includes bibliographical references and index.
 ISBN 978-1-4613-7468-8   ISBN 978-1-4615-5405-9 (eBook)
 DOI 10.1007/978-1-4615-5405-9
 1. Nervous system--Pathophysiology--Congresses.
 2. Neurochemistry--Congresses. 3. Nervous system--Degeneration--Congresses.   I. Teelken, Albert. II. Korf, Jaap. III. Title.
 RC347.E88 1996
 616.8'047--dc21   97-36254
                                                                 CIP

Proceedings of the 11th European Society for Neurochemistry Meeting,
held June 15–20, 1996, in Groningen, The Netherlands

ISBN 978-1-4613-7468-8

© 1997 Springer Science+Business Media New York
Originally published by Plenum Press, New York in 1997
Softcover reprint of the hardcover 1st edition 1997

http://www.plenum.com

10 9 8 7 6 5 4 3 2 1

All rights reserved

No part of this book may be reproduced, stored in a retrieval system, or transmitted in any form or by any means, electronic, mechanical, photocopying, microfilming, recording, or otherwise, without written permission from the Publisher

FOREWORD

20 Years of the European Society for Neurochemistry

The first meeting of the European Society for Neurochemistry (ESN), held in Bath, England, in September 1976, recognized the need at that time to bring together both basic and clinical neurochemists from all European countries during a period of considerable political uncertainty throughout the continent. The setting of that meeting, an informal atmosphere of some 350 delegates in a small university city, set the tone for many of the future biennial meetings, which tended to alternate between Western and Eastern Europe. The development of the society since that time is excellently reviewed in Chapter 1 in this volume by Herman Bachelard, a founding member of the society and host at that first meeting. I attended that first meeting very much as a junior neurochemist and was enthused by the science and determined to make my way in the discipline. However, my next appearance at an ESN meeting had to wait until the Budapest meeting in 1984, and since that time, I have been a regular contributor. I was fortunate to be elected to the council in 1985 and took over the role of secretary from George Lunt at the Goteborg meeting in 1988, assuming the position of ESN president in 1992. During my time as an officer for the society, volunteers for meeting organization were becoming increasingly thin on the ground as funding was becoming increasingly difficult to obtain and the work-load became more onerous! Planning was well under way for the 1994 Jerusalem ESN meeting by 1992, but a venue for 1996 was sought. Fortunately Albert Teelken and colleagues suggested their hometown, Groningen, as a possible venue, and a visit there revealed it to be a city with many of the traditional values we associated with ESN meetings. The date was set for June 1996 and the Martinihal Congress center was chosen as the best site for the meeting and all of its requirements. The local organizing committee and the program committee set about their tasks with enthusiasm, and the program committee met in Jerusalem in 1994 and at the International Society for Neurochemistry (ISN) meeting in Kyoto (July 1995) to finalize the content. The aim was to have a highly topical congress with emphases on both basic and clinical aspects and to have a program designed to attract young neurochemists. I hope that we achieved a lot of those ideals. The contents of this volume, reflecting many of the topics presented, should give the reader a flavor of the scientific content, if not the social success, of the meeting. A special element of the Groningen meeting was a reception to celebrate of the 20th anniversary of the society, attended by many of those who had contributed much to the development of the society over two decades.

In the scientific program, stimulating plenary lectures by Herman Bachelard, Elisabeth Bock, and Hans Stoof covered diverse and topical areas of neurochemistry, and the quality of the young scientists at the meeting was reflected in the choice of as many as three neurochemists as honorary young lecturers. In all, there were 432 registered scientists from 31 countries at the meeting, comparable to those attending the meeting in Bath and reflecting a constant and continuing interest in the subject. Among the important topics covered in the many symposia, workshops, and round tables and presented in 195 contributions in the proceedings were Alzheimer's and other neurodegenerative diseases, ischemia, neurotrophins, multiple sclerosis, mood disorders, and epilepsy. There was also a focus on new techniques and their applications in neurochemistry. There was a wonderful atmosphere at the meeting and enthusiastic participation by the delegates in the true traditions of ESN, and much credit must go to Albert Teelken and colleagues for putting together a gathering that was enjoyed by all who attended. The meeting was a truly international one with attendees from all parts of Western and Eastern Europe as well as from several other countries including North America.

ESN now moves into its third decade, and we all look forward to our next meeting, to be held in St. Petersburg, Russia, in 1998. This also promises to be an exciting event in a rapidly changing city and country with great scientific and cultural traditions. We must hope that the society will continue to thrive, to foster links between all parts of Europe, and, most of all, to welcome and encourage the new generation of neurochemists. In recent years ESN has also fostered closer links with other European neuroscience societies and has participated in meetings of the European Neuroscience Association in Amsterdam (September 1995) and Strasbourg (September 1996); we hope that these links continue and strengthen as neuroscience becomes an increasingly interdisciplinary subject area. We must also not forget our international links and the support that ISN has provided to ESN over the years. As a fitting end to the millennium, there will be the first joint meeting of ESN and ISN, to be held in Berlin in 1999, at the center of Europe. We hope to see you all there. Meantime, this volume should provide a topical and diverse guide to neurochemical advances and serve as a reminder of the wonderful time spent reflecting on neurochemistry at the 11th meeting of ESN in June 1996. Finally, I wish my successor as President, Pam Fredman, every success as she leads the society into the 21st century.

Tony Turner
ESN President 1992–1996
Leeds, March 1997

PREFACE

With the publication of the proceedings of the 11th ESN meeting a tradition of the earlier meetings of the European Society for Neurochemistry is reestablished. Including two contributions on the history of neurochemistry, 195 manuscripts have been accepted for publication.

The contributions come from nearly all countries in Europe, east and west. About 60% are from Western Europe (116), while 31% are from Eastern Europe (61) and 9% from other parts of the world. From Western Europe most of the papers came from Germany (27); second, The Netherlands (21); third, United Kingdom (17); and fourth, Italy (14). The other countries are France (8), Sweden (6), Denmark (5), Spain (4), Ireland, Finland and Norway (each 3), Belgium and Switzerland (each 2), and Austria (1). Most of the papers from Eastern Europe are offered by Poland (18); second is Russia (17), and third are equally Armenia, Hungary, and Slovakia (each 5). The other countries are Georgia (3), Yugoslavia (2), and Slovenia, Belorussia, Czechia, Estonia, Bosnia-Herzogowina (Sarajevo!), and the Ukraine (each 1). Outside Europe most of the papers are from the U.S.A. (9); second is Japan (6), and third, Israel (3).

Two papers treat the history of neurochemistry. One is from Herman Bachelard, former president of the ESN, and covers the history of the ESN, written in honor of its 20th anniversary. The other paper is on the history of neurochemistry in the Ukraine. Other contributions can be divided in three parts: (1) clinical aspects, (2) cellular functions, and (3) molecules and methods.

Very important in Part I are the papers on the role of neuroprotection and neurorescue in ischemia, stroke, and neuronal injury. Described are, among others, the role of calcium and calpain and the effect of hypothermia and deprenyl-like compounds. Much attention is paid to neurochemical mechanisms, and from these mechanisms, deduced potential therapeutic interventions for Alzheimer's disease, dementia, multiple sclerosis, Parkinson's disease, and epilepsy. Attention is also paid to mood disorders such as depression and anxiety and the modern views on treatment with antidepressants and antipsychotics. Finally, a chapter is devoted to the significance of cerebrospinal fluid examinations for diagnosis and treatment of neurological diseases.

Part II covers nutrient transport, energy metabolism, and metabolic trafficking in brain cells. The role of neurotrophic molecules is also described. Gene expression and new insights in the function of classical and peptide neurotransmitters are reported from the neuronal axon compartment. Described, among other topics, are the release and inactivation of neuropeptides by ecto-enzymes. Attention is also paid to the function of melatonin in brain and retina cells and to the molecular mechanisms in microglia cells.

Part III describes functional aspects of neuroactive compounds. Special attention is paid to acetylcholine, taurine, amino acid neurotransmitters, and neuropeptides. The role of lipid mediators in neurotransmission is also reported. Brain proteins, especially their glycoconjugates and neuronal proteins, are treated extensively. Further attention is paid to carnitine, free radicals, and NO. New insights in methodology are given in chapters on PET imaging, microdialysis, and capillary electrophoresis.

For a good and rapid communication with the contributors/presenting authors a complete list of addresses including telephone and fax numbers and, if available, e-mail addresses is included. The addresses of the organizers of the several symposia, colloquia, and workshops are also recorded for easy contact.

The editing of this book could not have been possible without the unceasing help of secretaries Margot Kuipers, Lies Plas, and Joke Venema. I am very grateful to Joke Hoekstra for correcting the English in many papers; to J. Froon, N. Barneveld-Schelling, and E. Kloens (Medical Library) for correcting references; and to Eddie Ligeon for solving many PC-disk problems. I am especially thankful to the coeditors of this book for editing the chapters related to their sessions.

Finally, I want to acknowledge the members of the local organizing committee of the 11th ESN meeting—Dop Bär, Paul Luiten, Fennie Radhakishun, Ben Westerink, and Jitty Jaarsma (Congress Office)—for creating the stimulating atmosphere that made the publication of this book possible.

Albert W. Teelken
Jakob Korf

CONTENTS

Section 1: History of Neurochemistry

1. A Brief History of the European Society for Neurochemistry 1
   Herman Bachelard

2. Development of Neurochemical Research in the Ukraine 9
   A. V. Nazarenko and Ya. V. Belik

Part I: Clinical Aspects and Pathology

Section 2: New Aspects of Alzheimer's and Other Neurodegenerative Diseases

3. Neuroinflammation and Alzheimer's Disease 15
   P. Eikelenboom

4. Neurochemical Derangements and Cognitive Deficits in Apolipoprotein E Deficient Mice following Closed Head Injury 21
   L. Lomnitski, Y. Chen, E. Shohami, R. Kohen, and D. M. Michaelson

5. The Role of Cytokines in Alzheimer's Disease 27
   M. Hüll, B. L. Fiebich, K. Lieb, B. Volk, M. Berger, and J. Bauer

6. Profiles of Brain Phospholipids and Fatty Acids in Alzheimer's Disease Brain .. 33
   E. Kienzl, K. Jellinger, L. Puchinger, B. M. Crandall, and G. W. Arendash

7. Synaptic Pathology in Alzheimer's Disease 39
   Pia Davidsson and Kaj Blennow

Section 3: Antidepressant and Anxiolytic Mechanisms of Serotonergic Autoreceptor Agonists, Antagonists, and Reuptake Inhibitors

8. 5-HT$_{1A}$ Partial Agonists as Anxiolytics and Antidepressants 47
   Jeffrey Sprouse

9. Effects of Repeated Administration of Dizocilpine on $5HT_{1A}$ Receptor Binding in the Rat Brain 55
   S. Rump and I. Jakowicz

10. A Pharmacological Characterization of Hippocampal $5-HT_{1D}$ Receptors in the Guinea Pig: A Microdialysis Study 59
    S. K. Long and A. I. Bosch

11. Antidepressant Drug Actions on Hypothalamic–Pituitary–Adrenal Axis in Olfactory Bulbectomized Male Rats 63
    A. Marcilhac, G. Anglade, F. Hery, and P. Siaud

12. Antagonist Affinity Estimates from Radioligand Binding and Functional Studies Vary at the Cloned Human $5-HT_{2C}$ Receptor 69
    M. D. Wood, T. L. Gager, D. R. Thomas, V. L. Holland, and A. M. Brown

Section 4: Mild Hypothermia in the Management of Stroke and Other Ischemic Brain Trauma; Basic Mechanisms and Clinical Experience

13. Glia and the Development of Brain Damage 73
    S. Knollema, S. V. van de Witte, and G. J. Ter Horst

14. L-deprenyl Reduces Brain Damage in the Caudate Putamen by a MAO-Dependent Effect 79
    S. Knollema, A. Wiersma, and G. J. Ter Horst

15. Homeostatic Effects of Adenosine on Potentially Neurotoxic Glial Cell Activation 83
    Peter Schubert, Tadanori Ogata, Cristina Marchini, and Kiyoshi Kataoka

16. Protective Effects of Mild Hypothermia on the Brain and Endothelium in Acute Embolic Stroke 91
    Hiroaki Naritomi, Takao Shimizu, Hidekazu Kinugawa, and Tohru Sawada

17. Prevention of Cerebral Thermo-Pooling, Free Radical Reactions, and Protection of A10 Nervous System by Control of Brain Tissue Temperature in Severely Brain Injured Patients 97
    Nariyuki Hayashi, Kosaku Kinosita, Akira Utagawa, Nario Jo, Takeo Azuhata, and Tadashi Shibuya

Section 5: Neurochemistry and Neuroimmunology of EAE; Implications for Therapy of MS

18. Experimental Allergic Encephalomyelitis: Blood-Borne Macrophages as Effector Cells 105
    Inge Huitinga, Jan Bauer, Sigrid R. Ruuls, and Christine D. Dijkstra

19. Autoreactive T Lymphocytes in Multiple Sclerosis: Pathogenic Role and Therapeutic Targeting 113
    Piet Stinissen, Jingwu Zhang, Caroline Vandevyver, and Jef Raus

20. Cytokines and Multiple Sclerosis: Implications for Treatment 121
    B. W. van Oosten and C. H. Polman

21. N-3 Polyunsaturated Fatty Acids: Biological Activities, in Relation to
    the Functioning of Immune Cells and Possible Modification of
    Induction and Course of Multiple Sclerosis 125
    J. de Vries

22. Suppression of Experimental Autoimmune Encephalomyelitis in the Rat
    by Oral Administration of Spinal Cord Protein Hydrolysate 137
    B. Kwiatkowska-Patzer, B. Baranowska, M. Barcikowska-Litwin, and
    A. W. Lipkowski

23. Selective Blood Brain Barrier Dysfunction after Intravenous Injections of
    RTNFα in the Rat 141
    Gert J. Ter Horst, Jan Gert Nagel, Mike J. L. De Jongste,
    and Ysbrand D. Van der Werf

Section 6: Neuroprotection in Model Studies for Dementia and Ischemia

24. Hypoxia, Ca^{2+} Homeostasis, and Development and Aging of the Brain 147
    Csaba Nyakas, Bauke Buwalda, and Paul G. M. Luiten

25. Glutamatergic Function in Alzheimer's Disease 153
    Paul T. Francis and Sas N. Dijk

26. Morphochemical Changes in the Brain of Rats under the Conditions of
    Dopamine System Dysfunction 161
    L. M. Gershtein and A. V. Sergutina

27. Neuroprotective Signal Transduction Pathways of sAPP and TNF 165
    Annadora J. Bruce, Katsutoshi Furukawa, and Mark P. Mattson

28. Combined Fructose 1,6-diphosphate-dimethyl Sulfoxide Treatment
    Reverses Chronic Memory Deficits in Brain Ischemic Rats 173
    Jack C. de la Torre, Nancy Nelson, and Robert J. Sutherland

29. Carnosine Protective Effect on Ischemic Brain *in Vivo* Study Using
    Proton Magnetic Resonance Spectroscopy 177
    D. Dobrota, I. Tkáè, V. Mlynárik, T. Liptaj, A. A. Boldyrev,
    and S. L. Stvolinsky

30. A Magnetic Resonance Imaging Study of the Brain and the Spinal Cord
    during Ischemia 183
    D. Dobrota, I. Tkáè, V. Mlynárik, and T. Liptaj

Section 7: Pathogenic Mechanisms in Mood Disorders

31. The Role of 5-hydroxytryptamine in Depression 187
    G. Curzon

32. The Role of Noradrenaline in Depression and Its Therapy: A Reappraisal of
    Its Importance 193
    Mike Briley and Chantal Moret

33. Monoaminergic and Glutamatergic Mechanisms in a Realistic Animal Model
    of Depression 197
    Mariusz Papp and Paul Willner

34. Effects of Ammonia and L-tryptophan Loading on Brain Extracellular 5-HT
    and 5-HIAA Levels in Chronic Experimental Hepatic Encephalopathy ... 201
    Peter B. F. Bergqvist, Stephan Hjorth, Robert M. Audet, Gustav Apelqvist,
    Roger F. Butterworth, and Finn Bengtsson

35. Dopamine Metabolism Trafficking in Genetically Defined Anxiety and
    Neuropathology 209
    N. K. Popova, D. F. Avgustinovich, E. M. Nikulina, and J. A. Skrinskaya

Section 8: Atypical Antipsychotics: The Neurochemical Relevance of Test Models

36. Differential Effects of 2-(ethylthio)apomorphine Enantiomers on Striatal
    Dopamine Overflow *in Vivo* 215
    F. Auth, I. Laszlovszky, B. Kiss, E. Kárpáti, S. Berényi, Cs. Csutorás, S.
    Makleit, and M. Lõw

37. Neurosynaptic Nets and the Induction of Consciousness:
    Relationships with Psychopathology 221
    J. S. Wassenaar

Section 9: Signalling and Myelin Gene Regulation during Development and Regeneration

38. Oligodendrocytes and Schwann Cells Respond Differently to
    a Transgene Induced Reduction of Axon Caliber 227
    Joël Eyer and Alan Peterson

39. Major Myelin Specific Genes Expression in pt PLP Gene Mutant Rabbit 233
    Joanna Sypecka and Krystyna Domañska-Janik

40. Growth Regulation and Myelin Protein Expression in a Newly Established
    Permanent Glial Cell Line 239
    C. Richter-Landsberg and W. Hoppenstedt

Section 10: Aspects of Pathogenesis and Therapy of Parkinson's Disease

41. The Role of Glutathione in the Pathogenesis of Parkinson's Disease 243
    B. Drukarch and J. C. Stoof

42. Sources of Cells for Transplantation and Gene Therapy in Parkinson's Disease .. 249
    S. B. Dunnett, L. E. Annett, A. L. Kendall, A. E. Rosser, C. Watts,
    and C. N. Svendsen

43. Characterization of a Signalling Pathway Underlying Neuronal Cell Death
    after Trophic Support Withdrawal or TNFα Treatment 259
    Nathalie Lambeng, Valentine France-Lanord, Bernard Brugg,
      Philippe Anglade, Patrick P. Michel, Yves Agid, and Merle Ruberg

Section 11: Neurochemistry of Epilepsy

44. Autoantibodies against Human Brain AMPA/Quisqualate Receptor Capable
    of Determining Persons with Brain Paroxysmal Activity and Epilepsy ... 265
    S. A. Dambinova and L. G. Gromova

45. GABA-Dependent Modulation of [^{3}H]TBOB Binding to the GABA$_A$ Receptor
    Complex: A Study with Thiopental 275
    Clementina M. van Rijn, Ris Dirksen, Jan Pieter C. Zwart, and
      Elly Willems-van Bree

46. Increased Incidence of CNS- and Phospholipid Antibodies in
    Epileptic Syndromes of Childhood 281
    W. Diener, R. Klein, K. Kast, U. Danneberg, M. Mohrmann, P. A. Berg,
      S. A. Dambinova, and R. Korinthenberg

47. Effect of Some Neuropeptides on MAO Activity under Experimental Epilepsy .. 287
    E. L. Dovedova

48. The Dynamics of Neurophysins, Exogenous Ferritin, Albumin in Pituitary
    (Neurohypophysis), and Blood Serum of Rats at Oxygen Epilepsia 291
    A. A. Sinichkin, V. A. Lepeohin, and E. A. Bardahchjan

49. The Anticonvulsive Drug Valproic Acid Affects Cell Morphology and
    Cytoskeletal Organization 295
    Vladimir Berezin, Peter Walmod, Anna Kawa, and Elisabeth Bock

Section 12: Neuroprotection and Neurorescue: Myths and Mechanisms

50. Possible Mechanisms of Neuroprotection in Response to Diverse Insults 299
    Keith Tipton

51. The Role of the Metabolism of (−)-Deprenyl in Neuroprotection 303
    K. Magyar

52. The Iron Chelator Desferrioxamine Protects against MPP$^+$ Neurotoxicity 309
    E. R. Matarredona, M. Santiago, A. Machado, and J. Cano

53. Production of Metabolites of L-deprenyl and pFluoro-L-deprenyl in
    Rat Brain after Central Administration 315
    A. Lajtha, H. Sershen, T. Cooper, A. Hashim, and J. Gaal

54. The Possible Involvement of Death-Related Genes in
    Methamphetamine-Induced Apoptosis 323
    Jean Lud Cadet

55. Methamphetamine Administration and Associated Neurotoxicity:
    Effects of Selegiline (*l*-Deprenyl) 327
    Steven R. Goldberg and Sevil Yasar

56. Long-Term Safety and Efficacy of Selegiline in the Treatment of
    Parkinson's Disease 331
    Outi Mäki-Ikola, Esa Heinonen, and Risto Lammintausta

57. The Protective Effect of Delta-Sleep Inducing Peptide under
    Hypokinetic Conditions 339
    A. Mendzeritsky, A. Matsionis, and A. Lysenko

58. Amiridine Delays Functional Degradation of Isolated Crayfish Neuron
    Occurring Spontaneously or under Energy Metabolism Inhibition 345
    A. B. Uzdensky, Y. V. Burov, and T. N. Robakidze

59. Combined Therapy with 8-OH-DPAT and Nimodipine against
    in Vivo NMDA-Induced Cell Death in Rat Nucleus Basalis 351
    B. T. Stuiver, C. M. Stienstra, B. J. Oosterink, C. Nyakas,
    and P. G. M. Luiten

60. Biological Effects of P_2 Purinoceptor Modulators in Cultured
    Primary Cerebellar Granule Neurons 357
    Cinzia Volonté and Daniela Merlo

Section 13: Calcium in Neuronal Injury

61. Generation of Ca^{2+} Signal in NMDA Receptors of Rat and Rabbit
    Hippocampus *in Vivo*: Ca^{2+} Influx and Mobilisation 361
    J. W. Łazarewicz, W. Rybkowski, E. Salinska, W. Gordon-Krajcer,
    E. Zieminska, A. Ziembowicz, M. Puka-Sundvall, and H. Hagberg

62. Transient Increase of Calcium/Calmodulin-Dependent Kinase II (CaM KII)
    in Its Autonomous State under Brain Ischemia 369
    Krystyna Domańska-Janik and Teresa Zalewska

63. Ischemia–Reperfusion Decreases Protein Levels of $InsP_3$ Receptor and PMCA
    but Not Organellar Ca^{2+} Pump and Calreticulin in Gerbil Forebrain 375
    Ján Lehotský, Peter Kaplán, Peter Račay, Luc Raeymaekers, and
    Viera Mézešová

64. Immunological and Functional Identification of Intracellular Ca^{2+} Store
    from Gerbil Forebrain 383
    Peter Račay, Peter Kaplán, Luc Raeymaekers, and Ján Lehotský

65. Type I Inositol (1,4,5) Trisphosphate Receptor and Phosphate-Activated
    Glutaminase Are Highly Enriched in Neurons as Compared to
    Glial Cells in Rat Neostriatum 389
    L. S. Haug, A. C. Østvold, I. Torgner, B. Roberg, L. Dvořáková, and
    S. I. Walaas

66. Quantification of Extracellular Calcium in Fish Brain after Optic Nerve Cut ... 395
    Petra Vöhringer, Karl-Heinz Körtje, and Hinrich Rahmann

67. Neuroprotective Approaches *in Vivo* against Ca^{2+}-Induced
    Neurodegeneration in Rat Nucleus Basalis 401
    B. J. Oosterink, C. M. Stienstra, B. T. Stuiver, B. R. K. Douma, J. Korf,
    and P. G. M. Luiten

Section 14: Role of Calpain in Alzheimer's Disease and Cerebral Ischemia

68. On the Mechanism of Calpain Activation under Ischemia 407
    Teresa Zalewska, Barbara Zablocka, Takaomi C. Saido, and
    Krystyna Domańska-Janik

69. Proteolytic Activity Is Increased in Lymphocytes from Patients with
    Alzheimer's Disease 415
    J.-O. Karlsson, K. Blennow, I. Janson, K. Blomgren, I. Karlsson,
    B. Regland, A. Wallin, and C. G. Gottfries

70. The Role of Delta-Sleep Inducing Peptide in Calpain Activity Regulation
    under Hypoxia 419
    A. V. Lysenko, N. I. Uskova, A. E. Matsionis, and P. E. Povilaitite

Section 15: Cerebrospinal Fluid Analysis

71. CSF Flow: Its Influence on CSF Concentrations of Brain-Derived and
    Blood-Derived Proteins 423
    Hansotto Reiber

72. Quality Assurance and Sample Handling in CSF Investigation 433
    Sten Öhman

73. Glutamine Synthetase in Cerebrospinal Fluid Is Brain-Derived and Elevated
    in Neurodegenerative Diseases 443
    H. Tumani, G. Q. Shen, and J. B. Peter

74. Glial Cell Pathobiology in Multiple Sclerosis Detected by CSF Markers 451
    A. R. Massaro, A. Carnevale, P. Tonali, and E. Bock

Section 16: Other Clinical Aspects

75. BB-Isoform of Creatine Kinase in Amniotic Fluid as a Marker of
    Brain Development Disorders 457
    A. V. Arutjunyan, N. G. Pavlova, N. N. Konstantinova, A. V. Pavlenko,
    T. K. Kascheyeva, L. V. Lyzlova, L. S. Kosina, E. I. Rusina,
    A. V. Mikhailov, and E. V. Shelayeva

76. Excitation–Contraction Coupling of Extensor Digitorum Longus Muscle
    of Dystrophic MDX Mouse 461
    A. De Luca, S. Pierno, and D. Conte Camerino

77. Graded Postischemic Reoxygenation, Phospholipids and Neuronal Damage in Rabbits after Ischemia 465
    N. Lukáčová, M. Marsala, and J. Marsala

78. Do Prostaglandins E_2, F_{2alfa}, and D_2 Have a Role in the Development of Idiopathic Trigeminal Neuralgia? 469
    Nebojsa Jovic and Radovan Jovic

79. Leg Movements Are Transiently Present in Fetuses and Infants with Spina Bifida Aperta 475
    D. A. Sival and J. H. Begeer

80. Histamine-Induced Vasogenic Brain Oedema Formation in Newborn Pigs: A Role for Endothelial Acid Phosphatase? 479
    Csilla Andrea Szabó, Mária Anna Deli, László Németh, István Krizbai, József Kovács, Csongor S. Ábrahám, and Ferenc Joó

81. Progressive Motor Neuron Impairment in an Animal Model of Familial Amyotrophic Lateral Sclerosis 485
    Mimoun Azzouz, Nathalie Leclerc, Mark Gurney, Jean-Marie Warter, Philippe Poindron, and Jacques Borg

82. The Influence of Hypokinesia on Synapsoarchitectonical Features of Rat's Lymbic and Extrapyramidal Structures 491
    Mzia G. Zhvania

Part II: Cellular Functions

Section 17: Neurotrophic Molecules in Development and Regeneration

83. Time Course of Trophic Responses of Astrocytes in Rat Hippocampus Induced by Trimethyltin Intoxication: An Immunocytochemical Study ... 497
    B. Oderfeld-Nowak, D. Koczyk, M. Skup, and M. Zaremba

84. Interleukin-1β Receptor Mediated NGF Synthesis and Secretion from Rat Cortical Astrocytes 501
    Marija Čarman-Kržan, Damijana M. Juric, and Vesna Pahor

85. Changes in Hippocampal PKCγ, mAChR- and trkB-Immunoreactivity after a Spatial Learning Paradigm 507
    B. R. K. Douma, E. A. van der Zee, and P. G. M. Luiten

86. Devascularizing Injury of the Rat Brain Neocortex Causes Different Neurotrophic Responses in Denervated Nucleus Basalis Magnocellularis and Thalamic Nuclei 513
    M. Skup, D. Torzewska, and M. Zaremba

87. Analysis of Perfusates Collected from Surviving Rat Brain Olfactory Cortex Slices 519
    Anatoly A. Mokrushin and Anton Y. Plekhanov

Section 18: Steroids and Nervous System

88. A Cellular Model for the Study of Estrogen Activity in Cells of Neural Origin .. 523
    Adriana Maggi, Paola Agrati, Cesare Patrone, Sabrina Santagati,
    Martine Garnier, and Elisabetta Vegeto

89. The Disturbances of the Hypothalamic Circadian Rhythm Regulation in
    Female Rats under the Influence of Xenobiotics 529
    A. V. Arutjunyan, M. G. Stepanov, A. V. Korenevsky, V. M. Prokopenko,
    and T. I. Oparina

Section 19: Molecular Mechanisms of Microglial Activation

90. Molecular Determinants of Microglial Activation Stages 535
    B. Küst, K. Appel, M. Buttini, A. Sauter, H. W. G. M. Boddeke,
    M. Berger, D. van Calker, and P. J. Gebicke-Haerter

91. Is Regulation of Cyclooxygenase-2 Important for the Neuroprotective and
    Neurotoxic Roles of Microglia? 541
    L. Minghetti, A. Nicolini, E. Polazzi, and G. Levi

92. The Response of Microglia to Axon Injury 549
    H. Aldskogius, L. Liu, N. P. Eriksson, J. K. E. Persson, and M. Svensson

Section 20: Metabolic Trafficking in the Brain

93. Role of Neuron–Glia Interactions in Coupling Neuronal Activity to
    Energy Metabolism: Implications for Functional Brain Imaging 555
    P. J. Magistretti, L. Pellerin, and P. G. Bittar

94. The Source of Metabolic Substrates for Neuronal Energy Metabolism 561
    Marianne Fillenz, Maria Demestre, Lesley K. Fellows, Manfred
    O. M. Berners, and Martyn G. Boutelle

95. Regulation of Glial Lipid Metabolism during Osmotic Stress Studied by
    ^{13}C-NMR Spectroscopy 571
    Ulrich Flögel, Wieland Willker, Josef Pfeuffer, and Dieter Leibfritz

96. Determination of Brain Extracellular Glucose in Vivo with
    an Implanted Biosensor 577
    John P. Lowry and Marianne Fillenz

Section 21: The Functional Role of Chromogranins and Related Peptides

97. Regulation of Parathyroid Secretion by Chromogranin A Amino
    Terminal Peptides 583
    Ruth Hogue Angeletti, Hassane Zouheiry, Thomas D'Amico, and
    John Russell

98. Pancreastatin Signaling in the Liver 589
    Víctor Sánchez-Margalet, José Santos-Alvarez, and Raimundo Goberna

99. The Functional Role of Bovine Adrenal Medullary New Cardioactive Peptides .. 595
R. Srapionian, Z. Paronian, and A. Galoyan

Section 22: Role of Melatonin in the Brain and Eye

100. Pharmacology of Melatonin as a Neural Antioxidant 599
Russel J. Reiter

101. Melatonin Biosynthesis in the Retina: Effect of UV-A Light 605
Jerzy Z. Nowak, Jolanta B. Zawilska, Jolanta Rosiak, and Tomasz Kalaóny

102. Melatonin Protects against Ischaemic-like Insults to Retinal Cells, *in Vitro* 611
Neville N. Osborne, Rukhsana Safa, Marta Ugarte, and Chantal Cazevieille

Section 23: Transporters for Essential Nutrients in Brain Cells

103. Molecular Identification of Proteins Involved in Transport of
Branched-Chain Amino Acids in Glial Cells 619
Stefan Bröer

104. Role of Arginine Transport in the Production of Nitric Oxide in
Cultured Neural Cells 625
A. Schmidlin, S. Fischer, P. Ricciardi-Castagnoli, and H. Wiesinger

105. Metabolic Trafficking in the Fetal Brain under Normal and Diabetic
Maternal Conditions, by ^{13}C NMR Isotopomer Analysis 631
S. Litvin and A. Lapidot

Section 24: Gene Expression in the Axon Compartment

106. Gene Expression in Axons and Nerve Endings 637
Antonio Giuditta, Marianna Crispino, and Carla Perrone Capano

107. Protein Synthesis in Brain Presynaptic Endings 643
Marianna Crispino, Rainer Martin, Juan Claudio Benech, Jaime Alvarez,
Barry B. Kaplan, and Antonio Giuditta

108. Long Term Survival of Isolated Axonal Segments as Revealed by *in Vitro* Studies 647
R. Oren, A. Dormann, D. Benbassat, and M. E. Spira

109. Direct Evidence for de Novo Protein Synthesis in Isolated Axons of
Identified Lymnaea Neurons 655
J. J. Bergman, J. Van Minnen, E. R. Van Kesteren, A. B. Smit,
W. P. M. Geraerts, and N. I. Syed

110. Ribosomes in Peripheral and Presynaptic Domains of Axons 661
Rainer Martin

111. Periaxoplasmic Plaques as Discrete Putative Translational Domains in
Mauthner Axons 667
Edward Koenig and Rainer Martin

112. The Major Protein of a Large Ribonucleoprotein Particle (Vault) Is Localized in Nerve Terminals 675
Walter Volknandt and Christine Herrmann

Section 25: Ecto-Enzymes in the Inactivation of Neural Messengers

113. New Insights into the Structure and Function of the Endopeptidase-24.11 Family of Ectoenzymes 683
Anthony J. Turner

114. Inactivation of Thyrotropin-Releasing Hormone (TRH) by the TRH-Degrading Ectoenzyme 691
H. Heuer, S. Turwitt, L. Schomburg, and K. Bauer

115. Ecto-Nucleotidases: Studies on the Diadenosine Polyphosphate Hydrolases from Neural and Endothelial Origin 695
Teresa Miras-Portugal, Pedro Rotllán, and Jesús Mateo

116. Functional Role of Ecto-5'-Nucleotidase in the Nervous System 701
Herbert Zimmermann, Alev Heilbronn, Norbert Braun, and Vera Maienschein

117. NCAM and Ecto-ATPases of Rat Synaptosomes 707
K. N. Dzhandzhugazyan and E. Bock

Section 26: Mechanism of Neurotransmitter Release

118. Quantal Transmitter Release at Botulinum-Treated Vertebrate Neuromuscular Junctions 713
Jordi Molgó, Frédéric A. Meunier, and Lawrence C. Sellin

119. Identification of Proteins Involved in Regulated Exocytosis 719
Alan Morgan and Robert D. Burgoyne

120. Catecholamines Levels in the Rat Brain and Plasma during Cardiac Arrest and after Resuscitation 729
Andrzej Kapuściński

121. Effect of Glutathione on [^{3}H]Dopamine Release from the Mouse Striatum Evoked by Glutamate Receptor Agonists 733
R. Janáky, V. Varga, A. Hermann, Z. Serfozo, R. Dohovics, P. Saransaari, and S. S. Oja

Section 27: Regulation of Acetylcholine Release in Brain

122. Compartmentation of Acetyl-CoA and Acetylcholine Metabolism in Pyrithiamine Encephalopathy 737
Hanna Bielarczyk, Maria Tomaszewicz, Agnieszka Jankowska, Yuri Kisielewski, and Andrzej Szutowicz

123. Enhancement of Hippocampal Acetylcholine Release by Local Ethanol Infusion: A Role for Phosphatidylethanol Formation? 743
C. Henn, K. Löffelholz, and J. Klein

Section 28: Mitochondrial Energy Metabolism

124. Effects of Inhibitors of the Mitochondrial Electron Transport Chain on Fatty Acid Beta-Oxidation in Rat Brain Cultured Astrocytes 751
Ali Esfandiari and Marion Paturneau-Jouas

125. Brain Metabolic Adaptation to Hypoxia Stress 757
Elena M. Khvatova, Elena I. Yerlykina, Murat R. Gaynullin, and Inessa I. Mickaleva

126. Changes in Contents of Phospholipids in Rat Brain under the Action of Noise .. 761
M. M. Melkonian, K. G. Karageuzyan, G. A. Hoveyan, A. Sargsian, and L. M. Hovsepian

127. Postnatal Development of Malic Enzyme Isoforms in Rat Brain 765
R. Vogel, H. Wiesinger, and B. Hamprecht

Part III: Molecules/Methods

Section 29: Neurotransmitter Transporters

128. Towards Overexpression of the Serotonin Transporter: An Overview 773
Friedrich M. Titgemeyer, Patrick Schloss, Cara R. Baker, D. Clive Williams, Sylviane Hilberer-Ehret, Arjen-Kars Boer, and George T. Robillard

Section 30: Structures of Neuronal Proteins in Rationalized Drug Design

129. Scorpion Toxins: Functional Variations on a Structural Theme 783
André Ménez and Marc Dauplais

130. A Novel Strategy of Effect-Directed Ligand Design for G-protein Coupled Receptors 791
Jaak Järv

131. Role of Some Biochemical Correlates in Mechanisms of LTP in Olfactory Cortex Slices 797
N. L. Izvarina, N. S. Nilova, T. S. Glushchenko, and A. V. Tokarev

Section 31: Progress in Cholinergic Neurotransmission

132. Presynaptic Nicotinic Receptors in the CNS: Function and Mechanism 801
S. Wonnacott and L. Soliakov

133. Muscarinic Receptors: Phosphorylation and Desensitization 807
   Tatsuya Haga, Kazuko Haga, Fumio Nakamura, Mariko Kato Hayashi,
   Kimihiko Kameyama, and Hirofumi Tsuga

134. Presence of Acetylcholine in the Blood and Its Production by
   Choline Acetyltransferase in Lymphocytes 813
   K. Kawashima, T. Fujii, H. Misawa, S. Yamada, S. Tajima, T. Suzuki,
   K. Fujimoto, and T. Kasahara

135. Metabolism of Acetyl-CoA and Cholinergic Neuropathies 821
   Andrzej Szutowicz, Hanna Bielarczyk, and Maria Tomaszewicz

136. Targeted Immunolesion of Cholinergic Neurons by 192 IgG-saporin:
   A Novel Tool to Study Neurochemical Events of Alzheimer's Disease ... 829
   Reinhard Schliebs, Steffen Roßner, Mechthild Heider, and Volker Bigl

137. The Relationships between Two Choline Uptake Components within
   Wide Range of pH 837
   O. V. Chumakova

138. Influence of Nicotinic Stimulation on $[Ca^{2+}]_i$ in Processes of Chick
   Sympathetic Neurons in Culture 843
   V. Dolezal, A. Schobert, and G. Hertting

139. Cholinergic Mechanisms in Inherited and Acquired Tolerance to Ethanol 847
   A. Jankowska, Y. Kisielevski, N. Oganesjan, M. Tomaszewicz, and
   A. Szutowicz

140. The State of M-Cholino- and Beta-Adrenoreception of Autonomic Ganglia
   in Spontaneously Hypertensive Rats 853
   V. V. Glinkina, L. A. Khyazeva, I. G. Charyeva, S. A. Pylaeva, and
   A. S. Pylaev

Section 32: Cell Adhesion in Neural Plasticity Induced by Experience

141. Structure and Function of the Neural Cell Adhesion Molecule NCAM 857
   Elisabeth Bock, Nina Pedersen, and Vladimir Berezin

142. Polysialic Acid (PSA) Associated with the Neural Cell Adhesion Molecule
   (N-CAM) May Play a Role in Spatial Learning and LTP in Rats 863
   Hans Welzl, Catherina G. Becker, Alain Artola, and Melitta Schachner

143. Regulated Expression of the CNS-Specific Cell Adhesion Molecule
   Ependymin after Acquisition of an Active Avoidance Behaviour
   Provides a Possible Mechanism for Memory Consolidation 869
   Rupert Schmidt

144. Time-Limited Roles for Cell Adhesion Molecule-Mediated Neuroplastic Events
   in the Commitment of Memory to Long Term Storage 877
   Keith J. Murphy, Gerard B. Fox, Alan W. O'Connell, and Ciaran M. Regan

145. Expression of α-2,8-polysialyltransferase during Retinoic Acid Induced
     Differentiation of a Neuroblastoma Cell Line 885
     Herbert Hildebrandt, Harald Rösner, Rita Gerardy-Schahn, and
     Hinrich Rahmann

146. Functional Characterization of the First Immunoglobulin-like Domain of
     the Neural Cell Adhesion Molecule 891
     Vladislav V. Kiselyov, Vladimir Berezin, Thomas E. Maar, Vladislav Soroka,
     Klaus Edvardsen, Arne Schousboe, and Elisabeth Bock

147. Expression of a Transmembrane Isoform of the Neural Cell Adhesion Molecule
     (NCAM) in Rat Glioma BT4Cn Cells Increases Their Motility 897
     Søren Prag, Nina Pedersen, Peter S. Walmod, Vladimir Berezin, and
     Elisabeth Bock

148. Brevican, a Conditional Proteoglycan from Rat Brain: Characterisation of
     Secreted and GPI-Anchored Isoforms 901
     Constanze Seidenbecher, Karin Richter, and Eckart D. Gundelfinger

149. Enhancement of Hippocampal Long-Term Potentiation *in Vitro* by
     Fucosyl-Carbohydrates 905
     H. Matthies, Jr., S. Staak, K. H. Smalla, and M. Krug

Section 33: Expression and Metabolism of Glycoconjugates

150. Glycoconjugates and Nuclear Membrane Lectin from Rat Brain Cell Nuclei ... 909
     N. G. Aleksidze, R. G. Akhalkatsi, and T. Bolotashvili

151. Glycosyltransferase Activities in 15 Human Meningiomas 913
     E. Sottocornola, I. Colombo, S. Rapelli, and B. Berra

152. Administration of Gangliosides Changes the Properties of Adenylate Cyclase
     System in the Sensorimotor and Limbic Structures of Rat Brain 919
     N. Nalivaeva, S. Plesneva, U. Chekulaeva, N. Dubrovskaya, and I. Zhuravin

153. Organ-Specific Expression of Membrane Lipids, Especially of Gangliosides,
     Depending on the State of Thermal Adaptation 925
     Hinrich Rahmann and Ute Balshüsemann

154. N-Linked Oligosaccharide Expression in Whole Tissue and
     Specific Glycoproteins of Rodent CNS 929
     D. R. Wing, Y-J. Chen, R. A. C. Clark, R. A. Dwek, and S. E. Zamze

155. Anti-Ganglioside Antibodies and Molecular Mechanism of Development
     of Guillain-Barre Syndrome 933
     Takao Taki, Nobuhiro Yuki, and Shizuo Handa

Section 34: CNS Taurine — A New Outlook

156. Neuroprotective Actions of Taurine 939
     Pirjo Saransaari and Simo S. Oja

157. The Volume Responses of Brain Cells during Osmotic Stress: How Important Is Taurine? 943
R. O. Law

158. Kainic Acid–Stimulated Release of Taurine from the Rat Substantia Nigra *in Vivo* 949
L. Bianchi, R. Zamfirova, M. Nerini, J. P. Bolam, and L. Della Corte

159. Kainic Acid–Stimulated Release of Taurine and GABA from the Rat Globus Pallidus and Substantia Nigra *in Vivo* 955
L. Bianchi, P. Bartolini, A. Colivicchi, J. P. Bolam, and L. Della Corte

160. Different Specificities for Taurine Analogues and Their Target Sites in Brain 959
M. Marangolo, D. Zisterer, D. C. Williams, K. F. Tipton, H. B. F. Dixon, and L. Della Corte

161. The Effect of Taurine and Saline on Neuroactive Amino Acid Levels in the Mouse Brain 963
M. Mijanovic, K. Valjevac, and Lj. Jozanc

Section 35: Free Radicals, NO, and Brain Pathology

162. Nitric Oxide in the Central Nervous System of Rats Suffering from Rabies Virus Infection or Experimental Allergic Encephalomyelitis 967
A.-M. Van Dam, S. R. Ruuls, C. Marquette, J. Bauer, C. J. A. de Groot, H. Tsiang, C. D. Dijkstra, and F. J. H. Tilders

163. The Effect of Glutamate, Antioxidants, and Gangliosides on Na^+,K^+-ATPase Activity in Rat Brain Cortex Synaptosomes 973
N. F. Avrova, V. A. Tyurin, I. O. Zakharova, T. V. Sokolova, and Y. Y. Tyurina

164. Nitric Oxide Level Dramatically Increases in the Rat Brain during Epileptiform Seizures 977
V. Bashkatova, A. Vanin, V. Mikoyan, G. Vitskova, V. Narkevich, and K. Rayevsky

165. Resistivity of Leech Retzius Nerve Cells to Long-Lasting Oxidant 983
Zorica Jovanovic and B. B. Beleslin

166. Effects of Nitric Oxide on the Catecholamine Release from Cultured Bovine Adrenal Chromaffin Cells 987
Ken Lee, Vladimír Dolezal, and Georg Hertting

167. Modification by Nitric Oxide of Acetyl-CoA and Acetylcholine Metabolism in Nerve Terminals 993
M. Tomaszewicz, H. Bielarczyk, A. Jankowska, and A. Szutowicz

Section 36: Molecular Mechanism of Lipid Mediators Action and Their Role in Neurotransmission and Signal Transduction

168. Role of the Phosphatidylinositol Transfer Protein in Intracellular Membrane Traffic 999
   Gerry T. Snoek, Klaas Jan de Vries, Philip I. H. Bastiaens, and Karel W. A. Wirtz

169. Lipid Metabolism in Human Neuroblastoma SK-N-BE Differentiated with Retinoic Acid 1005
   A. Petroni, N. Papini, P. La Spada, M. Blasevich, and C. Galli

170. The Role of Ca^{2+} and Protein Kinase C in Regulation of Phosphatidylserine Synthesis in Glioma C6 Cells 1011
   J. Barańska, M. Czarny, P. Sabała, and M. Wiktorek

171. The Possible Mechanisms Involved in New Opioid Agonist Fenaridin and Its Antagonist 1019
   M. I. Agadjanov and G. S. Vartanian

172. Lipid Metabolism Markers in Multi-Infarct Dementia 1025
   R. Shakarishvili, S. Tabagari, G. Thakhava, and M. Topuria

173. CMP-Dependent Degradation of Platelet-Activating Factor (PAF) by Rat Brain Microsomes 1029
   Ermelinda Francescangeli, Serena Porcellati, and Gianfrancesco Goracci

174. The Effect of Antidepressants on Phospholipase A2 Activity in the Brain of Rats   1035
   Andrzej Malecki, Krzysztof Kucia, Irena Krupka-Matuszczyk, and Henryk I. Trzeciak

Section 37: The Role of Carnitine in the CNS

175. A Brief History of Carnitine and Its Presence in the CNS 1039
   Rona R. Ramsay

176. Carnitine and the Maintenance of Cell Function 1047
   Victor A. Zammit

177. The Carnitine System and Acyl Trafficking in CNS: A New Task or a Biological Reappraisal? 1053
   Rita Ricciolini, Maurizio Scalibastri, Anna Floriana Sciarroni, Secondo Dottori, Menotti Calvani, Lluis Lligoña-Trulla, Roberto Conti, and Arduino Arduini

178. Carnitine Transport and Physiological Function in Neurones 1059
   Katarzyna A. Nałęcz, Agnieszka Wawrzenczyk, Joanna Mroczkowska, Urszula Berent, Nilolai A. Lobanov, and Maciej J. Nałęcz

179. Cerebral Metabolic Compartmentation as Revealed by ^{13}C NMR Analysis
 of [1-^{13}C]Glucose Metabolism 1065
 Tommaso Aureli, Maria Enrica Di Cocco, Caterina Puccetti, Rita Ricciolini,
 Giorgio Capuani, Menotti Calvani, and Filippo Conti

Section 38: Aminoacid Neurotransmitters

180. Age-Dependent Changes of the *in Situ* Protein Phosphorylation in
 Hippocampal Slices after Stimulation of Glutamate Receptors 1071
 Frank Angenstein and Sabine Staak

181. Regulation of Activity of Brain Phosphate Activated Glutaminase Isoforms 1077
 R. G. Kamalyan, A. V. Gyulkhandanyan, T. D. Karapetyan,
 E. R. Mikaelyan, and A. G. Vardanyan

182. Effect of GABA-ergic Substances on Cerebral Blood Flow and
 Intracellular Ca^{2+} in Hypokinesia 1083
 V. P. Hakopian, K. V. Melkonian, G. A. Kevorgian, and A. S. Kanayan

Section 39: Neuropeptides

183. Functional Evaluation of the Benzodiazepine Cholecystokinin Type-B
 Receptor Antagonists L-365,260, L-740,093, and YM022 1089
 John Dunlop, Neil Brammer, Non Evans, Ian Pass, and Chris Ennis

Section 40: Brain Proteins

184. A Role of the Phosphoproteins Dynamin and Rabphilin-3A in
 Neurotransmitter Release from Synaptic Vesicles 1095
 Else Marie Fykse

185. Synaptic Vesicle Proteins Are Rapidly Transported in the Optic Nerve 1101
 Jia-Yi Li and Annica Dahlström

186. Overexpression of B-50/GAP-43 Induces Formation of Filopodia in
 PC12 Cells 1107
 L. H. J. Aarts, H. B. Nielander, A. B. Oestreicher, L. H. Schrama,
 W. H. Gispen, and P. Schotman

187. Salt-Soluble Myelin Basic Protein Is Degraded by
 Myelin-Adsorbed Proteinases 1111
 U. Haas and H. H. Berlet

Section 41: Molecular Neurobiology

188. Suppression of Immediate Early Gene Expression by Intracerebrally
 Applied Antisense Oligonucleotides Impairs Mechanisms of
 Learning and Memory 1117
 Wolfgang Tischmeyer, Rita Grimm, Klaudia Lohmann, Horst Schicknick,
 and Eckart D. Gundelfinger

189. β-Alanine Behaves as γ-Aminobutyric Acid at *Xenopus* Oocytes Expressing
γ-Aminobutyric Acid Receptors 1123
Luis M. Orensanz and Luis C. Barrio

Section 42: Capillary Electrophoresis

190. Patch Clamp Detection of Neuroreceptor Modulators in
Capillary Electrophoresis 1131
Kent Jardemark, Owe Orwar, Ingemar Jacobson, Alexander Moscho,
Harvey A. Fishman, Anders Hamberger, Mats Sandberg,
Richard H. Scheller, and Richard N. Zare

Section 43: PET Imaging of Receptors; Experimental Models and Clinical Approaches

191. Neuroreceptor Mapping *in Vivo*: The Experimental Approach 1139
Adriaan A. Lammertsma, Susan P. Hume, and Ralph Myers

192. PET Imaging of Neurotransmission Systems in Neurology:
The Example of the Nigrostriatal Dopaminergic Pathway 1145
Eric Salmon

193. Measurement of the Cerebral Uptake and Metabolism of L-6-[^{18}F]
Fluoro-3,4-dihydroxy-phenylalanine in Newborn Piglets 1149
P. Brust, R. Bauer, R. Bergmann, B. Walter, J. Steinbach, F. Füchtner,
E. Will, H. Linemann, M. Obert, U. Zwiener, and B. Johannsen

Section 44: Multidisciplinary Application of Microdialysis

194. Dopamine Release in *N. accumbens* in a Conditioned Reward Paradigm 1157
K. P. Datla, R. G. Ahier, A. M. J. Young, J. A. Gray, and M. H. Joseph

195. Neuronal Activation Visualized by Fos-Expression after
Intracerebral Microdialysis of Drugs 1161
M. Feenstra, M. Bubser, E. Erdtsieck-Ernste, A. van der Wal,
M. Botterblom, and H. van Uum

Addresses of Presenting Authors, Organizers, and (Co)-Editors 1167

Author Index 1191

Subject Index 1197

PART III

Molecules/Methods

Section 29: Neurotransmitter Transporters

TOWARDS OVEREXPRESSION OF THE SEROTONIN TRANSPORTER

An Overview

Friedrich M. Titgemeyer,[1] Patrick Schloss,[2] Cara R. Baker,[3] D. Clive Williams,[3] Sylviane Hilberer-Ehret,[1] Arjen-Kars Boer,[1] and George T. Robillard[1]

[1]Department of Biochemistry
University of Groningen
Groningen, The Netherlands
[2]Department of Neurochemistry
Max-Planck-Institute for Brain Research
Frankfurt, Germany
[3]Department of Biochemistry
Trinity College, Dublin, Ireland

1. INTRODUCTION

Presynaptic Na^+-coupled neurotransmitter transporters are an essential element in neurotransmission and play a crucially important role in mental illnesses and drug abuse. In order to elucidate this role, we aim to solve the three-dimensional structure of one such transporter. This would greatly accelerate our understanding of the molecular mechanisms of transporter function and regulation, and may open new ways for therapeutic treatment and development of pharmaceuticals. A prerequisite of structure determination is the overexpression and purification of transporter protein in milligram quantities. In this overview, we present and discuss recent results on the heterologous expression of the serotonin transporter.

In the brain, the serotonin transporter (SERT) is located in the presynaptic cell membrane of serotonergic nerve terminals. It functions to terminate the action of the neurotransmitter serotonin (5-hydroxytryptamine) on postsynaptic serotonin receptors by re-uptake into the presynaptic cell. Serotonin is then recycled by storage in secretory vesicles via transport by the vesicular monoamine transporter or degraded by monoamine oxidase to 5-hydroxyindoleacetic acid, the urine secretion product (1, 2).

From a symposium organized by George Robillard (Groningen, The Netherlands), Clive Williams (Dublin, Ireland) and Baruch I. Kanner (Jerusalem, Israel). (Supported by SmithKline Beecham).

Medical modulation of SERT activity results in effective treatment of mental disorders like bipolar depression, alcoholism and drug abuse. Whenever SERT-mediated re-uptake is blocked or down-regulated, serotonergic neurotransmission is increased, leading to a relief of the pathogenic symptoms. The tricyclic antidepressant imipramine and most of its derivatives, and non-tricyclic antidepressants such as paroxetine, fluoxetine (Prozac™) and citalopram bind directly to SERT and inhibit transport (3, 4).

SERT belongs to a family of Na^+-dependent neurotransmitter transporters which are distinguishable by their pharmacological profile. Among those are the transport systems for γ-aminobutyric acid (GABA), norepinephrine, dopamine, proline, betaine and glycine. The biochemical characterisation of SERT started in the late 1960s (5). Accumulation of serotonin requires three ion gradients and occurs in a 1:1:1:1 fashion by co-transport of serotonin, Na^+, Cl^- and counter-transport of K^+ (3). Comprehensive analyses on substrate transport and inhibitor binding revealed that this strictly Na^+-dependent transporter has a K_m for serotonin of 300–600 nM and inhibitor constants (K_i) for SERT specific ligands in the range of 0.25–2000 nM (paroxetine > citalopram > imipramine > fluoxetine >> desipramine > cocaine) (3, 6–10).

In 1990, Guastella and co-workers reported the cloning of the first member of this family, the rat GABA transporter (11). This started the PCR-based identification of cDNAs encoding homologous transporters. The rat, human and *Drosophila melanogaster* SERT cDNAs have been cloned and their sequences reported (8–10, 12–14). The 630 amino acid residues of the rat and human SERT share 92% identity and both share 51% identity to the *D. melanogaster* protein. Until now, there are more than 80 members of this family known, which have been classified into the superfamily of Na^+-dependent solute transporters (15–18). They all have a common topology of 12 predicted transmembrane α-helices (TMH), exhibiting cytoplasmic N- and C-termini and an extracellular loop between TMH three and four containing one to four N-linked glycosylation sites (Fig. 1A, 19). The cytosolic N- and C-terminal ends show no conservation and are flexible in length. They often contain phosphorylation consensus sites for protein kinases such as protein kinase C (PKC) and protein kinase A (PKA), suggesting a regulatory function. This is supported by fully-retained transport function when parts of the N- and C-terminal sequences were deleted (20, 21). Furthermore, PKC- and PKA-mediated phosphorylation could be demonstrated *in vitro* when the N- and the C-terminal parts of SERT were purified as glutathione S-transferase (GST) fusion proteins, respectively (22). Stimulation of PKC *in vivo* resulted in twofold down-regulation of serotonin transport by a V_{max} reduction (23). Inhibition of PKA *in vivo* led to a fivefold up-regulation of SERT (24). However, some reports present contradictory results on transporter regulation or propose other regulatory mechanisms (3, 25). Taken together, these observations suggest that the N- and/or C-termini may comprise distinct

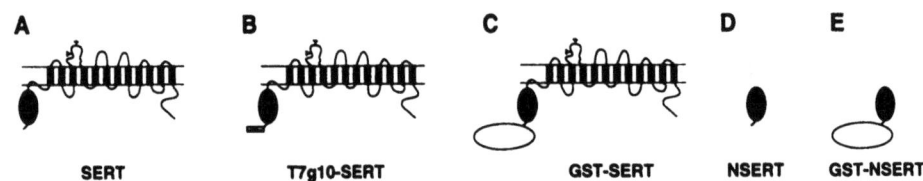

Figure 1. SERT expression constructs. Five SERT expression constructs are schematically displayed. The 12 TMHs are presented as black rectangles connected by hydrophilic loops. The N-terminal cytoplasmic SERT domain is cross-hatched, the fusion peptides are schown as chequered pattern for the T7 Gene 10 and as an oval circle for the GST protein. The two N-linked glycosylation sites are indicated by a fork within the extracellular loop between TMHs three and four. For further explanation see text.

regulatory domains which may trigger transporter activity depending on their phosphorylation state or their interaction with regulatory proteins. The membrane-embedded domain is obviously sufficient for proper transport function.

Many questions regarding the complex activity of the neurotransmitter transporters could be answered, if an expression system becomes available which allows the production of milligram quantities of transporter protein. In this overview, we focus on approaches towards SERT expression using *Escherichia coli* and compare to results obtained from yeast, baculovirus and mammalian cell expression systems.

2. BACTERIAL EXPRESSION

E. coli is the organism of choice for the overproduction of many proteins. It offers the broadest variety of genetic tools as well as fast and inexpensive protein production. A variety of bacterial membrane proteins have been overexpressed to amounts of up to 20 mg/l cell culture or up to 50% of membrane protein, respectively (26, 27). Yet, there are only three examples reported of overexpression of eukaryotic transporters in *E. coli*. The 6 TMH-type bovine mitochondrial oxoglutarate transporter was overexpressed in inclusion bodies to 15 mg/l culture (28). The 6 TMH-type yeast H^+-phosphate transporter was expressed to amounts of 2 mg/l culture and the 12 TMH-type yeast PHO84 protein, a high-affinity phosphate transporter, was expressed to 240 mg/l culture (29, 30). Nevertheless, these three transporters are evolutionary much closer related to bacterial organisms than mammalian transporters for which only two cases of functional expression at low levels have been reported. These are the mouse multidrug resistance protein and the human erythrocyte glucose transporter (31, 32).

Problems associated with heterologous membrane protein expression in *E. coli* include i) toxicity of the foreign gene product resulting in plasmid instability (33), ii) proteolytic degradation of the foreign protein induced via heat-shock response or other stress responsive factors (34), iii) low codon usage of the foreign gene for the *E. coli* host, especially when rarely used codons occur frequently at the beginning of the gene (35, 36), and iv) mRNA which is poorly translated by the prokaryotic ribosome (37). So far, there is no rule of thumb for a successful heterologous protein overproduction strategy.

Out of the family of the Na^+-dependent neurotransmitter transporters, SERT is an excellent candidate to attempt overexpression. It is biochemically well characterised and there are a variety of radioligands commercially available such as serotonin, imipramine, citalopram, paroxetine and RTI-55. Extensive work has been invested into the solubilisation, purification and reconstitution of SERT (38–41), and it has been demonstrated that the non-glycosylated form is active with respect to serotonin transport and ligand binding (42). Nevertheless, the codon composition still corresponds to poorly expressed *E. coli* genes, which have a codon adaptation index smaller than 0.3 (43). The SERT gene index is 0.258 and 0.199 for the first 86 codons of the cytoplasmic N-terminus, respectively.

We have made five genetic constructs to study SERT expression (Fig. 1). The full length rat SERT gene has been cloned behind the highly inducible T7 RNA polymerase promoter (SERT, Fig. 1A; 45). To improve translation initiation the rat SERT gene was fused at codon six to the first 16 codons of the highly expressible T7 gene 10 (T7g10-SERT; Fig. 1B). The human and the rat SERT gene were fused at codon six to the highly expressed GST protein of *Schistosoma japonicum* (GST-SERT, Fig. 1C; 46). The GST protein fusion expression system has been used to overproduce critical proteins in a less toxic form, resulting in reduced inclusion body formation and an increased yield (47). The

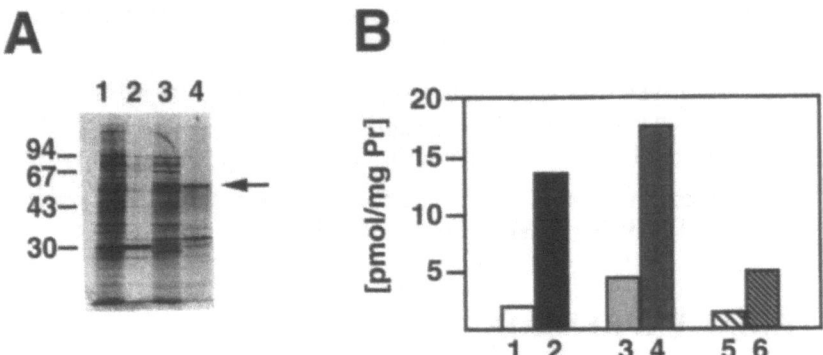

Figure 2. A-B: *In vivo* labelling and [^{3}H]imipramine binding of full-length SERT. A) An autoradiogram of a 10% polyacrylamide gel is shown. Plasmid pET3d-SERT (lanes 3, 4) and control plasmid pET3d (lanes 1, 2) were transformed into JM101/pGP1–2, respectively. Cells were radiolabelled with [^{35}S]methionine as described previously (44). In lanes 2 and 4 rifampicin was added to suppress radiolabelling of *E. coli* proteins. B) Whole cells of an OD$_{600}$ of 1.5 were incubated with 50 nM [^{3}H]imipramine for 30 min before rapid cell filtration and radioactivity determination (48). All constructs were analysed in strain BL21(DE3)/pLysE harbouring plasmids pET3d (column 1), pET3d-SERT (column 2), pRSET6b (column 3), pT7g10-SERT (column 4), pGEX-3X (column 5) and pGST-SERT (column 6).

system offers the convenience of single step purification via glutathione-agarose affinity chromatography and subsequent protease cleavage *i. e.* thrombin to prepare the desired protein. Two constructs were designed to study expression of the cytoplasmic N-terminal 86 residues of the human SERT, which comprise the putative regulatory domain hereafter termed NSERT. Construct four was engineered by introducing a stop codon at codon 87 and insertion of the gene fragment into a T7 promoter based expression plasmid (NSERT, Fig. 1D). Construct five was engineered by fusing the SERT codons 6 to 86 to the GST protein (GST-NSERT, Fig. 1E).

The *E. coli* strains JM101 and the Lon protease and OmpT periplasmic protease deficient strain BL21(DE3)/pLysE (49) were freshly transformed to analyse the three full length constructs, respectively, and cells were cultivated overnight at 30°C to prevent heat-shock response and to increase plasmid stability. Cells were then diluted and grown to OD$_{600}$ of 0.5–0.7 before protein production was induced for three to five hours by addition of IPTG. The non-fused SERT was labelled with [^{35}S]methionine and migrated on polyacrylamide gels at 58 kDa (Fig. 2A). The size agrees with that reported for the non-glycosylated protein (42, 50). Western blot analysis of cells expressing T7g10-SERT and the GST-SERT revealed molecular sizes of 57 kDa and 78 kDa, respectively. Cell fractionation showed that both proteins were found in inclusion bodies and in the membrane fraction (data not shown). Figure 2B shows specific binding of [^{3}H]imipramine cells expressing the three full length constructs. In contrast to native SERT, binding of imipramine was not Na$^+$-dependent. Isotherm saturation analysis of expressed T7g10-SERT revealed a K$_d$ of 73 μM and B$_{max}$ values between 0.9 and 3.7 nmol/mg protein (Fig. 3A). [^{3}H]imipramine binding was displaceable with non-labelled amitriptyline, desipramine, fluoxetine, imipramine, citalopram and serotonin (Fig. 3B and data not shown) with K$_i$ values ranging from 8 μM for amitriptyline to 6 mM for serotonin. Binding to [^{3}H]citalopram was also found to be specific and displaceable by the above mentioned ligands (Fig. 3C). Attempts to demonstrate [^{3}H]serotonin accumulation on whole cells or membrane vesicles failed to date.

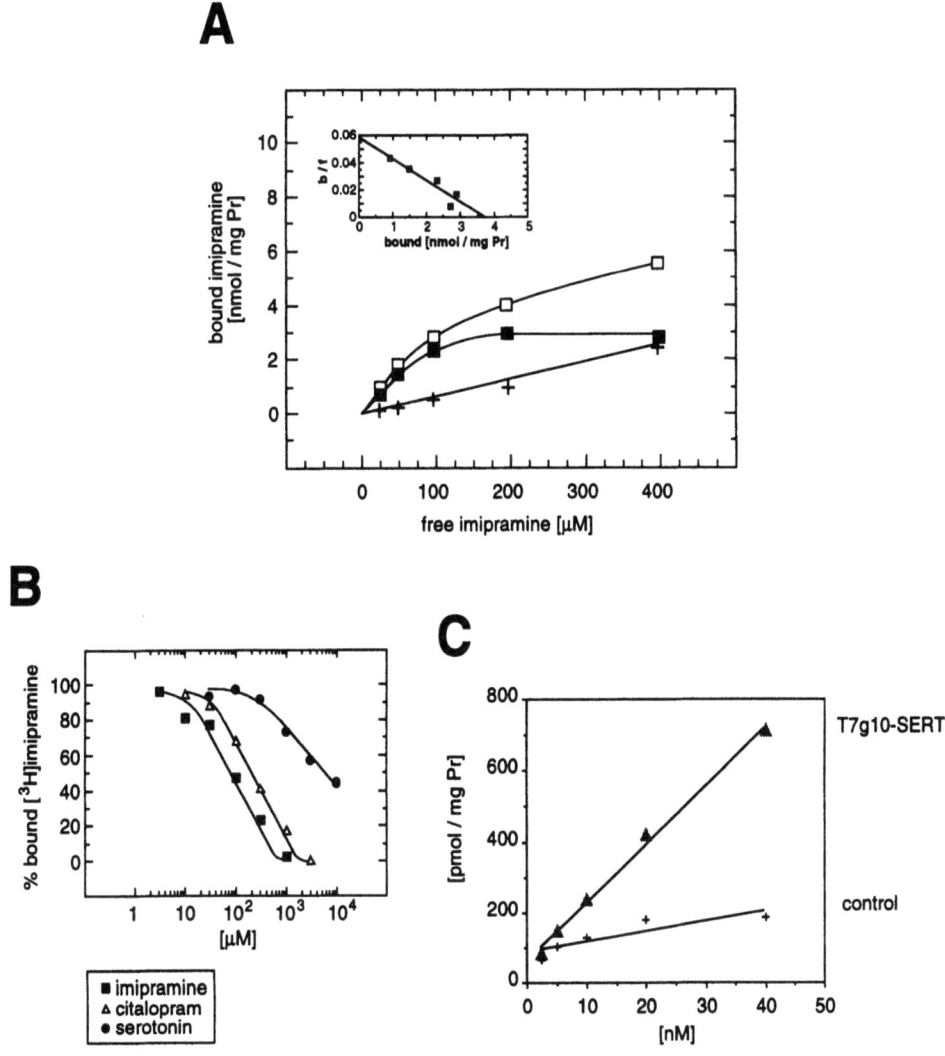

Figure 3. A–C: Binding of [^{3}H]imipramine and [^{3}H]citalopram to T7g10-SERT. A) Isotherm saturation analysis of [^{3}H]imipramine using whole cells. Open squares show total binding to cells expressing T7g10-SERT. Binding to control cells bearing the same plasmid (pRSET6b) without SERT gene insertion is shown by crosses. Closed squares resemble the specific binding. The inset shows the respective Scatchard plot. B) Displacement of 50 nM [^{3}H]imipramine with unlabelled imipramine, citalopram and serotonin. C) Binding of [^{3}H]citalopram to whole cells.

Expression of the two NSERT constructs revealed that non-fused NSERT was expressed at low levels, but that GST-NSERT was expressed at high levels. Figure 4 shows expression and purification of the 35 kDa GST-NSERT protein compared with the expression of the 26 kDa GST control protein. Similar amounts of protein production of about 20–50 mg/l cell culture were found (lanes 1, 2). The proteins were present in the soluble fraction (lanes 3, 4). After affinity purification both proteins were essentially pure (lanes 5, 6). The NSERT protein was cleaved off by thrombin treatment and purified with a yield of 3 mg/l culture. Pure NSERT was verified by N-terminal sequencing and mass spectroscopy. Experiments aiming to crystallise the GST-NSERT protein and aiming to study the structure of the 10 kDa NSERT domain by multidimensional NMR are currently in progress.

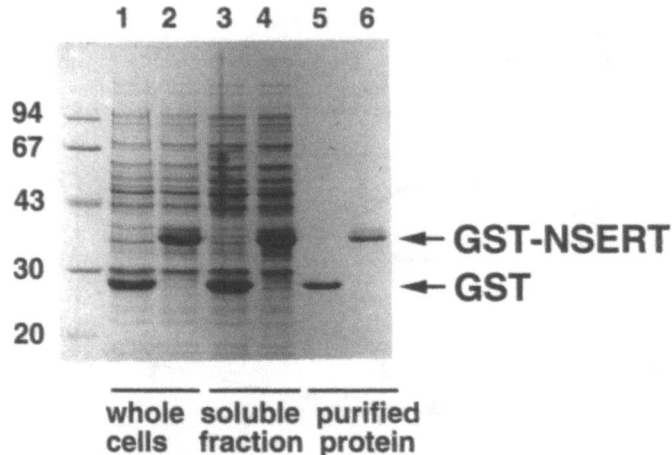

Figure 4. Overexpression of GST-NSERT. A 10% polyacrylamide gel stained with Coomassie Brilliant Blue is shown. Standard marker proteins expressed in kDa are shown in the left lane. For further explanation see text.

These results demonstrate that expression of the SERT and subdomains is feasible in *E. coli*. Expression levels of full-length SERT were up to 3.7 nmol/mg protein corresponding to about 500,000 imipramine binding sites per cell. These amounts should allow purification of milligram quantities from one litre of cell culture. To our knowledge the overexpression of SERT in *E. coli* represents the first example of a 12 TMH-type transporter originating from higher eukaryotes.

3. YEAST EXPRESSION

We have studied heterologous SERT expression in the methylotrophic yeast *Pichia pastoris*. The commercially available expression system is based on the chromosomal gene integration under the control of the methanol inducible promoter of alcohol oxidase, which can result in the production of 30% of total cell protein or up to several grams of protein per litre (51). Recently, the functional expression of the mouse 5-HT$_{5A}$ serotonin receptor has been reported as the first example for membrane protein expression in this system (52). 5-HT$_{5A}$ expression was shown to be significantly increased in a protease deficient strain.

We have integrated the rat SERT gene into the chromosome of *P. pastoris* wild-type and into a protease deficient strain, respectively. Southern blot analysis showed that integrants of single and multiple copies of the SERT gene were obtained. Northern blot analysis revealed the transcription of SERT gene mRNA upon growth on methanol. Specific [^{3}H]imipramine binding to membranes prepared from the producer strains was found to be of low affinity similar to that found in *E. coli*, and was displaceable by several SERT specific ligands. Western blot analysis of membranes showed a signal corresponding to a protein of about 58 kDa. These result demonstrate for the first time the possibility of functional SERT expression in *P. pastoris*.

4. INSECT CELL EXPRESSION

The baculovirus expression system has been used for the expression of numerous eukaryotic membrane proteins, most notable is the 12 TMH-type human glucose transporter GLUT1 which is overexpressed at 200 pmol/mg protein (reviewed in 27). Compared to yeast and *E. coli* expression systems, the baculovirus system has the advantage that the insect cell machinery confers all eukaryotic post-translational modifications and therefore resembles more closely the situation in the mammalian cell. The system is commercially available and can be scaled up to operate 100 litre fermentors. Tate and Blakely have successfully expressed the rat SERT in baculovirus (42). The transporter activity was comparable to brain SERT activity and expression levels were 9 pmol/mg membrane protein corresponding to 500,000 SERT copies per cell. They showed furthermore that affinity tags could be engineered at both ends of the protein allowing single step protein purification. The authors conclude that the expression level obtained in baculovirus may provide a suitable starting material for SERT purification in milligram quantities.

5. FUTURE PERSPECTIVES

SERT transporter has been heterologously overexpressed in *E. coli* to 3.7 nmol/mg total cell protein, in the yeast *P. pastoris*, in baculovirus to 9 pmol/mg membrane protein (42), in stable cell lines to 0.1 pmol/mg (53, 54) and in transiently transfected human embryonic kidney cells to 40 pmol/mg membrane protein (48). This compares to 2.8 pmol/mg in placenta brush-border membranes (55), to 2.5 pmol/mg platelet membranes (56) and 0.5 pmol/mg found in brain membranes (57).

Clearly, expression in *E. coli* produces the highest amounts of SERT. But this non-glycosylated form of SERT is only partially functional. It remains to be demonstrated if the produced membrane-embedded SERT or SERT protein formed in inclusion bodies can be renatured and reconstituted to restore full activity as has been demonstrated for other membrane protein (28, 30). Furthermore, an improvement of SERT expression could be achieved by growth medium supplementation with the osmoprotectants betaine and sorbitol (58; our unpublished results), and by expression in recently designed *E. coli* strains which tolerate membrane protein overproduction to higher levels (59). The baculovirus system provides functional SERT and is currently exploited to purify milligram amounts of SERT. Expression in the yeast *P. pastoris* has been achieved and future experiments will reveal expression levels and transporter functionality. Purification from mammalian cell culture expression systems seems not to be feasible, because the technology is expensive and difficult to scale up.

What kind of expression system finally will succeed to produce the milligram quantities of SERT required for the three-dimensional structure determination remains to be seen. The overviewed efforts directed towards SERT overexpression show that several promising strategies exist to achieve this goal and that these approaches may be applicable as well for other types of the biologically so important polytopic eukaryotic membrane proteins.

ACKNOWLEDGMENTS

We thank Chris Tate, Klaus-Peter Lesch and Klaas Nico Faber for stimulating discussions. This work was supported by the EC Programmes Biomed PL93110 and Human Capital and Mobility.

REFERENCES

1. Sirek, A., and Sirek, O.V., 1970, *C.M.A. Journal* **102**: 846–849.
2. Owens, M.J., and Nemeroff, C.B., *Clin. Chem.* **40**: 288–295.
3. Rudnick, G., and Clark, J., 1993, *Biochim. Biophys. Acta* **1144**: 249–263.
4. Marcusson, J.O., and Ross, S.B., 1990, *Psychopharmacol.* **102**: 145–155.
5. Sneddon, J.M., 1969, *Brit. J. Pharmacol* **37**: 680–688.
6. Meyerson, L.R., Ieni, J.R., and Wennogle, L.P., 1987, *J. Neurochem.* **48**: 560–565.
7. Phillips, O.M., Wood, K.M., and Williams, D.C., 1988, *Eur. J. Pharmacol.* **146**: 299–306.
8. Ramamoorthy, S., Bauman, A.L., Moore, K.R., Han, H., Yang-Feng, T., Chang, A.S., Ganapathy, V., Blakely, R., 1993, *Proc. Natl. Acad. Sci USA* **90**: 2542–2546.
9. Hoffman, B.J., Mezey, E., and Brownstein, M.J., 1991, *Science* **254**: 579–580.
10. Blakely, R.D., Berson, H.E., Fremeau, R.T., Jr., Caron, M.G., Peek, M.M. Prince, H.K., and Bradley, C.C., 1991, *Nature* **354**: 66–70.
11. Guastella, J., Nelson, N., Nelson, H., Czyzyk, L., Keynan, S., Miedel, M.C.,,Davidson, N., Lester, H.A., and Kanner, B.I., 1990, Science **249**: 1303–1306.
12. Mayser, W., Betz, H.., and Schloss, P., 1991, *FEBS Lett.* **295**: 203–206.
13. Lesch, K.-P., Wolozin, B.L., Estler, H.C., Murphy, D.L., and Riederer, P., 1993, *J. Neural. Transm.* **91**: 67–72.
14. Corey, J.L., Quick, M.W., Davidson, N., Lester, H.A., and Guastella, J., 1994, *Proc. Natl. Sci. Acad. USA* **91**: 1188–1192
15. Schloss, P., Mayser, W., and Betz, H. , 1992, *FEBS Lett.* **307**: 76–80.
16. Schloss, P., PŸschel, A.,and Betz, H.,1994, *Curr. Opin. Cell Biol.* **6**: 595–599.
17. Worrall, D.M., and Williams, D.C., 1994, *Biochem. J.* **297**: 425–436.
18. Reizer, J., Reizer, A., and Saier, M.H., Jr., 1994, *Biochim. Biophys. Acta* **1197**: 133–166.
19. Zafra, F., 1996, *J. Neurochem. (Suppl.2)* **66**: S42A.
20. Bendahan, A., and Kanner, B.I., 1993, *FEBS Lett.* **318**: 41–44.
21. Olivares,L., Aragòn, C., Gliménez, C., and Zafra, F., 1994, *J. Biol. Chem.* **269**: 28400–28404.
22. Quian, Y., Melikian, H.E., Moore, K.R., Duke, B.J., and Blakely, R.D., 1995, *Soc. Neurosci. Abstr.* **21**: 865
23. Anderson, G.M., and Horne, W.C., 1992, *Biochim. Biophys. Acta* **1137**: 331–337.
24. Ramamoorthy, J.D., Ramamoorthy, S., Papapetropoulos, A.,Catravas, J.D., Leibach, F.H., and Ganapathy, V., 1995, *J. Biol. Chem.* **270**: 17189–17195.
25. Sato, K., Betz, H., and Schloss, P., 1995, *FEBS Lett.* **375**: 99–102.
26. Henderson, P.J.F., 1991, *Curr. Opin. Struct. Biol.* **1**: 590–601.
27. Grisshammer, R., and Tate, C.G., 1995, *Quart. Rev. Biophys.* **28**: 315–422.
28. Fiermonte, G., Walker, J.E., and Palmieri, F, 1993, *Biochem. J.* **294**: 293–299.
29. Wohlrab, H., and Briggs, C., 1994, *Biochemistry* **33**: 9371–9375.
30. Berhe, A., Fristedt, U., and Persson, B.L., 1995, *Eur. J. Biochem.* **227**: 566–572.
31. Bibi, E., Gros, P., and Kaback, H.R., 1993, *Proc. Natl. Acad. Sci. USA* **90**: 9209–9213.
32. Sarkar, H.K., Thorens, B., Lodish, H.F., and Kaback, H.R., 1988, *Proc. Natl. Acad. Sci. USA* **85**: 5463–5467.
33. Harada, Y, Senda, T., Sakamoto, T., Takamoto, K, and Ishibashi, T., 1994, *J. Biochem*, **115**: 66–75.
34. Surek, B., Wilhelm, M., and Hillen, W., 1991, *Appl. Microbiol. and Biotechnol.* **34**: 488–494.
35. Goldman, E., Rosenberg, A.H., Zubay, G., and Studier, F.W., 1995, *J. Mol. Biol.* **245**: 467–473.
36. Kane, J.F., 1995, *Curr. Opin. Biotechnol.* **6**: 494–500.
37. Ferreira, G.C., and Pederson, P.L., 1992, *J. Biol. Chem.* **267**: 5460–5466.
38. Talvenheimo, J., and Rudnick, G., 1980, *J. Biol. Chem.* **255**: 8606–8611.
39. Biessen, E.A.L., Horn, A.S., and Robillard, G.T., 1990, *Biochemistry* **29**: 3349–3354.
40. Launay, J.-M., Geoffroy, C., Mutel, V., Buckle, M., Cesura, A., Alouf, J.E., and Da Prada, M., 1992, *J. Biol. Chem.* **2647**: 11344–11351.
41. Ramamoorthy, S., Cool, D.R., Leibach, F.H., Mahesh, V.B., and Ganapathy, V., 1992, *Biochem. J.* **286**: 89–95.
42. Tate, C.G., and Blakely, R.D., 1994, *J. Biol. Chem.* **269**: 26303–26310.
43. Sharp, P.M., and Li, W.-H., 1987, *Nucleic Acids Res.* **15**: 1281–1295
44. Titgemeyer, F., Jahreis, K., Ebner, R., and , Lengeler, J.W., 1996, *Mol. Gen. Genet.* **250**: 197–206.
45. Studier, FW., Rosenberg, A.H., Dunn, J.J., and Dubendorf, J.W., 1990, *Meth. Enzymol.* **185**: 60–89.
46. Smith, D.B., and Johnson, K.S., 1988, *Gene* **67**: 31–40.
47. Guan, K.L., and Dixon, J.E., 1991, *Analyt. Biochem.* **192**: 262–267.

48. Schloss, P., and Betz, H., 1995, *Biochemistry* **34:** 12590–12595.
49. Studier, F.W., 1991, *J. Mol. Biol.* **219:** 37–44.
50. Sur, C., Betz, H., and Schloss, P., 1996, *Neurosci.* **73:** 217–231.
51. Faber, K.N., Harder, W., AB, G., and Veenhuis, M., 1995, *YEAST* **11:** 1331–1344.
52. Weiβ, H.M., Haase, W., Michel, H., and Reiländer, H., 1995, *FEBS Lett.* **377:** 451–456.
53. Ramamoorthy, S., Cool, D.R., Mahesh, V.B., Leibach, F.H., Melikian, H.E., Blakely, R.D., and Ganapathy, V., 1993, *J. Biol. Chem.* **268:** 21626–21631.
54. Gu, H., Wall, S.C., and Rudnick, G., 1994, *J. Biol. Chem.* **269:** 7124–7130.
55. Ramamoorthy, S. Leibach, F.H., Mahesh, V.B., and Ganapathy, V., 1993, *Placenta* **14:** 449–461.
56. Wall, S.C., Innis, R.B., and Rudnick, G., 1993, *Mol. Pharmacol.* **43:** 264–270.
57. Graham, D., Esnaud, H., and Langer, S.Z., 1992, *Biochem. J.* **286:** 801–805.
58. Blackwell, J.R., and Horgan, R., 1991, *FEBS Lett.* **295:** 10–12.
59. Miroux, B., and Walker, J.E., 1996, *J. Mol. Biol.* **260:** 289–298.

Section 30: Structures of Neuronal Proteins in Rationalized Drug Design

SCORPION TOXINS

Functional Variations on a Structural Theme

André Ménez and Marc Dauplais

Département d'Ingénierie et d'Etudes des Protéines
CEA Saclay
Gif-sur-Yvette, France

1. INTRODUCTION

Scorpions form an order of animals in the phylum of Arthropods. Approximately 1400 species of scorpions, spread into 9 families, are distributed on all major land masses of the world, except Antartìca. Scorpions successfully survived over the past 450 million years, having undergone little morphological change since their appearance, despite a number of functional differences regarding their locomotion and respiration apparatus. As a result, scorpions are often considered as "living fossils", which carry a number of unique biological informations (1). Recent progresses have emerged in the literature regarding the molecular properties of venoms and toxins from scorpions and their ability to act on a wide diversity of prey. In particular, structural studies have shed light on the molecular basis associated with the functional diversity of scorpion toxins.

2. SCORPION TOXINS THAT ACT ON SODIUM CHANNELS

When injected in mice, the most potent scorpion toxins cause signs that are globally similar to those exerted by the venom, including convulsions and paralysis due to an increase of neurotransmitter release in the nerve endings. These toxins act selectively and with high affinity on sodium channels (2, 3) which are maintained in an open state (4). At least two categories of such toxins can be distinguished. A first category includes the α-toxins that act from the extracellular side of the membrane by slowing or blocking inactivation of sodium channels. Once these toxins are bound, sodium channels maintain an entrance of sodium flux and hence action potentials on nerve membranes are prolonged. These toxins bind in a voltage-dependent manner and structural changes of sodium channels occurs upon

From a symposium organized by Hermona Soreq (Jerusalem, Israel) and André Ménez (Saclay, France). (Supported by the IBRO and the Commissariat à l'Energie Atomique, Centre d'Etudes de Saclay.)

binding of such neurotoxins (5, 6). Toxins of the second category affect activation of sodium channels. These β-toxins also act from the extracellular side of the membrane, causing a shift of the voltage-dependent activation to more negative membrane potentials and slowing the onset of sodium current inactivation. In contrast to α-toxins, binding of β-toxins is not modified by membrane depolarization. In addition to these different types of actions, scorpion toxins often express a specificity for particular species. Thus, scorpion venoms contain toxins acting specifically on sodium channels of either mammals, insects or crustaceans, some toxins exerting cross-species actions. At least two groups of insect-selective toxins, the excitatory and depressant toxins (7), affect axonal sodium conductance in insects, like β-toxins affect sodium channels in vertebrates (8). Therefore, scorpion toxins exert various actions on sodium channels of various potential prey.

Scorpion toxins that act on sodium channels are small proteins of 60–70 amino acids and 4 disulfides. More than 45 different sodium channel-acting toxins have been isolated and their amino acid sequences have been identified. The three-dimensional structures of several scorpion toxins have been elucidated by X-ray diffraction and by NMR spectroscopy. Irrespective of whether they act as α-toxins or β-toxins in mammals, insects, crustaceans or else, all the toxins adopt the same general fold (9, 10, 11) that is shown in the figure 1 (top left).

This figure shows the structure of a toxin variant from *Centruroides sculpturatus* Ewing (CsE-v3) as elucidated by X-ray crystallography (9) This fold contains a short helix and a three-stranded β-sheet that are linked to each other by two disulfides. The rest of the chain adopts no canonical secondary structure.

3. SCORPION TOXINS THAT ACT ON POTASSIUM CHANNELS

Some scorpion toxins act on potassium channels, stopping potassium currents, and thus modify action potentials by preventing membrane repolarization (12). These toxins block various subtypes of voltage-dependent and/or calcium-activated potassium channels. Four families of such toxins have been identified, grossly mirroring the current classification of potassium channels (13).

Scorpion toxins that block potassium channels are small proteins with 30–40 amino acids and three disulfides. The architectures of various toxins of that type have been elucidated by NMR and modelling (14), revealing that irrespective of their subtype specificity, scorpion toxins acting on potassium channels adopt the same fold. As an illustration, Figure 1 (top right) shows the solution structure of charybdotoxin (15, 16), a toxin from the scorpion *Leiurus quinquestriatus hebreus* which binds to calcium-activated and voltage-gated potassium channels (17, 18). One can appreciate the compactness of this structure with a three-stranded β-sheet and a short helix associated to each other by two of the three disulfides.

4. A COMMON FOLD FOR SCORPION TOXINS

A close inspection of the folds of toxins acting on sodium or potassium channels (Figure 1, top left and right) reveals obvious similarities. Both folds possess the same general core composed of a small β-sheet strand and a short helix linked to each other by two disulfides. These two elements of secondary structure can be nicely superimposed in toxins acting on sodium or potassium channels. The average deviation along the polypeptide chains between the two toxins is equal to 1 Å. A similar calculation made with toxins from different phyla, acting on the same category of channels (potassium channels), would

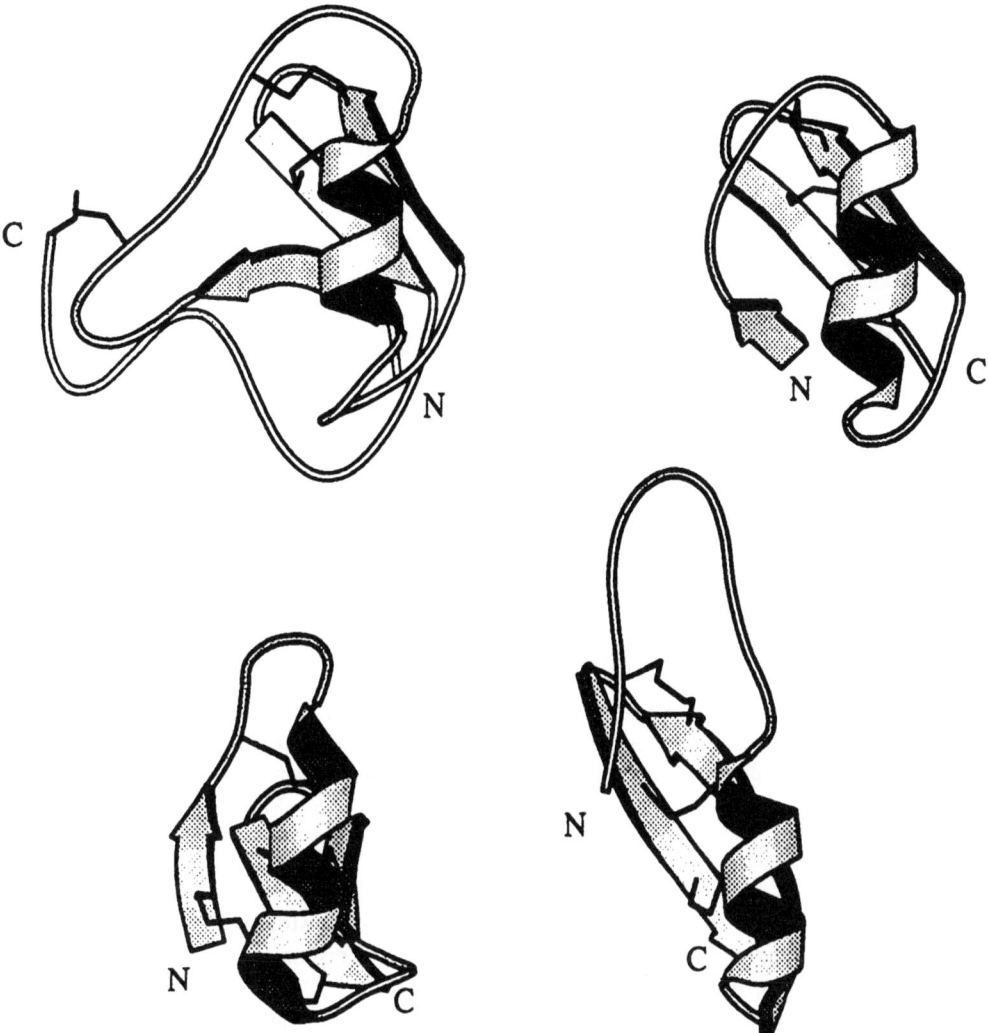

Figure 1. Three-dimensional structures of the polypeptide chains of three scorpion toxins and one insect defensin. The figure on top and left shows the typical architecture of a toxin that acts on sodium channels (9). The figure on top and right shows the structure of a toxin that acts on potassium channels (15, 16). On the bottom are shown, on the left the structure of a toxin that acts on chloride channels (19) and on the right the architecture of an insect defensin (20). One can appreciate the structural similarities between these functionally different proteins. The helix and β-sheet structures form a comparable compact core around which much variations occur.

yield a far larger deviation (approximately 7 Å). Other scorpion toxins may affect chloride channels. Such toxins are also small proteins with 36 amino acids and 4 disulfides and again, they fold as other scorpion toxins (19), as shown in figure 1 (bottom left).

To summarize the situation, all scorpion toxins acting on ion channels, irrespective of their general or subtype specificity, possess the same structural core that is decorated with a number of elements, more variable in size. These elements include the N-terminal region, the C-terminal region and loops. In other words, the functional diversity of scorpion toxins might be associated with a natural engineering of a core that is now identified. How such an engineering might occur remains unknown; however, the associated biochemical process must be exquisitely dicriminating since it preserves the spatial organiza-

Figure 2. Three-dimensional structure of BgK, a potassium-blocking toxin from the sea anemone Bunodosoma granulifera. The structure has been elucidated by NMR spectroscopy and modeling (44). This structure contains helical structure with no β-sheet, in contrast to potassium channel-blocking toxins from scorpions (see figure 1, top right).

tion of the core whilst its periphery can undergo considerable structural variability. Additional studies on the organization of the genes encoding scorpion toxins may help to clarify this fascinating problem.

Other proteins, not elaborated in venom glands, possess the structural core of scorpion toxins. These proteins, called defensins, act as antibacterial components. The solution structure of one of them (20) is illustrated in figure 2 (bottom, right). Defensins are found in the hemolymph of insects and of scorpions themselves (21). It is likely, therefore, that defensins and toxins are evolutionary linked. Though this is yet unprooven, it is remarkable that the same structural core is exploited to express a diversity of functions in histologically distinct regions of scorpions.

In summary, the functional prodigality of scorpion toxins is associated with a structural economy. Interestingly, the same general principle is adopted by other venomous animals, including snakes for which, however, more than a single fold is exploited (22). Such a practical scenario has suggested the possibility to engineer novel functions on conserved toxin cores (22). Preliminary successes have already been obtained in this domain (23, 24, 25), offering new potentialities in the frame of drug discovery.

5. HOW DO SCORPION TOXINS RECOGNIZE THEIR TARGETS?

Protein-protein interactions usually involve several amino acids on each partner. One way to identify interacting residues in such complexes is to modify one at a time, as many residues as possible on both partners. In this respect, site-directed mutational analyses appears to be particularly powerful. In 1991, preliminary results concerning site-directed mutagenesis experiments on a snake curaremimetic toxin were described (26, 27). More mutational data were subsequently reported on the same snake toxin (28, 29, 30) and on charybdotoxin (31, 32, 33), a scorpion toxin acting on potassium channels (17).

Charybdotoxin possesses between 4 and 7 functionally important residues, depending on the actual target, i.e. the channel subtype (31, 33, 34, 35). Among these residues is the lysine at position 27 which plays a predominant binding role. This residue is conserved in all scorpions toxins acting on potassium channels (10, 14, 15, 36, 37, 38). Some evidences have suggested that the lysine 27 enters into the pore by which potassium ions permeates through the membrane (32, 39, 40). The current view is that charybdotoxin interacts by the exposed face of its β-sheet with a vestibule located around the pore on the extracellular side of the channel (32, 40).

6. SCORPION TOXINS: JUST ONE SOLUTION AMONG OTHERS

Various venomous animals, other than the scorpions, can synthesize toxins acting on potassium channels. This is the case for snakes (41), cone shells (42) and sea anemones (43). Furthermore, some toxins elaborated by these animals compete with each other for binding to potassium channels. Thus, some scorpion toxins and sea anemone toxins inhibit the binding of dendrotoxin to brat brain synaptosomes. To understand the molecular basis associated with this mutually exclusive binding, we have solved the solution structure of a sea anemone toxin, BgK, from *Bunodosoma granulifera*, and compared it (44) to that of charybdotoxin. Although BgK contains 37 residues with 3 disulfides like charybdotoxin, its amino acid sequence as well as its disulfide pairings differ from those found in scorpion toxins (in preparation). Not surprisingly, the 3-D structure of BgK (Figure 2A), which is similar to that of another K$^+$ channel-blocking toxin from sea anemone (45), has no similarity with that of a scorpion toxin (Figure 1). BgK contains two nearly perpendicular helical stretches linked to each other by one disulfide. The N- and C-terminal regions are maintained in spatial proximity by another disulfide whereas the third disulfide brings the loops between residues 16–24 and 31–37 inside the center of the molecule, providing the toxin with a globular shape. Despite BgK and scorpion toxins have distinct architectures, they possess comparable flat surfaces, of approximately 200 Å^2 (18 × 12 Å and 21 × 9 Å, respectively). They are formed, in BgK, by the edge of the 9–16 helix and the two flanking loops (44) and, in scorpion toxins, by the exposed face of the b-sheet (14, 15).

We have identified functional residues of BgK by submitting the toxin to an alanine-scanning, elaborating all single point variants by chemical synthesis. Competitions between the variants and dendrotoxin on rat brain synaptosomes (in collaboration with A. L. Harvey) revealed the predominant role of a lysine, i.e. Lys25. This finding agrees with recent observations made with ShK, a homologous K$^+$ channel-blocking toxin from *Stichodactyla helianthus* (46). In BgK, two additional residues, i.e. Phe6 and Tyr26, also contribute to channel recognition. The other mutations caused either lower (Glu8, His13 and Ser23) or no affinity changes. Therefore, a lysine appears to be a major actor for both toxins from scorpions and sea anemones to recognize potassium channels. More strikingly, if one considers this lysine as a common center in BgK and charybdotoxin, their respective functionally important Tyr26 and Tyr36 nicely superimpose (44). The distance that separates the lysine from the aromatic side chain is 6.6 ± 1.0 Å, in both cases. The common capacity of the two toxins to recognize potassium channels seems therefore to be associated with the conservation of an identical diad of functional residues, a lysine and a close aromatic residue. This observation offers a molecular explanation as to how two toxins with unrelated architectures can compete for potassium channels. No superimposition could be achieved between other functional residues of scorpion and sea anemone toxins. Therefore, the scorpion and sea anemone toxins commonly share a lysine and an aromatic residue that constitute a kind of conserved functional core, assisted by more variable functional residues which might provide each toxin with a unique specificity toward given potassium channel subtypes.

7. CONCLUSION

We have seen that animal toxins can undergo both divergent and convergent functional evolutions. The divergent evolution is inferred from the finding that scorpion toxins with similar folds exert different biological functions. That a given scaffold can accept a wide diversity of functional surfaces offers new opportunities to engineer predetermined functions on a toxin fold. The convergent evolution of animal toxins is deduced from the observation that animals from distinct phyla can produce toxins with comparable functions though they have unrelated structures. Furthermore, two distinct folds with similar binding properties toward potassium channels possess an identical diad of functional residues organized in a comparable topographical manner. This finding suggests that targets, here the ion channels, may play an essential role during evolution to screen toxin functions. Also, it suggests that a predetermined binding function can be grafted on a diversity of host templates.

REFERENCES

1. Polis, G.A., 1990, *The Biology of Scorpions* (G.A. Polis, Ed), Stanford University Press, Stanford, California, USA.
2. Romey, G., Abita, J.P., Chicheportiche, R., Rochat, H., and Lazdunski, M., 1976, *Biochim. Biophys. Acta* **448**: 607–619.
3. Couraud, F., Jover, E. Dubois, J.M., and Rochat, H., 1982, *Toxicon* **20**: 9–16.
4. Caterall, W.A., 1992, *Physiol. Rev.* **72**: 15–48
5. Cestele, S., Ben Khalifa, R., Pelhate, M., Rochat, H., and Gordon, D., 1995, *J. Biol. Chem.* **270**: 15153–15161.
6. Gordon, D., 1996, *Cellular and Molecular Mechanisms of Toxin Action*, in press.
7. Zlotkin, E., Kadouri, D., Gordon, D., Pelhate, M., Martin, M.F., and Rochat, H., 1985, *Arch. Biochem. Biophys.* **240**: 877–887.
8. Stankiewicz, M., Ben Khalifa, R., Grolleau, F., Tomaszewski, R., Kadziela, W., and Pelhate, M., 1996, *Physiology and ecotoxicology of insects and mechanisms of adaptation in vertebrates* (Z. Bargiel, Ed). Scientific Society of Torun, Torun, Poland.
9. Fontecilla-Camps, J., Almassy, R.J., Suddath, F.L., Watt, D.D., and Bugg, C.E., 1980, *Proc. Natl. Acad. Sci. USA* **77**: 6496–6500.
10. Meunier S., Bernassau, J.M., Sabatier J.M., Martin-Eauclaire, M.F., Van Rietschoten, J., Cambillau, C., and Darbon, H., 1993, *Biochemistry* **32**: 11969–11976.
11. Lebreton, F., Delepierre, M., Ramirez, A.N., Balderas, C., and Possani, L.D., 1994, *Biochemistry* **33**: 11135–11149.
12. Carbone E., Wanke E., Prestipino G., Possani L.D., and Maelicke, A., 1982, *Nature* **296**: 90–91.
13. Miller, C., 1995, *Neuron* **15**: 5–10.
14. Dauplais, M., Gilquin, B., Possani, L.D., Gurrola-Briones, G., Roumestand, C., and Ménez, A., 1995, *Biochemistry* **34**: 16563–16573.
15. Bontems, F. Roumestand, C., Gilquin, B., Ménez, A., and Toma, F., 1991, *Science* **254**: 1521–1523.
16. Bontems, F., Gilquin, B., Roumestand, C., Ménez, A., and Toma, F., 1992, *Biochemistry* **31**: 7756–7764

17. Miller C., Moczydlowsi, E., Latorre R., and Philips M., 1985, *Nature* **313**: 316–318.
18. Sands, S.B., Lewis, R.S., and Cahalan, M.D., 1989, *J. Gen. Physiol.* **93**: 1061–1074
19. Lippens, G., Najib, J., Wodak, S.J., and Tartar, A., 1995, *Biochemistry* **34**: 13–21.
20. Hanzawa, H., Shimada, I., Kuzuhara, T., Komano, H., Kohda, D., Inagaki, F., Natori, S., and Arata, Y., 1990, *FEBS lett.* **269**: 413–420.
21. Cociancich, S., Goyffon, M., Bontems, F., Bulet, P., Bouet, F., Ménez, A., and Hoffmann, J., 1993, *Biochem. Biophys. Res. Comm.* **194**: 17–22.
22. Ménez, A., Bontems, F., Roumestand, C. Gilquin, B., and Toma, F., 1992, *Proc. of the Royal Soc. of Edimburgh* **99B**: 83–103.
23. Vita, C., Roumestand, C., Toma, F., and Ménez, A., 1995, *Proc. Natl. Acad. Sci. USA* **92**: 6404–6409.
24. Zinn-Justin, S., Guenneugues, M., Drakopoulou, E., Gilquin, B., Vita, C., and Ménez, A., 1996, *Biochemistry*, in press.
25. Drakopoulou, E., Zinn-Justin, S., Guenneugues, M., Gilquin, M., Ménez, A., and Vita, C., 1996, *J. Biol. Chem.*, in press.
26. Ducancel, F., Bouchier, C., Tamiya, T., Boulain, J.-C., and Ménez, A., 1991, *Snake Toxins*, (Harvey, A.L. Ed) Pergamon Press, New-York, PP. 385–414.
27. Ménez, A., 1991, *Toxicon* **29**: 1171.
28. Hervé, M., Pillet, L., Humbert, P., Trémeau, O., Ducancel, F., Hirth, C., and Ménez, A., 1992, *Eur. J. Biochem.* **208**: 125–131.
29. Pillet, L., Trémeau, O., Ducancel, F., Drevet, P., Zinn-Justin, S., Pinkasfeld, S., Boulain, J-C., and Ménez, A., 1993; *J. Biol. Chem.* **268**: 909–916.
30. Trémeau, O., Lemaire, C., Drevet, P., Pinkasfeld, S., Ducancel, F., Boulain, J-C., and Ménez, A. (1995) *J. Biol. Chem.* **270**: 9362–9369.
31. Park, C.S., and Miller, C., 1992, *Biochemistry* **31**: 7749–7755.
32. Goldstein, S.A.N., and Miller, C., 1993, *Biophy. J.* **65**: 1613–1619.
33. Goldstein, S.A.N., Pheasant, D.J., and Miller, C., 1994, *Neuron* **12**: 1377–1388.
34. Stampe, P., Kolmakova-Partensky, L.K., and Miller, C., 1994, *Biochemistry* **33**: 443–450.
35. Miller, C., 1992, *Biophys. J.* **62**: 8–9.
36. Johnson, B.A., and Sugg, E.E., 1992, *Biochemistry* **31**: 8151–8159.
37. Johnson, B.A., Stevens, S.P., and Williamson, J.M., 1994, *Biochemistry* **33**: 15061–15070.
38. Fernandez, I., Romi, R., Szendeffy, S., Martin-Eauclaire, M.F., Rochat, H., Van Rietschoten, J., Pons, M., and Giralt, E., 1994, *Biochemistry* **33**: 14256–14263.
39. MacKinnon, R., and Miller, C., 1988, *J. Gen. Physiol.* **91**: 335–349.
40. Park, C.S., and Miller, C., 1992, *Neuron* **9**: 307–313
41. Harvey, A.L., and Anderson, A.J., 1991, *Snake toxins*, (A.L. Harvey, Ed) Pergamon Press, PP 131–164.
42. Terlau, H., Shon, K-J., Grilley, M., Stoker, M., Stühmer, W., and Olivera, B.M., 1996, *Nature* **381**: 148–151
43. Aneiros, A., Garcia, I., Martinez, J.R., Harvey, A.L. Anderson, A.J., Marshall, D.L., Engström, A., Hellman, U., and Karlsson, E., 1993, *Biochim. Biophys. Acta* **1157**: 86–92.
44. Dauplais, M., Lecoq, A., Song, J., Cotton, J., Jamin, N., Gilquin, B., Roumestand, C., Vita, C., de Medeiros, C.L.C., Rowan, E.G., Harvey, A.L. and Ménez, A., 1996, *J. Biol. Chem.* in press.
45. Tudor, J.E., Pallaghy, P.K., Pennington, M.W., and Norton, R.S., 1996, *Nature, Structural Biology* **3**: 317–320.
46. Pennington, M.W., Mahnir, V.M., Krafte, D.S., Zaydenberg, I., Byrnes, M.E., Khaytin, I. Crowley, K., and Kem, W.R., 1996, *Biochem. Biophys. Res. Comm.* **219**: 696–701.

A NOVEL STRATEGY OF EFFECT-DIRECTED LIGAND DESIGN FOR G-PROTEIN COUPLED RECEPTORS

Jaak Järv

Institute of Chemical Physics
Tartu University
Tartu, Estonia

1. INTRODUCTION

Effectiveness of binding and the type of physiological activity are the two aspects of molecular recognition of drugs, which both should be quantitatively related to structure of these molecules in the case of the effect-directed ligand design. While different approaches have been worked out for quantitative analysis of the interrelationship between ligand structure and binding affinity (1), the formalised schemes for description of the second aspect of receptor specificity, i.e. for prediction of the agonistic, partially agonistic and antagonistic properties of ligands proceeding from their structure have not been proposed. We would like to fill this cap by introducing a non-exclusive model of ligand binding, which main features and possible implications in the effect-directed design of receptor ligands are discussed in this paper.

2. EXCLUSIVE AND NON-EXCLUSIVE LIGAND BINDING MODELS

There are two principally different ways to explain how antagonists block the agonist-induced receptor response. Initially, competition of agonist and antagonist molecules for the same receptor site was proposed in the classical receptor theory (2), assuming that antagonist occupies the receptor binding site and thus interferes with agonist binding and inhibits its effect. This model of exclusive ligand binding was further developed by assuming the presence of different binding sites for agonist and antagonist on the receptor molecule, while the inhibitory effect of antagonist should arise through allosteric exclusion of agonist from its binding site (for review see (3)).

From a symposium organized by Hermona Soreq (Jerusalem, Israel) and André Ménez (Saclay, France). (Supported by the IBRO and the Commissariat à l'Energie Atomique, Centre d'Etudes de Saclay.)

Neurochemistry, edited by Teelken and Korf
Plenum Press, New York, 1997

On the other hand, the possibility of simultaneous (non-exclusive) binding of two ligands with two distinct receptor sites, which are responsible for the agonistic and antagonistic effects, was proposed (4). Following this concept the ligand binding at one of these sites triggers the receptor response, whereas the other site elicits another response leading to the receptor blockade.

Thus, the main difference between the exclusive and non-exclusive ligand binding models is the possibility of simultaneous occupation of two receptor sites by two ligand molecules.

3. THE NON-EXCLUSIVE MODEL OF LIGAND BINDING

The model to be considered assumes that two binding sites exist simultaneously on G-protein coupled receptors (4,5). These sites, further denoted as α and β, interact independently with ligands and possess different ligand binding specificity. Binding of any ligand in the site α exerts the agonistic response, while ligand binding in the site β inhibits this response by blocking the conformational transition, which is responsible for transduction of the signal in the receptor molecule. Therefore, three different complexes of ligand A with the receptor R can be formed, including one ternary complex $A_\beta RA_\alpha$ as shown in Figure 1.

Under the excess of ligand over the receptor concentration ($R_0 \ll [A]$), this scheme can be described by 3 independent equilibria and by the mass conservation equation for the receptor, also defined in Figure 1. As the binary complex $_\beta RA_\alpha$ evokes the receptor response, the observed effect (E_{rel}) should be proportional to the ratio $[_\beta RA_\alpha]/R_0$. By appropriate algebraic manipulations of these equations the relationship of E_{rel} on A concentration can be derived (Figure 1), and the parameters of this equation can be calculated from the E_{rel} vs. ligand A concentration plots.

It is important to emphasise that differently from all exclusive ligand binding models the dose-response relationship includes quadratic term. This, in turn, leads to bell-shaped dose-response curves, which cannot be described by the exclusive ligand binding models, although the presence of the two distinct receptor sites α and β can still be assumed. This means that the formation of the receptor-ligand ternary complex, involving two ligands and one receptor molecule, is an obligatory condition for the deviation of the dose-response curve from hyperbolic plot. And on the contrary, the bell-shaped dose-response relationships give evidence for the non-exclusive ligand binding mechanism.

4. DUAL EFFECT OF AGONISTS

According to the non-exclusive model of ligand binding, the distinction between agonists and antagonists by the receptor is based on different affinity of these compounds for the agonistic and antagonistic binding sites. However, this distinction cannot be absolute if some common interaction mechanisms (hydrophobicity, ion-ion interactions, H-bonds etc.) are involved for molecular recognition of ligands in these sites. Therefore it can be predicted that agonists may also bind to the antagonistic site β at sufficiently high drug concentrations. This, in turn, should inhibit the receptor response elicited by the same ligand in site α and lead to the bell-shaped form of the dose-response curve in sufficiently wide agonist concentration interval. This situation is described by the quadratic term of the dose-response equation shown in Figure 1.

$$A\beta R\alpha \xleftrightarrow{K23} A\beta RA\alpha$$
$$\downarrow K30 \qquad\qquad \downarrow K21$$
$$\beta R\alpha \xleftrightarrow{K10} \beta RA\alpha \rightarrow \text{Response}$$

$$K10 = \frac{[A][\beta R\alpha]}{[\beta RA\alpha]}$$

$$K21 = \frac{[A][\beta RA\alpha]}{[A\beta RA\alpha]}$$

$$K30 = \frac{[A][\beta R\alpha]}{[A\beta R\alpha]}$$

$$R0 = [A\beta R\alpha] + [A\beta RA\alpha] + [\beta RA\alpha] + [\beta R\alpha]$$

$$E rel = \frac{[A]}{K10 + \left(1 + \dfrac{K10}{K30}\right)[A] + \dfrac{[A]^2}{K21}}$$

Figure 1. Non-exclusive model of ligand binding for two-site receptor $_\beta R_\alpha$. Site α is responsible for agonistic activity, site β blocks this activity, but does not interfere with ligand binding in site α.

Such dual effect of agonists was demonstrated in the case of the muscarinic receptor mediated inhibition of adenylate cyclase activity (6) and mobilization of intracellular calcium (7), but has also been observed in earlier pharmacological studies (8). In all these cases agonists evoked the response in dose-dependent manner, but the further increase in their concentration led to the bell-shaped form of the plots with dose-dependent downturn phase.

Following the non-exclusive model of ligand binding, it can be predicted that the inhibitory effect of agonist should be additive to the inhibitory effect initiated by antagonists, as in this case these ligands interact with the same receptor site. This situation was analysed experimentally by measuring mobilization of Ca^{2+} ions in the SH-SY5Y cells in the case of combined application of agonist and antagonist (unpublished data). Carbachol and atropine as classical and specific muscarinic ligands were selected for this experiment. The results of this study revealed that the bell-shaped dose-response curves could be observed also in the presence of low atropine concentrations, but in these cases the cell responses were smaller and a clear shift of the up-going phase of the dose-response curves towards higher carbachol concentrations was observed. On the other hand, the inhibitory phase of the dose-response curves was not shifted in these experiments, pointing that carbachol and atropine inhibited the receptor response by the same mechanism. In general, adequate description of the data was obtained by assuming that at least two ligand molecules can bind simultaneously and independently on the receptor and these interactions can be presented by the mass-action law.

Hence, within the non-exclusive ligand binding model the phenomenon of antagonism is not explained on the level of the ligand binding step, but should be related to some conformational transition of the receptor protein, responsible for the signal transmission. Therefore it is important to recapitulate that within the receptor concept under consideration the phenomena of agonism and antagonism arise from specific properties of the appropriate binding sites, and differentiation of the ligands by these sites determines the type of the pharmacological activity of the drug. Thus both carbachol and atropine can similarly antagonise the muscarinic receptor response while binding in the antagonistic receptor site β. As a result of that the inhibitory effects of these ligands were additive and the down-going phase of the bell-shaped dose response curve for carbachol was not shifted in the presence of atropine. But affinities of these ligands for this site were dramatically different (above 7 powers of magnitude).

5. KINETIC EVIDENCE FOR RECEPTOR-AGONIST-ANTAGONIST TERNARY COMPLEX FORMATION

The key point of the non-exclusive ligand binding model is formation of the ternary complex between receptor and two ligand molecules, which may bind in the agonistic and antagonistic binding sites. Evidence for simultaneous binding of agonist and antagonist with the same receptor molecule has been obtained in the case of muscarinic acetylcholine receptors by means of kinetic analysis of the receptor-ligand interactions. A typical muscarinic antagonist, [^{3}H]N-methylpiperidinyl benzilate, and two agonists, carbachol and oxotremorine, were used in these experiments (9), based on the fact that potent muscarinic antagonists interact with the receptor by a complex kinetic mechanism, involving fast formation of antagonist-receptor complex and the following "isomerization" of this complex into another slowly dissociating complex (10). It has been found that carbachol and oxotremorine had no effect on [^{3}H]N-methylpiperidinyl benzilate binding, but inhibited the "isomerization" step (9). This means that the antagonist binds equally well to the free receptor and to the receptor-agonist complex, suggesting formation of the ternary agonist-antagonist-receptor complex. At the same time inhibition of the "isomerization" step by agonists gives the impression of competition between the agonist and antagonist in the radioligand displacement experiments. For this reason the conventional equilibrium binding studies cannot reveal the formation of the agonist-antagonist-receptor ternary complexes (5,11).

6. TWO LIGAND-BINDING SITES ON MUSCARINIC RECEPTOR

The idea of agonist and antagonist binding in sterically different areas on muscarinic receptors, formulated in 1980 on the basis of kinetic analysis of ligand binding (9), has been supported later by site-directed mutagenesis experiments with the rat m3 muscarinic receptor (12,13) as well as by molecular dynamics calculations for the same receptor (14). In the former case it has been found that the replacement of N617(507) (asparagine residue 507 in the primary structure of the m3 muscarinic receptor, corresponding to the position 17 of the transmembrane helix 6) with A, S and D was not critical for agonist binding as well as for agonist-induced receptor activation, but dramatically reduced affinity of atropine-like ligands (12,13). Even if this asparagine residue is not directly interacting with agonists and only stabilises the binding site structure, the data suggest allosteric location of agonist and antagonist molecules on the receptor.

On the other hand, molecular dynamics simulations performed on the agonist-bound and antagonist-bound m3 muscarinic receptors, revealed different structural/dynamic perturbations of the receptor protein by the functionally different ligands. These results suggested that antagonists and agonists bind into different pockets formed by transmembrane helices 1–3 and 7 and 3–6, respectively (14). Although the role of the 6th transmembrane helix in agonist binding is in contrast with the site-directed mutagenesis studies cited above, there are good perspectives to elucidate the location and detailed structure of the agonistic and antagonistic binding sites on the receptor molecule.

7. PERSPECTIVES OF THE EFFECT-DIRECTED LIGAND DESIGN

The rational methods of the effect-directed ligand design can be developed by combining of the non-exclusive ligand binding model with the conventional methods of analysis of the structure-activity relationships, or ligand prediction proceeding from the structural properties of the binding sites (1). Following the receptor model under consideration, each ligand can independently interact with both agonistic and antagonistic receptor sites and thus can be characterised by the appropriate dissociation constants, arbitrarily denoted as K_{agon} and K_{antag}. The meaning of these parameters corresponds to K_{10} and K_{30} for a ligand A in Figure 1. It is evident that the ligands which bind more tightly to the agonistic site if compared with the antagonistic site ($K_{agon}/K_{antag}<1$) are recognised as agonists and other compounds which bind more tightly to the antagonistic site ($K_{agon}/K_{antag}>1$) are recognised as antagonists. This means that the ratio of K_{agon} and K_{antag} (or the difference between the appropriate pK values) is a quantitative descriptor of the type of drug activity. If specificity of these two sites can be presented by means of a relationship between the binding effectiveness (pK_{agon} and pK_{antag}) and ligand structure (the conventional QSAR approaches), difference of these relationships describes quantitatively the interrelationship between ligand structure and the type of drug activity and can be used for prediction of ligand structure like any conventional QSAR approach.

The model of the non-exclusive ligand binding can be extended to differentiate between full and partial agonists. The latter drugs should have close affinities for the agonistic and antagonistic binding sites ($K_{agon}/K_{antag}<1$) and therefore the inhibitory effect reveals before the maximal response is reached. Consequently, the activity of partial agonists can be quantified (and predicted) without using additional empirical parameters like efficacy or intrinsic activity. In the case of full agonists the constants K_{agon} and K_{antag} must differ more than two orders of magnitude, but in most cases the difference seems to be much larger and therefore the inhibitory effect can be seen at very high ligand concentrations. It is important to emphasise that the constants K_{agon} for antagonistic drugs cannot be directly determined from the dose-response data since these drugs already block the response at lower ligand concentrations. However, there are still possibilities to determine these parameters from binding experiments that may have practical importance for development of the methods of the effect-directed drug design.

REFERENCES

1. Dean, P.M., 1989, *Molecular foundations of drug-receptor interaction*, Cambridge University Press, Cambridge, New York, Port Chester, Melbourne, Sidney, Chapter 8.
2. Gaddum, J.H., 1936, *J. Physiol.* **89**: 7–9.
3. Ariens, E.J., and De Miranda, R., 1979, *Recent Advances in Receptor Chemistry* (F. Gualtieri, M. Giannella and C. Melchiorre, eds.), Elsevier/North Holland Biomedical Press, Amsterdam, PP.1–36.
4. Järv, J., 1995, *J. Theor. Biol.* **175**: 577–582.
5. Järv, J., 1992, *Selective Neurotoxicity. Handbook of Experimental Pharmacology*. Volume 102 (H. Herken, and F. Hucho, eds.) Springer-Verlag, Berlin, Heidelberg, New York, London, Paris, Tokyo, Hongkong, Barcelona, Budapest, PP. 659–680.
6. Järv, J., Toomela, T., and Karelson, E., 1993, *Biochem. Mol. Biol. Int.* **30**: 649–654.
7. Järv, J., Hautala, R., and Åkerman, K.E.O., 1995, *Eur. J.Pharmacol., Mol. Pharmacol. Section*, **291**: 43–50.
8. Ariens, E. J., 1954, *Arch. Int. Pharmacodyn.* **99**: 32–49.
9. Järv, J., Hedlund, B., and Bartfai, T., 1980, *J. Biol. Chem.* **225**: 2649–2651.
10. Järv, J., Hedlund, B., and Bartfai, T., 1979, *J. Biol. Chem.* **254**: 5595–5598.
11. Järv, J., 1991, *Per. Biologorum* **93**: 197–200.
12. Blüml, K., Mutschler, E., and Wess, J., 1994, *J.Biol.Chem.*, **269**: 18870–18875.
13. Wess, J., Blin, N., Mutschler, E., and Blüml, K., 1995, *Life Sci.*, **56**: 915–924.
14. Fanelli, F., Menziani, M.C., and De Benedetti, P.G., 1995, *Bioorg. and Med. Chem.*, **3**: 1465–1477.

Section 30: Structures of Neuronal Proteins in Rationalized Drug Design

ROLE OF SOME BIOCHEMICAL CORRELATES IN MECHANISMS OF LTP IN OLFACTORY CORTEX SLICES

N. L. Izvarina, N. S. Nilova, T. S. Glushchenko, and A. V. Tokarev

I.P. Pavlov Institute of Physiology
Russian Academy of Sciences
St. Petersburg, Russia

1. INTRODUCTION

According to modern views the long-term potentiation (LTP) phenomenon, i.e. long-term increase of neuronal activity after short-term tetanization, may serve as a model of learning and memory. LTP development is based on the changes in characteristics of the neurochemical membrane structures related to ionic fluxes. Postsynaptic signal transduction processes play an important part in all LTP phases, the most important role belonging the G-proteins conjugating the transmitter receptors to intracellular regulatory systems and to several types of ionic channels (3, 4).

The second aspect important for signal transduction is connected with the physic-chemical characteristics of the transducting medium, in particular the membrane lipid properties. One of the natural processes connected with the membrane lipid chemical changes is the lipid peroxidation (LP). LP processes change the hydrophylity, (viscosity), polarity, charge distribution and the conditions for functioning of receptor complexes, membrane bound enzymes and ionic channels (6).

Another important factor for the functioning of biological membranes especially in neurones is active Na^+ and K^+ transport. This transport is one of the characteristics of membrane potential and is carried out mainly by Na^+, K^+-ATPase (8).

On the basis of the considerations listed in the short preface above we studied the dynamics of G-proteins, LP and Na^+, K^+-ATPase activity and their interactions in the dynamics of LTP in rat piriform cortex slices.

From a symposium organized by Hermona Soreq (Jerusalem, Israel) and André Ménez (Saclay, France). (Supported by the IBRO and the Commissariat à l'Energie Atomique, Centre d'Etudes de Saclay.)

2. THE RESULTS AND DISCUSSION

As was shown in previous investigations the changes of population EPSPs in piriform cortex after 30 s tetanization had the following phases: the induction and transition from primary post-tetanic potentiation to LTP — 2–5 min; supporting phase — 0–25 min; and extinction to control value — 25–45 min (5).

It seems important to pay attention to expressed phase changes of all metabolic processes studied in LTP dynamics (Table 1).

The LTP induction trigger is neuronal depolarization accompanied by the increase of G-proteins GTPase activity, LP activation and the decrease of Na^+, K^+-ATPase activity. The increase of G-proteins function was probably related to excessive transmitter release and initiated their conjugation to multiple effector systems. It is well known that any depolarization is followed by LP activation. The increase of G-proteins and LP activity may, in their turn, be one of the causes of the Na^+,K^+-ATPase activity inhibition observed (2).

In the supporting phase of LTP the changes of all studied metabolic processes became different: GTPase activity of G-proteins decreased slightly towards the normal level accompanied by translocation of (-subunit from membrane to cytosolic fraction (7); the levels of LP products decreased sharply below control values; the activity of Na^+,K^+-ATPase increased towards the normal value. To all appearance in LTP supporting phase LP inhibition observed contributes to the decrease of membrane fluidity and its support at a constant level which is necessary for the conservation of synaptic transmission efficacy.

This modified physic-chemical state of membrane might condition the translocation of G-proteins (-subunit which in its turn might activate many important enzymes, for instance phospholipase C (9). It is possible that the decrease of G-proteins activity is the effect of protein kinase C activation described in this LTP phase (1), these proteins being target for the protein kinase C phosphorilating action. The decrease of G-proteins and LP system activity might be the cause of some disinhibition of Na^+,K^+-ATPase.

The LTP extinction phase is characterized by returning of neuronal bioelectric parameters to normal value. At the same time all the metabolic parameters investigated were elevated to a smaller or larger extent. Especially sharp was the increase of neuronal Na^+,K^+-ATPase activity, which may be connected to ionic balance restoration after intensive cellular bioelectric activity. Furthermore G-proteins activity continued to rise; the LP system product contents rose up to normal level and even slightly more.

In the special series of experiments we studied the influence of G-proteins on LP system in the LTP supporting phase using the nonhydrolizable GTP analogue — GTP-γ-S-(10^{-5} M) which is a G-proteins activator. The preincubation of slices in the media with addition of GTP-γ-S did not cause any changes of LP products as compared to control values. However the effect of LTP in presence of GTP-γ-S was characterized by deepening LP inhibition at the stage of conjugated dienes and by opposite changes of Shiff's bases level.

Probably previous G proteins activation resulted in additional membrane conformation changes under LTP conditions. Steric factor - the availability of fatty acid residues for attacks of active oxygen forms -might have an essential regulatory role on the initial steps of LP. That is what might be the explanation of additional LP inhibition in presence of G-proteins activator. At the same time the changes of membrane configuration under GTP-γ-S could cause the release of additional amino groups from proteins and lipids and these amino groups participated in Shiff's bases formation. It seems necessary to emphasize the fact that the change of membrane structure above took place only in conditions of G-proteins participating in signal transduction, i.e. in LTP.

Table 1. The dynamics of changes in processes of membrane structures metabolism in slices of rat brain piriform cortex in various LTP phases (in % from control values taken as 100 %)

Metabolic parameters		LTP phases		
		Induction (1-5 min)	Supporting (15-18 min)	Extinction (27-30 min)
GTPase activity of G-proteins	Total	+ 32 *	+ 23 *	+ 28 *
	Membrane		- 15 *	
	Cytosol		+ 31 *	
LP in LTP conditions	conjugate diens	+ 22 *	- 11 *	+ 5 *
	ketodienes	+ 15 *	- 9	+ 27 *
	Shiff's bases	0	- 17 *	+ 7
	isolated double bonds	+ 10	- 13 *	+ 8
LP under LTP + GTP-γ-S — LTP only	conjugated diens		- 17 *	
	Shiff's bases		- 13	
	isolated double bonds		- 16 *	
LP under LTP + GTP-γ-S — LTP + GTP-γ-S	conjugated diens		-29 ♦	
	Shiff's bases		+ 8 ♦	
	isolated double bonds		- 21 ♦	
Na^+, K^+ - ATPase activity in neurons		- 20 *	- 14 *	+ 36 *

The difference from control values are
* Statistically significant, $p < 0.05$
♦ The difference from the values, observed under LTP only

3. CONCLUSIONS

The results provide evidence on the interconnected responses under LTP conditions of all the three neurochemical systems studied: G-protein family, LP system and Na^+, K^+-ATPase. Such interrelationship may be considered a form of some universal reactions of cellular membranes involved in signal transduction processing.

REFERENCES

1. Bauer, S., Jakobs, K.H., 1986, *FEBS Lett.*, **198**: 43–46.
2. Bean, B.P., 1990, *J. Neurocsi.*, **10**: 1–10.

3. Bourne, H.R., Sonders, D.A., McCormick, M.S., 1991, *Nature*, **349**: 117–127.
4. Chou, J.C., Lee, E.H.J., 1995, *Neuroscience*, **64**: 5–15.
5. Emelyanov, N.A., Glushchenko, T.S., Izvarina, N.L. et. al., 1995, *Sechenov. Physiol. J.*, **81**: N 8, 29–33 (in Russian).
6. Faroqui, A.A., Horrocks, L.A., 1991, *Brain Res. Rev.*, **16**: 171–191.
7. Izvarina, N.L., Glushchenko, T.S., Tokarev, A.V., Emelyanov, N.A., 1994, *J. Neurosci. Methods*, **34**: A21-A28.
8. McGrail, K.M., Philips, Y.M., Sweadner, K.J., 1991, *J. Neurosci.*, **11**: 381–391.
9. O'Brien, C.A., Ward, N.E., Weinstein, J.B. et. al., 1988, *Biochem. Biophys. Res. Commun.*, **155**: 1374–1380.

PRESYNAPTIC NICOTINIC RECEPTORS IN THE CNS

Function and Mechanism

S. Wonnacott and L. Soliakov

School of Biology and Biochemistry
University of Bath
Bath, United Kingdom

1. INTRODUCTION

Neuronal nicotinic acetylcholine receptors (nAChR) are ligand-gated ion channels, sharing general structural features with other members of this gene superfamily, which includes $GABA_A$, glycine and 5HT-3 receptors (1). Since 1986, when the first cDNA clone coding a nicotinic α-subunit was isolated from vertebrate neurons, eight α-subunits (α2-α9) and three β-subunits (β2-β4) have been shown to be expressed by neurons (2). The enormous potential for heterogeneity of pentameric nAChR by "pick and mix" of these subunits poses a major challenge for understanding the functional roles of nAChR in the CNS. In contrast to muscle and ganglionic nAChR, which mediate fast excitatory transmission in the peripheral nervous system, no clear demonstration of nicotinic synaptic transmission within the brain has been presented so far. Instead, it has been postulated that neuronal nAChR modulate, rather than to mediate, synaptic transmission in the brain (2). This hypothesis derives from numerous neurochemical and electrophysiological experiments that have shown that nicotinic agonists stimulate the release of neurotransmitters from diverse brain areas, including noradrenaline release from hippocampus, acetylcholine release from hippocampal and cortical slices and synaptosomes, as well as glutamate and GABA release from rat prefrontal cortex and interpeduncular nucleus (3,4). However, the best studied example of nAChR-mediated transmitter release is that of dopamine release from the striatum. Here, we will review the recent evidence pertaining to the mechanisms underlying nAChR-mediated neurotransmitter release from nerve terminals. The results were derived by monitoring [^{3}H]dopamine ([^{3}H]DA) release from superfused rat striatal synaptosomes stimulated with nicotine or the potent nicotinic agonist, (±)anatoxin-a (AnTx).

From a symposium organized by Stanislav Tuček (Prague, Czechia) and Andrzej Szutowicz (Gdansk, Poland). (Supported by the British-American Tobacco Company Ltd., Staines, England.)

2. NICOTINIC MODULATION OF [³H]DOPAMINE RELEASE FROM STRIATAL SYNAPTOSOMES

2.1. Pharmacological Characterisation of Nicotinic Agonist-Evoked [³H]Dopamine Release

AnTx elicited [³H]DA release from rat striatal synaptosomes in a dose-dependent and mecamylamine-sensitive manner, with its maximum effect at 1μM (5). An EC_{50} of 0.11μM was determined, compared with an EC_{50} value of 0.5μM for nicotine-evoked striatal [³H]DA release, determined in parallel in this laboratory (6). Similarly, in the comparative study of Clarke & Reuben (7), they reported EC_{50} values of 0.05μM and 0.16μM for (+)AnTx- and nicotine-stimulated [³H]DA release, respectively, from striatal synaptosomes. Thus AnTx is second only to epibatidine in terms of rank order of nicotinic potency. [³H]DA release elicited by a maximally effective concentration of AnTx (1μM) and nicotine (10μM) was inhibited by 10μM mecamylamine by 78±2% and 77±2% respectively (5,8). Dihydroβerythroidine, methyllycaconitine and trimetaphan inhibited nicotine-evoked striatal [³H]DA release with IC_{50} values of 32nM, 38nM and 20μM respectively (7). Nicotine and AnTx were equiefficacious in releasing [³H]DA (7), but the nicotinic stimulation of release was only 20% of the maximum response elicited by KCl depolarisation (5). We have argued that this might reflect the presence of nAChR on only 20% of dopaminergic terminals or, alternatively, that the number or location of nAChR on the terminals limits their efficacy in releasing of [³H]DA. There is no additivity between AnTx and KCl concentrations of 15mM or above, with respect to evoked [³H]DA release (5). This may be explained by the effective voltage clamp imposed by KCl depolarisation, raising the neuronal potential to more positive potentials at which nAChR will gate little current, becoming essentially inactive. Thus the nicotinic modulation of transmitter release might be most effective under resting or hyperpolarised conditions.

2.2. Ionic Basis of nAChR-Mediated Transmitter Release

In agreement with previous studies (3), AnTx-evoked [³H]DA release was Ca^{2+}-dependent: in Ca^{2+}-free medium the agonist elicited only 7% of the release seen in control conditions (5). KCl-evoked release was similarly Ca^{2+}-dependent. These observations are consistent with Ca^{2+}-dependent exocytosis of neurotransmitter release. nAChR are ligand gated cation channels. In addition to permeability to Na^+, neuronal nAChR are characterised by a high permeability to Ca^{2+} (2,9). Thus the question arises whether presynaptic nAChR might themselves provide the Ca^{2+} for triggering exocytosis. In an attempt to answer this question, we designed experiments in low (10mM) external Na^+, in which Na^+ was largely replaced with N-methyl-D-glucamine. In this condition, AnTx-evoked [³H]DA release was almost totally abolished to about 7% of control release (5). This result suggests that the putative Ca^{2+} influx via the activated nAChR pore was unable to promote [³H]DA release. In other words, Na^+ influx through the open nAChR channel is necessary for depolarisation of the synaptic membrane, presumably leading to the opening of voltage-sensitive calcium channels (VSCC), coupled to neurotransmitter exocytosis. This was supported by the Cd^{2+} sensitivity of AnTx-evoked [³H]DA release (5,10).

2.2.1. Voltage-Sensitive Calcium Channels (VSCC) that Contribute to nAChR-Mediated [³H]DA Release. To further delineate the mechanism of nAChR-mediated transmitter release, we compared the sensitivity of AnTx- and KCl-evoked [³H]DA release to VSCC blockers (10). In addition to 25mM KCl, we also examined 15mM KCl which provoked the same amount of transmitter release as $1\mu M$ AnTx, suggesting that it produces a similar level of depolarisation. Examination of the sensitivity of evoked [^{3}H]DA release to Cd^{2+}, a nonselective antagonist of VSCC, showed that both nAChR-mediated and KCl-evoked release were blocked by Cd^{2+}. Indeed, AnTx-evoked release was more sensitive: maximum inhibition of 80% was achieved by $75\mu M$ Cd^{2+} compared with only 50% inhibition of KCl-evoked release (10). As Cd^{2+} should not directly block nicotinic receptor channels at this concentration (11), these results strongly support the involvement of VSCC in nAChR-mediated neurotransmitter release.

To identify the particular VSCC subtypes involved, we examined [^{3}H]DA release in the presence of selective blockers. AnTx-evoked [^{3}H]DA release was rather sensitive to blockade by ω-conotoxin GVIA, a highly selective blocker of N-type VSCC. At $1\mu M$, this toxin inhibited AnTx-evoked release by about 60%. However the release mechanism was resistant to ω-agatoxin IVA (90nM), a selective antagonist of P-type VSCC. In contrast, both toxins produced an equal decrease (about 30%) of KCl-evoked release. Another peptide blocker of VSCC, ω-conotoxin MVIIC ($1\mu M$), which is more selective for Q-type channels, was ineffective in producing any measurable inhibition of [^{3}H]DA release evoked by either AnTx or KCl (10). In comparison, the L-type VSCC blocker, nifedipine ($20\mu M$) significantly decreased AnTx-evoked [^{3}H]DA release (by about 28%), but had no significant effect on KCl-evoked release. Dihydropyridines have been found to interact directly with nAChR (12). To determine whether this might be the case for the presynaptic nAChR studied here, we utilised a $^{86}Rb^+$ efflux assay developed by Marks et al. (13) as a direct index of nAChR activation. Nifedipine ($20\mu M$) did indeed inhibit AnTx-evoked $^{86}Rb^+$ efflux whereas Cd^{2+} and ω-conotoxin GVIA had no significant effect (10). Taken together, these results suggest that N-type VSCC in rat striatal nerve terminals are coupled to nAChR-mediated neurotransmitter release while both N- and P-type VSCC are equally involved in KCl-evoked general depolarisation of the synaptic membrane followed by neurotransmitter exocytosis, a result consistent with other studies on neuronal preparations (14,15).

2.2.2. Do Voltage-Sensitive Na^+ Channels Contribute to nAChR-Mediated Transmitter Release? As already described, AnTx-evoked [^{3}H]DA release from striatal synaptosomes is almost totally dependent upon extracellular Na^+. Further experiments also implicated the involvement of voltage-activated Na^+ channels in the release process following agonist activation of nAChR. Experiments performed in the presence of tetrodotoxin (TTX, $0.1-2.0\mu M$) showed that AnTx-evoked [^{3}H]DA release is partially inhibited (about 40%) by the toxin and this could not be increased by raising the concentration of TTX (5). In contrast, TTX failed to affect KCl-evoked release whereas it produced a near total inhibition of [^{3}H]DA release evoked by veratridine (5). These results imply that in addition to VSCC, voltage-activated sodium channels also participate in AnTx-evoked [^{3}H]DA release. This is a rather unexpected finding because presynaptic nAChR have been considered to operate in a TTX-insensitive manner (3), although a similar inhibition of nicotine-evoked striatal $^{86}Rb^+$ has been reported by Marks et al. (16). In an attempt to clarify this issue, we undertook a set of comparative experiments in brain preparations of increasing complexity, using nicotine as the agonist. Nicotine ($10\mu M$) evoked [^{3}H]DA release from crude P2 and Percoll-purified synaptosomes, and this was inhibited by $1.5\mu M$ TTX by about 50% and 40% respectively (8). However, in striatal slices, nicotine-evoked [^{3}H]DA release was decreased to 30% of control, representing a total block of nAChR-mediated (mecamylamine-sensitive) release. In all of the brain preparations, veratridine-evoked [^{3}H]DA release was almost completely blocked by the same concentration of TTX (1.5μM) whereas KCl-evoked release was unaffected (8). *In vivo* microdialysis (with nicotine applied locally via the microdialysis probe to the striatal terminal regions) also demonstrated that nicotine-evoked dopamine release was completely blocked by addition of TTX (10μM) to the perfusion fluid (8). Taken together, these results support the involvement of voltage-activated sodium channels in nAChR-mediated neurotransmitter release from nerve terminals in rat striatum. The extent of TTX-sensitivity depended on the complexity of the tissue studied (from Percoll-purified synaptosomes to slices and whole animals). It is not at all clear why there should be so much discrepancy in the literature with respect to the TTX-sensitivity of nAChR-mediated neurotransmitter release (see Ref.8). However, the assumption that presynaptic nAChR are TTX-insensitive is likely to underestimate the true proportion of nAChR on terminals.

2.3. A Model for nAChR-Mediated [^{3}H]DA Release

Based on the results outlined above, a model can be constructed to summarise the events that may occur at a dopaminergic terminal stimulated with a nicotinic agonist (see Fig.1). Activation of presynaptic nAChR located on nerve terminals results in opening of the intrinsic cation channel permeable to Na$^+$ and Ca^{2+}. The increase in intracellular Na$^+$ sufficient to achieve a local depolarisation of the synaptic membrane results in activation of both VSCC and TTX-sensitive Na$^+$ channels. The VSCC (principally N-type) provide the Ca^{2+} that triggers exocytosis and the observed neurotransmitter release. Additional Na$^+$ influx through the activated voltage-sensitive Na$^+$ channels serves as an amplifying

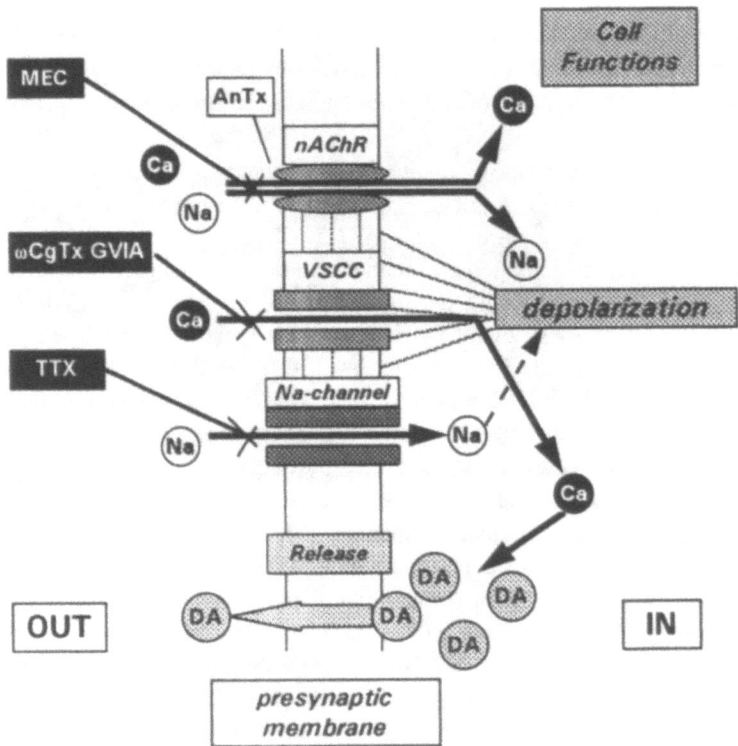

Figure 1. A model for nAChR-mediated [^{3}H]dopamine release from striatal nerve terminals.

mechanism, thus increasing the extent of membrane depolarisation and consequently augmenting the neurotransmitter release. Ca^{2+} influx via the activated nAChR channel does not appear to directly affect the neurotransmitter releasing machinery. However, this local Ca^{2+} influx may be involved in other neuronal functions and might exert an indirect modulation of neurotransmitter release via activation of numerous Ca^{2+}-dependent signal transduction pathways (17,18).

ACKNOWLEDGMENTS

The work carried out in this laboratory was supported by a grant from BBSRC. We are grateful to Professor T. Gallagher, University of Bristol, for the AnTx.

REFERENCES

1. Ortells, M.O., and Lunt, G.G., 1995, *Trends neurosci.*, **18**: 121–127.
2. McGehee, D.S., and Role, L.W., 1995, *Ann. Rev. Physiol.*, **57**: 521–546.
3. Wonnacott, S., Drasdo, A., Sanderson, E., and Rowell, P., 1990, In *Ciba Foundation symposium* **152**: The biology of nicotine dependence, Marsh J. Chichester: John Wiley and Sons pp 87–101.
4. Role, L.W., and Berg, D.K., 1996, *Neuron*, **16**: 1077–1085.
5. Soliakov, L., Gallagher, T., and Wonnacott, S., 1995, *Neuropharmacol.*, **34**: 1535–1541.
6. Whiteaker, P., Garcha, H.S., Wonnacott, S., and Stolerman, I.P., 1995, *Br. J. Pharmacol.* **116**: 2097–2105.
7. Clarke, P.B.S., and Reuben, M., 1996, *Brit. J. Pharmacol.*, **117**: 595–606.
8. Marshall, D., Soliakov, L., Redfern, P.R., and Wonnacott, S., 1996, *Neuropharmacol.*, in press.
9. Mulle, C., Choquet, D., Korn, H., and Changeux, J.-.P., 1992, *Neuron*, **8**: 135–143.
10. Soliakov, L., and Wonnacott, S., 1996, *J. Neurochem.*, **67**: 163–170.
11. Nutter, T.J., and Adams, D.J., 1995, *J. Gen. Physiol.*, **105**: 701–723.
12. Adam, L.P., and Henderson, E.G., 1990, *Pfluger's Arch. Eur. J. Physiol.*, **416**: 586–593.
13. Marks, M.J., Bullock, A.E., and Collins, A.C., 1995, *J. Pharmacol. Exp. Ther.*, **274**: 833–841.
14. Harsing, L.G., Sershen, H., Vizi, S.E., and Lajtha, A., 1992, *Neurochem. Res.*, **17**: 729–734.
15. Takahashi, T., and Momiyama, A., 1993, *Nature*, **366**: 156–158.
16. Marks, M.J., Farnham, D.A., Grady, S.R., and Collins, A.C., 1993, *J. Pharmacol. Exp. Ther.*, **264**: 542–552.
17. Vaughan, P.F.T., Kaye, D.F., Reeve, H.L., Ball, S.G., and Peers, C., 1993,, *J.Neurochem.*, **60**: 2159–2166
18. Coffey, E.T., Sihra, T.S., and Nicholls, D.G., 1993, *J.Biol.Chem.*, **268**: 21060–21065

MUSCARINIC RECEPTORS

Phosphorylation and Desensitization

Tatsuya Haga, Kazuko Haga, Fumio Nakamura, Mariko Kato Hayashi,
Kimihiko Kameyama, and Hirofumi Tsuga

Department of Biochemistry
Institute for Brain Research
Faculty of Medicine
University of Tokyo
Tokyo, Japan

1. INTRODUCTION

Muscarinic acetylcholine receptors are members of G protein-coupled receptor superfamily. All five subtypes (m1–m5) have putative seven transmembrane segments and are characterized with long third intracellular loops. The m1, m3, and m5 subtypes are linked with Gq family G proteins, and the m2 and m4 subtypes with Gi/Go family G proteins. Muscarinic receptors are known to be phosphorylated by cAMP-dependent protein kinase A (PKA), protein kinase C (PKC), and G protein-coupled receptor kinases (GRK) (see reviews (1,2)). The m2 receptor has been most extensively studied and is shown to be phosphorylated by PKA and GRK2, 3, 5 and 6. The m2 receptor of chicken heart is reported to be phosphorylated by PKC. Multiple sites are phosphorylated by each kinase. The m1 receptor is known to be phosphorylated by PKC. The m1 receptor is better substrate of PKC but poorer substrate of PKA compared with the m2 receptor. The m1 receptor was recently shown to be phosphorylated by GRK2, as described later. The m3 receptor is reported to be phosphorylated by GRK2 and 3 (3). Recently a novel kinase, which phosphorylates a fusion protein containing a peptide corresponding to a part of the third intracellular loop of the m3 receptor, was purified from cerebellum (4). No direct evidence is available for phosphorylation of m4 and m5 receptors, as far as we know.

Here we summarizes our recent studies on 1) phosphorylation of m1 receptors by PKC and GRK2, 2) phosphorylation of m2 receptors by GRK2, particularly regulation of GRK2, and 3) facilitation by GRK2 of internalization of m2 and other muscarinic receptors.

From a symposium organized by Stanislav Tuçek (Prague, Czechia) and Andrzej Szutowicz (Gdansk, Poland). (Supported by the British-American Tobacco Company Ltd., Staines, England.)

2. PHOSPHORYLATION OF M1 RECEPTORS BY PKC AND GRK2

Until very recently, we could not detect agonist-dependent phosphorylation of the m1 receptor by GRK2 under the same experimental conditions that the m2 receptor is phosphorylated in an agonist-dependent manner. Hosey and her group also reported that the m1 receptor is not phosphorylated in an agonist-dependent manner by an endogenous kinase of Sf9 cells, whereas the m2 receptor is phosphorylated in an agonist-dependent manner by the same kinase (5). We have extensively examined the purification procedure of m1 receptors and GRK2 preparations, and found that m1 receptors could be phosphorylated in an agonist dependent manner by GRK2 after removing unknown factor(s) that copurify with m1 receptors (6). Unknown factor(s) were removed by a procedure to precipitate m1 receptors with polyethylene glycol.

The phosphorylation of m1 receptors by GRK2 was dependent on the presence of agonist and was stimulated by G protein $\beta\gamma$ subunits as expected, in contrast with the phosphorylation by PKC which was independent of the presence of agonist or $\beta\gamma$ subunits. The amounts of phosphates incorporated into m1 receptors increased by adding one kinase to the other but was dependent on the order of addition of the two kinases, and were apparently greater when m1 receptors were phosphorylated by GRK2 at first and then by PKC. The amount of phosphates incorporated into m1 receptors was estimated to be 4.6, 2.8, 7.6, and 5.8 (mol/mol) for GRK2 alone, PKC alone, GRK2 and then PKC, and PKC and then GRK2, respectively. These results suggest that the phosphorylation sites by GRK2 and PKC are mutually exclusive and that phosphorylation by PKC suppresses the further phosphorylation by GRK2.

Major phosphorylation sites by GRK2 and PKC were located in 38 kDa fragment, which includes the third intracellular loop but not the carboxyterminal part, and in the 14 kDa carboxyterminal fragment, respectively. Serine and threonine residues in the sequence from 275 to 303 in the third intracellular loop are most likely to be phosphorylation sites by GRK2, because 1) the sequence contains basic amino acid residues only at both ends and the corresponding peptide with a molecular mass of 3 kDa is expected to be resistant to trypsin treatment, 2) a peptide corresponding to the sequence 279–293 was phosphorylated *in vitro* by GRK2, and 3) a sequence 286–291 fits to a consensus sequence for phosphorylation *in vivo* by GRK2, which was proposed from studies on $\alpha 2$ adrenergic receptors.

There are two serine and two threonine residues in the carboxyterminal tail and one serine and two threonine residues in the carboxyterminal portion of the third intracellular loop, which are franked by basic amino acids and are therefore good candidates for PKC phosphorylation sites.

GRK2 is known to be activated by agonist-bound receptors. Peptides corresponding to the second intracellular loop, the carboxyterminal end of the third intracellular loop and the carboxy terminal tail of m2 receptors activated GRK2 (7). All these peptides contain many basic amino acid residues. It is tempting to speculate that the phosphorylation by PKC of serine and threonine residues in the carboxyterminal tail or in the carboxyterminal end of the third intracellular loop of m1 receptors reduces the ability of the receptor to activate GRK2.

The stimulation by carbamylcholine of the formation of inositol 1,4,5 triphosphate (IP3) in the presence of GTP was observed in the reconstituted system composed of purified m1 receptors, one of three Gq family G proteins, GL1 (G11), GL2 (G14) or Gq, and phospholipase Cβ (8). Treatment of the reconstituted vesicle with PKC resulted in phosphorylation of m1 receptors (4 phosphate per mol) and phospholipase Cβ (0.6 phosphate per mol) but no phosphorylation of G protein α and β subunits. The agonist-stimulated formation of IP3 was not affected by this PKC treatment. There was a tendency that slightly higher concentrations of carbamylcholine were necessary for phosphorylated samples, but the difference was not significant. Phosphorylation of m1 receptors by GRK2 or by both GRK2 and PKC also did not affect the interaction of m1 receptors with Gq family G proteins.

There are several reports that the signal transduction from m1 receptors to Gq and PLC is desensitized by activation of PKC with phorbol ester. The present results indicate that the three components of m1, Gq, and PLC are sufficient for signal transduction from acetylcholine to IP3, but is not sufficient for the regulation of the signal transduction by PKC.

3. REGULATION OF PHOSPHORYLATION BY GRK2 OF M2 RECEPTORS

Prototypes of GRKs are GRK1 and GRK2 which were originally termed as rhodopsin kinase, and β adrenergic receptor kinase (βARK1), respectively (9). GRK 1 phosphorylates photoexcited rhodopsin, and GRK2, 3 phosphorylate agonist-bound β adrenergic, muscarinic and other G protein-coupled receptors. Recently GRK 4,5 6 were also identified. GRKs are characterized by the substrate specificity that they phosphorylate only light-activated or agonist-bound receptors. This strict specificity appears to be explained, at least partly, by the activation of kinase by substrates. In fact, GRK1 is known to be activated by photoexcited rhodopsin. GRK2 was also reported to be activated by agonist-bound β adrenergic receptors (10), but the extent of stimulation was limited probably because G protein βγ subunits were not included in this experiment. GRK2 and 3, but not other GRKs, are activated by G protein βγ subunits, and synergistically activated by both βγ subunits and agonist-bound receptors. The synergistic activation by βγ subunits and agonist-bound receptors was found for phosphorylation of m2 receptors by a muscarinic receptor kinase, which was purified from porcine brain and is the same as or very similar to GRK2 (6,11,12).

Phosphorylation sites in m2 receptors by muscarinic receptor kinase were located in the central part of the third intracellular loop by protein chemical methods (13). This result was confirmed by examining phosphorylation of a m2 receptor mutant lacking the central part of the third intracellular loop (I3-del m2) and a fusion protein with Glutathione S-transferase of a peptide corresponding to the phosphorylation sites (I3-GST). The I3-deleted mutant was not phosphorylated by GRK2 expectedly. Unexpectedly, however, I3-GST was a poor substrate (7). The ratio of Km/Vmax was 1,000-fold lower for the phosphorylation of I3-GST compared with the phosphorylation of m2 receptors. This result does not support a simple assumption that the phosphorylation sites in receptors are concealed in resting states and are exposed by activation of receptors, because the phosphorylation sites in the fusion proteins are thought to be exposed without activation.

The phosphorylation of I3-GST was found to be synergistically stimulated by mastoparan and G protein βγ subunits, and Vmax increase and Km decrease in the presence of both βγ and mastoparan (7). Mastoparan is a peptide in bee venom and is known to mimic the function of agonist-bound receptor and activate G proteins. In the presence of mastoparan, the ratio of Vmax to Km for phosphorylation of I3-GST is only 30-fold lower than the ratio for phosphorylation of m2 receptors. These results indicate that the difference between m2 receptors and I3-GST is largely due to the presence or absence of activation of the kinase, and support the assumption that the agonist-bound receptor activates the kinase. In fact, the phosphorylation of I3-GST was stimulated by carbamylcholine in the presence of m2 receptors or I3-deleted m2 mutants and G protein $\beta\gamma$ subunits (14). These results provide direct evidence that the carbamylcholine-bound m2 or m2 mutant interact with and activate the receptor kinase, and that the agonist-bound m2 receptors interact with the kinase at two different sites, the phosphorylation sites and the activation sites. Tentative activation sites were suggested to be intracellular regions adjacent to transmembrane segments, because synthetic peptides corresponding to these regions stimulated the phosphorylation by GRK2 of I3-GST.

GRK1 is known to be inhibited by s-modulin or recoverin in a Ca-dependent manner, and this inhibition is thought to explain the light adaptation of retina (15). S-modulin is a Ca-binding protein with three E-F hand domains. In analogy with GRK1, we have examined if GRK2 is also regulated by Ca-binding proteins, and found that GRK2 is inhibited by calmodulin or S100 protein. The inhibition was observed either in the absence or presence of $\beta\gamma$ subunits, but was suppressed by high concentrations of $\beta\gamma$ subunits. The dose-response curves shifted to the higher concentrations of $\beta\gamma$ subunits by addition of Ca-calmodulin, indicating that $\beta\gamma$ subunits and Ca-calmodulin compete to each other for interaction with GRK2 and/or agonist-bound receptors. The inhibitory effect by Ca-calmodulin or Ca S100 protein was also observed when photoexcited rhodopsin was used as a substrate. The inhibition, however, was not observed when I3-GST was used as substrate, irrespective of the presence or absence of $\beta\gamma$ subunits. On the other hand, the inhibition by Ca-calmodulin of the phosphorylation of I3-GST was clearly observed when the phosphorylation was activated by carbamylcholine in the presence of I3-deleted muscarinic receptors. These results indicate that Ca calmodulin does not affect catalytic activity of GRK2 directly but attenuate the kinase activity by inhibiting the activating effect of agonist-bound receptors.

4. FACILITATION BY GRK2 OF INTERNALIZATION OF MUSCARINIC RECEPTORS

The m2 receptor expressed in COS7 cells was phosphorylated in an agonist dependent manner, and the phosphorylation was further enhanced by coexpression of GRK2 and attenuated by coexpression of dominant-negative form of GRK2 (DN-GRK2). Sequestration or internalization of m2 receptors was assessed as loss of binding sites of the hydrophilic ligand [^{3}H]NMS from the cell surface. In the presence of 10^{-6} M carbamylcholine, thirty to forty per cent of m2 receptors were sequestered with a half life of 20 min from cells expressing GRK2, although virtually no sequestration was observed for control cells expressing m2 receptors alone or cells expressing m2 receptors plus DN-GRK2 (16). In the presence of higher concentrations of carbamylcholine, the sequestration was observed also for m2 receptors in control cells and was attenuated for m2 receptors in cells expressing DN-GRK2. These results indicate that the phosphorylation of m2 receptors by GRK2 facilitates the internalization of phosphorylated receptors. Recently we have found that internalization of other muscarinic receptor subtypes and dopamine D2 receptors is also facilitated by coexpression of GRK2. In addition, Caron's group reported that the internalization of β adrenergic receptors is also facilitated by GRK2 (17). It is likely that the phosphorylation of G protein-coupled receptors by GRK2 or other kinases is closely related to their internalization.

REFERENCES

1. DebNBurman, S.K., and Hosey, M.M. 1995, *Molecular mechanisms of muscarinic acetylcholine receptor function* (J. Wess, ed.), Springer, New York, PP. 209–226.
2. Haga, T., Kameyama, K., Haga, K. 1995, *Molecular mechanisms of muscarinic acetylcholine receptor function* (J. Wess, ed.), Springer, New York, PP. 227–248.
3. DebNBurman, S.K., Kunapuli, P., Benovic,J.L., and Hosey, M.M. 1995, *Mol. Pharmacol.* **47**: 224–233.
4. Tobin, A.B., Keys, B., and Nahorski, S.R., 1996, *J. Biol. Chem.* **271**: 3907–3916.
5. Richardson, R.M., and Hosey, M.M. 1992, *J. Biol. Chem.* **267**: 22249–22255.
6. Haga, K., Kameyama, K., Haga,T., Kikkawa, U., Shiozaki, K., and Uchiyama, H., 1996, *J. Biol. Chem.* **271**: 2776–2782.
7. Haga, K., Kameyama, K. and Haga, T. 1994, *J. Biol. Chem.* **269**: 12594–12599.
8. Nakamura, F., Kato, M., Kameyama, K., Nukada,T., Haga,T., Kato, H., Takenawa, T., and Kikkawa, U., 1995, *J. Biol. Chem.* **270**: 6246–6253.
9. Haga, T., Haga, K. and Kameyama, K. 1994, *J. Neurochem.* **63**: 400–412.
10. Chen, C.-Y., Dixon, S.B., Kim, C.M., and Benovic, J.L. 1993, *J. Biol. Chem.* **268**: 6886–6892.
11. Haga, K. and Haga, T. 1992, *J. Biol. Chem.* **267**: 2222–2227.
12. Kameyama, K., Haga, K., Haga, T., Kotani, K., Katada, T. and Fukada, Y. 1993, *J. Biol. Chem.* **268**: 7753–7758.
13. Nakata, H., Kameyama, K., Haga, K. and Haga, T. 1994, *Eur. J. Biochem.* **220**: 29–36.
14. Kameyama, K., Haga, K., Haga, T., Moro, O. and Sadée, W. 1994, *Eur. J. Biochem.* **226**: 267–276.
15. Kawamura, S. 1993, *Nature* **349**: 420–423.
16. Tsuga, H., Kameyama, K., Haga, T., Kurose, H. and Nagao, T. 1994, *J. Biol. Chem.* **269**: 32522–32527.
17. Ferguson, S.S.G., Ménard, L., Barak, L.S., Koch, W.J., Colapietro, A.-M., and Caron, M.G. 1995, *J. Biol. Chem.* **270**: 24782–24789.

PRESENCE OF ACETYLCHOLINE IN THE BLOOD AND ITS PRODUCTION BY CHOLINE ACETYLTRANSFERASE IN LYMPHOCYTES

K. Kawashima,[1] T. Fujii,[1] H. Misawa,[2] S. Yamada,[1] S. Tajima,[1] T. Suzuki,[1] K. Fujimoto,[1] and T. Kasahara[3]

[1]Department of Pharmacology
Kyoritsu College of Pharmacy, Minato-ku, Tokyo 105
[2]Department of Neurology
Tokyo Metropolitan Inst. for Neuroscience
Fuchu-city, Tokyo 183, Japan
[3]Department of Biochemistry
Kyoritsu College of Pharmacy, Minato-ku, Tokyo 105

1. INTRODUCTION

Acetylcholine (ACh), a classical neurotransmitter in both the central and peripheral nervous systems, is synthesized by choline acetyltransferase (ChAT, E.C. 2.3.1.6) from acetyl coenzyme A and choline (1). It has been suggested that the cholinergic system plays a role in the immune-neuroendocrine system (2, 3). In support of this, the presence of muscarinic receptors has been demonstrated in lymphocytes, using selective cholinergic ligands (2–6), Northern blot analysis (7), and reverse transcription-polymerase chain reaction (RT-PCR) (8–10). In addition, stimulation of muscarinic receptors in vascular endothelial cells by ACh induces vascular smooth muscle relaxation via the release of endothelium-derived relaxing factor (11). In view of these findings, we hypothesized that ACh is present in the blood of various animal species and is involved in various physiological functions including neuroimmunomodulation and local blood flow regulation.

In this paper we present a simplified overview of our recent studies demonstrating 1) the presence of ACh in the blood of various species of animals, using a sensitive, specific radioimmunoassay (RIA) (12–15), 2) the localization of ACh in mononuclear leukocytes (MNL) of humans (14), 3) the synthesis of ACh in human leukemic T-cell lines and stimulation of ACh synthesis by phytohemagglutinin (PHA) (16), and 4) the expression of mRNA for ChAT and the presence of ChAT protein in MOLT-3, a human leukemic T-cell line (17).

From a symposium organized by Stanislav Tuček (Prague, Czechia) and Andrzej Szutowicz (Gdansk, Poland). (Supported by the British-American Tobacco Company Ltd., Staines, England.)

2. DETERMINATION OF ACh IN THE BLOOD

2.1. Presence of ACh in the Blood and Plasma of Various Animal Species (12–15)

Venous blood samples (10 ml) from conscious humans, cattle, dogs, goats, horses, pigs, rabbits, and sheep were withdrawn into syringes containing 200 µl of a mixture of 12.5% EDTA-2NH$_4$ and 0.5 mM paraoxon, which produced rapid, complete inhibition of blood cholinesterase (ChE) activity. From conscious, unrestrained 10-week-old Wistar rats, blood samples (2.5 ml) were withdrawn via catheters inserted into the abdominal aorta three days previously. ACh extracts were prepared from whole blood and plasma samples. The ACh content in samples was determined by a RIA (18) with a limit of sensitivity of 3 pg (about 20 fmol)/tube, using [^{3}H]ACh and a rabbit antiserum against ACh. This RIA is specific for ACh and its levels of cross-reactivity with choline, phosphatidylcholine, and phosphorylcholine are less than 0.012%. The procedures used for the determination of ACh in whole blood and plasma by RIA are described in detail elsewhere (12–15).

Mean (± s.e.m.) blood and plasma ACh concentrations in various species of mammals are shown in Table 1 (12–15). The ACh levels in cattle and horse blood samples were more than 10 times higher than that in humans. The value for humans was comparable to those observed in the porcine and rabbit samples, but slightly higher than those in dog, goat, rat and sheep. Plasma ACh levels were extremely high in porcine samples, intermediate in humans and cattle samples, and low in samples from other species of animals. These findings provide the first direct evidence for the presence of ACh in the blood of mammals and for a greater concentration of ACh in blood cells than in plasma, except in pigs.

The reason why plasma ACh concentrations were higher in porcine samples than in those of other animals is unknown. We have previously demonstrated the synthesis and release of ACh by cultured vascular endothelial cells from bovine carotid artery (21) and porcine cerebral microvessels (22). In addition, the possibility of an endothelial origin for plasma ACh has been proposed on the basis of immunocytochemical localization of ChAT in vascular endothelial cells (23). Thus, one possible explanation for the higher plasma ACh concentrations in pigs is that ACh released from vascular endothelial cells contributes to the high plasma ACh concentration in this animal.

Table 1. Acetylcholine concentrations in blood and plasma of various species of mammals as determined by radioimmunoassay (12–15)

Species	ACh (pmol/ml)		P/B ratio
	Blood	Plasma	
Human	8.66 ± 1.02 (30)	3.12 ± 0.36 (32)	0.28
Cattle (7)	360.5 ± 59.7	7.14 ± 2.55	0.02
Dog (10)	1.37 ± 0.23	0.13 ± 0.04	0.02
Goat (5)	4.05 ± 0.23	0.22 ± 0.06	0.05
Horse (5)	93.8 ± 16.3	0.66 ± 0.23	0.01
Pig (5)	11.7 ± 1.58	15.4 ± 1.58	1.38
Rat (10)	1.43 ± 0.20	0.75 ± 0.11	0.57
Rabbit (7)	24.5 ± 6.14	0.96 ± 0.11	0.04
Sheep (5)	2.04 ± 0.20	0.27 ± 0.05	0.14

Values are means ± S.E.M. Numbers of samples are in parentheses.

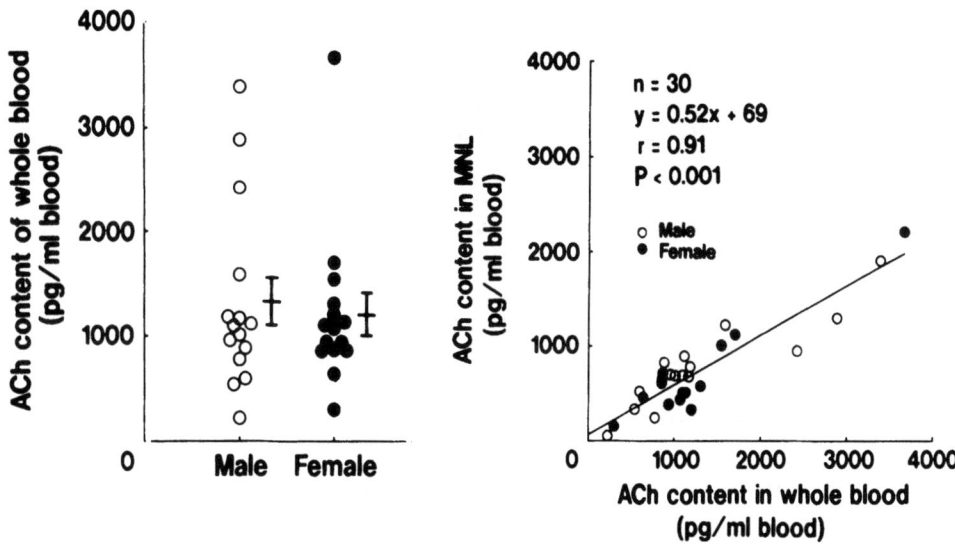

Figure 1. Left: ACh contents of whole blood from 15 male and 15 female subjects. Each bar represents a mean ± s.e.m. Right: Scatter plot showing the correlation between the ACh content of whole blood and that of MNL (14).

2.2. Localization of ACh in Circulating Human Mononuclear Leukocytes (MNL)

We also found release of ACh from blood cells into plasma in rabbits upon nicotine stimulation (13) and the presence of a major portion of rabbit blood ACh in the buffy coat layer (19). Therefore, focusing on leukocytes, we attempted to clarify the origin of blood ACh in humans by determining which type of blood cells contains the largest amount of ACh (14). Blood samples from 30 healthy volunteers were layered on Mono-Poly Resolving Medium and centrifuged to separate MNL and polymorphonuclear leukocytes (PMN) (20). The ACh contents in whole blood, the MNL and PMN fractions were determined by RIA. The ACh content of MNL was corrected for recovery of MNL and expressed as pg/ml blood.

The mean (± s.e.m.) blood ACh content in the 30 subjects was 1263.5 ± 149.0 pg/ml. Considerable inter-individual variation in the ACh content of whole blood was observed for both male and female subjects, although average ACh contents were equivalent in the two groups (Fig. 1, left). The cause of the large inter-individual variation remains to be determined. The mean (± s.e.m.) ACh content of MNL was 731 ± 85.8 pg/ml blood, corresponding to about 60% of the level of ACh in whole blood (Fig. 1, right). No ACh was detected in PMN. The dominant localization of blood ACh in MNL, and the significant correlation of ACh content of MNL with that of whole blood indicate that the major portion of plasma ACh originates from MNL, probably lymphocytes.

3. DETECTION OF ACh AND ACh SYNTHETIC ENZYME ACTIVITY IN HUMAN LEUKEMIC T-CELL LINES

The localization of ACh in MNL (14) and a paper by Rinner and Schauenstein reporting the presence of ChAT activity in lymphocytes and human leukemic cell lines (24) prompted us to determine the ACh content and ACh synthetic activity in various human leukemic cell lines (16).

Table 2. ACh content and ACh- and ACar-synthesizing activities in human leukemic cell lines (16)

Cell Line	ACh Content (pmol/10^6 Cells)	ACh Synthesis (pmol/mg protein per min)	ACar Synthesis (pmol/mg protein per min)
MOLT-3	79.6 ± 8.9 (3)	1.9 ± 0.1 (3)	53.3 ± 5.6 (3)
HSB-2	36.2 ± 3.5 (7)	1.9 ± 0.1 (3)	58.3 ± 15.3 (3)
CEM	9.5 ± 0.8 (4)	2.5 ± 0.1 (13)	22.8 ± 4.6 (7)
Jurkat	1.1 ± 0.1 (4)	0.7 ± 0.1 (12)	17.2 ± 1.9 (6)
Daudi	1.2 ± 0.1 (4)	NT	125.6 ± 44.2 (3)
YT.C3	0.9 ± 0.1 (3)	NT	NT
HL-60	0.9 ± 0.1 (4)	NT	NT
U-937	0.8 ± 0.4 (4)	NT	NT

Values are means ± S.E.M. Numbers of experiments are in parentheses. MOLT-3, HSB-2 and CEM, T cell lines; Jurkat, helper T cell line; Daudi, B cell line; YT.C3, NK cell line: HL-60, promyelocytic cell line; U-937, histiocytic cell line. ND, not detectable. NT, not tested.

3.1. Presence and Synthesis of ACh in Human Leukemic T-Cell Lines

In order to clarify the origin of ACh in human blood, we measured the ACh content and ACh synthetic activity in several human leukemic cell lines. Eight human leukemic cell lines were used: the human acute lymphoblastic leukemic T-cell lines, HSB-2, MOLT-3, CEM and Jurkat; a human promyelocytic leukemic cell line, HL-60; a human histiocytic mono-like lymphoma cell line, U-937; a human leukemic NK-cell line, YT.C3; and a human leukemic B-cell line, Daudi. The intracellular ACh content was determined by RIA. The ACh synthetic activity of the cell lines was determined by the method of Fonnum (25) with modifications. Since it has been suggested that carnitine acetyltransferase (CarAT, E.C. 2.3.1.7) is involved in ACh synthesis in the peripheral nervous system (26), we also determined acetyl-L-carnitine (ACar)-synthetic activity in human leukemic cell lines.

The ACh contents of three T-cell lines, MOLT-3, HSB-2 and CEM, were much higher than those of other cell lines (Table 2). The Jurkat cell line derived from helper T-cells, and the other leukemic cell lines tested, Daudi, HL-60, U-937 and YT.C3, had little ACh. ACh synthetic activity was found in all the T-cell lines tested, but was not detectable in Daudi, a B-cell line (Table 2). Although various degrees of ACar-synthetic activity were found in the various cell lines, the degree of this activity was not correlated with ACh content. Bromoacetylcholine (100 µM), a ChAT inhibitor (1, 26), and bromoacetyl-L-carnitine (100 µM), a CarAT inhibitor (26), decreased ACh synthetic activity in MOLT-3, HSB-2 and CEM, by about 50% and 30%, respectively (16, data not shown). These findings demonstrate that T lymphocytes have the potential to synthesize ACh, and suggest that ChAT, and to a lesser extent CarAT, may be involved in ACh synthesis in T-cells.

3.2. Stimulation of ACh Synthesis in MOLT-3 by PHA

The effect of PHA stimulation on the intracellular content and release of ACh into culture medium was investigated in MOLT-3, a T-cell line. MOLT-3 was incubated in the absence or presence of 3 or 10 µg/ml PHA for 48 h in a medium containing 10 µM DFP, an inhibitor of ChE (16).

PHA increased both the intracellular content and release of ACh into the culture medium, and increased ChAT activity in a dose-dependent manner (Fig. 2). These findings indicate that ACh released from activated T lymphocytes may act on the muscarinic receptors on the surface of various blood cells including lymphocytes, and suggest that ACh may function as a modulator of T-cell-dependent immune responses, as suggested by Maslinski (2) and Rinner and Schauenstein (3).

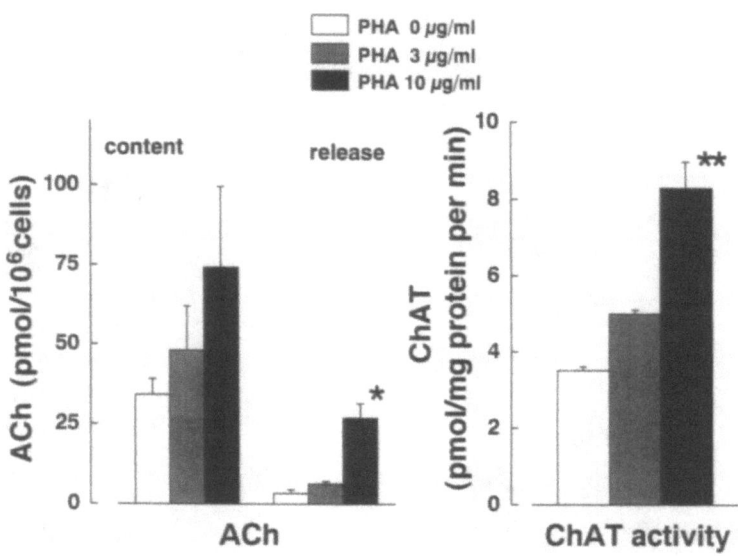

Figure 2. Effect of PHA treatment for 48 hr on the intracellular content and release of ACh and ChAT activity in MOLT-3 (15).

It has been shown that PHA activates the phospholipase C-inositol-1, 4, 5-trisphosphate system via the T-cell receptor/CD3 molecular complex (27). In addition, dibutyryl cAMP and phorbol 12-myristate 13-acetate have been found to increase ChAT activity via protein kinase A (28) and protein kinase C (29), respectively, in the nervous system. Therefore, the PHA-induced increase in intracellular content and release of ACh may have been at least in part due to enhanced production of ACh synthetic enzyme(s).

4. DETECTION OF ChAT mRNA AND ChAT PROTEIN IN MOLT-3

RT-PCR analysis of the MOLT-3 poly(A)+ RNA sample was performed to confirm the expression of ChAT mRNA using specific primers corresponding to positions 322–342 and 952–973 of human brain ChAT cDNA (17, 28, 30–32). A single transcript of RT-PCR products was observed on agarose gel (Fig. 3A). The position of the RT-PCR product obtained from poly(A)+ RNA of MOLT-3 was identical with that of the human brain sample used as a positive control, and the size of this product was equivalent to the calculated size of 652 base pairs on agarose gel. However, when reverse transcriptase was omitted from the reaction mixture, as a negative control, no DNA band was observed on the agarose gel. The sequence of the RT-PCR product of MOLT-3 was completely identical with nucleotide positions 322–973 of human brain ChAT cDNA (31).

The presence of ChAT protein in MOLT-3 was confirmed by Western blot analysis using a polyclonal antibody against human ChAT (17). The major band in MOLT-3 was observed at the same molecular weight (70 kDa) as human placental ChAT protein (Fig. 3B).

These findings provide the first direct evidence for the presence of ChAT mRNA and protein in T lymphocytes, and suggest that ACh is synthesized by ChAT in T lymphocytes. Since T-lymphocytes can directly contact their targets, such as various blood cells and vascular endothelial cells, even a small amount of ACh released from T lymphocytes should be able to interact with the receptors on the targets prior to its hydrolysis by ChE.

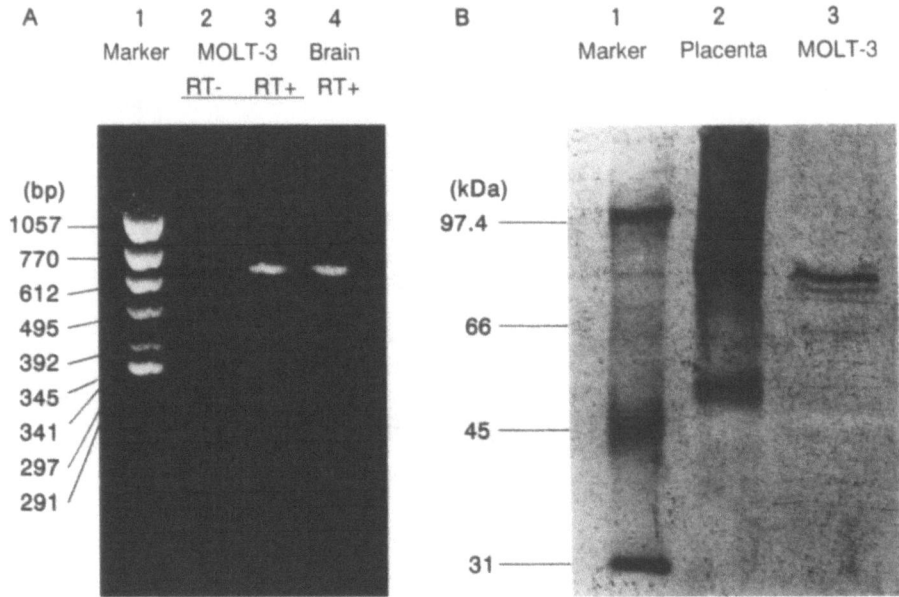

Figure 3. A: Expression of ChAT mRNA in the MOLT-3. Ethidium bromide-stained PCR products. Lane 1, the DNA size standards, with size listed in base pairs (bp); lane 2, MOLT-3 mRNA sample, with reverse transcriptase omitted from PCR mixture as a negative control; lane 3, MOLT-3 mRNA sample; lane 4, human brain mRNA sample as a positive control. B: Western blot analysis of MOLT-3 ChAT protein. Lane 1, molecular size markers with sizes listed in kDa; lane 2, human placental ChAT; lane 3, MOLT-3 sample (16).

5. CONCLUSION

We found that significant concentrations of ACh are detectable in the blood (1.4–426 pmol/ml) and plasma (0.1–15 pmol/ml) of various species of mammals including humans. ACh synthesized by ChAT in T lymphocytes is one of the origins of ACh in the blood, since most (about 60%) of human blood ACh was localized to MNL, probably lymphocytes, and the expression of ChAT mRNA and the presence of ChAT protein were confirmed in MOLT-3, a human leukemic T-cell line. The increases in both the synthesis and release of ACh in MOLT-3 induced by PHA, a stimulator of T lymphocytes, suggest that ACh in the blood may be involved in the regulation of T-cell-dependent immune responses. Since T lymphocytes can directly contact circulating blood cells and vascular endothelial cells, even a small amount of ACh released from T lymphocytes should be able to interact with receptors on the target cells prior to hydrolysis by ChE.

REFERENCES

1. Tucek, S., 1988, *Handbook of Experimental Pharmacology*, Volume **86**, (VP Whittaker ed.), Springer Verlag, Berlin, PP. 125–165.
2. Maslinski, W., 1989, *Brain Behav. Immun.* **3**: 1–14.
3. Rinner, I., and Schauenstein, K., 1991, *J. Neuroimmunol.* **34**: 165–172.
4. Costa, L.G., Kaylor, G., and Murphy, S.D., 1988, *Immunopharmacology* **16**: 139–149.
5. Eva, C., Ferrero, P., Rocca, P., Funaro, A., Bergamasco, B., Ravizza, L., and Genazzani, E., 1989, *Neuropharmacology* **28**: 719–726.

6. Genaro, A.M., Cremaschi, G.A., and Borda, E.S., 1993, *Immunopharmacology* **26**: 21–29.
7. Kaneda, T., Kitamura, Y., and Nomura, Y., 1993, *Mol. Pharmacol.* **43**: 356–364.
8. Costa, P., Castoldi, A.F., Traver, D.J., and Costa, L.,G., 1994, *J. Neuroimmunol.* **49**: 115–124.
9. Costa, P., Traver, D.J., Auger, C.B., and Costa, L.G., 1994, *Immunopharmacology* **28**: 113–123.
10. Costa, P., Auger, C.B., Traver, D.J., and Costa, L.G., 1995, *J. Immunopharmacol.* **60**: 45–51.
11. Furchgott, R.F., and Zawadzki, J.V., 1980, *Nature* (Lond.) **288**: 373–376.
12. Kawashima, K., Oohata, H., Fujimoto, K., and Suzuki, T., 1987, *Neurosci. Lett.* **80**: 339–342.
13. Kawashima, K., Oohata, H., Fujimoto, K., and Suzuki, T., 1989, *Neurosci. Lett.* **104**: 336–339.
14. Kawashima, K., Kajiyama, K., Fujimoto, K., Oohata, H., and Suzuki, T., 1993, *Biog. Amine.* **9**: 251–258.
15. Fujii, T., Yamada, S., Yamaguchi, N., Fujimoto, K., Suzuki, T., and Kawashima, K., 1995, *Neurosci. Lett.* **201**: 207–210.
16. Fujii, T., Tsuchiya, T., Yamada, S., Fujimoto, K., Suzuki, T., Kasahara, T., and Kawashima, K., 1996, *J. Neurosci. Res.* **44**: 66–72.
17. Fujii, T., Yamada, S., Misawa, H., Tajima, S., Fujimoto, K., Suzuki, T., and Kawashima, K., 1995, *Proc. Japan. Acad.* **71B**: 231–235.
18. Kawashima, K., Ishikawa, H., and Mochizuki, M., 1980, *J. Pharmacol. Methods* **3**: 115–123.
19. Kajiyama, K., Suzuki, T., Fujimoto, K., and Kawashima, K., 1991, *Japan. J. Pharmacol.* **55** (Suppl. 1): 194P.
20. Ferrante, A., and Thong, Y.M.,1981, *J. Immunol. Methods* **24**: 389–393.
21. Kawashima, K., Watanabe, N., Oohata, H., Fujimoto, K., Suzuki, T., Ishizaki, Y., Morita, I., and Murota, S., 1990, *Neurosci. Lett.* **119**: 156–158.
22. Ikeda, C., Morita, I., Mori, A., Fujimoto, K., Suzuki, T., Kawashima, K., and Murota, S., 1994, *Brain Res.* **655**: 147–152.
23. Parnavelas, J.G., Kelly, W., and Burnstock, G., 1985, *Nature* (Lond.) **316**: 724–725.
24. Rinner, I., and Schauenstein, K., 1993, *J. Neurosci. Res.* **35**: 188–191.
25. Fonnum, F., 1975, *J. Neurochem.* **24**: 407–409.
26. Tucek, S., 1982, *J. Physiol.* (Lond.) **322**: 53–69.
27. Imboden, B.J., Shoback, M.D., Pattison, G., and Stobo, D.J., 1986, *Proc. Natl. Acad. Sci. U.S.A.* **83**: 5673–5677.
28. Misawa, H., Takahashi, R., and Deguchi, T., 1993, *J. Neurochem.* **60**: 1383–1387.
29. Honegger, P., Du Pasquier, P., and Tenet, M., 1982, *Dev. Brain Res.* **29**: 217–223.
30. Oda, Y., Nakanishi, I., and Deguchi, T., 1992, *Mol. Brain Res.* **16**: 287–294.
31. Kengaku, M., Misawa, H., and Deguchi, T., 1993, *Mol. Brain Res.* **18**: 71–76.
32. Misawa, H., Ishii, K., and Deguchi, T., 1992, *J. Biol. Chem.* **267**: 20392–20399.

METABOLISM OF ACETYL-CoA AND CHOLINERGIC NEUROPATHIES

Andrzej Szutowicz, Hanna Bielarczyk, and Maria Tomaszewicz

Department of Clinical Biochemistry
Medical University of Gdańsk
Gdansk, Poland

1. INTRODUCTION

Pyruvate derived from glucose is a main source of acetyl-CoA and energy in all cells of mammalian brain. However in cholinergic neurons acetyl-CoA additionally serves as a source of acetyl moiety for acetylcholine synthesis. It implies that cholinergic neurons may be, to the higher degree than others, dependent on appropriate oxidative metabolism of glucose(1).

There are several encephalopathies in which lesions are preferentially located in cholinergic neurons. They include both primary and secondary disorders developing in the course of Alzheimer's disease (AD), dialysis encephalopathy (DE), thiamine deficiency (TD) and alcoholism, inherited pyruvate dehydrogenase deficiencies and diabetes.

Several typical symptoms (memory loss, agnosia, hyperexcitability, muscular weakness) and their improvement after treatment with cholinomimetic drugs as well as autopsy data showing loss of cholinergic markers, provide an ultimate evidence on the existence of cholinergic lesions in these diseases (2). In addition, another common feature of these pathologies is concomitant suppression of uptake and oxidative metabolism of glucose in the brain (3, 4, 5).

One must stress however that the destruction of cholinergic neurons can not be itself responsible for this phenomenon, since pyruvate dehydrogenase (PDH) and other glycolytic and TCA cycle enzymes, are evenly distributed throughout the brain (6). The source of this discrepancy may explain some *post mortem* studies of AD brain showing marked decrease of PDH activity at unchanged expression of the complex proteins (3). In addition, the inhibition of glucose utilization in AD brain has been shown to preceede development of cholinergic degeneration (2). It indicates that functional disturbances can be caused by a prolonged inhibition of acetyl-CoA metabolizing enzymes by still poorly defined pathogens.

From a symposium organized by Stanislav Tuček (Prague, Czechia) and Andrzej Szutowicz (Gdansk, Poland). (Supported by the British-American Tobacco Company Ltd., Staines, England.)

One has to remember that acetyl-CoA used for ACh synthesis must be transported from the site of its synthesis in nerve terminal mitochondria to cytoplasm which is the site of neurotransmitter synthesis (1, 7). Our previous studies have shown that it may be a critical step limiting acetyl-CoA availability for ACh synthesis particularly under pathologic conditions (1, 8). Therefore, in this article we review the data which show how recognized neurotoxic factors might contribute to the dysfunction and damage of cholinergic neurons through the interference with acetyl-CoA metabolism.

2. ACETYL-CoA SUPPLY FOR ACh SYNTHESIS

All acetyl-CoA utilized for ACh synthesis is synthesized in mitochondria of cholinergic terminals. Main source of acetyl-CoA is pyruvate derived from glucose. Under some physiologic and pathologic states β-hydroxybutyrate and lactate may serve as supplementary sources of acetyl units. Acetyl-CoA is mainly utilized in TCA cycle and only 1% of its pool is used for ACh synthesis in cholinergic nerve ending cytoplasm (1). Mitochondrial membrane in resting terminals is poorly permeable for acetyl-CoA. Therefore it has to be transported by mitochondrial membrane indirectly using citrate and acetylcarnitine as carriers. These metabolites in synaptoplasm and inner mitochondrial membrane are converted back to acetyl-CoA by ATP-citrate lyase and carnitine acetyltransferase, respectively. Several concordant data show that about 30% of acetyl-CoA for ACh synthesis is provided *via* ATP-citrate lyase pathway (1, 7). Data concerning acetylcarnitine are less clear and demonstrate rather beneficial effect of this metabolite on cholinergic transmission than quantitative assessment (1, 9). It sems however that very important source of cytoplasmic acetyl-CoA is its direct transport by mitochondrial membrane, which is stimulated by the rise in cytoplasmic Ca^{2+} taking place during functional depolarization of the terminal plasma membrane (1, 8). Such mechanism would couple directly increased demand for acetyl-CoA resulting from post release stimulation of ACh synthesis, with complementary direct transport of this precursor to cytoplasm of depolarized cholinergic terminals. It would assure provision of cytoplasmic acetyl-CoA adequate to actual demand for ACh synthesis.

3. ALUMINUM

Aluminum (Al) is widely investigated but still a controversial factor in the AD. On the contrary, its involvement in DE is unequivocal (10). In both encephalopathies preferential loss of cholinergic neurons is evident. Al may be transported into the brain and neurons as an Al-transferrine complex (11). The process is activated by low pH which is likely to occur during hypoxic episodes common in aged brains or due to metabolic acidosis in uremic patients (11).The Al levels in AD brains were found to be from 60% to several times higher than in nondemented ones (10). This cation is thought to facilitate formation of neurotoxic β-amyloid aggregates. On the other hand, in brains of DE victims the Al levels were even higher than in AD although not accompanied by β-amyloid lesions. It indicates that Al itself is capable to exert independent neurotoxic influences. They might be due to its inhibitory interactions with Ca-dependent metabolism of cholinergic neurons. However, we found that Al had no effect on pyruvate utilization in K-depolarized synaptosomes. It seems to exclude this cation as a potential source of marked inhibition of pyruvate dehydrogenase activity seen in AD brains (4). On the other hand, Al

Table 1. Effect of different experimental *in vivo* and *in vitro* conditions on acetyl-CoA and ACh metabolism in nerve terminals

Conditions	Pyruvate utilization	Synaptosomal Acetyl-CoA Mitochondria	Synaptosomal Acetyl-CoA Cytoplasm	Acetylcholine Release Nonquantal	Acetylcholine Release Quantal	Ref.
in vitro						
AlCl$_3$ 0.1-0.25 mM	100	147	65	138	36	12
SNP 0.2 mM	97	109	133	99	178	17
Br Pyr 0.25 mM	33	0	16	29*		8
in vivo						
Pyrithiamine 15 days	74	88	76	159	58	5
Diabetes 10 days	147	130*		156	144	25
Ethanol 6 days	95	90*		106	211	26

Presented data are expressed as a percent of respective controls calculated from original data given in references. Abbreviations: Br Pyr, Bromopyruvate; SNP, sodium nitroprusside. * Acetyl-CoA in whole synaptosomes or total ACh synthesis.

was found to inhibit Ca-evoked ACh synthesis, linked with its quantal release, without affecting a nonquantal one. In addition, Al caused a 35% decrease of acetyl-CoA content in synaptoplasm and its 40% rise in synaptosomal mitochondria (Tab. 1)(12). It may indicate that Al inhibits Ca-dependent, direct transport of acetyl-CoA from mitochondria to synaptoplasm (12). These data substantiate the theory that Al might aggravate existing cholinergic deficits in AD and DE brains by simultaneous suppresion of acetyl-CoA provision to synaptoplasm and ACh quantal release in terminals of cholinergic neurons which avoided degeneration.

4. NITRIC OXIDE

NO and derived free radicals are postulated to mediate excitotoxic mechanisms in the brain caused by excessive release of glutamate from glutamatergic neurons, which was reported to take place during ischemia and hypoxia as well as in encephalopathies such as AD, Parkinson's or Huntington's disease (13). The rise of cytoplasmic Ca^{2+}, evoked by overactivation of NMDA receptors, was found to increase NO synthesis. Inhibitors of NO synthase decreased glutamate and NMDA toxicity both *in vivo* as well as in tissue slices and neuronal cell cultures (14). The NO level was found to increase in several brain pathologies (13). However, reports on its direct and free radical dependent neurotoxicity are contradictory. NO is known to suppress both glycolytic and oxidative energy metabolism by inhibition of aconitase as well as other enzyme activities. It rised the suggestion that NO is more likely to cause cell injury by an inhibition of energy metabolism than by activation of free radicals formation (14).

NO was also found to stimulate ACh release from cortical and striatal neurons (15). This may increase demand of cholinergic neurons for energy and acetyl-CoA necessary for restoration of the neurotransmitter pool. It is however in conflict with inhibition of energy metabolism evoked by NO. Several reports demonstrated that inhibition of glucose utilization or pyruvate oxidation by wide range of metabolic inhibitors always caused suppression of ACh synthesis (16). Our data provided a direct evidence that it was due to the shortage of acetyl-CoA in nerve ending cytoplasm (5). Therefore in NO-treated terminals some additional factors must exist which increase provision of acetyl-CoA to nerve ending cytoplasm despite its decreased formation in mitochondria.

Our recent experiments indicate that sodium nitroprusside (SNP) caused 34% inhibition of pyruvate utilization and nonproportionally higher 55% drop of acetyl-CoA content in whole brain mitochondria (17). On the contrary, in synaptosomes SNP caused no change of pyruvate consumption but increased level of acetyl-CoA in synaptoplasm (Tab. 1)(17). Under these conditions no changes in nonquantal ACh release were observed in K-depolarized synaptosomes in the absence of Ca^{2+}. However, after addition of Ca^{2+}, the SNP caused 78% stimulation of quantal ACh release (Tab. 1) (17). These data indicate that NO is capable to cause a selective increase of mitochondrial membrane permeability for acetyl-CoA, which in turn would possibly activate quantal ACh release in cholinergic synaptosomes. Increased consumption of acetyl-CoA for ACh synthesis in connection with its decreased utilization for energy production could, at least in part, explain preferential loss of cholinergic neurons in states in which the increase of brain NO level takes place.

5. THIAMINE DEFICIENCY

TD encephalopathy in animals and humans is casually linked with loss of cholinergic functions. The decrease of brain ACh level seen in these conditions was thought to be caused by suppression of PDH and oxoglutarate dehydrogenase (OGDH) activities yielding insufficient production of acetyl-CoA and energy in this tissue. However, in animal models of thiamine deprivation or pyrithiamine encephalopathy, no significant changes in PDH activity were observed in homogenates of most brain regions (18). Significant loss of activities of thiamine dependent enzymes was observed mainly in peripheral tissues.

On the contrary, in our studies we found marked reductions of PDH and OGDH activities in synaptosomal and mitochondrial fractions isolated from forebrains of pyrithiamine treated rats (5). It indicates that these changes can not be limited to small brain regions. In addition, no changes in choline acetyltransferase and high affinity choline uptake activities were found, what would indicate that structural integrity of cholinergic neurons in this pathology was well preserved (16).

On the other hand, recent autopsy studies on brains of alcohol encephalopathy and hepatic coma victims showed a significant decreases of PDH, OGDH and transketolase activities in regions responsible for memory, associative and motor functions (19). Reports concerning acetyl-CoA content in TD brains are scarce and contradictory (1).

Our studies have shown 47% decrease of acetyl-CoA content in isolated brain mitochondria and 24% in cytoplasmic compartment of synaptosomes of TD rats incubated with pyruvate in the presence of l-malate (Tab. 1) (5). This finding was in agreement with decreased oxidation of pyruvate. TD increased Ca-indepedent nonquantal release of ACh for 58%, and inhibited Ca-dependent quantal release for 42% (Tab. 1) (5). On the contrary, TD caused no changes in synaptoplasmic acetyl-CoA and quantal ACh release when synaptosomes were incubated in medium without l-malate. In these conditions one step pyruvate utilization took place and ATP-citrate lyase pathway was not operative. It let us to think that TD impairs provision of cytoplasmic acetyl-CoA via ATP-citrate lyase pathway (5).

Hence, changes in ACh synthesis linked with its nonquantal release were apparently not related to shifts in synaptoplasmic acetyl-CoA level. The activation of nonquantal ACh release in TD synaptosomes was likely to be caused by energy deficits and depletion of thiamine triphosphate from their plasma membranes (20). This compound has been shown to control sodium gating in excitable membranes. The rise of nonquantal ACh release could explain presence of neuronal hyperexcitability and depletion of transmitter pool in brains of TD rats (1, 16). On the other hand, TD-evoked inhibition of quantal ACh release might be linked with the decreased level of synaptoplasmic acetyl-CoA and presumably could be responsible for clinical symptoms of cholinergic deficiency such as paralysis or memory loss.

6. INHERITED PDH DEFICIENCIES

This is heterogenic group of inherited deficiencies concerning various enzymes of PDH complex. Depending of the severity of deficit different degrees of brain lesion and lactic acidosis were observed (18). The link of PDH deficiency with cholinergic hypofunction was confirmed by many groups. Several unrelated compounds were found to exert inhibitory effect on PDH activity, which was accompanied by a proportional suppression of ACh synthesis (8, 16). Our studies proved that the direct cause of inhibition of ACh synthesis by 3-bromopyruvate, an inhibitor of PDH, is decreased level of acetyl-CoA in nerve terminal cytoplasm (8).

The dichloroacetate (DCA), a potent inhibitor of PDH kinase was used in treatment of inherited enzyme defects, with aim to increase proportion of active dephospho form of PDH (21). In some cases improvement of clinical status was observed after application of DCA (21). However PDH in the brain was found to be almost completely dephosphorylated. Therefore, DCA had no effect on pyruvate oxidation in brain preparations (22). In addition, PDH in cultured fibroblasts from majority of affected individuals failed to respond to DCA (23). These contradictions point out the existence of another site(s) of DCA action in the brain. We suppose that DCA might overcome supression of ACh synthesis increasing acetyl-CoA supply to synaptoplasm by activation its direct Ca-dependent transport through the mitochondrial membrane (22).

7. DIABETES

High demand of the brain for glucose is met thanks to high concentration of insulin-independent Glut1 transporter on the plasma membranes of blood-brain barrier and glial cells. The density of this carrier is reciprocally regulated by glucose concentration to keep the rate of glucose supply on constant level (24). On the other hand, neurons possess mainly Glut3 transporter which does not seem to be regulated by glucose levels (24). Both hyperglycemia and ketonemia were found to activate cholinergic activity, demonstrated by reversal of suppressory effect of quinuclidynyl benzilate on ACh level in the striatum or by increase of scopolamine-induced ACh efflux in hippocampus (1, 9). Also insulin itself was found to exert a direct suppressory effect on cholinergic activity in the brain (1, 24). Thus, in diabetic brain one can expect activation of cholinergic activity in by two coexisting conditions, hyperglycaemia/ketonemia combined with hypoinsulinemia. Our studies have shown that streptozotocin diabetes lasting 10 days brought about concordant increases of pyruvate oxidation (47%), acetyl-CoA level (30%) as well as ACh synthesis linked with its nonquantal and quantal release (about 40%) (Tab. 1) (25). Also utilization of β-hydroxybutyrate and its contribution to synaptosomal acetyl-CoA and ACh was activated under these pathologic conditions for 80, 48 and 75%, respectively (25). These data indicate that cholinergic hyperactivity seen in diabetes is likely to be evoked by an adaptative stimulation of pyruvate and β-hydroxybutyrate utilization by cholinergic terminals. Excessive activity of cholinergic neurons can make them particularly vulnerable to hypoxic insults caused by angiopathy which frequently developes in diabetic brain.

8. ETHANOL

Narcotic effect of ethanol is thought to be, at least in part, caused by its inhibitory effect on ACh release. However, few day administration of ethanol to rats was found to bring about development of full tolerance, which was accompanied by the increase of PDH and acetyl-CoA synthetase activities and over two-fold activation of quantal ACh release at no changes in synaptosomal acetyl-CoA (26). This adaptative overactivation of cholinergic neurons could be due to ethanol-evoked increase of phosphatidyl ethanolamine content in terminal plasma membranes (27). In connection with decreased acivity of purinergic neurons it could be an important factor in long-term adaptation to ethanol (26). Increased activity of cholinergic neurons in brains of chronic alcoholics could increase their sensitivity to thiamine deficiency and other toxic insults (19, 28). This subsequently could lead to irreversible loss of cholinergic neurons seen in late stages of alcoholism.

9. CONCLUSIONS

The data presented in this review indicate that disorders in acetyl-CoA metabolism in nerve terminals are likely to play a primary role in development of several brain pathologies. The inhibition of ACh metabolism was shown to be accompanied by decreased level of acetyl-CoA in synaptoplasm. In AD, DE, TD and other brain pathologies several neurotoxic agents may aggravate existing cholinergic deficits by inhibition of acetyl-CoA provision to cytoplasm. On the contrary, excessive activation of ACh release in depolarized synaptosomes could be linked with increased concentration of acetyl-CoA in terminal cytoplasm. Such conditions could increase natural susceptibility of cholinergic neurons to various pathogens. It may be due to the fact that cholinergic neurons, unlike others, utilize acetyl-CoA not only for energy production and lipid synthesis but also for ACh syntheis. The same applies to choline which in noncholinergic neurons is utilized for structural lipids formation, while in cholinergic ones is additionally used for the transmitter synthesis (27). The recognition of mechanisms leading to disturbances in acetyl-CoA metabolism in cholinergic neurons should be considered as a key task for understanding their preferential vulnerability in several encephalopaties.

REFERENCES

1. Szutowicz, A., Tomaszewicz, M., and Bielarczyk, H., 1996, *Acta Neurobiol. Exp.* **56**: 323–339.
2. Weinstock, M., 1995, *Neurodegeneration* **4**: 349–356.
3. Kalaria, R.N., and Harik, S.I., 1989, *J. Neurochem.* **53**: 1083–1088.
4. Sheu, K.F.R., Kim, Y.T., and Blass, J.P., 1985, *Ann. Neurol.* **17**: 551–558.
5. Bielarczyk, H., Tomaszewicz, M., Jankowska, A., Kisielevski, Y., and Szutowicz, A., 1997, *Neurochemistry* (A.W. Teelken, and J. Korf, eds.), PlenumPress, New York, This volume.
6. Szutowicz, A., Stępień, M., Bielarczyk, H., Kabata, J., and Łysiak, W., 1982, *Neurochem. Res.* **7**: 798–910.

7. Tucek, S., 1993, *Prog. Biophys. Molec. Biol.* **60**: 59–69.
8. Bielarczyk, H., and Szutowicz, A., 1989, *Biochem. J.* **262**: 377–380.
9. Ricny, J., Tucek, S., and Novakova, J., 1992, *Brain Res.* **576**: 215–219.
10. Meiri, H., Banin, E., Roll, M., and Rousseau, A., 1993, *Progr. Neurobiol.* **40**: 89–121.
11. Shi, B., and Huang, A., 1990, *J. Neurochem.* **55**: 551–558.
12. Szutowicz, A, and Bielarczyk, H., 1992, *Neurochem. Int. Suppl.* **21**: A13
13. Davson, V.L., and Davson, T.M., 1996, *Proc. Soc. Exp. Biol. Med.* **211**: 33- 40.
14. Lees, G.J., 1993, *Neuroscience* **54**: 287–322.
15. Guevara-Guzman, R., Emson, P.C., and Kendrick, K.M., 1994, *J. Neurochem.* **62**: 807–810.
16. Gibson, G.E., Barclay, L., and Blass, J,P., 1982, *Ann. New York Acad. Sci.* **387**: 382–403.
17. Tomaszewicz, M., Bielarczyk, H., Jankowska, A., and Szutowicz, A., 1997, *Neurochemistry* (A.W. Teelken, and J. Korf, eds.), Plenum Press, New York, This volume.
18. Butterworth, R.F, 1985, *Cerebral EnergyMetabolism and Metabolic Encephalopathy* (D. W. McCandless ed.). Plenum Press, New York 121–141.
19. Lavole, J., and Butterworth, R.F., 1995, *Alcoholism* **10**: 1073–1077.
20. Bettendorff, L.L., 1994, *Metab. Brain Dis.* **9**: 193–209.
21. Stackpoole, P.W., 1989, *Metabolism* **38**: 1124–1144.
22. Szutowicz, A., Bielarczyk, H., and Skulimowska, H., 1994, *Neurochem. Res.* **19**: 1107–1112.
23. Kuroda, Y., Naito, E., Takeda, E., Yokota, I. and, Miyao, M., 1987, *Enzyme* **38**: 108–114.
24. Wozniak, M., Rydzewski, B., Baker, S.P., and Raizada, M.K, 1993, *Neurochem. Int.* **22**: 1–10.
25. Szutowicz, A., Tomaszewicz, M., Jankowska, A., and Kisielevski, Y., 1994, *Neuroreport* **5**: 2421–2424.
26. Jankowska, A., Kisielevski, Y., Oganesjan, N., Tomaszewicz, M., and Szutowicz, A., 1997, *Neurochemistry* (A.W. Teelken, and J. Korf, eds.), Plenum Press, New York, This volume.
27. Klein, J., Chalifa, V., Liscovitch, M., and Loffelholz, K., 1995, *J. Neurochem.* **65**: 1445–1453.
28. Montoliu, C., Valles, S., Renau-Piqueras, J., and Guerri C., 1994, *J. Neurochem.* **63**: 1855–1862.

TARGETED IMMUNOLESION OF CHOLINERGIC NEURONS BY 192 IgG-SAPORIN

A Novel Tool to Study Neurochemical Events of Alzheimer's Disease

Reinhard Schliebs, Steffen Roßner, Mechthild Heider, and Volker Bigl

Paul Flechsig Institute for Brain Research
University of Leipzig
Medical Faculty
Jahnallee 59, D-04109 Leipzig, Germany

1. INTRODUCTION

The basal forebrain cholinergic system is known to play an important role in cortical arousal and normal cognitive function. Cortical cholinergic dysfunction has been implicated in cognitive deficits that occur in Alzheimer's disease, and the cholinergic projection from the nucleus basalis of Meynert (Nbm) to areas of the cerebral cortex is the pathway that is most early and severely affected in brains from Alzheimer patients. Investigations on the functions of the central cholinergic system require adequate animal models to produce specific cholinergic deficits in vivo. This would allow for a detailed evaluation of the neurochemical, neuropathological, and behavioural sequela as well as functional implications of plastic repair mechanisms following cholinergic hypofunction, and provide information that cannot or only partially be obtained in humans. At present there is no adequate animal model available which could mimic all the biochemical, behavioural, and histopathological abnormalities as observed in patients with Alzheimer's disease. However, partial success can be achieved with so called "isomorphic models" (1) representing partial parallelism between model and some human conditions. The value of such models is to delineate mechanisms underlying the pathological processes as well as to test for new potential therapeutic strategies. In the last few years an increasing number of studies have applied neurotoxins including excitotoxins or cholinotoxins by stereotaxic injection into the Nbm to produce reductions in cortical cholinergic activity. One of the most serious limitations of these lesion paradigms is the fact that basal forebrain choliner-

---

From a symposium organized by Stanislav Tuček (Prague, Czechia) and Andrzej Szutowicz (Gdansk, Poland). (Supported by the British-American Tobacco Company Ltd., Staines, England.)

Neurochemistry, edited by Teelken and Korf
Plenum Press, New York, 1997

gic neurons are always intermingled with populations of non-cholinergic cells and that the cytotoxins used are far from being selective to cholinergic cells. Recently, a novel approach for neuronal lesioning has been introduced by Wiley et al.(2) by using immunotargeting of unspecific cytotoxins. Cholinergic neurons of the basal forebrain possess nerve growth factor (NGF) receptors whereas other neurons in this region including the cholinergic cells in the nearby striatum do not express detectable levels of NGF receptors (3,4). It was demonstrated that a well-characterized monoclonal antibody to the low-affinity NGF receptor, 192IgG, accumulates bilaterally exclusively in cholinergic neurons of the basal forebrain following intracerebroventricular administration (see e.g. ref. 5). Employing these properties of 192IgG, a cholinergic immunotoxin was developed by chemical linking of 192IgG to the ribosome inactivating protein saporin (192IgG-saporin; see refs. 2,6; for details of preparation, see ref. 7). Here we demonstrate the usefulness of 192IgG-saporin as a powerful tool for producing an animal model with selective and specific basal forebrain cholinergic lesions in rats which can be applied to study the impact of reduced cortical cholinergic input on neurochemical events in cholinoceptive target regions as well as to test therapetic strategies to compensate for cortical cholinergic dysfunction.

2. EXPERIMENTAL METHODS

Male Wistar rats received a stereotaxic injection of 4 µg of the cholinergic immunotoxin 192IgG-saporin into the left lateral ventricle (for details, see ref. 8). 7, 15, 30, and 90 days following lesion brains were rapidly removed and subjected to histochemical (acetylcholinesterase-AChE, choline acetyltransferase-ChAT, NGF-receptor, parvalbumin, micro-and astroglia) and biochemical (choline uptake, acetylcholine release, receptor-mediated inositol phosphate production, protein kinase C) analysis. For receptor autoradiography, serial coronal sections were cryocut, thaw-mounted on gelatin coated slides and subjected to receptor binding applying receptor-specific radioligands (cholinergic, glutamatergic, GABAergic receptor subtypes). In situ hybridization to detect muscarinic acetylcholine receptor mRNA was performed using S-35 labeled oligonucleotide probes. Autoradiograms were obtained by exposing the slides to ^{3}H-Hyperfilm (Amersham) for various periods of time. Quantitative analysis of the autoradiograms was done on a video camera-based, computer assisted imaging device using the autoradiographic software package MCID 4.0.

3. RESULTS

Intracerebroventricular administration of 4 µg of 192IgG-saporin conjugate resulted in substantial reductions in the activity of ChAT in widespread areas of the cortex and hippocampus and in a nearly complete disappearance of ChAT-positive, NGF receptor-immunoreactive neurons in the medial septum, in both the vertical and horizontal limbs of the nucleus of the diagonal band of Broca and in the nucleus basalis magnocellularis, whereas cholinergic interneurons in the striatum are not affected (9). Seven days following injection of the immunotoxin there was a dramatic loss of acetylcholinesterase staining by up to 90% in frontal, parietal, piriform, temporal and occipital cortices, hippocampus and olfactory bulb, but not in the striatum and cerebellum (8). The number of parvalbumin-containing GABAergic projection neurons in the septum-diagonal band of Broca complex and nucleus basalis magnocellularis was not reduced following intraventricular 192IgG-saporin application (9). Corresponding to the topographic location of cholinergic neurons in the basal forebrain a dramatic lesion-induced increase in microglia has been demonstrated (9). As summarized in the table, already seven days following immunolesion cortical high-affinity choline uptake sites were decreased by up to 45%, whereas M_1-and M_2-muscarinic receptor binding sites were increased by about 20% in a number of cortical regions including parietal, temporal and occipital cortex reflecting receptor supersensitivity (10,11). The lesion-induced increases in cortical muscarinic acetylcholine receptors were complemented by corresponding alterations in m1 and m4 muscarinic acetylcholine receptor mRNA as revealed by semiquantitative in situ hybridization (10). In the basal forebrain, however, immunolesion caused about a 40% decrease in m2 muscarinic receptor mRNA which might be due to the loss of cholinergic cells (10). The immunolesion-induced increase in cortical M_1-muscarinic receptor density did not alter muscarinic receptor sensitivity as measured by carbachol-stimulated inositol phosphate production or phorbol ester binding to membrane-bound protein kinase C. However, the K^+-stimulated release of [^{3}H]acetylcholine from cortical and hippocampal slices from immunolesioned rats was found to be markedly decreased one week after injection (11).

One week after a single intracerebroventricular injection of 4 µg of 192IgG-saporin, NMDA receptor binding was markedly reduced in cortical regions displaying a reduced activity of acetylcholinesterase and high-affinity choline uptake sites as a consequence of cholinergic lesion, whereas AMPA and kainate binding sites were significantly increased in these regions (12).

Muscimol binding to $GABA_A$ receptors was increased in the caudal portions of frontal and parietal cortices as well as occipital and temporal cortex as compared to the corresponding brain regions from vehicle-injected control rats. Binding levels of benzodiazepine receptors were not affected by the lesion in any of the cortical regions studied (12).

α_1-A-drenoceptor binding sites were not affected by cholinergic immunolesion in any of the cortical regions studied, whereas the levels of α_2-and β-adrenoceptor binding were decreased in a number of cortical regions and hippocampal formation after lesion. 5-HT2-serotonin receptor binding was markedly reduced in cortical regions displaying a reduced activity of acetylcholinesterase, while 5-HT1_A receptor binding was found to be transiently reduced in some anterior cortical regions but not in the hippocampal areas following immunolesion.

To prove the capability of chronic nerve growth factor (NGF) treatment to induce recovery of cholinergic markers after partial lesion of rat basal forebrain cholinergic system, rats received intracerebroventricular transplants of either Swiss 3T3 cells retrovirally transfected to secrete NGF or Swiss 3T3 cells containing the retrovirus alone (control) for eight weeks (13). The levels of NGF in the cerebrospinal fluid from immunolesioned rats that received transplants of NGF-producing fibroblasts for eight weeks, were strikingly higher as compared to those detected in control groups. Eight weeks after NGF treatment of immunolesioned rats, the activity of choline acetyltransferase, and the sodium-dependent high-affinity choline uptake in the parietal cortex and hippocampus have increased reaching corresponding values which are detectable in untreated control rats. Histochemistry revealed that the density of acetylcholinesterase-positive cells in the basal forebrain as well as fibers in cerebral cortex and hippocampus were reduced by up to 60% in immunolesioned rats, but was completely restored after chronic NGF treatment. The remaining cholinergic cells following partial immunolesion displayed a higher intracellular staining for NGF-like immunoreactivity and the cell size was increased by up to 25% as compared to unoperated controls (13).

4. DISCUSSION

Administration of 192IgG-saporin conjugate was shown to result in substantial reductions in ChAT activity in widespread areas of the cortex and hippocampus and in a nearly complete disappearance of ChAT-positive neurons in the basal forebrain complex, which is in agreement to the findings of other similar studies (14–16). Non-cholinergic septal neurons containing parvalbumin and non-cholinergic substantia innominata neurons containing calbindin-D_{28K} or NADPH-diaphorase were not affected by 192IgG-saporin (16). Similarly, the number of parvalbumin-containing GABAergic projection neurons in the septum-diagonal band of Broca complex and Nbm was not reduced following intraventricular 192IgG-saporin application (9, 17,18). Moreover, 192IgG-saporin did not destroy neurotensin, galanin, somatostatin, or neuropeptide neurons within the Nbm (19). Corresponding to the topographic location of cholinergic neurons in the basal forebrain a dramatic increase in microglia has been demonstrated (9), suggesting that the immunotoxin is lethal to cholinergic cells in the Nbm rather than suppressing the expression of cholinergic markers (e.g. ChAT) in these cells (20).

It was found that 192IgG-saporin affects two neuronal groups outside of the basal forebrain which express p75NGF receptors: NGF-reactive cerebellar Purkinje cells after intraventricular injection and cholinergic striatal interneurons after injections into the substantia innominata (16). There are ChAT-positive, but NGF-receptor negative neurons in the rat Nbm-substantia innominata complex innervating the amygdala and parts of the rhinal paralimbic areas (see e.g. refs.21,22) which are spared or only partially affected by the immunotoxin (16). Similarly, cholinergic neurons in the ventral pallidum and sublenticular substantia innominata not expressing p75NGF receptors are not affected by the immunotoxin.

Behavioural studies have shown that complete cholinergic lesion by 192IgG-saporin did not produce any deficit in the Morris water maze task (23). Despite the high depletion in cortical ChAT activity by 192IgG-saporin acquisition, performance of the delayed alternation or passive avoidance tasks were not impaired by the lesions suggesting that selective loss of cholinergic cells is not sufficient to produce functional impairments (19). In contrast, other authors reported that intracerebral administration of 192IgG-saporin induced dose-dependent (ranging between 1 and 10 μg) impairments in the water maze task and passive avoidance retention, but only weak effects on locomotor activity (17,24). Intracerebroventricular injections of 192IgG-saporin severely affected spatial and cued navigation (14,25). However, an almost 90% reduction in ChAT activity is needed to produce substantial behavioural deficits (24).

In this study, receptor autoradiography and AChE staining were performed in adjacent brain sections, which allows to simultanously detect the consequences of lesions on various parameters in a distinct cortical area, and thus provides an appropriate tool to reveal correlations between cortical cholinergic hypoactivity and lesion-induced adaptive response in distinct cholinoceptive target regions. Seven days after immunolesion we found significantly increased $GABA_A$ but not benzodiazepine binding sites in the frontal and parietal cortices suggesting an up-regulation of postsynaptically localized $GABA_A$ receptors as an adaptive response to the reduced GABAergic input as measured by Gomeza et al. (26). In a current study alterations in dendritic morphology of cortical neurons after basal forebrain lesions have been described (27) suggesting that the cholinergic input plays an important modulatory role in cortical function and plasticity. The changes in the number of cortical NMDA receptors following lesion could be considered as a loss and/or down-regulation of NMDA receptor sites. In contrast, the increased kainate and AMPA

Table 1. Changes in histochemical, biochemical, in situ hybridization and receptor autoradiographic markers in the cerebral cortex, hippocampus, and basal forebrain nuclei of adult rat seven days after an intracerebroventricular administration of 4 µg 192IgG-saporin

Parameters	Changes in percentage over control		
	Cerebral cortex	Hippocampus	Cholinergic basal forebrain
Histochemistry			
AChE	-80%	-90%	n.d.
ChAT-ir	n.d.	n.d.	Nearly complete loss
Parvalbumin-ir	n.d.	n.d.	No change
GFAP-ir	No change	Weak increase	No change
Microglia	No change	Weak increase	Very strong increase
Biochemistry			
ChAT	-36%	-74%	n.d.
Choline uptake	-36%	-42%	n.d.
ACh-release	-46%	-57%	n.d.
Inositol phosphate accumulation	No change	n.d.	n.d.
Protein kinase C	No change	n.d.	n.d.
M1-mAChR (Bmax)	+27%	n.d.	n.d.
In situ hybridization			
m1-mAChR mRNA	+20%	n.d.	n.d.
m2-mAChR mRNA	No change	n.d.	-40%
m3-mAChR mRNA	No change	n.d.	n.d.
m4-mAChR mRNA	+25% (VC only)	n.d.	n.d.
c-fos mRNA	No change	n.d.	Increase
c-jun mRNA	Increase (VC,PC only)	n.d.	Increase
Receptor autoradiography			
Choline uptake binding sites	-40%	-50%	n.d.
M1-mAChR	+20%	+20%	n.d.
M2-mAChR	+15%(PC only)	No change	n.d.
nAChR	No change	No change	n.d.
NMDA-receptor	-20%	-30%	n.d.
AMPA-receptor	+20%	No change	n.d.
kainate receptor	+25%	+20%	n.d.
GABAA receptor	+20%	Weak increase	n.d.
Benzodiazepine receptor	No change	No change	n.d.
5-HT1A	-50% (FC, PC only)	No change	n.d.
5-HT2	-30%(FC,VC only)	-35%	n.d.
α1-adrenoceptor	No change	No change	n.d.
α2-adenoceptor	-20%	-25%	n.d.
β-adrenoceptor	-30%	-35%	n.d.

AChE, acetylcholinesterase; ChAT, choline acetyltransferase; GFAP, glial fibrillary acidic protein; mAChR, muscarinic acetylcholine receptor; nAChR, nicotinic acetylcholine receptor; n.d., not determined; PC, parietal cortex; VC, visual cortex.

binding following lesion should be considered as up-regulation of receptor sites. Up-regulation of postsynaptic glutamate receptors is assumed to compensate for reduced presynaptic input. This would suggest that cholinergic terminals directly affect glutamate transmission on presynaptic glutamatergic elements. However, MK-801 is assumed to bind to a site within the NMDA receptor ion-channel and thus can also be considered as a marker of the agonist-bound, open state of the channel (28). Therefore, the immunotoxin-induced decline in MK-801 binding also indicates a lower amount of glutamate bound to the NMDA receptor channel. This supports the suggestion that cholinergic hypofunction reduces cortical glutamatergic activity by less release of glutamate from presynaptic elements presumably due to enhanced inhibition by GABA. However, regardless of possible interpretations the immunotoxin-induced differential changes in glutamate and GABA receptor subtypes in cortical regions displaying reduced cholinergic activity clearly demonstrate that cortical glutamatergic and GABAergic markers are partially driven by cholinergic activity. Moreover, it is interesting to note that the same sort of alterations in glutamate and GABA receptor subtypes observed in rat cortex following basal forebrain cholinergic immunolesion have been detected in cortical brain areas from patients with Alzheimer's disease (29). This supports the suggestion that the receptor changes observed might indicate compensatory mechanisms due to presumably cholinergic degenerative events. However, these data further support a glutamatergic strategy which might be therapeutically potential in treating Alzheimer's disease (30–32).

Cortical cholinergic deficits induced by 192IgG-saporin might also be useful to study the effect of neurotrophins on the recovery of the cholinergic transmission. Our data on chronic NGF treatment of cholinergic immunolesioned rats suggest that NGF affects the remaining basal forebrain cholinergic cells by increasing both the size, activity of choline acetyltransferase activity and the accumulation of NGF through enhanced low-affinity NGF receptor density, and induces regenerative mechanisms in cholinoceptive target regions to restore cholinergic markers as well as to establish or remodel new synaptic contacts. The data further support the usefulness of a neurotrophic strategy to treat cholinergic deficits in demential disorders like Alzheimer's disease and the capability of transfected cell lines to continouesly deliver trophic factors into lesioned nervous tissue.

In conclusion, the detailed characterization of 192IgG-saporin suggests that cholinergic immunolesion by 192IgG-saporin exhibits a valuable tool to produce specific cholinergic deficits in rats, which can be used as model to study both the impact of reduced cortical cholinergic input on cortical neurotransmission and the efficiency of novel therapeutic strategies.

ACKNOWLEDGMENT

This work was partly supported by a grant of the Bundeministerium für Forschung und Technik to R.S., no. FKZ 01 ZZ 9103/2.8.

REFERENCES

1. Fisher, A., and Hanin, I., 1986, *Ann. Rev. Pharmacol. Toxicol.*, **26**: 161–181.
2. Wiley, R.G., Oeltmann, T.N., and Lappi, D.A., 1991, *Brain Res.*, **562**: 149–153.
3. Gage, F.H., Batchelor, P., Chen, K.S., Chin, D., Deputy, S., Rosenberg, M.B., Higgins, G.A., Koh, S., Fischer, W., and Björklund, A., 1989, *Neuron*, **2**: 1177–1184.
4. Yan, Q., and Johnson, jr., E.M., 1989, *J. Comp. Neurol.*, **290**: 585–598.
5. Thomas, L.B., Book, A.A., and Schweitzer, J.B., 1991, *J. Neurosci. Meth.*, **37**: 37–45.
6. Wiley, R.G. ,1992, *Trends Neurosci.*, **15**: 285–290.
7. Wiley, R.G., and Lappi, D.A., 1993, *Neurosci. Protoc.*, 93–020–02–01–12.
8. Roßner, S., Perez-Polo, J.R., Wiley, R.G., Schliebs, R., and Bigl, V., 1994, *J. Neurosci. Res.*, **38**:282–293.
9. Roßner, S., Härtig, W., Schliebs, R., Brückner, G., Brauer, K., Perez-Polo, J.R., Wiley, R.G., and Bigl, V., 1995, *J. Neurosci. Res.*, **41**: 335–346.
10. Roßner, S., Schliebs, R., Perez-Polo, J.R., Wiley, R.G., and Bigl, V., 1995, *J. Neurosci. Res.*, **40**: 31–43.
11. Roßner, S., Schliebs, R., Härtig, W., and Bigl, V., 1995, *Brain Res. Bull.*, **38**:371–381.
12. Roßner, S., Schliebs, R., and Bigl, V., 1995, *Brain Res.*, **696**: 165–176.
13. Roßner, S., Yu, J., Pizzo, D., Werrbach-Perez, K., Schliebs, R., Bigl, V., and Perez-Polo, J.R., 1996, *J. Neurosci. Res.*, in press.

14. Berger-Sweeney, J., Heckers, S., Mesulam, M.-M., Wiley, R.G., Lappi, D.A., and Sharma, M., 1994, *J. Neurosci.*, **14**: 4507–4519.
15. Book, A.A., Wiley, R.G., and Schweitzer, J.B., 1992, *Brain Res.*, **590**: 350–355.
16. Heckers, S., Ohtake, T., Wiley, R.G., Lappi, D.A., Geula, C., and Mesulam, M.M., 1994, *J. Neurosci.*, **14**:1271–1289.
17. Leanza, G., Nilsson, O.G., Wiley, R.G., and Björklund, A., 1995, *Eur. J. Neurosci.* **7**: 329–343.
18. Lee, M.G., Chrobak, J.J., Sik, A., Wiley, R.G., and Buzsáki, G., 1994, *Neuroscience*, **62**: 1033–1047.
19. Wenk, G.L., Stoehr, J.D., Quintana, G., Mobley, S., and Wiley, R.G., 1994, *J. Neurosci.*, **14**: 5986–5995.
20. Book, A.A., Wiley, R.G., and Schweitzer, J.B., 1994, *J. Neuropath. Exp. Neurol.*, **53**: 95–102.
21. Bickel, U., and Kewitz H., 1990, *Dementia*, **1**: 146–150.
22. Woolf, N.J., Gould, E., and Butcher, L.L., 1989, *Neuroscience*, **30**: 143–152.
23. Torres, E.M., Perry, T.A., Blokland, A., Wilkinson, L.S., Wiley, R.G., Lappi, D.A., and Dunnett, S.B., 1994, *Neuroscience*, **63**: 95–122.
24. Waite, J.J., Chen, A.D., Wardlow, M.L., Wiley, R.G., Lappi, D.A., and Thal, L.J., 1995, *Neuroscience*, **65**: 463–476.
25. Nilsson, O.G., Leanza, G., Rosenblad, C., Lappi, D.A., Wiley, R.G., and Björklund A, 1992, *Neuroreport*, **3**: 1005–1008.
26. Gomeza, J., Aragón, C., and Giménez, C., 1992, *Neurochem. Res.*, **17**: 345–350.
27. Wellman, C.L., and Sengelaub, D.R., 1995, *Brain Res.*, **669**: 48–58.
28. Seeburg, P.H., 1993, *Trends Pharmacol. Sci.*, **14**: 297–303.
29. Nordberg, A., 1992, *Cerebrovasc. Brain Met. Rev.*, **4**: 303–328
30. Advokat, C., and Pelligrini, A.I., 1992,. *Neurosci. Behav. Rev.*, **16**:13–24.
31. Burney, R.N., 1994, *Neurobiol. Aging*, **15**: 271–273.
32. Carlson, M.D., Penney jr., J.B., and Young A.B., 1993, *Neurobiol. Aging*, **14**: 343–352.

THE RELATIONSHIPS BETWEEN TWO CHOLINE UPTAKE COMPONENTS WITHIN WIDE RANGE OF pH

O. V. Chumakova

Institute of Biochemistry
50 BLK Grodno, Belarus

1. INTRODUCTION

The transfer of choline through neuronal membranes for acetylcholine (Ach) synthesis is a specific, high affinity and sodium-dependent process (HACU)(8,10,19).

Having high affinity for choline (0.4–4 μM) and specifically related to Ach biosynthesis, HACU is carried out by a transporter localized in the neuronal membrane (15,19). The transporter is probably a polypeptide with molecular weight of 80 kDa (3). HACU is one of the two presynaptic cholinergic "markers", choline acetyltransferase synthesizing Ach (E.C. 2.3.1.6.) being the second "marker" (15).

In addition to the high affinity transporter (K_m=1.4 μM, for rat striatum), the low affinity transporter (K_m= 93 μM for rat striatum) is wide-spread in brain tissue (21,24). They differ in localization and functional significance. HACU is located in nerve terminals and it is a rate-limiting and regulatory step in Ach synthesis. The low affinity system is not associated with efficient synthesis of Ach and is in the membranes of the majority of cells. It is possibly needed for choline transfer by diffusion (21).

Many aspects of the HACU process — its ionic and energy dependencies and relationship to Ach synthesis — have been described (8,10,16,17,22). In neurons and their models, synaptosomes, high-affinity but Na^+-independent choline uptake proceeds simultaneously and concurrently to Na^+-dependent transport (10,16). It has been shown that the HACU-inhibitor (N-allyl-3-quinuclidinol) inhibited the Na^+-dependent (IC_{50}=0.9 μM) and Na^+-independent(IC_{50}= 680 μM) transport in different ways (19). This fact seems to provide evidence for relative independence of the two processes.

Na^+-independent uptake may play a role in a number of processes taking place during impulse-flow in synapse.

The important condition for the synapse function is the acid-alkaline homeostasis involving ionic correlation, many of these ions (Na^+,Ca^{2+},Cl^-) are necessary participants of

---

From a symposium organized by Stanislav Tuček (Prague, Czechia) and Andrzej Szutowicz (Gdansk, Poland). (Supported by the British-American Tobacco Company Ltd., Staines, England.)

Neurochemistry, edited by Teelken and Korf
Plenum Press, New York, 1997

the impulse-flow. It was shown that pH value is an important bioregulator. The enzyme activities were found to increase and the rates of many membrane-associated reactions to alter after changing pH (2,7,11,14).

The aim of the study was to describe the relationships between the two processes of choline uptake within a wide range of pH values.

2. MATERIAL AND METHODS

To obtain the fraction enriched with synaptosomes, we used the striata of male albino rats (weighing 140–160 g, obtained from the Rappolovo breeding colony of the Academy of Medical Sciences of Russia).

Brains rapidly excised in cold were placed in an ice-cold solution of 0.32 M sucrose (pH 7.4) containing EDTA. After 1 min striata were removed (6). The fraction containing synaptosomes was isolated according to Weiler et al (22). Its pellet was resuspended in the Na^+-free Krebs-Ringer solution to restore the synaptosome ultrastructure. The synaptosome suspension, which was diluted to obtain the necessary amount of protein (from 90 to 560 μg) was added to the flasks.

The procedure of HACU determination was carried out according to Simon and Kuhar (18). To estimate the intensity of the Na+-dependent process, 100 μl of the suspension was added to each 800 μl sample of the Krebs-Ringer phosphate buffer (pH 7.4) containing 122 mM NaCl, 4.9 mM KCl, 1.3mM $CaCl_2$, 15.8 mM Na_2HPO_4, 1.2 mM $MgSO_4$ and 11 mM glucose. The intensity of Na^+-independent HACU was determined under the same conditions, but in the Na^+-free Krebs-Ringer solution(pH 7.4) NaCl was replaced by 252 mM sucrose and Na_2HPO_4 by 15.8 mM Tris phosphate. For the buffer solution of pH<7 we used Tris-HCl, and for those of pH>7 we used Tris-OH.

After 5-min preincubation at 37° C the reaction was started by addition of 100 μl of ^{14}C-choline. After 4-min incubation the reaction was stopped by the addition of 2 ml of ice-cold Krebs-Ringer buffer of the appropriate ion composition and pH. The samples were immediately placed on an ice-bath and after 1 min their contents were passed through CF/F filters which were washed twice with 5 ml of ice-cold saline and transferred to scintillator-containing flasks. The rate of choline uptake processes was expressed as pmoles choline/min/mg protein.

The pH-dependence of the HACU rats at the choline concentration of 1, 2, 4 and 5 M was studied. The kinetic properties of the processes were estimated at the choline concentrations of 0.010; 0.030; 0.035; 0.105; 0.310; 0.94 and 2.84 μM.

Two-tailed Student's t-test was used to assess possible differences.

3. RESULTS

Figure 1 summarizes the results of 4 independent experiments. The choline concentration was 4 μM at different amounts of protein (from 90 to 300 mg/tube). The aim of the experiments was to assess a generality. The figure shows three areas. 1. At pH>7.0 the process of sodium-dependent uptake prevails over the sodium-independent uptake. 2. The region of 6 to 7 pH values is the area of inversion which shows crosses. 3. In the pH region of <6.0 the prevalence of Na^+-independent uptake over Na^+-dependent uptake was observed.

Experiments using different substrate concentrations (1,2 and 5 μM) and constant protein content in the samples were designed to support this regularity. The pH values were

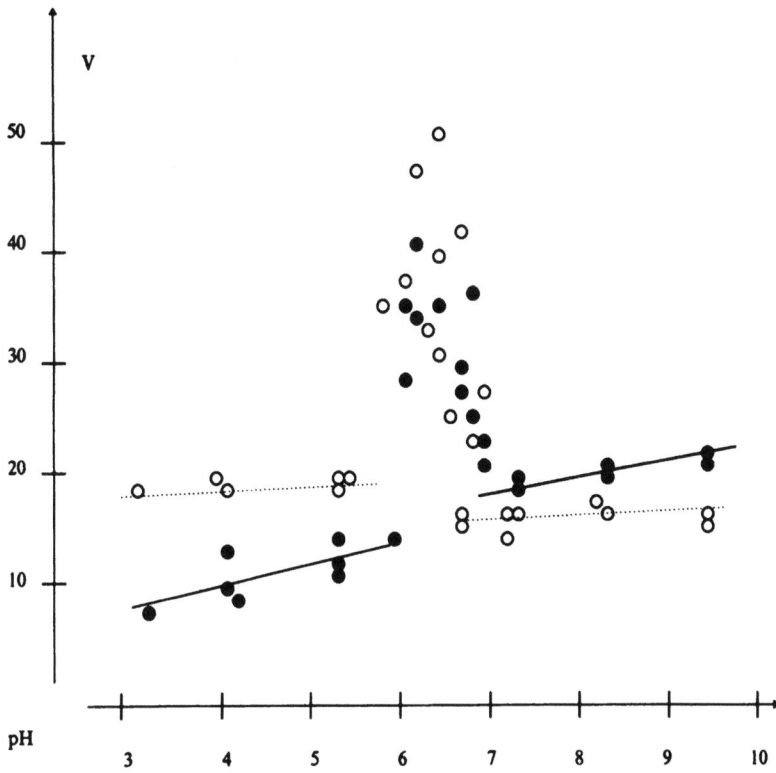

Figure 1. Dependence of HACU reaction rate (V, pmoles choline/min/mg protein) on pH of the reaction medium. The three regions of relationships. (Na$^+$-dependent uptake ●, Na$^+$-independent uptake ○).

within the range of 5.1 to 8.4. Figure 2 shows that the 3 relationships described above have remained. The character of relationships changes with the choline concentrations. More interesting are probably the low unsaturate substrate concentrations (1 and 2 μM) because they make it possible to distinguish between two reactions. The 5 μM choline concentration is saturated and the curves are parallel and almost coincide in the alkaline region.

The kinetic parameters were calculated by linearization of the dependence of the initial reaction rate on the substrate concentration in Lineweaver-Burk coordinates.

The K_m and V_{max} values were calculated for three different pH values: 7.4 is the optimum pH for Na$^+$-dependent uptake; at 5.17 — a clear-cut prevalence of the Na$^+$-independent uptake over the Na$^+$-dependent uptake was observed; at 9.09 a clear prevalence of the Na$^+$-dependent uptake was noted. The K_m and V_{max} values are listed in the table.

Table 1. Kinetics of choline uptake at various pH (n=3)

Series	pH	Ions Na$^+$ dependence	K_m, μM	V_{max}	V_{max}/K_m
1	5.17	Na$^+$-dependent	2.11±0.29	24.36±6.34	11.39±1.77
		Na$^+$-independent	4.05±0.69	58.3±20.9	13.68±2.68
2	7.4	Na$^+$-dependent	0.68±0.09[a]	116.6±8.35[a]	175.55±14.2[a]
		Na$^+$-independent	1.8±0.36[a]	70.1±8.02	39.6±3.30[a]
3	9.09	Na$^+$-dependent	0.50±0.028[a]	68.27±1.57[ab]	137.13±5.63[a]
		Na$^+$-independent	0.75±0.04[ab]	51.8±1.84	69.61±5.21[ab]

Significant differences-[a]compared with series 1; [b] compared with series 2.

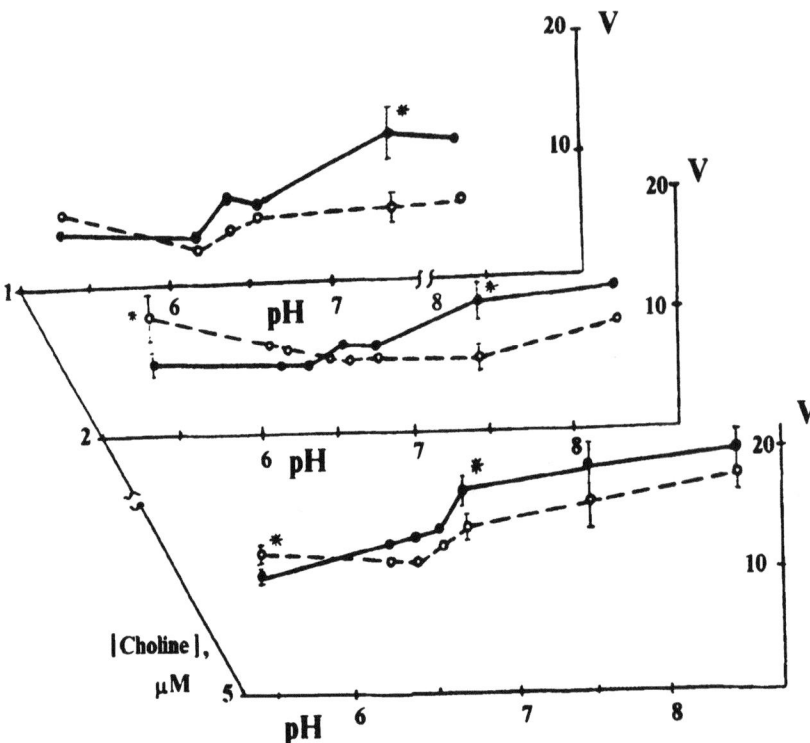

Figure 2. pH-Dependence of HACU reaction rate (V, pmoles choline/min/mg protein) in the medium with Na^+ (●) and Na^+-free medium (○) on various choline concentrations.

The table shows that the K_m Na^+-dependent uptake is significantly higher in the acid medium in comparison with the optimal and alkaline ones, which suggests a decrease of Na^+-dependent transporter affinity for choline at pH 5.17. The K_m of the Na^+-independent process significantly increases at pH 5.17 and subsequently decreases by nearly 2-fold in neutral and alkaline media (also about 2-fold). The highest V_{max} value characterizes Na^+-dependent uptake in the optimal medium, whereas it decreases in the acid (4.7-fold) and alkaline (1.7-fold) media. The variations in V_{max} of the Na^+-independent uptake at the pH points studied have not been shown.

4. DISCUSSION

Various membranes (chloroplast and myocardium membranes (13,14)) show pH-induced conformational transitions, therefore the pH-dependent changes in the choline transport across neuronal membranes are quite interpretable and governed by membrane reactions.

It is assumed that each of the three pH regions described may have a certain functional sense. In the first (pH>7.0) and third (pH<6.0) regions the choline uptake depends on the pH medium. Na^+-dependent transport is more sensitive to the acid medium as compared to the alkaline because at pH<6 it is slowed down and the V_{max} decreases more than 4-fold. The second pH region (6.0–7.0) may be characterized as a region of unequilibrium. In this case, the choline uptake does not only depend on pH, but also on some other parameter associated with the content of synaptosomal protein in the sample. It is possible that the ratio of the protein or proteolipid components to the transferred substrate is of importance in this pH region. The kinetic analysis of the properties of the choline transport also demonstrates the differences in Na^+-dependent and Na^+-independent processes. In various pH media, the changes in the K_m and V_{max} values may depend on protonation/deprotonation reactions of the membrane protein, including protein-transporters (5). The effect of screening of the surface protein charges at high and low pH can also be significant (9). The role of protonation of the enzyme ionogenic groups was described by three fundamental variants of the classical kinetics in the catalytic act. If we consider the transporter-choline interaction as obeying the Michaelis-Menten kinetics at the initial steps, then these variants can be used as models. The variants are as follows, at pH variation: 1. The K_m value and the V_{max}/K_m ratio can be changed 2. Both the V_{max} and V_{max}/K_m values can be changed 3. Both the K_m and V_{max} change but the V_{max}/K_m remains unchanged (4,5).

Thus, the Na^+-dependent choline transport is characterized by changing the K_m and V_{max} values in the acid medium and the V_{max} values only in the alkaline medium. The Na^+-independent uptake shows K_m and V_{max}/K_m changes in the acid and alkaline media.

Thus, the process of the Na^+-dependent uptake deviations does not fit the classical scheme. In terms of its primary importance in the cholinergic function, the regulation of this transport can be mediated through a delicate mechanism of pH changes. The Na^+-independent uptake may be described as the first variant under these conditions. This means that protonation/deprotonation of the transporter can change the share of reactive molecules and thus affect the affinity for substrate. The substrate binding decreases ionization without any changes in the V_{max} value, but the K_m value remains unaltered.

Since both the Na^+-dependent and Na^+-independent uptakes have different kinetical characteristics, they may be suggested to have different uptake mechanisms and, possibly, independent channels and carriers.

But on the basis of the biological principle of plastic economy in the cell, the existence of one transporter is more probable. But the changed intensity of its functions and its dependence on Na^+ ions are governed by choline requirements of different reactions : intracellular (Ach synthesis) and intramembranous.

Recently it has been shown that the contents of Ach, phosphatidylcholine and choline in brain cells are interrelated and that phosphatidylcholine in neuronal membranes may be a source of synthesized ACh (23). With regard to this viewpoint the existence of two interrelated processes of high-affinity choline uptake in nerve terminals is possible.

As Na^+-dependent uptake is a source of choline for neurotransmitter synthesis, Na^+-independent uptake may be related to the choline pool in neuronal membrane. Then the switching over of two processes may be possible at a certain ratio of choline and protein-transporter in the pH range from 6.0 to 7.0. Na^+ ions are not modifiers of these transitions due to their high and practically constant concentrations outside the cell (about 145 μ) (24). However, the local and transitory pH changes in the synaptic cleft may result from some ionic fluxes and induce reciprocal transitions of HACU process as a regulatory factor of neuronal activity.

REFERENCES

1. Alberts, B., Bray, D., Lewis, J. et al, 1986, *Molecular Biology of the cell. Moscow: "Mir"*, P. 99–109.
2. Barskyi, E.,L., Nazarenko, A.I., Samuilov, V.D., Khakimov, C.A., 1989, *Biol. membranes.*, 6: 720–723.
3. Breer, H., Knipper M., Kahle C., 1989, *J. Protein Chem.* 8: 372–374.
4. Dixon, M., Webb, E., Enzymes, N.Y. et al., 1979, *Longman Group Ltd.*
5. Fedosova, N.U., Berdashkevich, E.A., Fedosov, S.N., Boldyrev, A.A., 1993, *Biochemistry*, 58: 1077–1084.

6. Glowinsky, J., Iversen, L.L. , 1966, *J. Neurochem.*, **13**: 655–669.
7. Iijuma, T., Ciani, S., Hagiwara, S., 1986, *Proc. Nat. Acad. Sci.*, **83**: 654–658.
8. Jope, R.S., Weiler, M.H., Jenden, D.J., 1978, *J. Neurochem.*, **30**: 949–954.
9. Khazipov, R.N., Khamitov, Kh.S., Giniatullin, R.A., Garaev, R.S., 1990, *Biol.membranes*, **7**: 1001–1007.
10. Kuhar, M.J., Murrin, L.C., 1978, *J. Neurochem.*, **30**: 15–21.
11. Leontyeva, O.V., Kutuzova, G.D., Turovetsky, V.B., Ugarova, N.N., 1994, *Biochemistry*, **59**: 113–117.
12. Murrin, L.C., Kuhar, M.J., 1976, *Mol. Pharmacol.*, **12**: 1082–1090.
13. Nakipova, O.V., Kokoz, Yu.M., Povzun, A.A., Lazarev, A.B., 1989, *Biol. membranes*, **6**: 1285–1295.
14. Opanasenko, V.B., 1993, *Biochemistry*, **58**: 716–723.
15. Patel, P.J., Messer, W.J., Hudsson, R.A., 1993, *J. Med. Chem.*, **36**: 1893–1901.
16. Rea, M.A., Simon, J.R., 1981, *Brain Res.*, **219**: 317–326.
17. Roskoski, R.J., 1978, *J. Neurochem.*, **30**: 1357–1361.
18. Simon, J.R., Kuhar, M.J., 1976, *J. Neurochem.*, **27**: 93–99.
19. Sterling, G.H., Doukas, F.H., Ricciardi, F.J., O'Neill, J.J., 1991, *J. Pharm. Sci.*, **80**: 785–789.
20. Takano, Y., Kohjimoto, Y., Uchimura, K., Kamiya, H.-O., 1980, *Brain Res.*, **194**: 583–587.
21. Tuchek, S., 1981, Acetylcholine synthesis in neurons, "Mir", P. 128–151.
22. Weiler, M.H., Gundersen, C.B., Jenden, D.J., 1981, *J. Neurochem.*, **36**: 1802–1812.
23. Wurtman, R., Ulis, I.H., Bluszlajn, J.K. et al., 1989, *J. Neurochem.*, **52**: 1.
24. Yamamura, H.I., Snyder, S.H., 1973, *J. Neurochem.*, **21**: 1355–1374.

INFLUENCE OF NICOTINIC STIMULATION ON $[Ca^{2+}]_i$ IN PROCESSES OF CHICK SYMPATHETIC NEURONS IN CULTURE

V. Dolezal,[1] A. Schobert,[2] and G. Hertting[2]

[1]Institute of Physiology
Czech Academy of Sciences
Prague, Czech Republic
[2]Department of Pharmacology
Med. Fac. of the University
Freiburg, Germany

1. INTRODUCTION

The transmission of impulses from sympathetic nerve terminals to target cells is mediated by noradrenaline. The action potentials open voltage-dependent calcium channels (VDCCs) in the nerve terminals and the resulting increase of the free intraterminal concentration of calcium ($[Ca^{2+}]_i$) causes calcium-dependent liberation of noradrenaline. The action potential-evoked release of transmitters is subjected to modulation by the transmitter itself and by other neurotransmitters that can occur at the level of the cell bodies or at the level of the nerve terminals (9). The effect on the cell bodies usually affects the propensity of neurons to fire action potentials whereas that on the nerve terminals is often a consequence of a changes in calcium influx.

Sympathetic ganglion neurons in vivo are mainly excited by acetylcholine acting via nicotinic receptors located on neuronal cell bodies. However, nicotinic receptors are also believed to be present on sympathetic nerve terminals, and the stimulation of these receptors is thought to bring about the extracellular calcium-dependent release of noradrenaline (7,11). The stimulation of presynaptic nicotinic receptors might permit the influx of calcium either directly through the open pores of these receptors or indirectly by the depolarization of the nerve terminals and the resulting opening of VDCCs.

Changes of $[Ca^{2+}]_i$ in neuronal terminals can be measured by microfluorimetry in the cells grown in culture. Chick sympathetic neurons in culture represent a convenient preparation for the study of synaptic transmission. They accumulate noradrenaline and release it in

---

From a symposium organized by Stanislav Tuček (Prague, Czechia) and Andrzej Szutowicz (Gdansk, Poland). (Supported by the British-American Tobacco Company Ltd., Staines, England.)

a calcium-dependent manner from their dense neuronal processes (6). The stimulation of their nicotinic receptors evokes calcium-dependent and largely tetrodotoxin-sensitive liberation of noradrenaline (1,2,4). We employed this preparation to study the nicotinic stimulation-evoked changes of the intracellular concentration of calcium in sympathetic nerve terminals in order to clarify the pathway(s) of calcium entry into the nerve terminals.

2. METHODS

Neuronal cultures were prepared from 12-day-old chicken embryos as described (1). Cultures were used after 3–6 days. Cells were incubated for 30 min in a medium with 5 mol/l FURA 2-AM and then transferred into a closed superfusion chamber on the stage of an inverted fluorescence microscope. The chamber was superfused at 20–23 °C with a physiological salt solution at a flow rate of 0.7 ml/min. A place in the area of the process network which gave a clear, fast response to a single electrical pulse (2 msec width, 30 mA) was chosen as the synaptic area for the measurements (10). Sequential light excitation was at 340/360/380 nm and light emission was recorded at 510–560 nm. The frequency of sampling at each wavelength was 10 Hz, and the calculated $[Ca^{2+}]_i$ was averaged over 1 s intervals. The changes in $[Ca^{2+}]_i$ evoked by either nicotinic (1 min pulse of 100 μmol/l DMPP) or electrical (2 msec monopolar pulses, 36 pulses at 3 Hz, 30 mA) stimulation were calculated as described (5).

3. RESULTS

Under control conditions, the increases of $[Ca^{2+}]_i$ evoked both by the stimulation of nicotinic receptor (100 μmol/l DMPP for 1 min) and by the electrical stimulation (36 pulses at 3 Hz) are diminished by L- and N-type calcium channel blockers (+) Bay k 8644 (1 μmol/l) and ω-conotoxin (CTX; 100 nmol/l), but are not influenced by P-type calcium channel blocker ω-agatoxin (300 nmol/l). The contribution of the N-type calcium channel to the increase of $[Ca^{2+}]_i$ is $55 \pm 6\%$ (n=6) and $66 \pm 1\%$ (n=3) for nicotinic stimulation and electrical stimulation, respectively, and the contribution of the L-type calcium channel is $39 \pm 7\%$ (n=7) and $15 \pm 5\%$ (n=10), respectively.

Tetrodotoxin (TTX; 300 nmol/l) blocks the increases of $[Ca^{2+}]_i$ evoked by electrical stimulation (Fig. 1), which indicates that the propagation of action potentials and the subsequent opening of VDCCs are absolutely necessary for the electrically evoked $[Ca^{2+}]_i$ increase. In contrast, DMPP in the presence of TTX still evokes increases of $[Ca^{2+}]_i$ amounting $27 \pm 1\%$ (n=10) of control responses in the absence of TTX. These increases are nicotinic in nature because they are blocked by the nicotinic antagonist mecamylamine (10 μmol/l), although not by α-bungarotoxin (BTX; 125 nmol/l). The response evoked in the presence of TTX ($27 \pm 1\%$, n=10) is only slightly reduced by (+) Bay k 8644 (to $22 \pm 2\%$, n=3), by CTX (to $20 \pm 3\%$, n=7) or by combination of (+) Bay k 8644 and CTX (to $18 \pm 1\%$, n=4). The L-type calcium channel activator (-) Bay k 8644 potentiates DMPP induced response in the presence of TTX (to $54 \pm 6\%$, n=8).

4. DISCUSSION

Similarly to the electrical stimulation, the nicotinic stimulation of sympathetic neurons (3) performed under control conditions causes an influx of calcium into nerve termi-

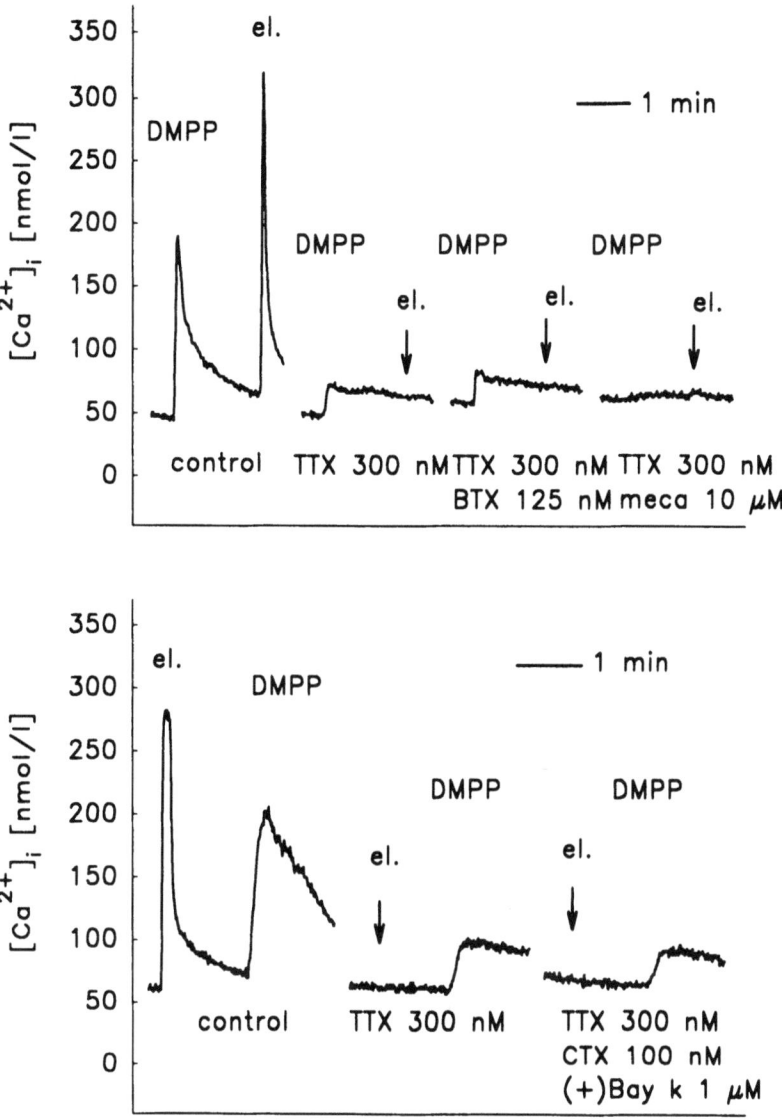

Figure 1. Sympathetic neurons were repeatedly stimulated by DMPP (DMPP; 100 μmol/l) or electrically (el.; 36 pulses at 3 Hz) at 24 min time intervals at first under control conditions and then in the presence of TTX, in the presence of TTX and nicotinic antagonists (upper panel), or in the presence of TTX and calcium channel blockers (lower panel). Each panel represents a different preparation. Ordinate: $[Ca^{2+}]_i$ in nmol/l. Time bar 1 min.

nals mostly through N- and L-type calcium channels that are activated by action potentials. The contribution of the L-type calcium channels seems to be somehow bigger than when the influx is activated by electrical stimulation (39% vs 15%).

The extracellular calcium-dependent increase of $[Ca^{2+}]_i$ in the nerve terminals evoked by DMPP is strongly diminished, but not blocked, by TTX. This finding can be explained either by an incomplete inhibition of the fast sodium channels or by a local action of the nicotinic agonist on nerve terminals that does not need impulse propagation. The block of the increase of $[Ca^{2+}]_i$ evoked by electrical stimulation in the presence of the same concentration of TTX excludes the first explanation (Fig 1). The local activation of presynaptic nicotinic receptors might bring about the influx of sodium ions leading to local depolarization and to an opening of VDCCs. Alternatively the influx of calcium ions might occur directly through the pores of nicotinic receptors. Substantial permeability for calcium of various types of nicotinic receptors on cell bodies has been reported (8). The results of our measurements on the nerve terminals favour the latter possibility. The influx of calcium into nerve terminals evoked by DMPP in the presence of TTX is not substantially decreased by calcium channel blockers at concentrations which do profoundly inhibit calcium influx under control conditions.

In conclusion, we have demonstrated that the stimulation of presynaptic nicotinic receptors activates calcium influx that is independent of the propagated electrical activity, and that is not blocked by blockers of VDCCs. We conclude that calcium ions enter the terminals through the pores of nicotinic receptors.

ACKNOWLEDGMENT

This work was supported by the Deutsche Forschungsgemeinschaft (SFB 325). V.D. was supported by EC grant ERB CIPA-CT92-3014 and by grant A7011506 of the Grant Agency of the Academy of Sciences of the Czech Republic.

REFERENCES

1. Dolezal, V., Schobert, A., Heldt, R., and Hertting, G., 1994, *Neurosci. Lett.* **180**: 63–66.
2. Dolezal, V., Schobert, A., and Hertting, G., 1995, *Neurochem. Res.*, **20**: 17–23.
3. Dolezal, V., Schobert, A., and Hertting, G., 1995, *J. Neurochem.* **65**: 1874–1879.
4. Greene, L.A., and Rein, G., 1978, *J. Neurochem.*, **30**: 579–586.
5. Grynkiewicz, G., Poenie, M., and Tsien, R.Y., 1985, *J. Biol. Chem.*, **260**: 3440–3450.
6. Przywara, D.A., Bhave, S.V., Chowdhury, P.S., Wakade, T.D., and Wakade, A.R., 1993, *Neuroscience* **52**: 973–986.
7. Rand, M.J. and Story, D.F., 1991, *Presynaptic Release: A Handbook* (Faigenbaum J.and Hanani, M. eds), Freund Publishing House, Ltd., Tel Aviv, pp. 1033–1070.
8. Sargent, P.B., 1993, *Ann. Rev. Neurosci.* **16**: 403–443.
9. Starke, K., Göthert, M. & Kilbinger, H., 1989, *Physiol. Rev.* **69**: 864–989.
10. Toth, P.T., Bindokas, V.P., Bleakman, D., Colmers, W.F. & Miller, R.J., 1993, *Nature* **364**: 635–638.
11. Wonnacott, S., Irons, J., Rapier, J., Thorne, B., and Lunt, G.G., 1989, *Progress in Brain Res.* **79**: 157–164.

CHOLINERGIC MECHANISMS IN INHERITED AND ACQUIRED TOLERANCE TO ETHANOL

A. Jankowska, Y. Kisielevski, N. Oganesjan, M. Tomaszewicz, and
A. Szutowicz

Department of Clinical Biochemistry
Medical University of Gdańsk
80-211 Gdańsk, Poland

1. INTRODUCTION

It is known that rats, which in free choice test prefer ethanol instead of water, differ in number of behavioral and metabolic parameters from siblings which choose water on their own accord. Each group may be identified on the basis of short (SS) or long (LS) duration of narcotic sleep after intraperitoneal (i.p.) administration of ethanol, respectively (1)(section 2.). Alcohol preferring (SS) rats displayed slower rates of glycolysis and metabolism of some aminoacids and lipids in their brains, than the LS ones (2). It gave rise to the hypothesis that inherited inclination to alcohol might be linked with relative deficiency of acetyl-CoA and energy in SS brains which could be supplemented by ethanol-derived acetate through the acetyl-CoA synthetase reaction (3). One has to remember that in brain cholinergic neurons acetyl-CoA is additionally utilized for acetylcholine (ACh) synthesis. This process has been shown to be strongly dependent on the rate of pyruvate oxidation which was found to regulate supply of acetyl-CoA to cytoplasm of cholinergic terminals (4,5). On the other hand, alcohol does not seem to be a likely source of acetyl-CoA for ACh synthesis since acetate has been found to be incorporated to brain lipids but not to the neurotransmitter pool. One must also stress, that theory of acetyl group deficiency-dependent alcoholims has not been substantiated by relevant experimental data.

Inherited sensitivity to ethanol may be modified by its consecutive applications. They cause quick development of resistance to ethanol narcotic action. The mechanism of this phenomenon is not clear. It has been claimed that it might be due to induction of ethanol metabolizing enzymes or suppression of adenosine and other inhibitory transmitter systems (6). We assumed that ethanol may evoke adaptative responses in acetyl-CoA metabolic pathways and thereby influence cholinergic activities. Inhibitory effects of chronic

---

From a symposium organized by Stanislav Tuček (Prague, Czechia) and Andrzej Szutowicz (Gdansk, Poland). (Supported by the British-American Tobacco Company Ltd., Staines, England.)

alcohol administration on ACh metabolism in whole brain preparations were reported. However neither long nor short term studies of ethanol effects on acetyl-CoA and ACh metabolism were performed in nerve terminals. Studies of isolated synaptosomes seem to be important for understanding this phenomenon since the transmitter pool of ACh is synthesized exclusively in this compartment of the brain.

2. MATERIALS AND METHODS

Male Wistar rats weighing 100–120 g were used. SS and LS rats were selected by i.p. injection of ethanol 3.0 g/kg of body weight as 10% solution and observation of time of postinjection sleep. About 15% animals with sleep time shorter than 80 min was qualified as SS, while another 15% with sleep time longer than 120 min was included into the LS group. After 14 days recovery animals were used for studies of acetyl-CoA and ACh metabolism. Rats with intermediate sleep time were used for short term ethanol tolerance studies. They were treated with daily i.p. injections of ethanol, 3 g/kg of body weight. Control group received i.p. injections of glucose. 24 hours after sixth injection their brains were taken for experiments. For studies of distribution of enzymes of acetyl-CoA metabolism brain regions were collected by free hand dissection. Enzyme assays were performed in homogenates as described elsewhere (7). For metabolic studies synaptosomes were isolated from pooled homogenates from 5 forebrains (8). They were incubated in the depolarizing medium containing in a final volume of 2.0 ml: 20 mM Na-HEPES, 1.5 mM Na-phosphate buffer (final pH. 7.4), 90 mM NaCl, 30 mM KCl, 2.5 mM Na-pyruvate, 2.5 mM Na-L-malate, 0.01 mM choline chloride, 0.01 mM eserine sulfate, 32 mM sucrose and 2.5 - 3.0 mg of synaptosomal protein. Quantal ACh release was stimulated by addition of 1.0 mM $CaCl_2$. Incubation was started by addition of synaptosomal suspension and carried out for 30 min at 37°C in water bath with continuous shaking 100 cycles/min. Incubation was stopped by centrifugation at 15.000 x g for 3 min in Eppendorf tubes. Pelet and supernatant were deproteinized and taken for metabolite assays (8). Results are means ± S.E.M. from 5–7 experiments done in duplicate. Statistical comparisons were made by Students or paired t tests.

3. RESULTS

3.1. Enzyme Activities in Brain Regions of LS and SS Rats

Pyruvate dehydrogenase activity in cortex and striatum of SS was found to be 22–47% lower than in LS rats (Tab. 1). On the other hand, ATP-citrate lyase (ACL) activities appeared to be lower in SS hypothalamus and medulla oblongata for 14 and 36%, and those of carnitine acetyltransferase (CaAT) for 18 and 15% higher, respectively than in the same regions of LS animals (Tab. 1). Higher activities of oxoglutarate dehydrogenase and acetyl-CoA synthetase were found in single brain regions of SS rats (not shown). No differences were detected for choline acetyltransferase activities (Tab. 1).

These findings indicate that in some brain regions of SS rats, the synthesis of intramitochondrial acetyl-CoA might proceed in slower rate than in LS siblings. In addition, differences in ACL, CaAT and acetyl-CoA synthetase activities suggest existence of variations in acetyl-CoA transporting pathways in both groups of animals.

Table 1. Enzyme activities in brain regions of rats of different susceptibility to narcotic effect of ethanol

Enzyme		TCtx	FCtx	Str	Hp	Ht	MO
PDH	LS	26.5±1.8	26.1±1.5	24.6±2.0	19.5±1.1	20.2±2.4	15.8±1.0
	SS	20.8±1.2*	20.4±1.3*	13.1±1.4*	19.3±0.9	18.8±1.1	16.0±1.3
ACL	LS	6.8±0.4	7.7±0.4	7.3±0.3	6.2±0.4	10.3±0.6	13.7±0.6
	SS	7.6±0.5	8.1±0.4	8.1±0.2*	7.3±0.7	8.9±0.43*	8.8±0.5*
CaAT	LS	24.8±1.3	24.9±1.7	23.5±1.4	26.1±1.8	31.2±1.9	31.3±2.5
	SS	24.4±1.6	26.0±1.8	24.2±0.9	26.5±1.2	36.9±0.9*	36.1±1.1*
CAT	LS	0.51±0.03	0.60±0.03	1.35±0.11	0.45±0.02	0.69±0.10	0.82±0.06
	SS	0.55±0.03	0.62±0.06	1.29±0.10	0.55±0.02	0.55±0.02*	0.86±0.04

Enzyme activities in nmols/min/mg of protein. Abbreviations: LS, long sleep; SS, short sleep rats; PDH, pyruvate dehydrogenase; ACL, ATP-citrate lyase; CaAT, carnitine acetyltransferase; CAT, choline acetyltransferase; TCtx, temporal cortex; FCtx, frontal cortex; Str, striatum; Hp, hippocampus; Ht, hypothalamus; MO, medulla oblongata.
*Significantly different from long sleep group, p<0.05.

3.2. Acetyl-CoA Metabolism in Synaptosomes of SS and LS Rats

Rates of pyruvate consumption by SS and LS synaptosomes were similar both in the presence and in the absence of Ca^{2+} in the medium. On the contrary, levels of acetyl-CoA, and rate of resting (nonquantal) ACh release in SS were found to be 28 and 58%, respectively lower than in LS (Tab. 2). These data are concordant with our earlier observations that the rate of ACh release and related synthesis is correlated with the acetyl-CoA contents in nerve terminals (8 for review).

The addition of Ca^{2+} increased total ACh release by the same value in both groups (Tab. 2). It suggests that functions of cholinergic neurons which depend on the release of quantal ACh pool remain the same in alcohol preferring and nonpreferring rats (9). On the other hand, the excitatory threshold of cholinoceptive neurons, could be probably lower in LS rats, due to higher Ca-independent (nonquantal) ACh release in this group (Tab. 3.) (9). Therefore we think that the activity of brain cholinergic system is not involved in differentiation of inherited susceptibility of SS or LS rats to narcotic effects of ethanol.

Table 2. Pyruvate utilization, acetyl-CoA levels and ACh release in synaptosomes of SS and LS rats

Conditions/parameter (*units*)	Long sleep (LS)	Short sleep (SS)
No Ca^{2+}		
Pyruvate *(nmols/min/mg protein)*	10.1 ± 0.6	8.1 ± 0.8
Acetyl-CoA *(pmols/mg protein)*	35.7 ± 3.2	25.9 ± 2.4^{++}
ACh *(pmols/min/mg protein)*	14.6 ± 1.7	6.1 ± 0.1$^+$
1mM Ca^{2+}		
Pyruvate *(nmols/min/mg protein)*	6.5 ± 0.6*	6.4 ± 0.6
Acetyl-CoA *(pmols/mg protein)*	33.7 ± 2.6	24.4 ± 1.1$^+$
ACh *(pmols/min/mg protein)*	22.2 ± 2.0*	13.4 ± 2.2$^+$ (7.2)

Numbers in parentheses indicate Ca-evoked ACh release. *Significantly different from respective no Ca^{2+} conditions, p<0.05; $^+$significantly different from long sleep group, p<0.01; $^{++}$significantly different from long sleep group, 0.01<p<0.05.

Table 3. Effect of six day ethanol adaptation on pyruvate utilization, acetyl-CoA content and ACh release in rat brain synaptosomes

Conditions/parameters (units)	Control	Ethanol 6 days
No Ca^{2+}		
Pyruvate (nmols/min/mg protein)	12.9 ± 0.5	11.4 ± 1.6
Acetyl-CoA (pmols/mg protein)	34.9 ± 2.8	33.7 ± 3.8
ACh (pmols/min/mg protein)	15.0 ± 0.6	15.9 ± 2.3
1mM Ca^{2+}		
Pyruvate (nmols/min/mg protein)	8.5 ± 0.8*	8.1 ± 0.4
Acetyl-CoA (pmols/mg protein)	29.8 ± 3.0	26.9 ± 2.1*
ACh-total (pmols/min/mg protein)	21.1 ± 3.0*	29.0 ± 1.1*+
ACh-quantal (pmols/min/mg protein)	6.2 ± 1.8	13.1 ± 1.6+
Ado - in. (pmols/mg protein)	810 ± 100	403 ± 71*
Ado - ext. (pmols/mg protein)	3013 ± 267	2930 ± 196

Abbreviations: Ado-in, synaptosomal adenosine content; Ado-ext, adenosine released.
*Significantly different from respective no Ca^{2+} conditions, $p<0.05$; +significantly different from respective controls, $p<0.05$.

3.3. Effect of Short Term Ethanol Administration on Acetyl-CoA Metabolism in Synaptosomes

Rats with intermediate sensitivity to narcotic action of ethanol, after third consecutive daily i.p. administration of 3g of ethanol/kg of body weight, developed full resistance to this effect. This phenomenon was not likely to be due to increased metabolic turnover of ethanol since previous reports did not show it to be the case (10). Six day alcohol administration caused no significant changes in pyruvate utilization, acetyl-CoA level and resting (nonquantal) ACh release in synaptosomes (Tab. 3). On the contrary, in the presence of Ca^{2+} synaptosomes from ethanol treated rats showed 37% and 110% higher rates of total and quantal (Ca-dependent) releases of ACh, respectively (Tab. 3). In addition, ethanol caused no change in release but 50% decrease in adenosine content in synaptosomes (Tab. 3).

4. DISCUSSION

Shifts in synaptosomal metabolism, found here, prove that ethanol ingestion may cause adaptative changes in ACh synthesizing compartment. Unlike in several other experimental neuropathological models, the alcohol-evoked activation of quantal ACh release can not be linked with increased production of acetyl-CoA (Tab. 3.). It might be brought about by ethanol-evoked changes in composition of brain lipids which facilitate fusion of synaptic vesicles with presynaptic membranes (11). The increase of quantal ACh release in ethanol treated synaptosomes, found here (Tab. 3), is concordant with this hypothesis. These data would indicate that increase of cholinergic activity is apparently involved in development of acquired resistance to narcotic effects of ethanol.

It has been reported that ethanol activates adenosine efflux in brain, which is known to inhibit ACh release. On the other hand, decrease in activity of purinergic neurons during chronic ethanol administration, was postulated to play a role in development of ethanol desensitization (10). The decrease of adenosine content in ethanol treated synaptosomes in concordant with this view (Tab. 3). Thus decreased purinergic activity in connection with increased activity of cholinergic system could facilitate development of acquired desensitization to ethanol, reported here. These data also prove that mechanisms of inherited and acquired resistance to narcotic actions of ethanol may be entirely different.

ACKNOWLEDGMENTS

This work was supported by KBN grant No 6 P04A 013 10 and Medical University of Gdańsk project W-113.

REFERENCES

1. Burov, Y.V., Borysenko, S.A., 1979, *Pharmacol. Toxicol.* **3**:291–293.
2. Segovia-Requelme, N., Campos, J., Solodkowska, W., 1964, *Med. Exp.* **11**:185–190.
3. Atrens, D., Harfaing-Jallat, P., LeMangen, J., 1983, *Biochem. Behav.* **19**:571–575.
4. Gibson, G.E., Barclay, L., Blass, J.R., 1982, *Ann. New York Acad. Sci.* **387**:382–403.
5. Szutowicz, A., Tomaszewicz, M., Bielarczyk, H., 1996, *Acta Neurobiol. Exp.* **56**:323–339.
6. Smolen, T. N., Smolen, A., 1991, *Alcohol* **8**:123–130.
7. Szutowicz, A., Stêpieñ, M., Bielarczyk, H., Kabata, J., Lysiak, W., 1982, *Neurochem. Res.* **7**:799–810.
8. Szutowicz, A., Bielarczyk, H., Skulimowska, H., 1994, *Neurochem. Res.* **19**:1107–1112.
9. Van der Kloot, W., 1988, *Neuroscience* **24**:1–7.
10. Nurmi, M., Khanmaa, K., Sinclair, J.I., 1994, *Alcohol* **11**:315–321.
11. Montoliu, C., Valle, S., Renau-Piqueras, J., Guerri., *J. Neurochem.* **63**:1855–1862.

THE STATE OF M-CHOLINO- AND BETA-ADRENORECEPTION OF AUTONOMIC GANGLIA IN SPONTANEOUSLY HYPERTENSIVE RATS

V. V. Glinkina, L. A. Khyazeva, I. G. Charyeva, S. A. Pylaeva, and A. S. Pylaev

Russian State Medical University
Moscow, Russia

1. INTRODUCTION

The experimental model of the genetically determined hypertensive state in rats has gained acceptance in spite of some discordance between this state in rats and the human hypertensive disease. The reason is that the spontaneously hypertensive rats (SHR) are a suitable object for a study of ACh-NA interaction in the nervous system under conditions of the changed balance of classic transmitters (1) and co-transmitters (2). It is known that both central and peripheral nervous systems, especially their autonomic divisions, are involved in the development of hypertensive status. Catecholamines produced by the adrenals (3) along with corresponding receptor systems (4) are naturally thought to play a leading role in the genesis of spontaneous hypertension. At the same time the high local tissue concentrations of vasoconstrictor substances are mainly produced by the endings of sympathetic neurons. These neurons in turn receive a cholinergic synaptic input characterized by a marked muscarinic component (5).

All the above considerations led us to investigate what role the changed reception in autonomic ganglia plays in triggering and maintaining increased blood pressure. All ganglionic receptors are perceived to be available for active substances of synaptic and blood origin. To study the transmitter system of autonomic ganglia involved in the development of spontaneous hypertension we traced the formation of M-cholinoreception and beta-adrenoreception in the ganglion nodosum (GN), cardiac ganglia (CG), lumbar ganglia of the sympathetic chain (LG), and main pelvic ganglion (MPG) in SHR at different periods of the postnatal ontogenesis (the animals aged 4, 11, and 19 weeks).

From a symposium organized by Stanislav Tuček (Prague, Czechia) and Andrzej Szutowicz (Gdansk, Poland). (Supported by the British-American Tobacco Company Ltd., Staines, England.)

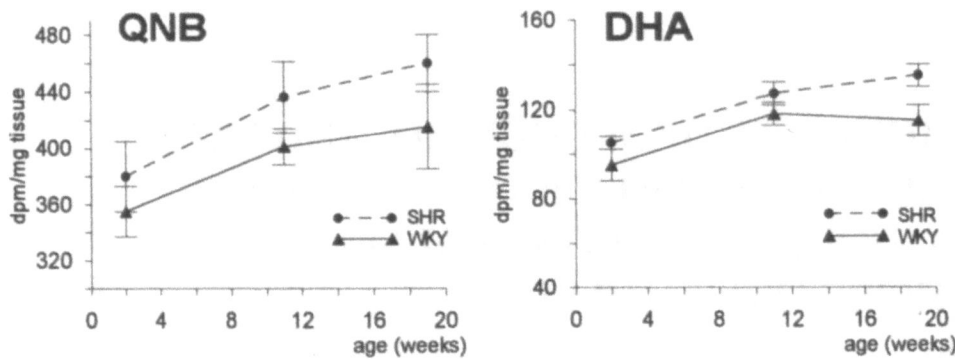

Figure 1. Binding levels for ^{3}H-quinuclidinyl-benzilate and ^{3}H-dihydroalprenolol in lumbar ganglia of the sympathetic chain.

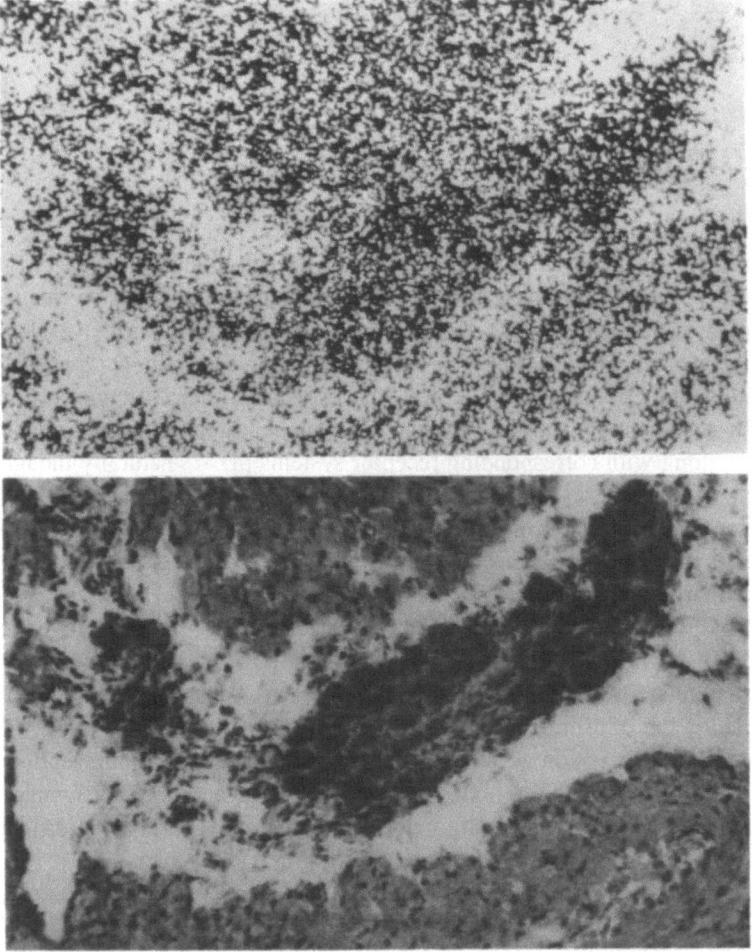

Figure 2. Radioautograph and micrograph of cryostate section of cardiac ganglion. Bar-50 μm.

2. METHODS

Wistar Kyoto rats (WKR) served as a control. The levels of reception (relationship between the number of binding sites and affinity) were determined in terms of binding of ^{3}H-quinuclidinylbenzylate (QNB) and ^{3}H-dihydroalprenolol (DHA). Incubation of cryostat sections of the labelled ligands was performed according to standard protocol. In the case of CG, preparations were covered with dry emulsion and the optical density was measured in radioautographs after exposure and routine photographic processing (Fig. 1). Cryostat sections of other ganglia were placed in scintillation vials on small glasses. The level of specific binding was calculated as the difference between the optical densities of autographs in preparations with total binding and in the preparations with displacement of the binding ligand by the "cold" ligand.

3. RESULTS

The results of our study showed no reliable difference in the level of M-cholino- and beta-adrenoreception in GN and CG tissue of SHR as compared to normotensive WK rats. The levels of binding of QNB and DHA labelled with tritium were similar for MPG tissue of SHR and WKR at all times studied. Coincidence of the reception levels in SHR and WKR attests to the absence of vasotonic neurons in autonomic ganglia studied. Previously, the changes in binding of ligands were demonstrated on the brain slides of SHR for those brain regions which are involved in the regulation of arterial pressure (6).

The analysis of M-cholinoreception levels in the LG tissue revealed a slight rise of the scintillation count for ^{3}H-QBN binding in SHR. The general trend of the reception formation was, however, similar to that in GN and MPG. The higher number of ligand binding sites was more pronounced in 19-week-old SHR as compared to the younger animals. ^{3}H-DHA binding assays gave similar results with statistically reliable differences at later stages of the experiment (Fig. 2).

4. DISCUSSION

The idea arises that slight hypersensibilization is peculiar to M-cholinoreceptors just in LG. This assumption seems to be justified. The lumbar sympathetic ganglia are known to innervate somatic vessels, mainly vessels of the posterior extremities, and our findings attest that they play an important part in the development of systemic vascular reactions. Interpretation of the fact that the beta-adrenoreceptor level in the LG tissue of SHR exceeds the corresponding parameter in WKR, which is statistically reliable, involves difficulties. However, several speculations can be made. The presence of catecholaminergic terminals in neuropil of autonomic ganglia points to the presence of a target substrate for synaptic and extrasynaptic effects of sympathetic transmitter. Thus, it may be concluded that ganglionic tissue is prone to hypertensive reaction similar to other tissues innervated sympathetically. The levels of reception at the different periods of hypertension development correspond to the extent of involvement of the peripheral parts of cardiovascular system needed to maintain high values of the arterial pressure.

REFERENCES

1. Roeske, W.R., et al., 1983, *Molecular pharmacology of neurotransmitter receptors* (T. Segava et al eds.), Raven press, New York.
2. Bucinskaite, V., et al, 1995, *Neuroscience Letters* **192** (2): 93–96.
3. Floras, J.S., 1992, *Hypertension* **19** (1): 1–18.
4. Castellano, M., et al, 1993, *Hypertension* **11** (Suppl.5): S64-S65.
5. Koval, L.M., et al, 1982, *J. of Autonomic Nervous System*, **6**: 37–46.
6. Le Fur, G., Guilloux, F., Uzan, A., 1983, *Clin. Exp. Hypertens.*, Volume 5, **9**: 1537–1542.

Section 32: Cell Adhesion in Neural Plasticity Induced by Experience

STRUCTURE AND FUNCTION OF THE NEURAL CELL ADHESION MOLECULE NCAM

Elisabeth Bock, Nina Pedersen, and Vladimir Berezin

The Protein Laboratory
Panum Institute
University of Copenhagen
Denmark

1. INTRODUCTION

Cell-cell adhesion molecules (CAMs) comprise a broad range of membrane-associated glycoproteins characterized by affinities for constituents of the extracellular matrix and cell-surface molecules expressed by other cells. Based on structural similarities and related functions, CAMs can be grouped into several families: the immunoglobulin (Ig) superfamily (IgSF), the cadherin superfamily, the integrin family and the selectin family (1). The neural cell adhesion molecule, NCAM, initially described as a synaptic membrane protein termed D2 (2), is the most extensively studied CAM. NCAM, mediating cell-cell as well as cell-substratum interaction and promoting neurite outgrowth, plays a key role in nervous system development, regeneration and learning (1,3).

2. ORGANIZATION OF THE NCAM GENE

In the eukaryotic genome NCAM is encoded by a single gene which comprises at least 25 exons (see, e.g., 1). Alternative splicing of precursor mRNA produces a variety of mature mRNA species. Exons 1 to 10 encode five Ig-like domains in the N-terminal part of NCAM, and the optional insertion of a 30 bp exon (VASE) may take place between exons 7 and 8 which compose the fourth Ig-like domain. Exons 11 to 14 encode two fibronectin type III (F3) repeats and optional insertions of four other exons, a (15 bp), b (48 bp), c (42 bp) and an AAG triplet, may occur between exons 12 and 13. When exons a, b, c, and AAG are expressed together, the combination is termed the muscle specific domain (MSD). In brain mainly exons a and AAG are expressed. The mode of attachment of NCAM to the plasma membrane and the size of the intracellular NCAM domains are de-

From a colloquium organized by Rupert Schmidt (Giessen, Germany) and Ciaran Regan (Dublin, Ireland).

Neurochemistry, edited by Teelken and Korf
Plenum Press, New York, 1997

termined by exons downstream to exon 14. The optional choice of exon 15 introduces a stop-codon and two polyadenylation sites, and the subsequently generated mRNA species are translated into a glycosylphosphatidyl inositol (GPI) anchored 120 kDa isoform (NCAM-C). Exon 16 encodes a transmembrane part of NCAM, while exons 17, 18 and 19 encode the intracellular domain of a 180 kDa isoform (NCAM-A). Exon 18 is optional and not present in the cytoplasmic part of the 140 kDa isoform (NCAM-B).

3. STRUCTURE OF NCAM DOMAINS

The five Ig-like domains of NCAM have, based on the amino acid sequence, been classified as belonging to the C2-set. However, the three-dimensional structure of the first Ig domain of mouse NCAM has recently been determined by NMR spectroscopy (4) and identified as belonging to the intermediate set (I-set), which contains two β-sheets with the A, B, D and E strands arranged in one β-sheet and the A', C, C', F and G strands in the other. The strands are antiparallel, except the A' strand and the C-terminal part of the G strand which are parallel (see Fig. 1). The two sheets are connected by a cysteine bridge between Cys 24 in strand B and Cys 79 in strand F. The A' strand and the B strand are connected by a type II β-turn.

Based on the three-dimensional structure determined using NMR spectroscopy and X-ray crystallography, three other members of the I-set have so far been identified: the first domain of the vascular cell adhesion molecule, VCAM-1 (5); the M5 domain in the gigantic protein titin, located in the thick filaments of vertebrate muscle (6), and telokin, an acidic protein identical to the C-terminal portion of smooth muscle light chain kinase (7). Alignment of structurally equivalent residues of the known members of the I-set and those found in the Ig-like domains of other proteins has allowed the prediction that other IgSF domains of neural CAMs and proteoglycans also belong to the I-set (4). These include the third Ig domain of human axonin-1, the third Ig domain of mouse contactin, the third and the fifth domains of mouse L1, the fifth domain of fruit fly neuroglian and the fourth, fourteenth and fifteenth domains of mouse perlecan. The third Ig-like domain of NCAM is also predicted to belong to the I-set, and, finally, the second Ig domain of

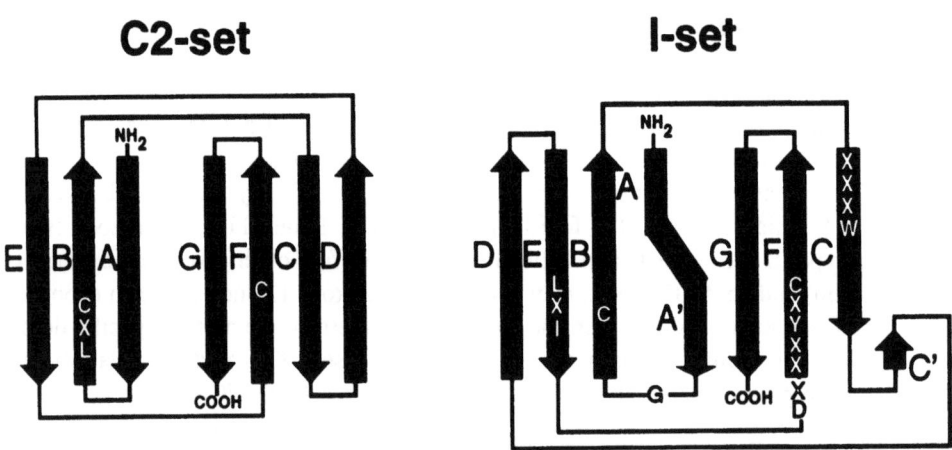

Figure 1. The structure of the C2-set and the I-set of IgSF.

NCAM contains most of the I-set sequence profile characteristics and therefore may also exhibit this tertiary structure (4).

The three-dimensional structure of the two F3 domains of NCAM has not yet been determined.

4. FUNCTION OF NCAM

It has been demonstrated that NCAM mediates cell-cell interactions by a Ca^{2+} independent homophilic binding mechanism (8) and cell-substratum interactions by heterophilic interactions, binding various collagens (9), heparin/heparan sulphate and chondroitin sulphate proteoglycans (10,11). NCAM also binds to an oligomannosidic glycan present on another CAM, L1 (12), and the NCAM-A isoform, but not NCAM-B or -C, is able to interact with the actin cytoskeleton associated protein, spectrin (13). A tight association between an ecto(Ca^{2+}-Mg^{2+})-ATPase activity and NCAM has been demonstrated (14). This variety of binding capabilities probably reflects the diversity of NCAM functions. Some structural determinants of NCAM function have recently been identified (see Fig. 2).

It has been shown that the first Ig domain is involved in cell adhesion and cell spreading (15), and that it contains a heparin-binding site and is able to bind the second Ig domain (Kiselyov et al., submitted) which also contains a heparin-binding site (10). The first Ig-like domain has also been shown to interact with the fifth domain and the second domain supposedly interacts with the fourth. Finally, the third domain has been demonstrated to carry a homophilic binding site (16). The fourth Ig domain may or may not contain an insert of 10 amino acid residues (VASE) which reduces the neurite outgrowth promoting activity of NCAM (17). The fourth domain is also believed to carry a lectin-like oligomannosidic binding site which allows NCAM to be involved in "assisted" homophilic binding with another CAM, L1 (18). The fifth Ig domain contains three N-linked glycosylation sites and may carry extreme amounts of sialic acid residues in a polymer form (polysialic acid, PSA) which modulates the homophilic binding and neurite outgrowth (17). PSA is predominantly expressed early during development, and NCAM is probably the only mammalian protein carrying this specific carbohydrate (see 1). The NCAM-F3 domains have been demonstrated to induce fasciculation (15) and to be involved in cell-cell adhesion and cellular spreading, with a proline-rich insertion optionally encoded by exons a and AAG having a modulatory effect on F3 mediated cellular spreading and cell aggregation (19).

The mechanisms by which NCAM on one cell interacts with NCAM on another cell is poorly understood. Since it has been demonstrated that the third Ig domain is able to self-aggregate and that the first and second Ig domains bind to the fifth and fourth Ig domains, respectively, it has been suggested that all five Ig domains are involved in NCAM homophilic adhesion, see Fig. 3A (16). However, when highly hydrated and negatively charged long chains of PSA are present, this type of interaction may be inhibited. Since the first Ig domain also binds the second Ig domain, it may be hypothesized that when polysialylated NCAM is expressed during nervous system development, regeneration or transient synaptic contact rearrangements, the homophilic NCAM mediated adhesion may take place via the first and the second Ig domains in a symmetrical double reciprocal binding, see Fig. 3B (modified from Kiselyov et al., submitted). The recently determined three-dimensional structure of the first Ig domain and the surface charge map of the assembled two N-terminal domains of NCAM in a computer simulation support this model (4).

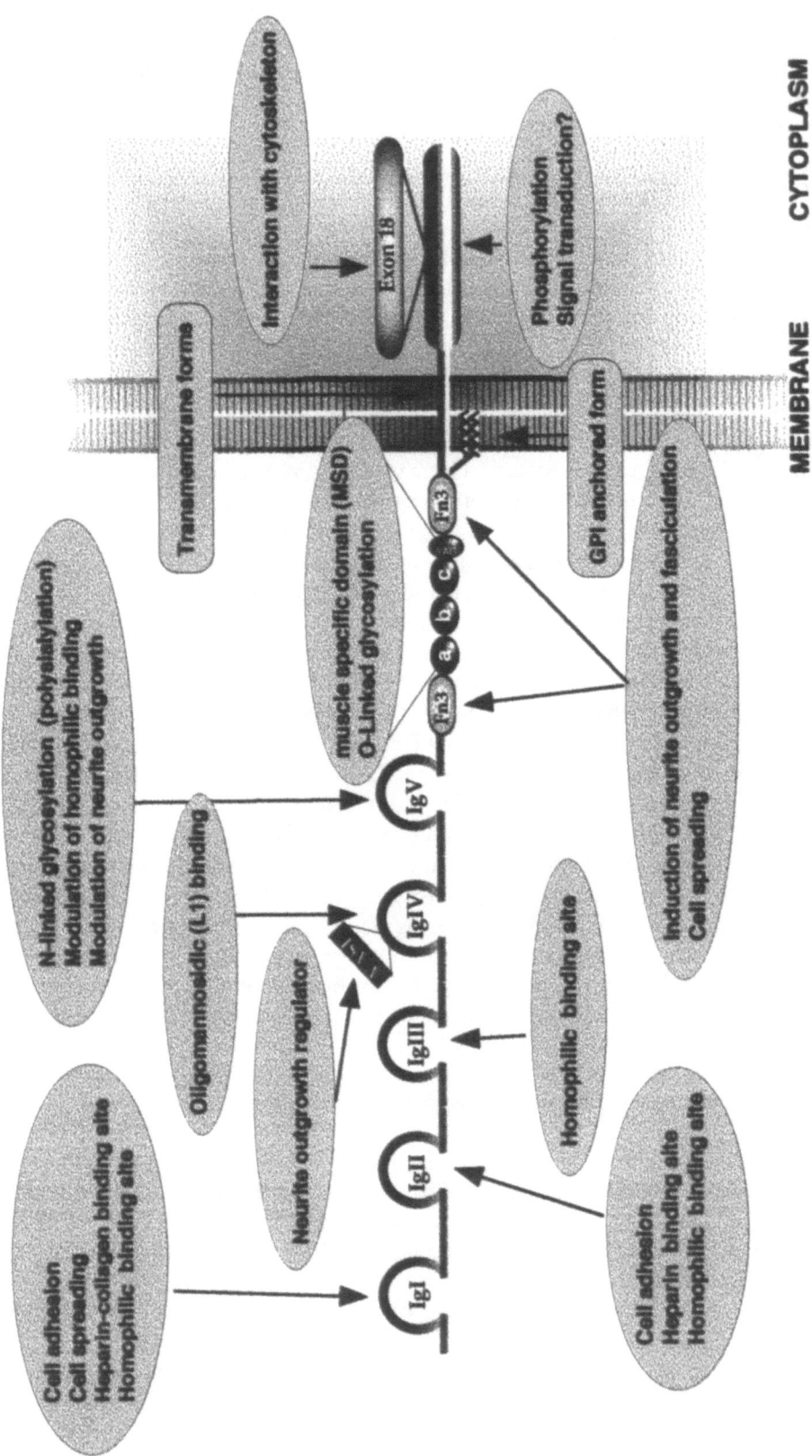

Figure 2. The putative functions of NCAM domains.

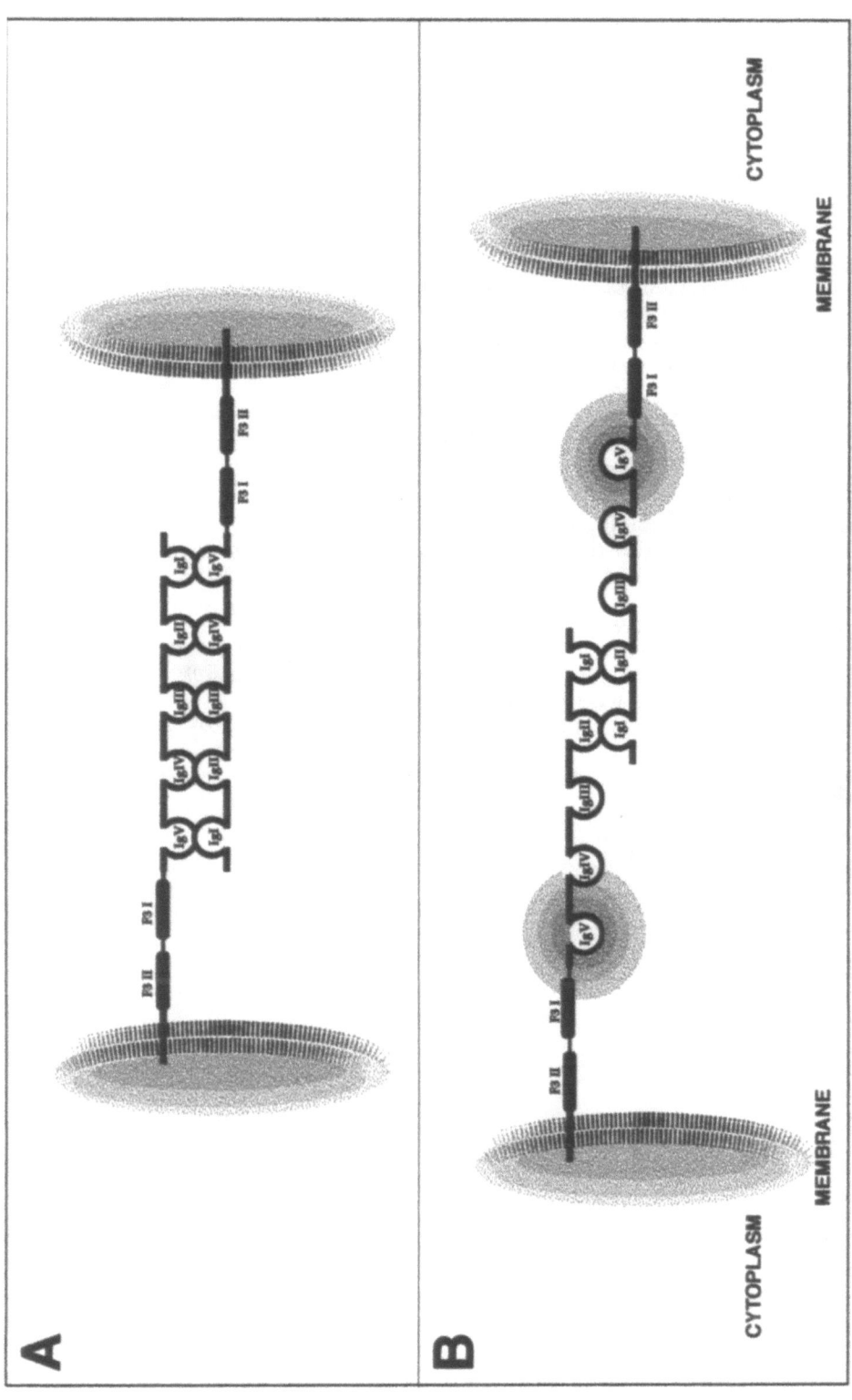

Figure 3. Schematic presentation of antiparallel models for NCAM homophilic binding. A) A model of homophilic binding of non-sialylated NCAM involving all five Ig domains. B) A model of homophilic binding of highly polysialylated NCAM involving the first and the second Ig domains in a symmetric double reciprocal interaction.

ACKNOWLEDGMENTS

The financial support of the Danish Medical Research Council, the Danish Cancer Society, and the Lundbeck Foundation is gratefully acknowledged.

REFERENCES

1. Gegelashvili, G., and Bock, E., 1996, *Biomembranes* **3**: in press.
2. Jørgensen, O.S., and Bock, E., 1974, *J. Neurochem.* **23**: 879–880.
3. Rønn, L.C., Bock, E., Linnemann, D., and Jahnsen, H., 1995, *Brain Res.* **677**: 145–151.
4. Thomsen, N.K., Soroka, V., Jensen P.H., Berezin, V., Kiselyov, V.V., Bock, E., and Poulsen, F.M., 1996, *Nature Structural Biology* **3**: 581–585.
5. Wang, J.H., Pepinsky, R.B., Stehle, T., Liu, J.H., Karpusas, M., Browning, B., and Osborn, L., 1995, *Proc. Natl. Acad. Sci. U.S.A.* **92**: 5714–5718.
6. Pfuhl, M. and Pastore, A., 1995, *Structure* **3**: 391–401.
7. Holden, H.M., Ito, M., Hartshorne, D.J., and Rayment, I., 1992, *J. Mol. Biol.* **227**: 840–851.
8. Rutishauser, U., Hoffman, S., and Edelman, G.M., 1982, *Proc. Natl Acad. Sci. U.S.A.* **79**:685–689.
9. Probstmeier, R., Kühn, K., and Schachner, M., 1989, *J. Neurochem.* **53**: 1794–1801.
10. Nybroe, O., Moran, N., and Bock, E., 1989, *J. Neurochem.* **52**: 1947–1949.
11. Grumet, M., Flaccus, A., and Margolis, R.U., 1993, *J. Cell Biol.* **120**: 815–824.
12. Kadmon, G., Kowitz, A., Altevogt, P., and Schachner, M., 1990, *J. Cell Biol.* **110**: 209–218
13. Pollerberg,G.E., Burridge, K., Krebs, K., Goodman, S.R., and Schachner, M., 1987, *Cell Tissue Res.* **250**: 227–236.
14. Dzhandzhugazyan, K., and Bock, E., 1993, *FEBS Let.* **336**: 279–283.
15. Frei, T., von Bohlen und Halbach, F., Wille, W., and Schachner M., 1992, *J. Cell Biol.* **118**: 177–194.
16. Ranheim, T.S., Edelman, G.M., and Cunningham, A., 1996, *Proc. Natl. Acad. U.S.A.* **93**: 4071–4075.
17. Doherty, P., Fazeli, M.S., and Walsh, F.S., 1995, *J. Neurobiol.* **3**: 437–446.
18. Horstkorte, R., Schachner, M., Magyar, J.P., Vorherr, T., and Schmitz, B., 1993, *J. Cell biol.* **121**: 1409–1421.
19. Kasper, C., Stahlhut, M., Berezin, V., Maar, T.E., Edvardsen, K., Kiselyov, V.V., Soroka, V., and Bock E., 1996, *J. Neurosc. Res.* **45**: in press.

Section 32: Cell Adhesion in Neural Plasticity Induced by Experience

POLYSIALIC ACID (PSA) ASSOCIATED WITH THE NEURAL CELL ADHESION MOLECULE (N-CAM) MAY PLAY A ROLE IN SPATIAL LEARNING AND LTP IN RATS

Hans Welzl,[1] Catherina G. Becker,[2] Alain Artola,[2] and Melitta Schachner[2]

[1]Laboratory of Behavioral Biology
[2]Laboratory of Neurobiology
ETH Zürich
Zürich, Switzerland

1. INTRODUCTION

In the developing organism, adhesion molecules play an important role in cell migration, neurite outgrowth and axon fasciculation. The neural cell adhesion molecule N-CAM — a member of the immunoglobulin superfamily — is involved in all three processes (1). N-CAM exists in 3 major isoforms which are mainly expressed on glia (120 kDa form of N-CAM), glia or neurons (140 kDa form of N-CAM) or neurons only (180 kDa form of N-CAM) and form Ca^{2+}-independent homo- and heterophilic bonds (2).

The adhesive properties of N-CAM are determined by the presence or absence of attached α-2,8-linked polysialic acid (PSA) side chains (3). There are 3 attachment sites for PSA on the fifth immunoglobulin domain of N-CAM. The long linear PSA chains (up to at least 100 sialic acid residues per chain) create clusters of negative charges. As a consequence, the adhesiveness of N-CAM is decreased and also a number of other types of cell-cell interactions are affected (4).

The importance of N-CAM sialylation for developmental processes in the brain is demonstrated by several studies. Enzymatic removal of PSA prevents or alters neurite outgrowth (5). Further, an increase in PSA content in the postnatal cerebral cortex coincides with the initial period of synapse formation (6). Thus, controlled sialylation of N-CAM seems to be part of the process of axon guidance and synapse formation.

In adult rats N-CAM and PSA are expressed in discrete areas; e.g., N-CAM 180 is associated with synaptic membranes in the adult hippocampus (7). In addition, modifications of synapses accompany lesion-induced plastic changes (e.g., 8) and memory formation (e.g., 9) in adult animals. The continuous presence of N-CAM and — more specifically — polysialy-

From a colloquium organized by Rupert Schmidt (Giessen, Germany) and Ciaran Regan (Dublin, Ireland).

lated N-CAM into adulthood suggests that in the adult PSA probably also serves processes of modification at synapses. Indeed, the role of N-CAM and related adhesion molecules in memory formation is supported by several lines of research carried out in a variety of animal species. Interfering with the normal function of N-CAM by application of specific antibodies attenuates the acquisition of spatial memory (10) and passive avoidance learning in rats (11). The same antibodies also prevent long-term potentiation (LTP; 12, 13) — an activity-dependent enhancement of synaptic transmission — in hippocampal CA1 (for review on LTP in this area, see 14). Similar interventions impair active avoidance learning in chicks (15). N-CAM-related adhesion molecules in fish (16) and *Aplysia californica* (17) have also been shown to be involved in modifications of synapses during memory formation. The assumption of an important role of PSA in adult plasticity is supported by results demonstrating an increase of PSA in parallel to lesion-induced plastic changes (including processes such as sprouting) in specific areas of the brain (18,19,20). In addition, the PSA content in the hippocampal formation rises as a consequence of avoidance training in rats (21,22).

To collect further experimental support for an involvement of PSA in memory formation, we investigated the role of PSA in spatial learning as measured in the Morris water maze. In addition, since it is generally accepted that long-lasting, activity-dependent changes in synaptic strength are of fundamental importance for information storage in the brain, we also tested whether PSA is implicated in the form of LTP elicited in the Schaffer collateral/commissural pathway. PSA was removed by the PSA-specific enzyme endoneuraminidase NE (endoN; 23). These data have already been published in more detail (24).

2. METHODS AND RESULTS OF SPATIAL LEARNING EXPERIMENTS

In a circular pool filled with water (diameter 1.6 m), rats had to locate a hidden platform in order to escape from the water. The platform was hidden below the water surface and, therefore, invisible to the rats; no other maze-internal cues indicated its position. Thus, in order to locate the platform animals were forced to use maze-external spatial cues distributed throughout the experimental room. Each trial started at one of four different positions at the wall of the pool. A set of four trials constituted a session, and on each of the 2 days of acquisition rats were submitted to 4 sessions. On the third day the platform was removed and the time was measured the animal spent searching for the platform in the quadrant of the pool where it had been during acquisition (retrieval session). EndoN (9.5 μU/0.5 μl) or vehicle solution (0.5 μl) was intracranially injected via chronically implanted cannulae into both hippocampi 14 hr before the beginning of acquisition training. Animals were sacrificed immediately after the end of the experiment, and the extent of PSA removal was verified using a monoclonal antibody (25).

PSA was removed to a large degree in the rostral part of the dorsal hippocampus covering at least 3 mm of rostro-caudal extension of the hippocampus. An area of varying size of the ventral parts of the hippocampal formation and the overlying neocortex were similarly affected. Removing PSA attenuated the acquisition of the spatial task (ANOVA $F(1,21) = 9.49$, $p = .0057$) and impaired memory retrieval for the spatial location during the transfer test (ANOVA $F(1,21) = 5.24$, $p = .033$) in comparison to vehicle injected animals (Figure 1). In contrast, PSA removal did not affect the ability of rats to escape onto a visible platform, i.e., a platform that was raised above water level and painted white; ANOVA $F(1,21) = 0.44$, $p = $ n.s.). The latter result assures that differences between treatment groups cannot be attributed to disturbed visual and/or motor abilities.

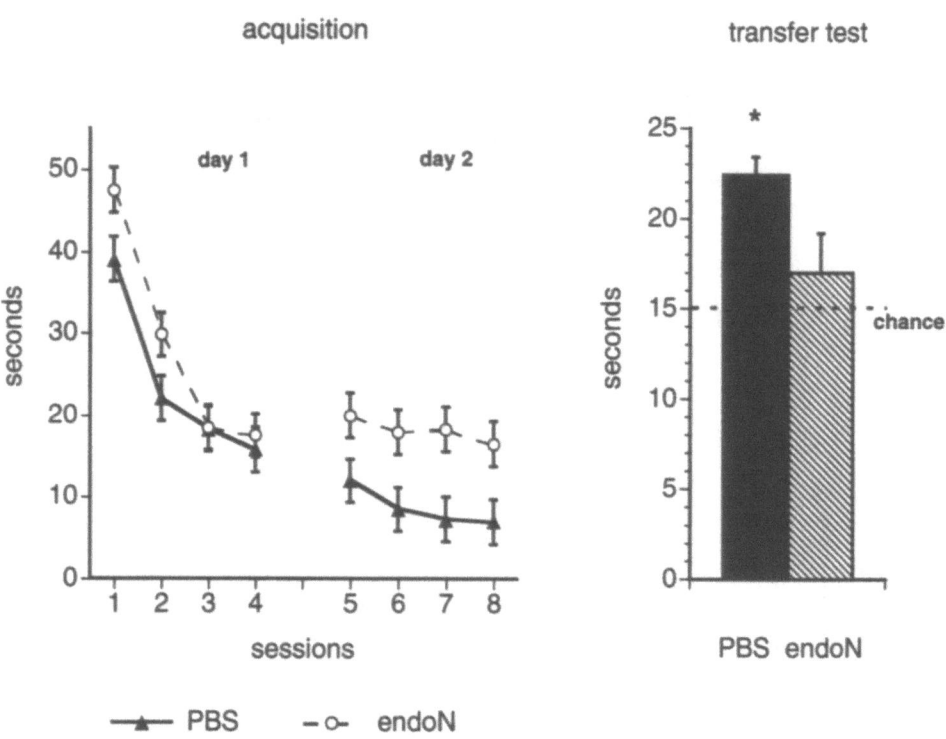

Figure 1. Escape latencies (in seconds) during the 2 days of acquisition (4 sessions/day) are plotted on the left. PBS-injected animals (controls) demonstrated better acquisition of the spatial task than endoN-injected animals (PSA removed from N-CAM), i.e., controls needed less time to escape onto the hidden escape platform than endoN-injected animals during the 4 sessions on day 2. On the right, the time spent in the quadrant where the platform had been during acquisition is given (transfer test). Controls showed a significantly greater preference for the training quadrant than endoN-injected animals (* p = 0.033; modified after 24).

Results were similar when distance swum instead of escape latencies (acquisition, visual platform) or time in platform quadrant (transfer test) in seconds was taken as a measurement of learning efficiency.

3. METHODS AND RESULTS OF LTP EXPERIMENTS

To remove PSA, hippocampal slices were incubated for 3 h in the bathing medium (Ringer's solution) containing endoN (0.2 μU). This was enough to suppress PSA from the slices as checked after each experiment using a monoclonal antibody specific to α-2,8-linked PSA (mab 735; 25). Intracellular recordings were then obtained from CA1 neurons in these slices and a single high-frequency stimulus (tetanus; 2 seconds, 100 Hz) was applied to attempt to induce LTP. This tetanus which produced a robust LTP in control slices (interleaved experiments) failed to induce any LTP in endoN treated slices. Only posttetanic and short-term potentiations could be observed in these treated slices (see Figure 2). This suppression of LTP was independent of any other detectable modification of synaptic transmission and of passive membrane properties.

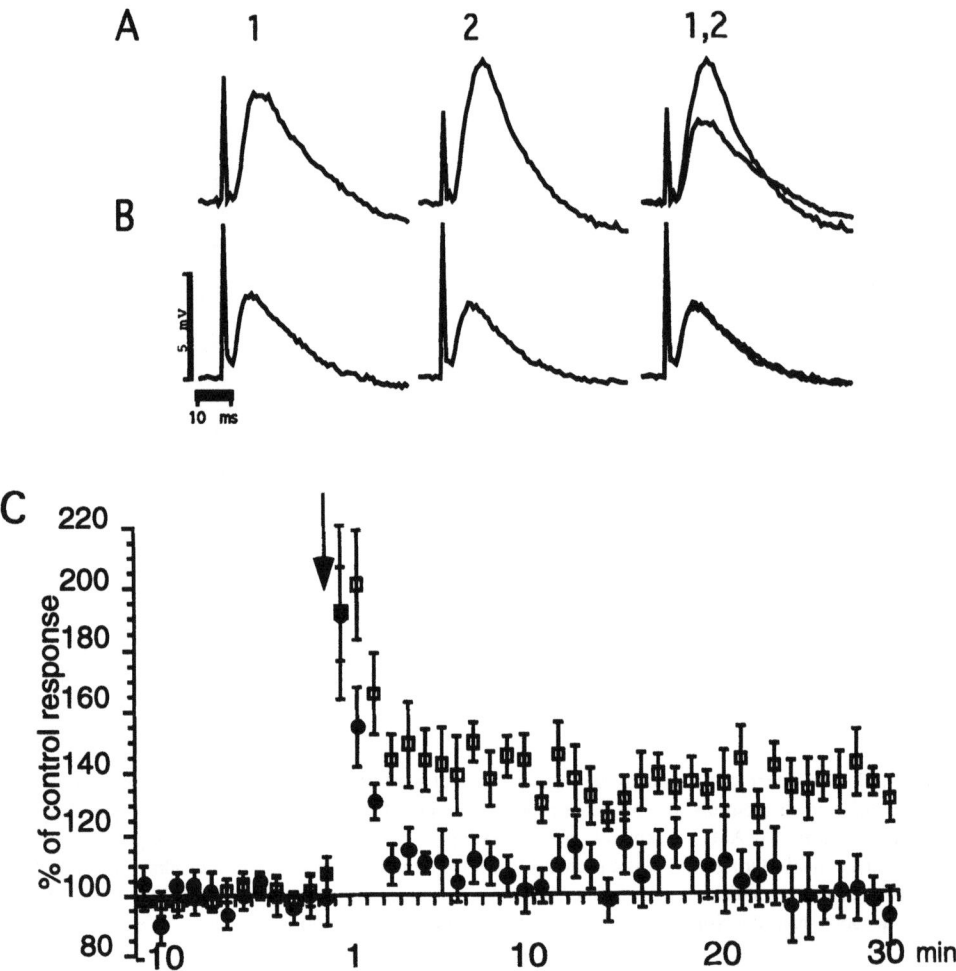

Figure 2. Time-course of the PSA-dependent synaptic potentiation. A,B: Synaptic responses to stimulation of the Schaffer collateral-commissural pathway recorded before (1) and 35 min after tetanus (100 Hz, 2 sec) (2) in two different cells, A (control slice) and B (endoN treated slice). Responses 1 and 2 are superimposed on the right (1,2). Each trace is the average of 5 successive responses (0.06 Hz). Vmr (mV) is -70 in A and -69 in B. C: Averaged, normalized EPSP slopes measured before and after tetanus (indicated by an arrow) in endoN treated slices (filled circles, n=6) and non-treated control slices (empty squares, n=6) are shown (modified after 24).

4. DISCUSSION

Specific removal of PSA was shown to impair spatial learning in the water maze — a learning paradigm known to be hippocampus-dependent (26) — and the induction of LTP in hippocampal slices. The specificity of the effect on LTP is supported by the fact that other electrophysiological measures (input resistance, resting membrane potential, short-term potentiation) were unaffected by treatment of the slice with endoN. These data also provide further evidence that memory formation in a spatial learning paradigm (i.e., the Morris water maze) shares common elements (i.e., modulation of N-CAM by polysialic acid) with the induction of LTP in vitro.

PSA has also been reported to be present on sodium channels (27). It is unlikely, however, that the effects of PSA removal by endoN on LTP are attributable to PSA removal from sodium channels, as changing the function of sodium channels would have affected basic electrophysiological measures (see above) as well as LTP. However, neither post-tetanic potentiation nor short-term potentiation were affected by endoN treatment of the slices.

The underlying neuronal mechanisms of memory formation are modifications of synaptic transmission, and these modifications are induced by increased synaptic activity at the time memory is acquired. These assumptions are supported by, e.g., the observation that LTP develops as a consequence of increased synaptic activity during tetanic stimulation. In cultured cortical neurons the translocation of polysialylated N-CAM to the cell surface was dependent on depolarization and the presence of extracellular Ca^{2+} suggesting that this process is also dependent on electrical activity (28); these data are, thus, in line with the hypothesis that a similar increase in polysialylated N-CAM expression takes places during memory formation in vivo when electrical activity is increased.

There is ample evidence that several types of learning lead to structural changes at the level of the synapse (9) including spatial learning in the water maze (29). It is likely, therefore, that N-CAM 180 — which is involved during development in neurite outgrowth and the establishment of synaptic contacts — continues to be involved in the maintenance of synaptic structures in the adult animal. A likely first step in changing synaptic structures as a consequence of a learning experience is to reduce the adhesiveness at the synapse to allow structural changes to take place. Adding polysialylated N-CAM 180 to the membrane or replacing existing N-CAM with polysialylated N-CAM 180 provides one mechanism of reducing adhesion and/or signal transduction.

An increase in N-CAM 180 and PSA expression in the hippocampus 24 hours after a passive avoidance task has been demonstrated recently (21,22). The 24 hour time lag phase correlates with our observed strong effects of PSA removal on spatial learning in the water maze on the second day (acquisition) and third day (transfer test) of training.

Interestingly, the effect of PSA removal on the formation of LTP is immediate, whereas its effect on spatial learning becomes significant only on the second day of acquisition training. This points to the possibility that PSA removal affects two different cellular processes which are involved in memory formation. In addition to a possible participation in structural rearrangements at a later stage, PSA might play a critical role — at an earlier time point — in cell-cell signaling mediated by N-CAM 180 and L1 (30,31).

On the basis of these results and results from other laboratories we would like to speculate that N-CAM 180 and the regulation of its adhesive properties by polysialylation is an important step in memory formation.

REFERENCES

1. Rutishauser, U., 1986, *Trends Neurosci.* **9**: 374–378.
2. Hoffman, S., and Edelman, G.M., 1983, *Proc. Natl. Acad. Sci. USA* **80**: 5762–5766.
3. Rutishauser, U., Acheson, A., Hall, A.K., Mann, D.M., and Sunshine, J., 1988, *Sciene* **240**: 53–57.
4. Rutishauser, U., 1993, *Polysialic Acid* (Roth, J., Rutishauser, U., and Troy II. F.A., eds.) Birkhäuser, Basel, PP. 215–228.
5. Rutishauser, U., and Landmesser, L., 1991, *Trends Neurosci.* **14**: 528–532.
6. Regan, C.M., 1993, *Polysialic Acid* (Roth, J., Rutishauser, U., and Troy II. F.A., eds.) Birkhäuser, Basel, PP. 299–312.
7. Persohn, E., Pollerberg, G.E., and Schachner, M., 1989, *J. Comp. Neurol.* **288**: 92–100.
8. Steward, O., 1986, *The Hippocampus*, Volume 3 (Isaacson, R.L., and Pribram, K.H., eds.), Plenum Publ. Co., New York, PP. 65–111.

9. Bailey, C.H., and Kandel, E.R., 1993, *Ann. Rev. Physiol.* **55**: 397–26.
10. Arami, S., Jucker, M., Schachner, M., and Welzl, H., 1996, *Behav. Brain Res.*, in press.
11. Doyle, E., Nolan, P.M., Bell, R., and Regan, C.M., 1992, *J. Neurochem.* **59**: 1570–1573.
12. Lüthi, A., Laurent, J.P., Figurov, A:, Muller, D., and Schachner, M., 1994, *Nature* **372**: 777–779.
13. Ronn, L.C.B., Bock, E., Linnemann, D., and Jahnsen, H., 1995, *Brain Res.* **677**: 145–151.
14. Bliss, T.V., and Collingridge, G.L., 1993, *Nature* **361**: 31–39.
15. Scholey, A.B., Rose, S.P.R., Zamani, M.R., Bock, E., and Schachner, M., 1993, *Neurosci.* **55**: 499–509.
16. Schmidt, R., 1995, *Behav. Brain Res.* **66**: 65–72.
17. Mayford, M., Barzilai, A., Keller, F., Schacher, S., and Kandel, E.R., 1992, *Science* **256**: 638–644.
18. Miller, P.D., Styren, S.D., Lagenaur, C.F., and DeKosky, S.T., 1994, *J. Neurosci.* **14**: 4217–4225.
19. Aubert, I., Ridet, J.-L., and Gage, F.H., 1995, *Curr. Opinion Neurobiol.* **5**: 625–635.
20. Fryer, H.J.L., and Hockfield, S., 1996, *Curr. Opinion Neurobiol.* **6**: 113–118.
21. Doyle, E., Nolan, P.M., Bell, R., and Regan, C.M., 1992, *J. Neurosci. Res.* **31**: 513–523.
22. Fox, G.B., O'Connell, A.W., Murphy, K.J., and Regan, C.M., 1995, *J. Neurochem.* **65**: 2796–2799.
23. Finne, J., Mäkelä, P.H., 1985, *J. Biol. Chem.* **260**: 1265–1270.
24. Becker C.G., Artola, A., Gerardy-Schahn, R., Becker, T., Welzl, H., and Schachner, M., 1996, *J. Neurosci. Res.*, in press.
25. Frosch, M., Görgen, I., Boulnois, G.J., Timmis, K.N., and Bitter-Suermann, D., 1985, *Proc. Natl. Acad. Sci. USA* **82**: 1194–1198.
26. Morris, R.G.M., Garrud, P., Rawlins, J.N.P., and O'Keefe, J., 1982, *Nature* **297**: 681–683.
27. Zuber, C., Lackie, P.M., Catterall, W.A., and Roth, J., 1992, *J. Biol. Chem.* **267**: 9965–9971.
28. Kiss, J.Z., Wang, C., Olive, S., Rougon, G., Lang, J.C., Baetens, D., Harry, D., and Pralong, W.F., 1994, *EMBO J.* **13**: 5284–5292.
29. Moser, M.-B., Trommald, M., and Andersen, P., 1994, *Proc. Natl. Acad. Sci. USA* **91**: 12673–12675.
30. Williams, E.J., Furness, J., Walsh, F.S., and Doherty, P., 1994, *Neuron* **13**: 583–594.
31. Schuch, U., Lohse, M.J., and Schachner, M., 1989, *Neuron* **3**: 13–20.

Section 32: Cell Adhesion in Neural Plasticity Induced by Experience

REGULATED EXPRESSION OF THE CNS-SPECIFIC CELL ADHESION MOLECULE EPENDYMIN AFTER ACQUISITION OF AN ACTIVE AVOIDANCE BEHAVIOUR PROVIDES A POSSIBLE MECHANISM FOR MEMORY CONSOLIDATION

Rupert Schmidt

Biotechnology Centre
Justus-Liebig-University
Leihgesterner Weg 217, D 35392 Giessen, Germany

1. INTRODUCTION

Many phenomena of short term adaptations of the CNS have been explained with changes in the conductance of ion channels, that are induced by cascades of second transmitter systems ultimately leading towards ion channel phosphorylation. On the other hand, it is the long-term consolidation of memory that appears to be vulnerable, in particular in the context of age-related disseases of which Alzheimer's is but the most prominent example. A few years ago, cell adhesion molecules were considered just sticky agglutinants, but now it is well accepted that they are accurately regulated glycoproteins involved in ligand-receptor relationships triggering intracellular second messenger cascades. They are known to guide axonal growth and the migration of neurectodermal cells during epigenetic differentiation. Also during CNS regeneration cell adhesion molecules are involved.

2. INVOLVEMENT OF EPENDYMIN IN CNS REGENERATION

Damaged central tracts in higher vertebrates regenerate only in the embryonic central nervous system (CNS), a feature that is lost after the embryonic stage of life. This change severely reduces the therapeutical possibilities in humans. Lower vertebrates, however, retain the ability to generate and regenerate central pathways throughout their life. In

---

From a colloquium organized by Rupert Schmidt (Giessen, Germany) and Ciaran Regan (Dublin, Ireland).

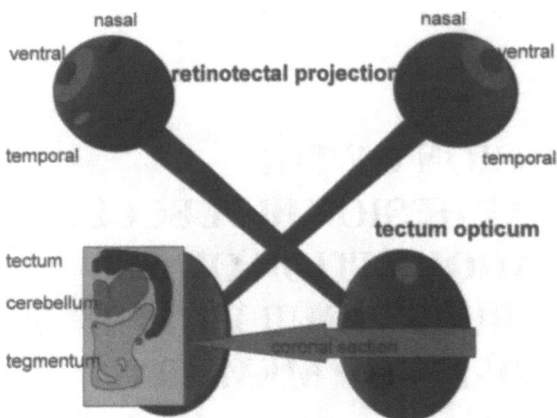

Figure 1. Retinotectal projection in teleosts. Regenerating retinal ganglion cell axons invade the optic tectum and make synaptic contacts in a topically correct manner, i.e., from the nasal part of the retina in the caudal tectum, from the temporal retina in the rostral tectum etc.

adult teleosts, e.g., the retina gowths concentrically, whereas its projection area, the optic tectum, growths by neuronal mitoses at its caudal rim. In order to preserve a matched retinotectal projection, therefore, central fibres and their synapses must continously be disconnected and reestablished with new target cells (Fig. 1).

When cut, the axons of goldfish retinal ganglion cells regenerate and reinnervate their projection area in the contralateral tectum within three weeks in a topically correct manner. However, receptive fields recorded in the optic tectum are initially enlarged. During this period synthesis of a cell adhesion molecule, named ependymin, is increased, as has been confirmed independently by several groups. Three to six weeks after optic nerve crush, receptive fields are sharpened by means of the synchronous activity in neighbouring ganglion cell axons. Eventually they reach their physiological value of 11°. However, when anti-ependymin antibodies were infused into the tectal ventricles, sharpening was blocked, and the receptive fields remained enlarged at 31° (1).

3. MOLECULAR AND CELL-BIOLOGICAL PROPERTIES OF EPENDYMIN

When given a choice between stripes of different concentrations of ependymin deposited on glass, regenerating ganglion cell axons from explanted goldfish retinae grow into the ependymin gradient (2). Ependymin also facilitates their fasciculation.

Ependymin was isolated and purified, and the primary sequence was deduced from cDNA cloning (3): Typical of a secretory protein, a lipophilic leader sequence is followed by a recognition site for cleavage by signal peptidase. Goldfish have a duplicated genome with two ependymin genes. There are many point mutations in some parts of the sequence, whereas other parts are well conserved, also with regard to the sequences of ependymins (4) from several other fish, including rainbow trout, atlantic salmon, pike, carp, zebrafish and herring. Most conservatively treated is the domain around two N-glycosylation sites that give rise to mono- and bi-N-glycosylated variants in *cyprinidae* (31 and 37 kDa, respectively, in the goldfish). In goldfish, 4 cysteine-residues form two loops and a fifth cysteine provides for an additional disulfide bond linking the monomers to homo- and hetero-dimeric molecules (5). The bi-N-glycosylated form is very rich in sialic acid. The carbohydrate moieties are of the complex type, and they carry the HNK-1 cell adhesion epitope (6) with terminal 3-sulfated glucuronic acid, characteristic of the immunoglobulin superfamily.

Monomeric ependymin binds radioactive calcium to ion-specific, saturable, high affinity binding sites on the carbohydrate chains (7), and calcium-binding induces a conformational change as indicated by circular dichroism measurements (8). Ependymin also binds zinc at independent binding sites on the peptide chain (7). It is the most prominent zinc binding protein in goldfish brain. Binding is stoichiometric, every two ependymin subunits chelating one Zn^{2+} cation. Apparently, calcium ions stabilize the monomers, whereas zinc stabilizes the dimeric forms. Removal of the metal cations was found to induce formation of high molecular weight association products, possibly polymer ependymins (9) or association products including other molecules of the extracellular matrix such as collagen and laminin (2).

Using biotinylated and radioactive oligodeoxynucleotide probes complementary to ependymin mRNA for in situ hybridization, biosynthesis was demonstrated only in the leptomeninx with its deep invaginations into the brain, in particular around blood vessels (10). In teleosts the endomeninx is composed of three layers. Ependymin is synthesized in reticular shaped fibroblasts of the inner layer, probably the homologue of the arachnoid in mammals. These fibroblasts have long been known for their prominent secretory activity. Despite ependymin synthesis is restricted to leptomeningeal fibroblasts, the protein is observed in various regions of the CNS (11). With fluorescent antibodies ependymin is detected in the neural parenchyma and the proximal portion of the optic nerve, but it is absent from its distal parts and from other tissues outside the CNS (Fig. 2). The protein is recognized most prominently in the meninx, but also in distinct neuronal populations and in the ependymal zone. Here, it was first localized immunocytochemically (12) and, therefore, named ependymin.

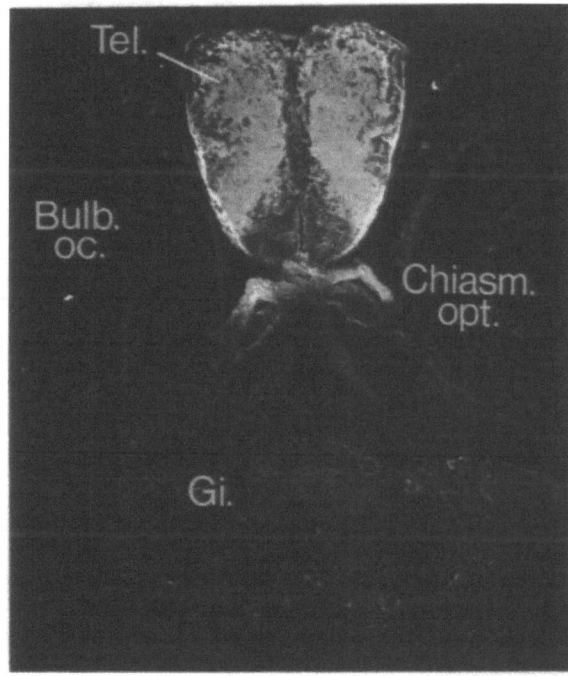

Figure 2. CNS-specificity of ependymin. A coronal section of a goldfish head was incubated with specific rabbit anti-ependymin antiserum, followed by FITC-conjugated goat anti-rabbit IgG. Note prominent staining of the telencephalon (Tel.) and the proximal part of the optic nerve. The eyes (Bulb. oc.) and gills (Gi.) are almost devoid of ependymin immunoreactivity.

4. ROLE OF EPENDYMIN IN BEHAVIOURAL PLASTICITY

Ependymin is a cell adhesion glycoprotein involved not only in CNS regeneration, but also in CNS reorganisation after learning: Goldfish were trained on an active shock-avoidance conditioning, in which a light signal precedes electric shocks administered alternatingly at one side of a shuttle-box. In 20 trials, goldfish learn to swim into the dark compartment in order to avoid the punishment. The learning-criterion was set to 80% correct responses. Three days after learning, the fish remembered the task (13). By in situ hybridization a marked increase in ependymin mRNA was detected after the avoidance-learning. In situ hybridization signals were analysed semi-quantitatively (10) and compared (Fig. 3) with steady-state concentrations of the translated protein measured by a very specific and sensitive RIA (5). The maximal concentration of intracellular ependymin was reached 5 h after acquisition. Afterwards the translation product was recovered in the extracellular and perimeningeal brain fluids (11, 14). At the time of maximal secretion at 9 h after acquisition, expression of ependymin mRNA had already returned to baseline levels. Apparently the amount of circulating ependymin provides a negative feed-back signal to the fibroblasts to stop the synthesis of this protein, possibly mediated by the dimeric, zinc-binding conformation (15). Feed-back regulation is also suggested by studies in which anti-ependymin antisera were injected into goldfish brain ventricles: Inactivation of circulating ependymin molecules resulted in an increase in

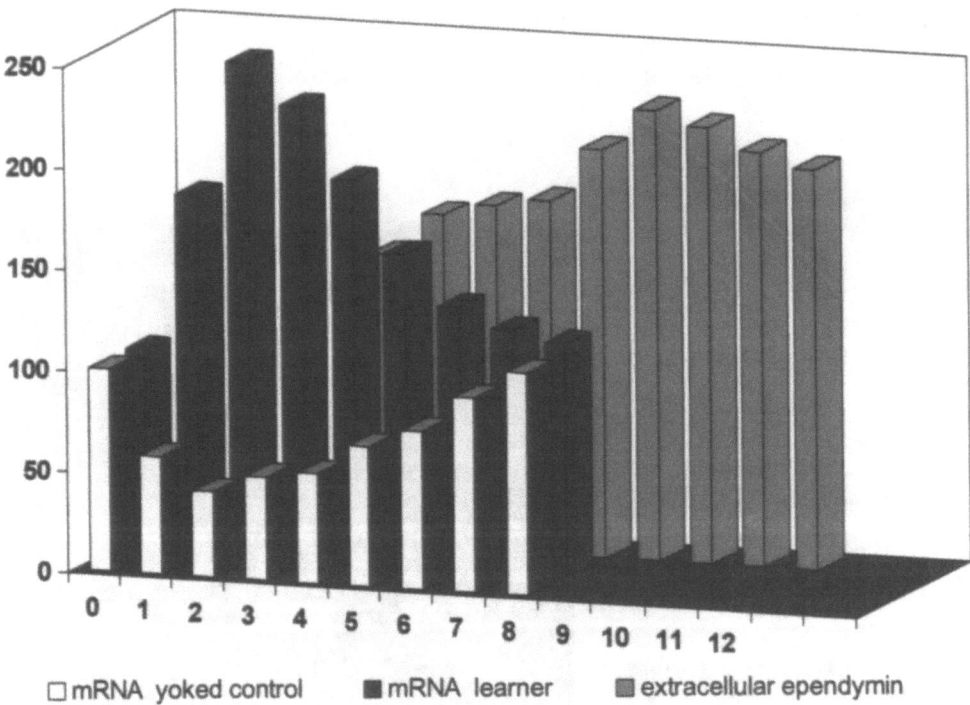

Figure 3. Relative amounts of ependymin mRNA and of its translation product. Ependymin mRNA expression after learning of the shuttle-box task was analysed semi-quantitatively on autoradiograms obtained from in situ hybridization as described in (10) and compared with mRNA expression following a stressful pseudo-conditioning (yoked control). Extracellular ependymin steady-state concentrations were measured in learning animals by a specific RIA (relative amounts, ordinate). Note, that mRNA decreases in learning fish, when the extracellular ependymin concentration increases (3–9 h, abscissa). Ependymin mRNA is temporarily depressed in yoked controls.

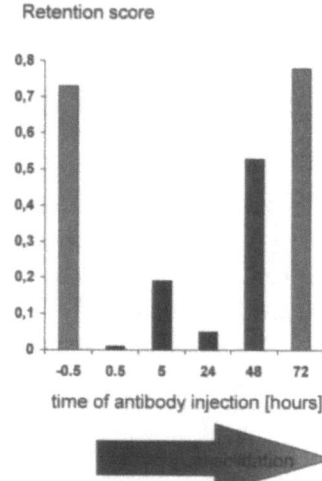

Figure 4. Induction of amnesia depends on the time of anti-ependymin injection. Goldfish trained on the active shock avoidance paradigm were intracerebroventricularly injected with anti-ependymin antibody at various times before (−0.5 h) or after acquisition (0.5–72 h). Retention scores were calculated according to the algorithm described in (13). Anmesia is induced by anti-ependymin injections between 0.5 and 24 h after learning.

ependymin mRNA expression. Trained, but non-learning, fish did not exhibit the increase in ependymin mRNA or protein synthesis (10, 14). Furthermore, yoked controls, that were stressed by exposure to a randomized sequence of light and electroshock stimuli, even displayed decreased hybridization signals below the control level. Experiments with yoked controls suggest that the expression of ependymin is blocked by some aspect of the physiological stress reaction (10, 15).

In order to test, whether the increase in this calcium-binding cell adhesion molecule is causally related to learning and memory formation, or whether we measured an epiphenomenon of CNS metabolism after behavioural adaptations, we have tried to interfere with learning by intracerebroventricular injection of anti-ependymin antisera (13, 14). Fish injected 0.5 h before training learned and remembered the task (Fig. 4). When the antisera were injected 0.5 h after learning, however, fish did not recall the avoidance response in the test session on day 4. Still, after clearance of the antibodies from brain, they learnt the task again and then also remembered it. Amnesia was observed, when secreted ependymins were inactivated up to 24 h after acquisition. Injected antisera did not interfere with the retrieval, when injected before the test, nor with the performance of the shuttle-box task as such, tested by injection into the brains of overtrained goldfish (13, 14).

The limited time window after learning during which the antibodies interfere with memory consolidation suggested that only the newly synthesized ependymin molecules are functionally involved in the consolidation process. To test this possibility, 18mer phosphorothioate derivatives of oligodeoxynucleotide probes against ependymin mRNA, which are stable against degradation by endogenous RNAse activities in vivo, were injected into the perimeningeal fluid, twice, at 18 and 1 h before training (16). Goldfish tolerate rather high water temperature. Therefore, the whole animal was kept and trained at 35°C to facilitate hybridization and to inhibit de novo ependymin synthesis without interference with pre-existing ependymin molecules. They learned the shuttle-box task, but in the test session an amnesic effect was measured, identical to that induced by the antisera. As a control, FITC-labelled oligodeoxynucleotide probes were injected, and indeed, they were incorporated into the meningeal fibroblasts. Further controls were matched, radomized sequences which did not interfere with memory consolidation. Overtrained fish were used as behavioural controls (16). These were the first experiments to demonstrate a direct effect of antisense probes on a specific behaviour.

5. REDISTRIBUTION OF EPENDYMIN TO NEURONAL TARGET SITES

From the antisense experiments, we conclude that only newly synthesized ependymin participates in the molecular cascades of memory consolidation, possibly because this cell adhesion molecule is rapidly bound to cell membranes of, and incorporated into, activated neurons (15).

The cytological architecture of the teleost tectum has thoroughly been analysed by comparative neuroanatomists. They described a conspicuous lamination in which each sensory modality projects on a specific class of neurons in one respective layer (17). The myelinated retinal ganglion cell axons, e.g., project on the long apical dendrites of large type I projection neurons. Actinopterygian fish possess an additional marginal layer in the optic tectum, invaded by unmyelinated marginal fibres from the longitudinal torus that make synaptic („S1") contacts with spines on the distal portion of the apical dendrites of the same type I neurons. The marginal fibres are supposed to conduct information about the timing of expected events. Therefore, the type I neurons are well suited to integrate the stimuli involved in our active avoidance conditioning, and, although these type I neurons exhibited no ependymin mRNA in the in situ hybridization studies, they are specifically recognized by immunogold anti-ependymin antibodies (10, 15, 18). Immunostaining was most prominent on dendrites and in the immediate proximity of the synapses projecting on their spines. After shuttle-box conditioning a marked increase in ependymin immunoreactivity was observed in the tectum, and in particular, an increase in ependymin precipitates at S1 synapses from marginal fibres (15).

6. TOWARDS A POSSIBLE MECHANISM FOR MEMORY CONSOLIDATION

Fig. 5 is a summarizing cartoon on ependymin redistribution. Ependymin is synthesized in the reticular shaped fibroblasts of the leptomeninx. But the protein is also observed in neurons of the optic tectum. Morris stimulated the marginal fibres form the longitudinal torus tetanically and recorded long term potentiation at the level of type I neurons. Furthermore, he measured that the calcium concentrations in the surrounding extracellular fluid decreased. Both, the potentation and the decrease in calcium were more pronounced, when the cells had first been stimulated by a conditioning stimulus via the optic nerve (19). Possibly, the local drecrease in extracellular calcium around activated synapses triggers the in vivo formation of high molecular weight ependymin association products in the extracellular matrix, later used to induce morphological alterations at activated synapses. Although most changes in metal ion concentrations induced by synaptic activity are temporary, also long-lasting consequences in synaptic calcium binding capacities during plastic CNS adaptations have been documented (20). Functional implication of ependymin in memory consolidation has been questioned by some workers in the field (8), but recently much evidence was compiled for the involvement of various cell adhesion molecules in the molecular cascades of learning and memory formation (21, compare also 22, 23, 24).

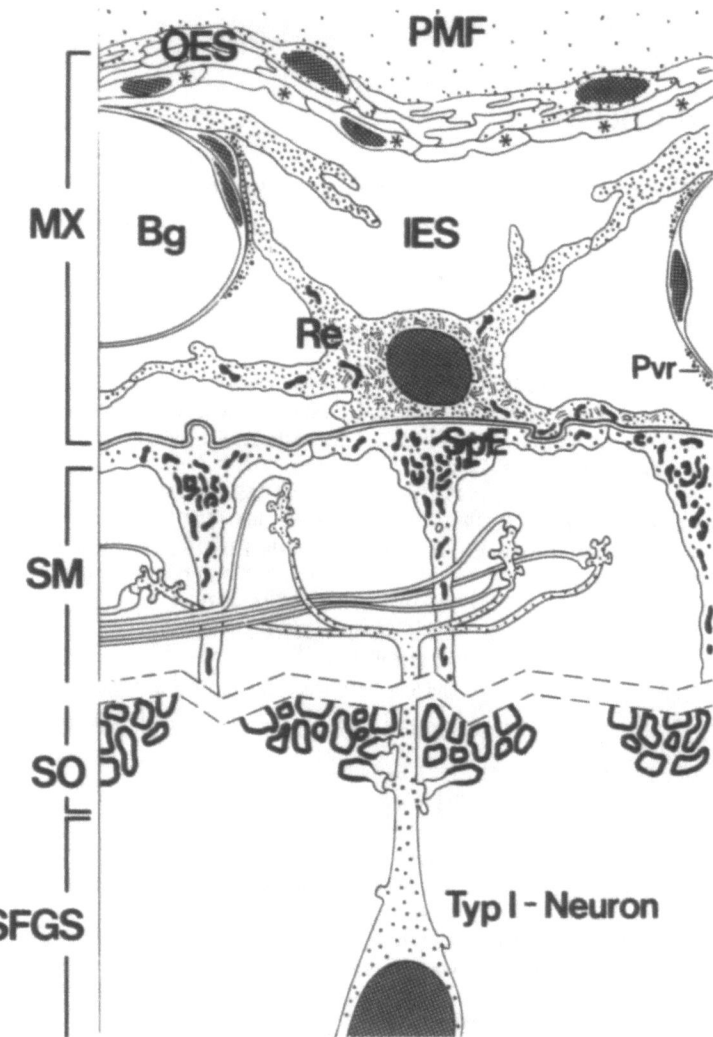

Figure 5. Synthesis and redistribution of ependymin in goldfish brain. Strongly stained reticular shaped fibroblasts (Re) of the inner endomeningeal layer (IES) synthesize ependymin. Their extensions contact blood vessels (Bg). Note staining of perivascular space (Pvr). Type I neurons of the stratum fibrosum et griseum superficiale (SFGS) incorporate ependymin. On dendritic spines they obtain synaptic input from marginal fibres of the longitudinal torus (in the marginal layer, SM) and from optic fibres (in the optic layer, SO, and the SFGS). Subpial endfeet (SpE) of radial glial cells display week staining. MX, meninx; OES, outer endomeningeal layer; PMF, perimeningeal fluid; asterisks, intermediate cell layer.

7. CONCLUSIONS

We suggest two intercalated regulatory mechanisms for memory formation. On the one hand, stress hormones, may evaluate the preceding acquisition with respect to its biological importance and determine, when and how much of cell adhesion molecules, such as ependymin, are synthesized and secreted. Local concentrations of metal cations in the neuronal micro-environment with its specific synaptic connections, on the other hand, reflect the semantic aspect of the preceding activity and may, therefore, direct the cell adhesion molecule to — and deposit it at — those synapses, that have to be consolidated in order to improve their efficacy for future use.

ACKNOWLEDGMENT

Support by grants from the Deutsche Forschungsgemeinschaft is gratefully acknowledged.

REFERENCES

1. Schmidt, J.T., and Shashoua, V.E., 1988, *Brain Res.* **446**:269–284.
2. Schmidt, J.T., Schmidt, R., Lin, W., Jian, X., and Stuermer, C.A.O., 1991, *J. Neurobiol.* **22**:40–54.
3. Königstorfer, A., Sterrer, S., Eckerskorn, C., Lottspeich, F., Schmidt, R., and Hoffmann, W., 1989, *J. Neurochem.* **52**:310–312.
4. Hoffmann, W., and Schwarz, H., 1996, *Interntl. Rev. Cytol.* **165**:121–158.
5. Schmidt, R., and Shashoua, V.E., 1981, *J. Neurochem.* **36**:1368–1377.
6. Shashoua, V.E., Daniel, P.F., Moore, M.E., and Jungalwala, F.B., 1986, *Biochem. Biophys. Res. Commun.* **138**:902–909.
7. Schmidt, R., and Makiola, E., 1992, *Life Sci. Advances* **10**:161–171.
8. Ganss, B., and Hoffmann, W., 1993, *Eur. J. Biochem.* **217**:275–280.
9. Shashoua, V.E:, 1988, *Neurochem. Res.* **13**:649–655.
10. Rother, S., Schmidt, R., Brysch, W., and Schlingensiepen, K.-H., 1995, *J. Neurochem.* **65**:1456–1464.
11. Schmidt, R., 1989, *Progr. Zool.*, Vol. 37 (Rahmann, H., ed.), Gustav Fischer Verlag, Stuttgart, New York, , PP. 327–339.
12. Benowitz, L.I., and Shashoua, V.E., 1977, *Brain Res.* **136**:227–242.
13. Piront, M.-L., and Schmidt, R., 1988, *Brain Res.* **442**: 53–62.
14. Schmidt, R., 1987, *J. Neurochem.* **48**:1870–1878.
15. Schmidt, R., 1995, *Behav. Brain Res.* **66**:65–72.
16. Schmidt, R., Brysch, W., Rother, S., and Schlingensiepen, K.-H., 1995,*J. Neurochem.* **65**:1465–1471.
17. Meek, J., 1981, *J. Comp. Neurol.* **199**:149–173.
18. Schmidt, R., Rother, S., Schlingensiepen, K.-H., and Brysch, W., 1992, *Progr. Brain Res.* **91**:7–12.
19. Morris, M.E., Ropert, N., and Shashoua, V.E., 1986, *Annals New York Acad. Sci.* **481**: 375–377.
20. Körtje, K.-H., and Körtje, D., 1992, *J. Microscopy* **166**:34–358.
21. Bailey, C.H., and Kandel, E.R., 1993, *Ann. Rev. Physiol.* **55**: 397–426.
22. Murphy, K.J., O'Connell, A.W., and Regan, C.M., 1997, *this volume*.
23. Welzl, H., Becker, C., Artola, A., and Schachner, M., 1997, *this volume*.
24. Mileusnic, R., and Rose, S.P.R., 1997, *this volume*.

Section 32: Cell Adhesion in Neural Plasticity Induced by Experience

TIME-LIMITED ROLES FOR CELL ADHESION MOLECULE-MEDIATED NEUROPLASTIC EVENTS IN THE COMMITMENT OF MEMORY TO LONG TERM STORAGE

Keith J. Murphy, Gerard B. Fox, Alan W. O'Connell, and Ciaran M. Regan

Department of Pharmacology
University College, Belfield
Dublin, Ireland

1. INTRODUCTION

When arrangements of peripheral targets are perturbed in adult animals by amputation or nerve section the cortical representation of the body undergoes substantial reorganisation (1). This demonstrates the remarkable capacity of the brain for immediate functional plasticity coupled with long-lasting structural change. These changes may depend on axonal growth of synaptic terminals into new zones or that a large number of potential synaptic connections already exist. The earliest indication that remodelling of nerve endings occurs among neuronal targets was the observation that axonal branches sometimes appeared to be degenerating or growing (2). Such remodelling appears to depend on functional adaptation to the external milieu as animals reared in complex environments exhibit change in synapse number suggestive of altered connectivity pattern through synapse formation or retraction as well as through modification of pre-existing connections (3). Similar changes are observed in animals subjected to maze or motor skill training suggesting that the nervous system exhibits both subtle and specific neural change in response to stimuli such as learning (4,5).

Although the literature on the neurobiology of memory is extensive, much of what is known can be summarised to suggest that memory has stages and is continually changing and that long-term memory may be represented by plastic changes localised in multiple brain regions (6,7). Thus, the current consensus for information storage is that long-term memory is stored in the brain in terms of change in synaptic connectivity pattern within discrete ensembles of neurons (8). The fundamental concept that neuronal growth and change occurs during the formation of permanent memory brings the neurobiological study of memory into contact with developmental neurobiology. Behaviour emerges

---

From a colloquium organized by Rupert Schmidt (Giessen, Germany) and Ciaran Regan (Dublin, Ireland).

gradually as the brain develops. Initially this is under the control of genetic and developmental programmes. Influences of the environment begin to exert their effect *in utero* and become a prime importance after birth. Epigenetic influenc es which control plasticity arise within the organism and from the external environment. The former involve intracellular signals that consist of diffusable factors and surface receptors such as adhesion molecules. The latter provides nutritive factors, sensory and social experiences and learning, which mediate their effects through changes in neural activity.

2. THE ROLE OF CELL ADHESION MOLECULES IN MEMORY FORMATION

The neural cell adhesion molecule (NCAM) is one of the most abundant adhesion molecules on the surface of neural cells. It exists in multiple protein forms derived from extensive post-translational modifications of a single gene and the attachment of α2,8 linked polysialic acid (PSA) chains (9). The molecular mechanisms by which NCAM regulates neural structuring remain to be established however, evidence exists to suggest an important role for NCAM prevalence and polysialylation in neuritogenesis and neurite path finding (10,11,12). In the adult, NCAM polysialylation has been associated consistently with areas that undergo or can be induced to undergo activity-dependent morphological remodelling in response to physiological and/or experimental stimulation (13,14, 15). Thus, NCAM plays an ongoing role in processes of neuronal plasticity.

2.1. Time-Dependent Role of NCAM Function

Memory-associated neuroplastic change is now recognised to be critically dependent on the intact functioning of NCAM. Antibody inter ventive studies have demonstrated a specific role for this cell adhesion molecule in a defined 6–8 h post-training period following avoidance training in both the chick and rat (16,17). This strict time-dependent effect cannot be attributed to a slow penetration of the antibody as similar studies indicate anti-amyloid precursor protein to be maximally effective during training and in the immediate 0–2 h posttraining period (18). In contrast to NCAM, the L1 cell adhesion molecule produces amnesia in the chick model of avoidance learning when administered either during training or in the 6–8 h posttraining period (19). Thus, there appears to be a role for L1 cell adhesion molecule function in the molecular events underlying memory acquisition and consolidation whereas that of NCAM is r estricted to consolidation of processes entrained during learning. The periods of L1 and NCAM function correspond to times in which protein synthesis inhibitors are effective in inducing amnesia in the chick passive avoidance response (20), implying the requirement for their increased biosynthesis which may relate to synapse formation and/or remodelling. This process may not cease at the 6–8 h posttraining time, as the amnesic action of anti-NCAM emerges gradually and only becomes apparent at recall times in excess of 24 hours (16). Thus, the time-limited role for neuroplastic events involved in the commitment of memory to long term storage must exceed that implied by the above interventive studies. Posttranslational modifications of NCAM appear to contribute to these processes.

2.2. Modulations of NCAM Polysialylation State

Intracerebroventricular administration of labelled N-acetyl-D-mannosamine, the sialic acid precursor, to rodents during passive avoi dance training, and at increasing time points thereafter, has revealed a transient increase in the sialylation of immunoprecipitated hippocampal NCAM at 12–24h following the initial learning trial (21). This is specific to learning as no increase is observed in animals rendered amnesic with scopolamine. Analysis of the relative incorporation of precursor into the various isoforms of NCAM indicates the label to be associated specifically with the 180 kDa form. The N-acetyl-D-mannosamine precursor appears to be incorporated as polysialic acid, as immunoprecipitation with anti-PSA isolates only the 180 kDa form. As this NCAM form is enriched in the postsynaptic density of differentiated neurons (22), it suggests that the learning-associated change in NCAM polysialylation state may dominate plastic events associated with the synapse. However, the extent of polysialylation may not be similar to that observed in development. Western blotting procedures reveal none of the charateristic 'smearing' patterns observed in samples of embryonic NCAM. However, a discrete band of 210 kDa has been observed occasionally (23). This may reflect reduced polysialylation or a uniquely spliced NCAM transcription product. A similar isoform has been reported previously in adult animals (24).

Identification of a role for NCAM polysialylation in memory consolidation provided a mechanism to explore directly the morphological substrate involved in memory-associated neuroplasticity. Immunohistochemical studies using a monoclonal antibody which specifically recognises polysialic acid (15), have provided direct evidence for learning-associated change in hippocampal NCAM polysialylation in a defined population of granule-like cells at the hilar border of the granule cell layer of the adult rat dentate gyrus (25). Following training in a passive avoidance paradigm a two-fold increase in the frequency of these is observed at the 10–12 h post-training time (Fig. 1). This does not occur in animals rendered amnesic with scopolamine, suggesting that this frequency change is specific to learning. Their rapid activation is consistent with *in vitro* studies which indicate PSA delivery to the cell surface to occur within one hour of synthesis (26) and with *in vivo* studies which demonstrate NCAM-PSA expression to be dependent on neural activity (14). Coincident with the learning-associated increase in NCAM polysialylation a tonic disinhibition of dopaminergic function occurs at 10–12 h following passive avoidance training (23). It has been suggested that a significant proportion of these polysialylated dentate neurons are *de novo* granule cell precursors (27) however, immunohistochemical studies employing bromo-deoxyuridine immunohistochemistry have failed to detect a significant difference in the small number of heterogenously distributed dentate neurons in animal trained in a passive avoidance paradigm (25). Thus these dentate cells appear to be capable of reactivating their polysialylation status within a discrete post-training time period requiring a neuroplastic response.

The learning-associated change in NCAM polysialylation status is not specific to non-spatial forms of learning. Precisely the same transient and temporal changes occur following training in the Morris water maze, which is a spatial form of learning (Fig. 1; 28). These increases are activity-dependent and specific to learning as they are not observed in animals required to locate a visible platform, swim freely in the maze or in those rendered amnesic with scopolamine. Animals trained in further water maze sessions exhibit similar transient and repetitive increases in dentate polysialylated neuron frequency. These are of a similar magnitude and occur at the same 12 h post-training time, despite improved performance as evidenced by significantly decreased escape latencies between the first and subsequent training sessions (28). This suggests that activation of NCAM-dependent events is required for processing information in response to task-associated sensory stimuli rather than retrieval from or contribution to previously stored, task-associated memory. These results are consistent with an earlier study in which mice homozygous for disrupted NCAM gene function were shown to exhibit spatial learning

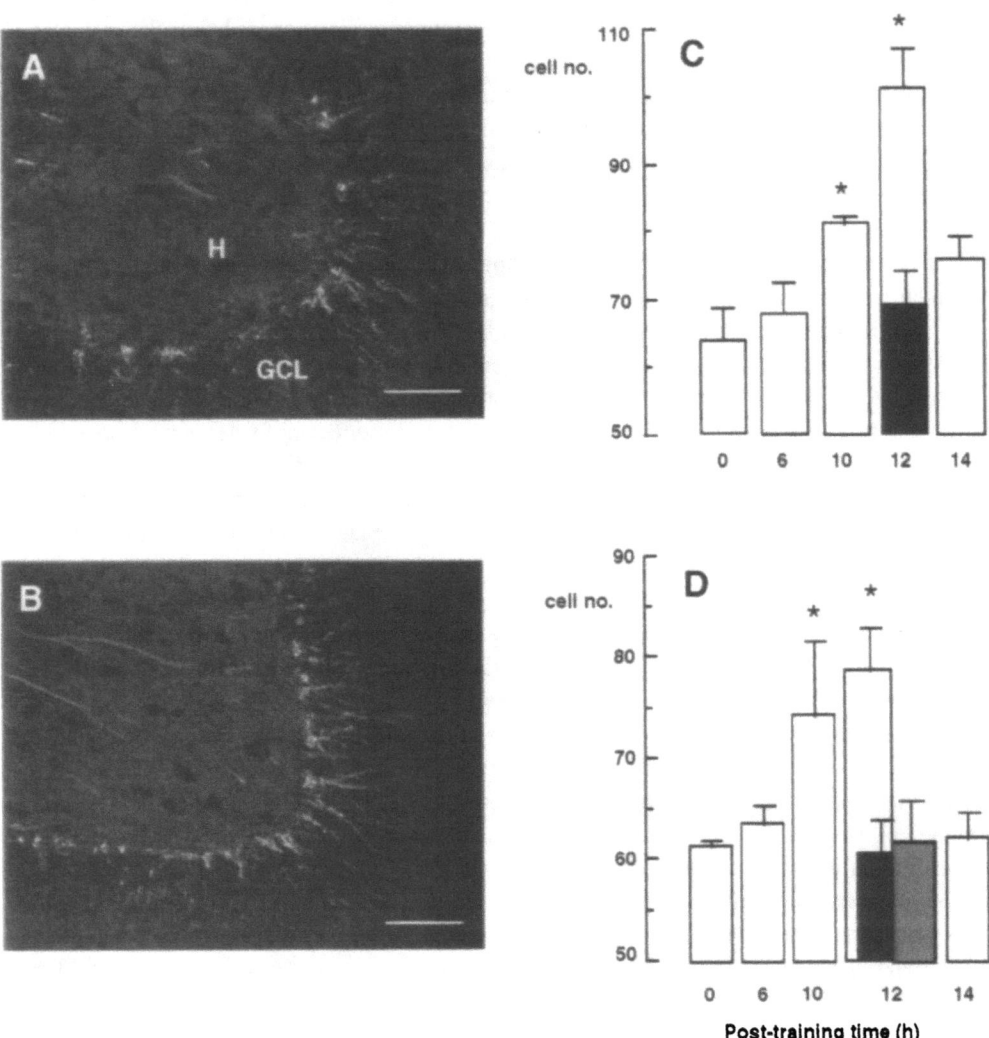

Figure 1. Time-dependent modulation of NCAM polysialylation state in the dentate gyrus following spatial and non-spatial forms of learning. Panels *A* and *B* illustrate PSA immunoreactivity in the rat hippocampal neurons at the granule cell layer (GCL)/hilar (H) border of the dentate gyrus at 0 h and 12 h following a single 5 trial training session in the Morris water maze. The scale bar represents 100 μm. Panels C and D demonstrate the time-dependent increase in the frequency of dentate polysialylated neurons following passive avoidance and Morris water maze training, respectively. No increase in frequency was noted in the scopolamine-treated (filled column) or passive (stippled column) controls. All values are the mean±SEM number of cells/0.15mm^2 of the granule cell layer (3<n<9). Those significantly different (p<0.05) from the 0 h time are indicated with an asterisk. The data is based on previously published observations in which details of the training protocols and immunohistochemical procedures may be obtained (Fox et al., 1995; Murphy et al., 1996).

deficits (29), an effect consistent with lack of PSA modulation, for which NCAM is the major carrier (15). Thus the increase in NCAM-PSA-positive cells which is observed at 10 and 12 h following training appears to be a universal learning response that is critical to both spatial and non-spatial paradigms.

The activation of neuronal polysialylation in the dentate following passive avoidance and water maze training is of considerable interest as the relative contributions of the hippocampus in these separate forms of memory function is controversial. The major parameters proposed to be analysed by this brain structure include environmental spatial cues (30), the contextual significance of stimuli (31) and the recruitment of working memory (32). These roles may not be mutually exclusive as the relative contribution of the hippocampus to spatial or non-spatial learning appears to be dependent on the representational demands of the task (33, 34, 35, 36). Furthermore, these findings directly identify a morphological substrate of memory in which synaptic remodelling may occur. In the antibody interventive studies described above, administration of anti-NCAM results in amnesia at the 48 h recall time but not at 12–24 h when NCAM sialylation is maximal. This suggests that memory-associated NCAM polysialylation may play a role in the selective stabilisation of synaptic contacts produced in the learning process, as the novel synaptic contacts/modifications would appear to be preserved at 24 h but fail to compete or mature and the ensuing amnesia results (37).

3. LTP INVOLVES CELL ADHESION MOLECULE FUNCTIONS

Similar cell adhesion molecule-mediated neuroplastic mechanisms appear to underlie the induction and maintenance of longterm potentiation (LTP). LTP occurs in many brain regions but is studied most widely and best understood in the hippocampus where it has been used as a model for the early stages of memory formation and learning. Induction of LTP has been associated with rapid synaptic remodelling as evidenced by an increased number of synapses onto dendritic shafts as compared to low-frequency stimulated control groups (8). Cell recognition systems, such as the integrins, L1 cell adhesion molecule and NCAM, may contribute to this process as their functional blockade prevents the induction and maintenance of LTP (38, 39, 40). These effects not only apply to the early phases of LTP. Fazeli and colleagues (41) have observed a longterm transient increase in the release of soluble NCAM into hippocampal perfusates at 1.5–2.5 h following the induction of LTP. This has been suggested to arise from increased proteolytic activity involved in the removal of 'stabilizing' forms of NCAM and their replacement by 'growth promoting forms' (42), as has been suggested previously by others (43, 44). Thus the role of cell adhesion molecules in the induction and maintenance of LTP has many parallels to those observed in behavioural models of animal learning, most notably, the presence of time-restricted events in the development of enduring longterm structural modifications.

4. CONCLUSIONS

These results demonstrate that a neuroplastic continuum is maintained within the period predicted to be necessary for the eventual commitment of memory to long term storage. The time required for learning-induced modulations of NCAM polysialylation is lengthy but within the consolidation period proposed by recent investigations. Cognition-enhancing and nootropic agents would appear to operate within this time-frame as their memory improving effects become apparent only after the 16–24 h post-training time, despite their differing primary modes of action (45, 46). These agents would appear to augment NCAM-mediated neuroplastic events as co-administration of nefiracetam during training reverses scopolamine-induced amnesia by preserving the learning-associated modulations of polysialylation at the 12 h post-training time (47), further confirming that the acquired memory trace remains plastic for considerable lengths of time.

REFERENCES

1. Merzenich, M.M., Kaas, J.H., Wall, J.T., Nelson, R.J., Sur, M., and Felleman D.J., 1983, *Neurosci.* **8**: 3–55.
2. Sotelo, C., and Palay, S.L., 1971, *Lab. Invest.* **25**: 653–671.
3. Turner, A.M., and Greenough, W.T., 1985, *Brain Res.* **329**: 195–203.
4. Black, J.E., Isaacs, K.R., Anderson, B.J., Alcantara, A.A., and Greenough, W.T., 1990, *Proc. Natl. Acad. Sci. USA* **87**: 5568–5572.
5. Greenough, W.T., Juraska, J.M., and Volkmar, F.R. 1979, *Behav. Neural Biol.* **26**: 287–297.
6. Squire, L.R., 1987, *Memory and Brain*, Oxford Press, New York.
7. Alvarez, P., and Squire, L.R. 1994, *Proc. Natl. Acad. Sci. USA* **91**: 7041–7045.
8. Bailey, C.H., and Kandel, E.R., 1993, *Annu. Rev. Physiol.* **55**: 397–426.
9. Regan, C.M., 1991, *Int. J. Biochem.* **23**: 513–523.
10. Doherty, P., Cohen J., and Walsh, F.S., 1990, *Neuron* **5**: 209–219.
11. Doherty, P., Fruns, M., Seaton, P., Dickson, G., Barton, C.H. Sears, T.A., and Walsh, F.S., 1990, *Nature* **343**: 464–466.
12. Tang, J., Landmesser, L., and Rutishauser, U., 1992, *Neuron* **8**: 1031–1044.
13. Bonfanti, L., Olive, S., Poulain, D., and Theodosis, D., 1992, *Neurosci.* **49**: 419–436.
14. Kiss J., Wang C., and Rougon, G., 1993, *Neurosci.* **53**: 213–221.
15. Rougon G., 1993, *Eur. J. Cell Biol.* **61**: 197–207.
16. Doyle, E., Nolan, P., Bell, R., and Regan, C.M., 1992, *J. Neurochem.* **59**: 1570–1573.
17. Scholey, A.B., Rose, S.P., Zamani, M.R., Bock, E., and Schachner, M., 1993, *Neurosci.* **55**: 499–509.
18. Doyle, E., Bruce, M.T., Breen, K.C., Smith, D.C., Anderton, B., and Regan, C.M. 1990, *Neurosci. Lett.* **115**: 97–102.
19. Scholey, A.B., Mileusnic, R., Schachner, M., and Rose, S.P.R., 1995, *Learning and Memory* **2**: 17–25.
20. Freeman, F.M., Rose, S.P.R., and Scholey, A.B., 1995, *Neurobiol. Learning and Memory* **63** 291–295.
21. Doyle, E., Nolan, P., Bell, R., and Regan, C.M. 1992, *J. Neurosci. Res.* **31**: 513–523.
22. Persohn, E., Pollerberg, G.E., and Schachner, M., 1989, *J. Comp. Neurol.* **288**: 92–100.
23. Doyle, E., and Regan, C.M., 1993, *J. Neural Transm.* **92**: 33–49.
24. Key, B., and Akeson, R.A., 1990, *J. Cell Biol.* **110**: 1729–1743.
25. Fox, G.B., O'Connell, A.W., Murphy, K.J., and Regan, C.M., 1995, *J. Neurochem.* **65**: 2796–2799.
26. Kiss, J.Z., Wang, C., Olive, S., Rougon, G., Lang, J., Baetens, D., Harry, D., and Pralong, W-F., 1994, *EMBO J.* **13**: 5284–5292.
27. Seki, T., and Arai, Y., 1993, *J. Neurosci.* **13**: 2351–2358.
28. Murphy, K.J., O'Connell, A.W., and Regan, C.M., 1996, *J. Neurochem.* **67**: 1268–1274.
29. Cremer, H., Lange, R., Christoph, A., Plomann, M., Vopper, G., Roes, J., Brown, R., Baldwin, S., Kraemer, P., Scheff, S., Barthels, D., Rajewsky, K., and Willie, W., 1994, *Nature* **367**: 455–459.
30. O'Keefe, J., and Nadel, L., 1978. *The hippocampus as a cognitive map*, Oxford University Press. Oxford.
31. Solomon, P.R., 1980, *Physiol. Psychol.* **8**: 254–261.
32. Olton, D.S., Becker, J.T., and Handelman, G.E., 1980, *Physiol. Psychol.* **8**: 239–246.
33. Morris, R.G.M., Garrud, J., Rawlins, N.P., and O'Keefe, J., 1982, *Nature* **297**: 681–683.
34. Bunsey, M., and Eichenbaum, H., 1996, *Nature* **379**: 255–257.
35. Eichenbaum, H., Mathews, P., and Cohen, N.J., 1989, *Behav. Neurosci.* **103**: 1207–1216.
36. Eichenbaum, H., Stewart, C., and Morris, R.G.M., 1990, *J. Neurosci* **10**: 331–339.
37. Doyle, E., Nolan, P.M., Bell, R., and Regan, C.M., 1992, *Network* **3**: 89–94.
38. Luthi, A., Laurent, J.P., Figurov, A., Muller, D., and Schachner, M., 1994, *Nature* **372**: 777–779.
39. Ronn, L.C.B., Bock, E., Linnemann, D., and Jahnsen, H., 1995, *Brain Res.* **677**: 145–151.
40. Staubli, U., Vanderklish, P., and Lynch, G., 1990, *Behav. Neural Biol.* **53**: 1–5.
41. Fazeli, M.S., Breen, K.C., Errington, M.L., and Bliss, T.V.P., 1994, *Neurosci. Lett.* **169**: 77–80.
42. Doherty, P., Fazeli, M.S., and Walsh, F.S, 1994, *J. Neurobiol.* **26**: 437–446.

43. Bailey, C.H., Chen, M., Keller, F., and Kandel E.R., 1992, *Science* **256**: 645–649.
44. Sheppard, A., Wu, J., Rutishauser, U., and Lynch, G., 1991, *Biochim. Biophys. Acta* **1076**: 156–160.
45. Mondadori, C., Hengerer, B., Ducret, T., and Borkowski, J. 1994, *Proc. Natl. Acad. Sci. USA* **91**: 2041–2045.
46. Doyle, E., O'Boyle, K.M., Shiotani, T., and Regan, C.M., 1996, *Neurochem Res* **21**: 649–652.
47. Doyle, E., Regan, C.M., and Shiotani, T. 1993, *J Neurochem.* **61**: 266–272.

**EXPRESSION OF
α-2,8-POLYSIALYLTRANSFERASE DURING
RETINOIC ACID INDUCED DIFFERENTIATION
OF A NEUROBLASTOMA CELL LINE**

Herbert Hildebrandt,[1] Harald Rösner,[1] Rita Gerardy-Schahn,[2] and Hinrich Rahmann[1]

[1]Inst. für Zoologie
University Hohenheim
Stuttgart, FRG
[2]Inst. für medizinische Mikrobiologie
Medizinische Hochschule Hannover
Hannover, FRG

1. INTRODUCTION

Neuroblastoma is a neural crest-derived tumor predominantly composed of undifferentiated neuroblast-like cells. The human neuroblastoma cell line SH-SY5Y is widely used as a model system for neuronal differentiation (1). Upon addition of retinoic acid the proliferation rate of the heavily migrating neuroblast like cells decreases and the cells differentiate morphologically. About 20–30% of the cells differentiate into a flat, substrate-adherent cell type (S-cells), while the majority of the cells develop the characteristic phenotype of differentiated neurons with long neuritic extensions and spindle-shaped perikarya (neuron-like, NL-cells). Like many other neuroendocrine tumors (2, 3) SH-SY5Y cells exhibit strong immunoreactivity with an antibody that specifically recognizes a homopolymer of 8 or more α-2,8-linked sialic acid residues (polysialic acid, PSA) (4).

As demonstrated for adult rat brain, PSA is expressed on the sodium channel α subunit (5) but in embryonic tissues as well as on a variety of tumor cells it is an abundant modification of the neural cell adhesion molecule (NCAM) (2, 6). PSA decreases the adhesive properties of NCAM and therefore interferes with cell-cell interactions (6, 7). It has been suggested to promote axonal growth (8, 9), cell migration (10, 11) and invasiveness of tumor cells (3). In addition, a crucial role of PSA in cell division has been implied (12). The sialylation state of NCAM is controlled by the activity of developmentally regulated

From a colloquium organized by Rupert Schmidt (Giessen, Germany) and Ciaran Regan (Dublin, Ireland).

Golgi sialyltransferases (13) and recently a hamster α-2,8-polysialyltransferase (PST-1) was cloned that induces PSA synthesis in N-CAM expressing cell lines (14). Northern blot analysis suggested that PST-1 activity is regulated on the level of the mRNA since expression is high in embryonic but low in adult brain. Subsequent cloning of human PST revealed an almost identical amino acid sequence (97.8 %) and transfection studies demonstrate the ability of PST to influence neuritic outgrowth (15). More recently it has been demonstrated that a second, closely related sialyltransferase (STX) is also able to determine PSA synthesis (16).

In order to examine the role of polysialylation in differentiation and neuritic outgrowth of SH-SY5Y cells we compared occurrence of PSA and the expression of PST during retinoic acid induced differentiation. Here we provide first evidence that PSA immunostaining correlates with PST expression in SH-SY5Y cells.

2. MATERIALS AND METHODS

SH-SY5Y cells were seeded at a density below $2 \times 10^3/cm^2$ and grown in 35 mm petri dishes in the presence or absence of 10 μM retinoic acid as described (4). For immunostaining cells were fixed in 3.8% paraformaldehyde/PBS for 20 min, washed, and incubated for 3h with PSA-specific monoclonal antibody (mAb) 735 (17) in 2% BSA/PBS at 4°C. As secondary antibody fluoresceine-labelled anti-mouse IgG was coupled for 1 h. To obtain antisense RNA probes 2 μg of plasmid (pBluescript), linearized with Hind III, were transcribed with T3 RNA-polymerase in the presence of 70 nmoles DIG-UTP (Boehringer). RNA fragments app. 300 nucleotides in length were obtained by alkaline hydrolysis, collected by ethanol precipitation, centrifuged and resuspended in hybridization buffer containing 50% formamide (Amersham). For in situ hybridization cells in petri dishes were sequentially treated with 4% formaldehyde/PBS (5 min), 200mM HCl (10 min), 1% Triton X-100 (2 min), dehydrated in a graded series of ethanol (60, 80, 95, 100, 100%, 1 min each) and stored in 95% ethanol at 4°C. Cells were incubated with 3 ng antisense RNA in 10 μl hybridization buffer under a coverslip. Hybridization was carried out overnight at 55°C and cells were washed with 0.1 × SSC at 60°C (2 × 30 min). DIG-labeling was detected by alkaline phosphatase-conjugated anti-DIG-antibody (Boehringer, 1:750, 30 min at 37°C). and BCIP/NBT as substrate (30°C, overnight). Bright field illumination was used for photographic documentation of the hybridization signals because cells were grown on plastic and therefore Nomarski-optics could not be applied.

3. RESULTS AND DISCUSSION

Immunostaining was performed with fixed, not detergent permeabilized SH-SY5Y cells and therefore only antigen on the cell surface was assessed. For undifferentiated, strongly proliferating cells detection of PSA with mAb 735 resulted in a moderate, apparently homogeneous surface labeling of virtually all cells (Fig. 1a). But as shown for one of the frequently observed protrusions (Fig. 1b), higher magnification reveals a punctate distribution of the staining. After retinoic acid treatment, staining of the flat substrate adherent S-cells was clearly reduced (Fig. 1c), and a punctate pattern distributed over the whole cell surface was obtained. In contrast, surface concentration of PSA was high on cell bodies and neurite-like extensions of NL-cells. Thus, PSA appears to be expressed at all stages of differentiation of SH-SY5Y cells, but NL- and S-cells differ strongly in PSA sur-

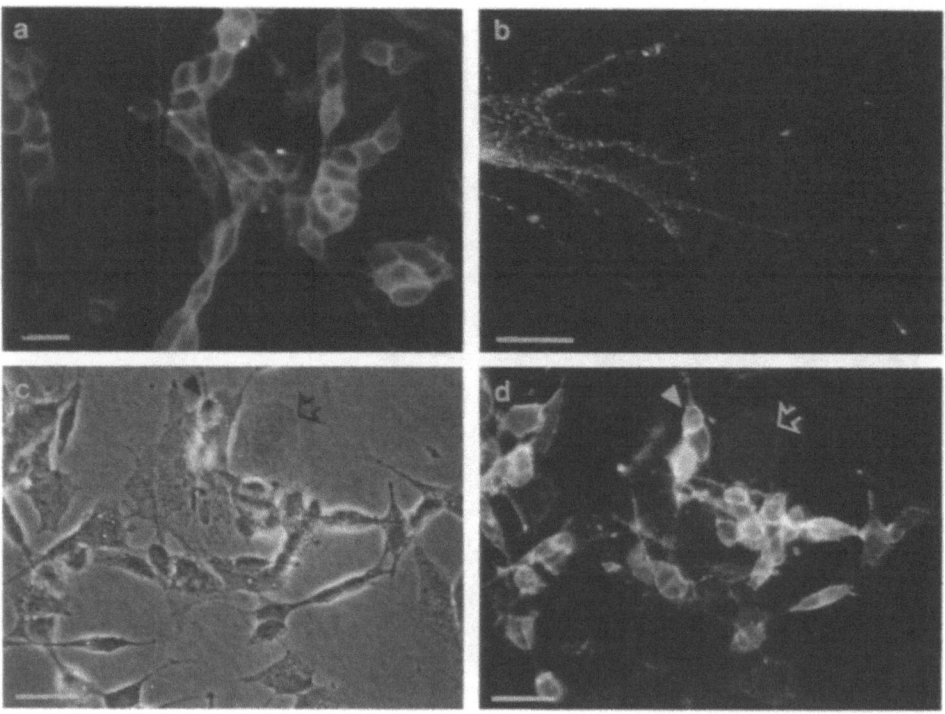

Figure 1. Fluorescence-immunostaining (a, b, d) and phase contrast image (c) of SH-SY5Y cells grown in the absence (a, b) or presence of retinoic acid (c, d). Arrowhead: Neuron-like cell type (NL-cell). Open arrow: substrate adherent cell type (S-cell). Calibration bars: 25 μm (a, c, d) and 10 μm (b).

face labeling. A punctate pattern of PSA staining has been frequently detected on tissue sections and cultured cells (10, 12, 18, 19). For staining of live cells (18) this pattern was interpreted as a common artifact due to cross-linking by the antibodies, which in the case of NCAM immunostaining could be prevented by aldehyde fixation (20). Since in the present study staining was performed on fixed cells it appears unlikely, that the patchy staining occurs due to cross-linking by the antibodies. It rather seems that on the surface of SH-SY5Y cells PSA is distributed in clusters.

Expression of PST in SH-SY5Y cells was assessed by in situ hybridization with a probe directed against the cDNA of PST-1. Due to the largely identical nucleotide sequences of hamster PST-1 and human PST (90% for the coding region), hybridization could be performed under high stringency conditions. Strong signals were obtained in all undifferentiated SH-SY5Y cells (Fig 2a, b). Upon retinoic acid induced differentiation the NL-cells were still strongly labeled, although the intensity of the signals appears somewhat reduced if compared to the undifferentiated cells. On the other hand, staining was clearly reduced in S-cells (Fig. 2c, d). Control hybridizations with a probe directed against a rat olfactory receptor did not give any detectable signals (not shown).

The high expression of PST in undifferentiated and neuron-like SH-SY5Y cells indicated by the in situ hybridization correlates to the strong surface expression of PSA in these cell types. Likewise, the apparent down-regulation of PST expression in S-cells corresponds to reduced PSA immunostaining. The results therefore imply that in SH-SY5Y cells PST activity and PSA expression are regulated at least in part on the level of the mRNA. Never-

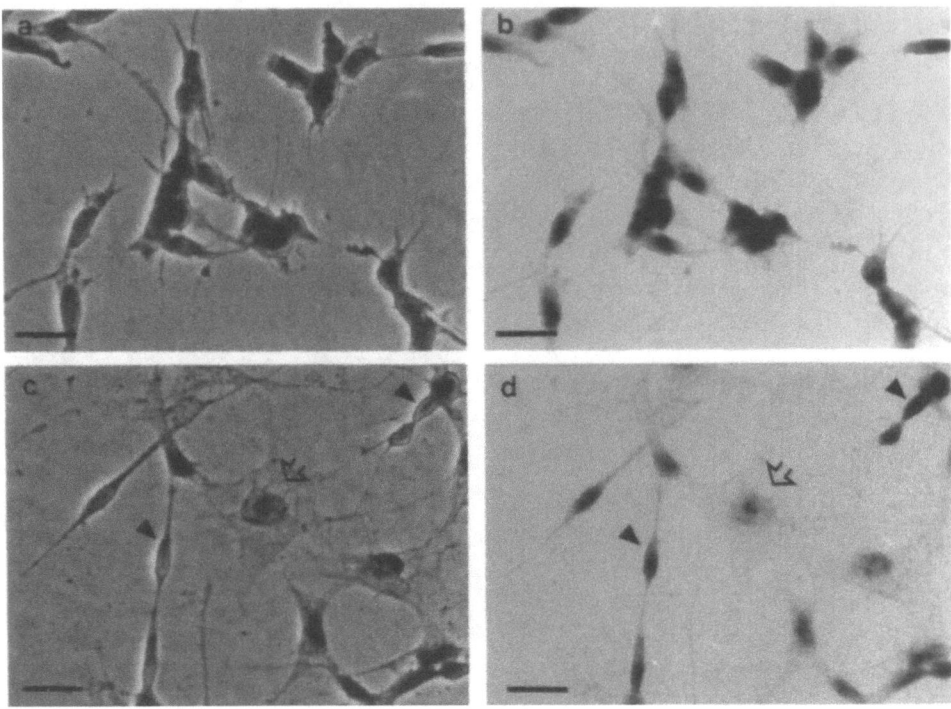

Figure 2. In situ hybridization analysis of PST expression of SH-SY5Y cells grown in the absence (a, b) or presence of retinoic acid (c, d). Phase contrast reveals cell morphology (a, c) whereas under bright field illumination (b, d) only hybridization signals appear black. Arrowhead: Neuron-like cell type (NL-cell). Open arrow: substrate adherent cell type (S-cell). Calibration bars: 25 μm.

theless, it will be of great interest to examine the expression of STX, the other PSA-promoting sialyltransferase. Taken together, high expression of PSA and PST in the undifferentiated, strongly proliferating and migrating cells as well as in the NL-cells that form neuritic extensions appears consistent with the suggested roles of PSA (6–12). Thus, the PST sequence should provide a useful tool to study the functions of PSA in SH-SY5Y cells.

ACKNOWLEDGMENT

We thank Dr. J. Strotmann, Hohenheim, for introduction into in situ hybridization and his support in RNA probe preparation.

REFERENCES

1. Pahlman, S., Mamaeva, S., Meyerson, G., Mattson, M.E.K., Bjelfman, C., Örtoft, E., and Hammerling, U., 1990, *Acta Physiol. Scand.* **592** (Suppl.):25–37.
2. Figarella-Branger, D.F., Durbec, P.L., and Rougon, G.N., 1990, *Cancer Res.* **50**: 6364–6370.
3. Roth, J., Zuber, C., Komminoth, P., Scheidegger, E.P., Warhol, M.J., Bitter-Suermann, D., and Heitz, P.U., 1993, *Polysialic Acid* (J. Roth, U. Rutishauser, and F.A. Troy, eds.), Birkhäuser Verlag, Basel, PP. 335–348.
4. Rebhan, M., Vacun, G., Bayreuther, K., and Rösner, H., 1994, *Neuroreport* **5**: 941–944.

5. Zuber, C., Lackie, P.M., Catterall, W.A., and Roth, J., 1992, *J. Biol. Chem.* **267**:9965–9971.
6. Rougon, G., 1993, *Eur. J. Cell Biol.* **61**:197–207.
7. Rutishauser, U., 1993, *Polysialic Acid* (J. Roth, U. Rutishauser, and F.A. Troy, eds.), Birkhäuser Verlag, Basel, PP. 215–227.
8. Doherty, P., and Walsh, F.S., 1992, *Curr. Opin. Neurobiol.* **2**:595–601.
9. Doherty, P., Fazeli, M.S., and Walsh, F.S., 1995, *J. Neurobiol.* **26**:437–446.
10. Wang, C., Rougon, G., and Kiss, J.Z., 1994, *J. Neurosci.* **14**:4446–4457.
11. Hu, H., Tomasiewicz, H., Magnuson, T., and Rutishauser, U., 1996, *Neuron* **16**:735–743.
12. Bonfanti, L., and Theodosis, D.T., 1994, *Neuroscience* **62**:291–305.
13. Breen, K.C., and Regan, C.M., 1988, *Development* **104**:147–154.
14. Eckhardt, M., Mühlenhoff, M., Bethe, A., Koopman, J., Frosch, M., and Gerardy-Schahn, R., 1995, *Nature* **373**:715–718.
15. Nakayama, J., Fukuda, M.N., Fredette, B., Ranscht, B., and Fukuda, M., 1995, *Proc. Natl. Acad. Sci. USA* **92**:7031–7035.
16. Scheidegger, E.P., Sternberg, L.R., Roth, J., and Lowe, J.B., 1995, *J. Biol. Chem.* **270**:22685–22688.
17. Frosch, M., Görgen, J., Boulnois, G.J., Timmis, K.M., and Bitter-Suermann, D., 1985, *Proc. Natl. Acad. Sci. USA* **82**:1194–1198.
18. Rougon, G., Dubois, C., Buckley, N., Magnani, J.L., and Zollinger, W., 1986, *J. Cell Biol.* **103**:2429–2437.
19. Kiss, J.Z., Wang, C., Olive, S., Rougon, G., Lang, J., Baetens, D., Harry, D., and Pralong, W.F., 1994, *EMBO J.* **13**:5284–5292.
20. Van den Pol, A.N., Ellisman, M., and Deerinck, T., 1989, *Colloidal Gold. Principles, Methods and Applications* (M.A. Hayat, ed), Academic Press, San Diego, PP. 451–487.

Section 32: Cell Adhesion in Neural Plasticity Induced by Experience

FUNCTIONAL CHARACTERIZATION OF THE FIRST IMMUNOGLOBULIN-LIKE DOMAIN OF THE NEURAL CELL ADHESION MOLECULE

Vladislav V. Kiselyov,[1] Vladimir Berezin,[1] Thomas E. Maar,[1] Vladislav Soroka,[1] Klaus Edvardsen,[1] Arne Schousboe,[2] and Elisabeth Bock[1]

[1]Protein Laboratory
Panum Institute
University of Copenhagen
Denmark
[2]Royal Danish School of Pharmacy
Copenhagen, Denmark

1. INTRODUCTION

The neural cell adhesion molecule (NCAM) is a cell surface glycoprotein belonging to the immunoglobulin (Ig) superfamily. NCAM is known to mediate cell-cell and cell-substratum interactions (1). In this study, we functionally characterized the IgI NCAM domain using a bacterially produced fusion protein of the IgI domain with maltose binding protein (MBP) as a fusion partner and a recombinant protein of the IgI domain produced in an eukaryotic expression system of the yeast *Pichia pastoris*. For comparison, other Ig-like NCAM domains were also produced as fusion proteins. Both the IgI and IgII domains inhibited aggregation and neurite formation of mouse cerebellar neurons, whereas other Ig-like domains of NCAM did not. Immobilized IgI domain promoted cell attachment of NCAM transfected L-cells, and antibodies against the IgI NCAM domain inhibited aggregation of NCAM transfected BT4Cn cells.

2. EXPERIMENTAL PROCEDURES

Production of bacterially expressed fusion proteins of the IgI, IgII, IgIII, IgIV and IgV NCAM domains and a recombinant protein of the IgI domain expressed in an eukaryotic expression system of the yeast *Pichia pastoris* is described (Kiselyov, et. al., submitted). The fusion proteins were designated IgN-MBP (Ig-like domain number N produced in *E. coli* as a

From a colloquium organized by Rupert Schmidt (Giessen, Germany) and Ciaran Regan (Dublin, Ireland).

fusion protein with MBP as a fusion partner). The IgI domain of NCAM was also produced as a recombinant protein in *Pichia pastoris*. The recombinant protein was designated as IgI *P. pastoris* (Ig-like domain I produced in *P. pastoris*). The various NCAM fragments consisted of the following amino acids: IgI *P. pastoris* — residues 20–118, IgI-MBP — residues 20–110, IgII-MBP — residues 115–209, IgIII-MBP — residues 215–307, IgIV-MBP — residues 309–405 (including exon VASE (ASWTRPEKQE) between residues 354 and 355), IgV-MBP — residues 411–497. The numbering of amino acid residues is according to GenBank/EMBL under accession number A44290. Number 1 refers to the first amino acid of the secretion signal.

Polyclonal rabbit antibodies were raised against IgI-MBP as described (2). The antibodies were used as an anti-serum. A preimmune serum from the same rabbit was used as a control.

A mouse L 929 cell line (L-cells) was provided by the European Collection of Animal Cell Cultures. Transfection of L-cells and a rat glioma cell line, BT4Cn, with full-length cDNA encoding the human NCAM-B isoform inserted in pHβ-Apr-1-neo was described (3,4). Control L-cells were transfected with the expression vector alone. Cells were routinely grown and passaged in Dulbecco's modified Eagle's medium (DMEM) (Gibco BRL, Denmark) supplemented with 10% heat-inactivated FCS, penicillin 100 units/ml, and streptomycin 100 μg/ml.

Mixed primary microwell cultures of dissociated mouse cerebellum were prepared as described (5). Briefly, a suspension of dissociated cells from cerebella of 6-day-old mice was plated onto an uncoated 60-well microtiter plate (Nunc, Denmark) at a density 45,000 cells/well in a volume of 10 μl. Cells were grown in a culture medium consisting of DMEM supplemented with 19 mM KCl, 25 mM glucose, 0.2 mM glutamine, 0.1 I.U./l insulin, 0.15 mM p-aminobenzoate, 100 I.U./ml penicillin and 10% (v/v) FCS. Protein constructs were added to the culture medium before plating at 10 μg/ml. The number of aggregates and processes was counted per field after 72 h. Aggregates were defined as clusters of more than 50 cells. Processes connecting two aggregates were counted if their length exceeded 100 μm.

96-well tissue culture plates (Nunc, Denmark) were coated overnight (100 μl/well) with nitro-cellulose dissolved in methanol (1 cm^2 of BA 85 nitro-cellulose, Schleicher & Schuell, The Netherlands, was dissolved in 48 ml of methanol). The dishes were washed 3 times with PBS, incubated for 1 h at 37 °C with 10 μg/ml IgI-MBP or MBP, washed 3 times with PBS and blocked with 10% FCS in DMEM for 1 h. L-cells were grown for 4 days, trypsinized for 10 min with 0.05% w/v trypsin, 0.02% w/v EDTA solution (Gibco BRL, Denmark), seeded in a 96-well plate at a density of 30,000 cells/well in DMEM containing 1% BSA and incubated at 37 °C for the indicated time periods. Cells were washed 3 times with PBS, fixed with 3% paraformaldehyde, stained with 0.5% crystal violet in 20% methanol for 15 min and washed 5 times with distilled water. Finally, the residual dye was dissolved with 0.1 M sodium citrate pH 4.2, 50% ethanol, and the absorbance at 540 nm was measured.

Confluent BT4Cn cells transfected with the human NCAM-B isoform were treated with 7 mM EDTA in DMEM for 20 min and plated onto a 24-well plate (Nunc, Denmark) coated with 1 ml per well 0.75% Agar Noble (Difco Laboratories, USA) in DMEM at a density of 10^6 cells/ml (100 μl per a well). Anti-serum against IgI-MBP or preimmune anti-serum or PBS were added to 100 μl cells at a dilution 1:50. Cells were incubated at 37 °C for 180 min on a shaking platform (120 rpm). Aggregation was stopped by adding 1 ml 0.1% formaldehyde in PBS either immediately after plating or after 180 min. After incubating cells with formaldehyde for 10 min, cells were transferred to 30 ml PBS containing 0.1% formaldehyde and counted using a Coulter Multisizer (Coulter Electronics Ltd., England).

3. RESULTS AND DISCUSSION

To asses the effect of the IgI NCAM domain on neural cell aggregation and process formation, a mixed primary cell culture of mouse cerebellar neurons was employed. Dissociated cells from cerebella of 6-day-old mice were plated on a microtiter plate and grown in the presence of various protein constructs. From Figs. 1a it appears that addition of both IgI-MBP, IgII-MBP and IgI *P. pastoris* inhibited the formation of aggregates and therefore led to a decrease in the size of aggregates and, as a result, to an increase in the number of smaller aggregates. IgIII-MBP and IgV-MBP did not have any effect when compared to MBP or PBS, and IgIV-MBP had only a slight effect. Furthermore, addition of IgI-MBP, IgII-MBP and IgI *P. pastoris* inhibited the formation of processes (Fig. 1b), whereas IgIII-MBP and IgV-MBP did not have any effect when compared to MBP or PBS, and IgIV-MBP had a slight effect.

This indicates that the IgI domain binds to neural cell surface molecules and thereby interferes with cell aggregation and inhibits the formation of neuronal processes.

In the previous experiment we showed that the IgI domain could bind to neural cell surface molecules. To test whether this binding is homophilic or heterophilic, we studied cell attachment of NCAM and vector transfected L-cells. From Fig. 2 it appears that NCAM and vector transfected L-cells hardly attached to the control matrix, MBP, whereas they attached readily to IgI-MBP. This means that the IgI domain may be involved in heterophilic interactions with components on the surface of L-cells. However, the attachment of NCAM transfected L-cells to IgI-MBP was twice that of the vector transfected L-cells,

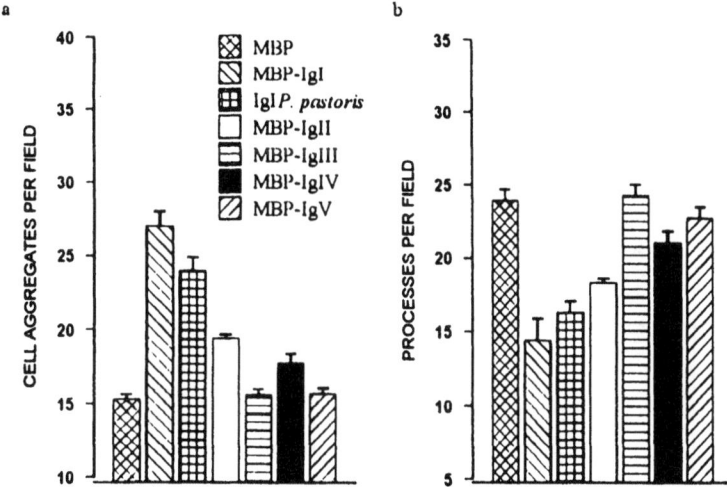

Figure 1. Effect of the Ig-like NCAM domains on mixed primary cell cultures of mouse cerebellar neurons. IgI-MBP, IgI *P. pastoris*, IgII-MBP, IgIII-MBP, IgIV-MBP, IgV-MBP and MBP were added to the culture medium before plating at a concentration of 10 μg/ml. *a*) Number of cell aggregates. *b*) Number of processes. Three independent experiments, each in ten replicates, were performed. The data from different experiments were pooled. The results are shown as mean ± SEM.

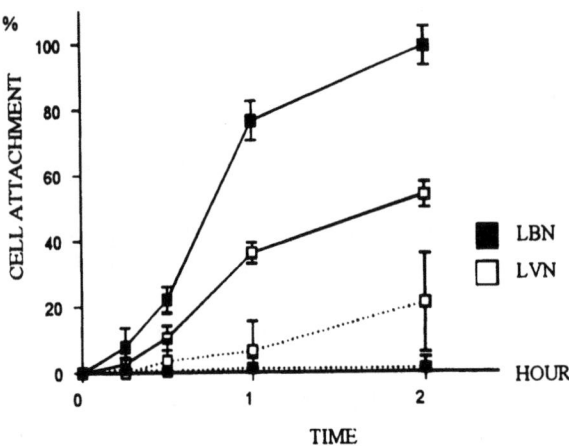

Figure 2. Attachment kinetics of L-cells to immobilized IgI-MBP or MBP. LBN and LVN stands for NCAM and vector transfected L-cells, respectively. Cell attachment to the IgI-MBP is represented by solid curves, and to MBP — by dotted curves. Attachment at various time points is shown as the percentage of the signal obtained after 4 h of incubation of the respective cell line (attachment complete). Mean values ± SEM are from three independent experiments, each carried out in quadruplicates.

indicating that the IgI domain of NCAM probably also is involved in homophilic binding to NCAM.

If the IgI domain is involved in cell-cell adhesion, then antibodies against it may be expected to interfere with the adhesion process. It has been demonstrated that aggregation of NCAM-B transfected BT4Cn cells is NCAM dependant (6). From Fig. 3 it appears that antibodies against the IgI domain inhibited cell aggregation of NCAM-B transfected BT4Cn cells, whereas control antibodies did not, supporting the notion that the IgI domain of NCAM participates in NCAM mediated cell-cell adhesion.

Thus, our data indicate that the IgI NCAM domain is involved in cell-cell adhesion and cell-substratum interaction. This is in agreement with our recent results demonstrating that the IgI domain binds the IgII domain and heparin in a surface plasmon resonance analysis assay (Kiselyov, et. al., submitted) and with the results of other researchers who showed that the IgI and IgII NCAM domains, synthesized in a bacterial expression sys-

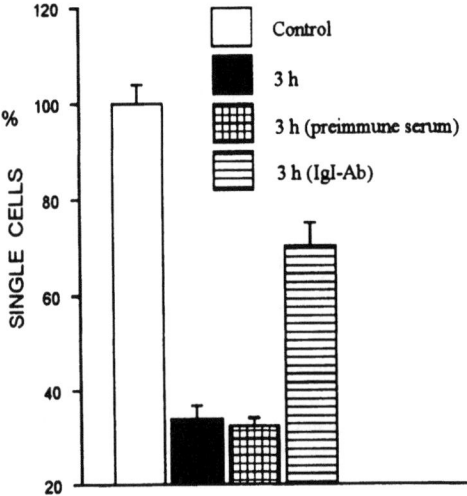

Figure 3. Inhibition of cell aggregation of BT4Cn cells transfected with NCAM-B. Column 1 (from left to right) — aggregation of cells was stopped immediately after the beginning. Column 2 — aggregation was stopped after 3 h. Column 3 — aggregation in the presence of a preimmune serum, stopped after 3 h. Column 4 — aggregation in the presence of an antiserum against the IgI domain, stopped after 3 h. The number of single cells in the control was taken as 100%. Three independent experiments were made, each in six replicates. The results are given as mean ± SEM. A representative experiment is shown.

tem, when used as immobilized substratum, clearly promoted cell adhesion of NCAM expressing cells, whereas other NCAM domains hardly had any adhesion effects (7). Our data are also in accordance with reports that a 25-kDa N-terminal NCAM fragment containing the IgI and IgII NCAM domains is involved in cell adhesion (8,9).

ACKNOWLEDGMENT

The financial support of the EU-Programme on Neuronal Development and Regeneration (contract No. BIO2-CT-93–0260), the Danish Medical Research Council, the Danish Biotechnology Programme, the Danish Cancer Society and the Lundbeck Foundation is gratefully acknowledged.

REFERENCES

1. Gegelashvili, G., and Bock, E., 1996, *Treatise on Biomembranes, in press.*
2. Harboe, N., and Ingild, A., 1973, *Scand. J. Immunol.* Suppl. **2**: 161–164.
3. Kasper, C., Stahlhut, M., Berezin, V., Maar, T. E., Edvardsen, K., Kiselyov, V. V., Soroka, V., and Bock, E., 1996, *J. Neurosci. Res., in press.*
4. Edvardsen, K., Pedersen, P. -H., Bjerkvig, R., Hermann, G. G., Zeuthen, J., Laerum, O. D., Walsh, F. S., and Bock, E., 1994, *Int. J. Cancer.* **58**: 116–122.
5. Trenkner, E., and Sidman, R. L., 1977, *J. Cell Biol.* **75**: 915–940.
6. Edvardsen, K., Chen, W., Rucklidge, G., Walsh, F. S., Öbrink, B., and Bock, E., 1994, *Proc. Natl. Acad. Sci. U. S. A.* **90**: 11463–11467.
7. Frei, T., von Bohlen und Halbach, F., Wille, W., and Schachner, M., 1992, *J. Cell Biol.* **118**: 177–194.
8. Cole, G. J., and Glaser, L., 1986, *J. Cell Biol.* **102**: 403–412.
9. Frelinger III, A. L., and Rutishauser, U., 1986, *J. Cell Biol.* **103**: 1729–1737.

Section 32: Cell Adhesion in Neural Plasticity Induced by Experience

EXPRESSION OF A TRANSMEMBRANE ISOFORM OF THE NEURAL CELL ADHESION MOLECULE (NCAM) IN RAT GLIOMA BT4Cn CELLS INCREASES THEIR MOTILITY

Søren Prag, Nina Pedersen, Peter S. Walmod, Vladimir Berezin, and Elisabeth Bock

The Protein Laboratory
Panum Institute
University of Copenhagen, Denmark

1. INTRODUCTION

Cell movement is a fundamental part of basic biological and pathological phenomena such as neural development, wound healing, and cancer metastasis. Integrins are involved in interactions of cells with components of the extracellular matrix (ECM), formation of focal adhesion complexes and rearrangement of the actin cytoskeleton, and therefore play a pivotal role in cell motility (1). The role of other cell adhesion molecules on cell motility is less characterized. It has been demonstrated that the expression of a transmembrane form of the neural cell adhesion molecule NCAM-B (NCAM 140 kDa) in rat glioma BT4Cn cell lines inhibits their penetration of Matrigel *in vitro* and reduce their invasive behaviour and growth *in vivo* (2, 3). Here, we show that NCAM-B transfected rat glioma cell lines grown on various culture substrata exhibited increased rates of random cell movement compared to control vector transfected cell lines.

2. MATERIALS AND METHODS

2.1. Cell Culture

The BT4C rat glioma cell line was obtained from fetal BD IX rat brain cells after exposure to ethyl-nitrosourcea *in vivo*. The BT4Cn cell line is a cloned subline of the BT4C cell line (4). BT4Cn cell lines transfected with human NCAM-B cDNA and control cell lines transfected with the expression vector alone have been characterized previously (2). Four NCAM-B transfected cell lines and four control lines were used. Cells were grown in

From a colloquium organized by Rupert Schmidt (Giessen, Germany) and Ciaran Regan (Dublin, Ireland).

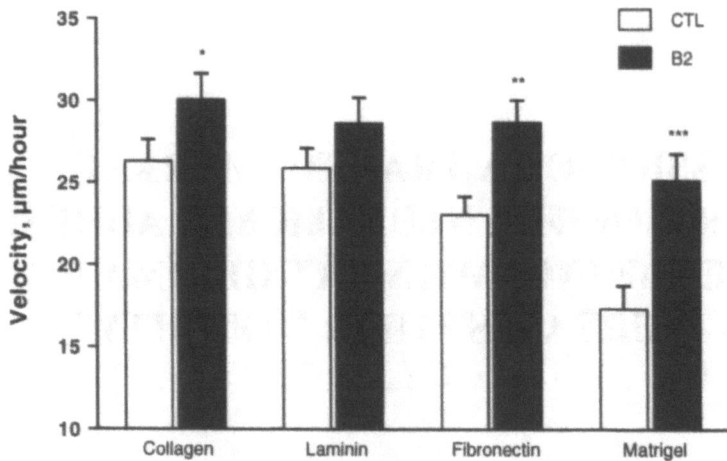

Figure 1. Effects of NCAM-B expression and culture substrata on cell motility. A control cell line (CTL) and NCAM-B expressing cell line (clone B2) were plated in dishes pre-coated with collagen type I, laminin, fibronectin or Matrigel and recorded 4 hours after plating with a 5 min interval between observations for 30 min. The number of cells observed varied from 90 to 110 in the individual experiments. Results are given as mean ± SEM. *, ** and *** -indicate that the mean values were significantly different from the values shown for control cells ($p < 0.05$, $p < 0.005$ and $p < 0.001$, respectively).

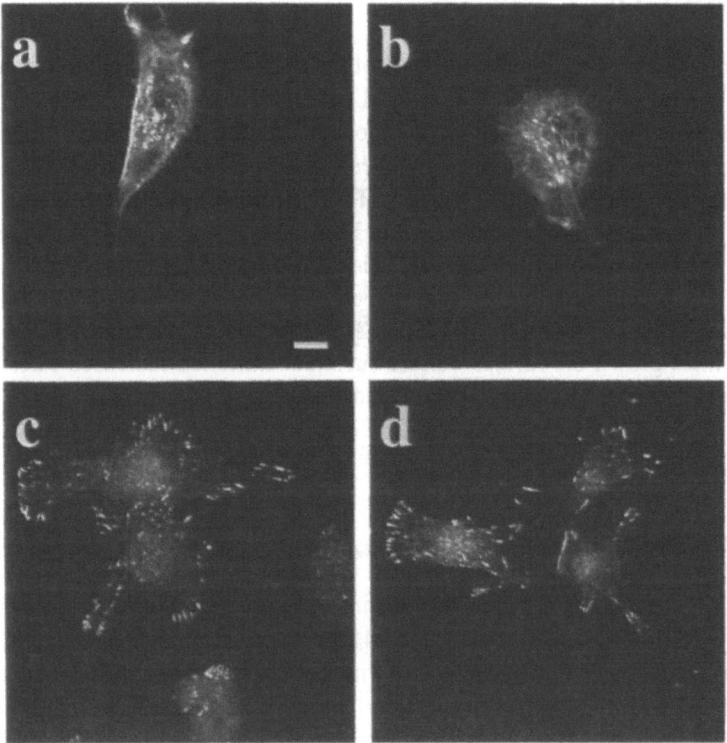

Figure 2. Fluorescence micrographs of control cells (a, c) and NCAM-B expressing cells (b, d). Actin filaments are stained with Texas Red-conjugated phalloidin (a, b) and are projections of a section series. Focal adhesions are visualized by vinculin stained by indirect immunofluorescence (c, d). Bar, 10 μm.

DMEM (Gibco BRL) supplemented with 10% fetal calf serum at 37°C in a humidified atmosphere with 5% CO_2. In order to study cell motility, cells were dislodged with 5 mM EDTA in Dulbecco's PBS and replated at a density of 5×10^4 cells/cm^2 on 35 mm diameter tissue culture dishes pre-coated with 1 μg per cm^2 of fibronectin (from bovine plasma, Sigma), laminin (from mouse sarcoma basement membrane, Sigma), Matrigel (purified basement membrane components from murine Engelbreth-Holm-Swarm tumor, kindly provided by H. K. Kleinman, NIH, Bethesda), or collagen type I (prepared from rat tail tendon (5)), in the presence of 50% conditioned medium collected from confluent cultures of mouse fibroblastoid L-cells (L929). The cells were plated 4 hours prior to video recording.

2.2. Video-Recording and Estimation of Cellular Motility

The cells were observed in a Nikon Diaphot 300 inverted microscope equipped with phase contrast optics, and video recording of cultures of the live cells placed on a thermostatically controlled stage was made at 34°C using a black and white CCD video camera (Burle, USA). Using an image analysis system (LARPOP, Image House, Denmark), automated 512 × 512 pixel images were acquired at 5 minute intervals for 15 min. Images were processed by means of thresholding and binary transformation using the software PRIMA developed at the Protein Laboratory. Centroid coordinates of the outlined cells from consecutive video frames were utilized to generate tracks of cell movement and for calculation of cellular velocity.

2.3. Fluorescence Staining and Microscopy

Cells grown on fibronectin coated LabTek slides (Nunc, Denmark) were fixed in 3% paraformaldehyde, permeabilized with Triton X-100, and stained for filamentous actin (Texas Red X-conjugated phalloidin, Molecular Probes, USA) and vinculin (monoclonal Ab against human vinculin, Sigma followed by rabbit anti-mouse Ig-fluorescein conjugate from DAKO, Denmark). Stained cells were scanned using a Multi-Probe 2001 Laser Scanning Confocal Microscope with an argon/krypton laser (Molecular Dynamics, USA).

3. RESULTS AND DISCUSSION

The effect of NCAM-B expression on random cell motility was estimated by determination of the rate of centroid translocation. From Fig. 1 it can be seen that control cells moved faster when grown on collagen type I and laminin than on fibronectin and Matrigel. Since Matrigel mostly consists of laminin and proteoglycans, it may be assumed that proteoglycans in the ECM modulate cell motility. Random cell movement of NCAM-B expressing cells was statistically significantly higher than that of the control cells when measured 4 hours after plating on collagen I, fibronectin, or Matrigel coated dishes. NCAM expression in glioma cells caused a dramatic increase in velocity of cells plated on Matrigel, but not on laminin. These results were confirmed using three other NCAM-B transfected glioma cell lines and three control lines (not shown). This points toward a possible modulatory effect of NCAM-B on cell motility, possibly due to direct interactions of NCAM with heparan proteoglycans of the Matrigel.

Since the dynamics of focal adhesion complexes and rearrangements of the actin cytoskeleton are critical for cell motility, we examined whether NCAM-B expression changed the pattern of distribution of focal contacts and filamentous actin in NCAM expressing glioma cells. From Fig. 2 it appears that the distribution of vinculin and the appearance of the actin cytoskeleton did not change in NCAM-B expressing cells compared to the control cells 4 hours after plating on fibronectin. It may therefore be hypothesized that NCAM-B modulates cell motility via a direct interaction with components of the ECM, but not via changes in organization of the focal contacts or actin cytoskeleton.

4. CONCLUSION

In conclusion, this study demonstrates that NCAM-B expression not only influences tumor cell growth and invasive capacity (2, 3), but also modulates the velocity of random cell movement.

ACKNOWLEDGMENTS

This study was supported by The Danish Cancer Society.

REFERENCES

1. Schmidt, C.E., Horwitz, A.F., Lauffenburger, D.A., and Sheetz, M., 1993, *J. Cell Biol.* **123**: 977–991.
2. Edvardsen, K., Brunner, N., Spang-Thomsen, Walsh, F., and Bock, E., 1993, *Int. J. Devl Neurosci.* **11**: 681–690.
3. Edvardsen, K., Pedersen, P.H., Bjerkvig, R., Hermann, G.G., Zeuthen, J., Laerum, O.D., Walsh, F., and Bock, E., 1994, *Int. J. Cancer* **58**: 116–122.
4. Laerum, O.D., Rajewsky M.F., Schachner, M., Stavrou, D., Haglid, K.H., and Haugen, A., 1977, *Z. Krebsforsch.* **89**: 273–295.
5. Poez, K.A.,1967, In: *Treatise on Collagen*, Ed. G.N. Ramachandran, **vol 1**: 207–252.

BREVICAN, A CONDITIONAL PROTEOGLYCAN FROM RAT BRAIN

Characterisation of Secreted and GPI-Anchored Isoforms

Constanze Seidenbecher,[*] Karin Richter, and Eckart D. Gundelfinger

Institute for Neurobiology
Magdeburg, Germany

1. PROTEOGLYCANS OF THE AGGRECAN/VERSICAN FAMILY

The extracellular matrix of the brain is a macromolecular aggregate composed of a great variety of structurally diverse glycoproteins and proteoglycans serving as a microenvironment for neurons and glial cells. Proteoglycans are large molecules consisting of a protein core and glycosaminoglycan (GAG) side chains. During morphogenesis and neural differentiation, the matrix undergoes dramatic changes in its composition and structure. Up to now, more than 20 different proteoglycan core proteins have been described to occur in the brain, most of them being developmentally regulated in their patterns of expression (1). Some of the brain proteoglycans are considered members of the aggrecan/versican family, a group of chondroitin sulphate (CS) bearing secreted molecules with modular structures (for review see ref. 2). Fig. 1A gives a schematic overview of this proteoglycan family. Members of the family are aggrecan, known as the main aggregating proteoglycan of cartilage (3), versican, originally identified in fibroblasts (4), and the two neural-specific molecules neurocan (5) and brevican (6). All these proteoglycans share common structural features including N-terminal globular domains called proteoglycan tandem repeats (PTR) that are capable of binding hyaluronic acid (HA) and link protein. The central part of the molecules has a stretched architecture and carries the GAG side chains. In the C-terminal part, epidermal growth factor (EGF)-like domains, C-type lectin domains and a complement-regulatory protein (CRP)-motive are to be found. The putative HA-binding protein BEHAB (7) is identical to the N-terminus of brevican. Using antisera against a synaptic protein preparation from rat brain we have cloned rat brevican from a

---

[*] Present address: Institute for Medical Psychology, University "Otto von Guericke", Magdeburg, Germany.

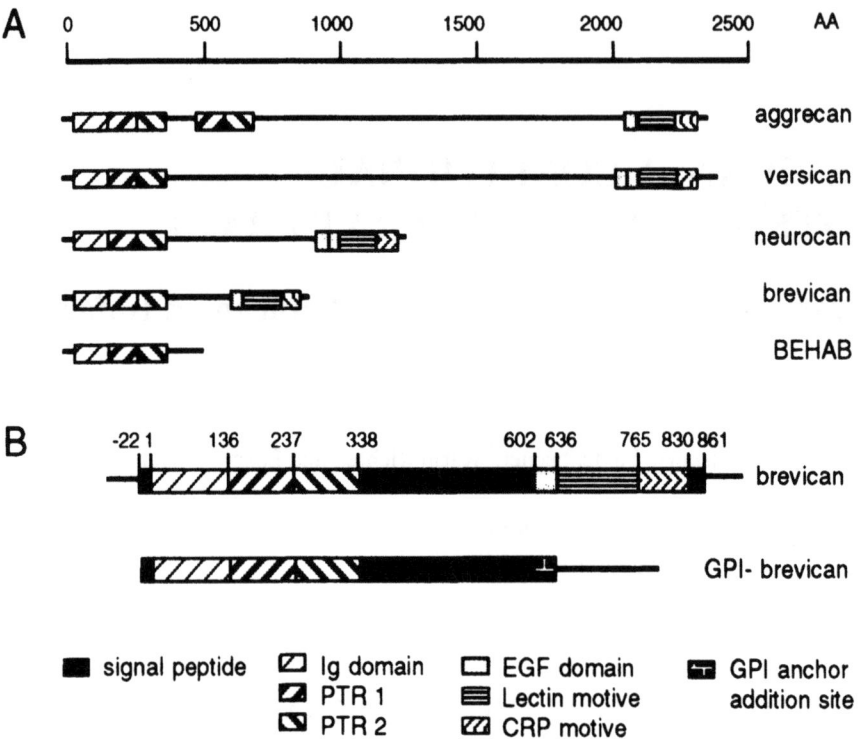

Figure 1. (A) Structural scheme summarising the domain organisation and homology of the members of the aggrecan/versican family of proteoglycans. (B) Comparison of the secreted and the GPI-anchored brevican isoforms. Numbering of amino acids indicates domain borders.

cDNA expression library (8). With polyclonal antisera developed against a bacterially expressed glutathione S-transferase(GST)-brevican fusion construct we could show the existence of multiple brevican isoforms of 145, 125 and 80 kDa with the latter being a hypothetical proteolytic product generated by a specific aggrecanase-like protease (9).

2. PROTEIN ISOFORMS OF BREVICAN

The cDNA cloning revealed two distinct rat brevican cDNAs that differ in their 3'-terminal regions. On Northern blots the corresponding transcripts of 3.6 and 3.3 kilobases (kb) can be detected exclusively in brain RNA. The larger mRNA was shown to encode secreted brevican. As was reported for bovine brevican (6), the secreted isoform in the rat has a molecular weight of 145 kDa. From digestions with chondroitinase ABC we concluded that it occurs as a conditional proteoglycan. This means that a subpopulation of brevican molecules bears CS side chains whereas another subpopulation does not. This could have some importance for the physical properties of the ECM, because a high content of sulphated sugar side chains results in a highly charged matrix. The gene product of the 3.3 kb transcript encodes a putative glycosylphosphatidylinositol (GPI)-anchored isoform (Fig. 1B). Indeed, the membrane protein fraction contains 140 and 125 kDa brevican anchored via a phosphatidylinositol-specific phospholipase C (PI-PLC)-cleavable GPI anchor. This was also shown by experiments with eukaryotic cells stably transfected with the 3.3 kb brevican transcript yielding GPI-brevican comparable to the isoforms found in the brain (8). This is a new and important finding because it is the first example of glypiation in the aggrecan/versican family. All the other members so far reported to be expressed in the brain are secreted in the extracellular space and appear in the soluble protein fraction. Only versican is reported to be capable of binding to cell surfaces via adhesive C-terminal structures (10). What could be the biological significance of the GPI anchoring? As was reported for other glypiated molecules, these anchors may serve as a targeting signal delivering the proteins to specialised cellular subdomains. A controlled release by endogenous phospholipases is possible, and a connection with signal-transducing protein tyrosine kinases is discussed in the literature (for review see ref. 11). Therefore, GPI-brevican could serve as a cell-surface receptor for the extracellular matrix. In contrast to neurocan, which is expressed at high amounts during early developmental stages and then declines to low levels (12) the expression of both secreted and glypiated brevican increases during postnatal development indicating a possible function of the molecules in the terminally differentiating and the adult nervous system.

3. BREVICAN LOCALISATION IN RAT BRAIN

For the immunohistochemical detection of brevican on rat brain sections, three different polyclonal antisera were used that gave identical results. All antisera were developed against the same fusion construct. They can not discriminate between secreted and GPI-anchored brevican, therefore, the immunohistochemical pattern represents both isoforms. The immunoreactivity is widely distributed throughout the CNS in both grey and white matter with the most intense labeling of brain areas containing large neurons. The Purkinje cells of cerebellar cortex and the magnocellular neurons of deep cerebellar nuclei are "coated" by brevican immunostaining. In the hippocampus, both pyramidal and granule cells are strongly labeled. Fig. 2A demonstrates the localisation in the CA3 hippocam-

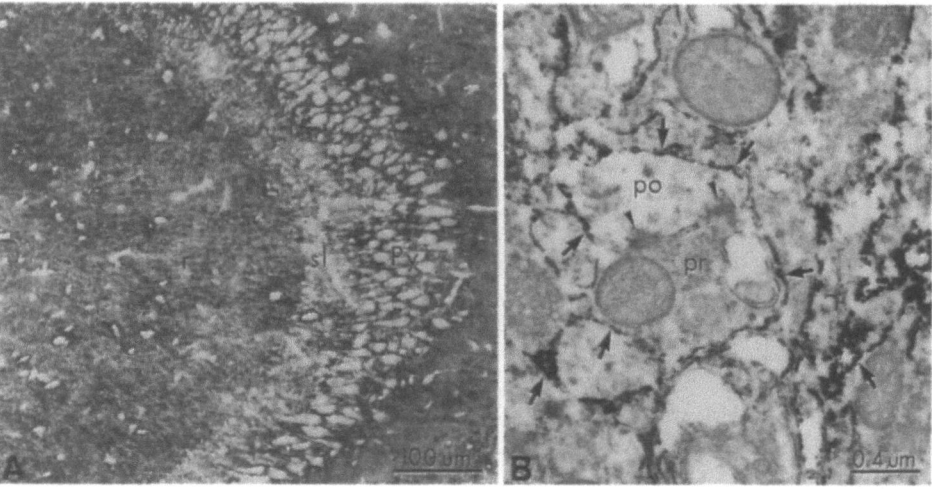

Figure 2. (A). Hippocampal CA3 region. Immunostaining of horizontal vibratome sections (50μm) from perfusion-fixed (4% paraformaldehyde/ 0.1% glutaraldehyde) brains of 30-day-old rats was performed with a 1:20 diluted affinity-purified rabbit anti-brevican antiserum. For visualisation the biotin-avidin-peroxydase method was used. Py= pyramidal cell layer, sl= *stratum lucidum*, r= *stratum radiatum*. (B) Electronmicrograph showing brevican immunoreactivity in the CA3 *stratum radiatum* by means of the pre-embedding peroxydase technique. Arrows indicate accumulated reaction product, small arrowheads mark postsynaptic densities (PSDs). pr= presynaptic element, po= postsynapse.

pal region. Pyramidal cell bodies and their primary dendrites show strong positivity. Staining was never observed to be intracellular and seems not to be restricted to neurons.

Analysis at the electron microscopical level has revealed that the antigen is localised in the extracellular space and along the surfaces of neurons but is excluded from regions of synaptic contact (Fig. 2B). Such a perisynaptic localisation was previously reported for the aggrecan-related brain proteoglycan CAT-301 (13), for HA and versican (10) and is supposed to be characteristic for perineuronal nets, a specialised form of brain ECM (14, 15). Different functions have been discussed to be performed by the perineuronal nets including the stabilisation of synapses, the local concentration of growth factors and a barrier-like function hindering the formation of new cell-cell contacts (reviewed in 14). Up to now, only a few constituents of perineuronal nets have been identified, amongst them hyaluronic acid and CS proteoglycans. One can assume that brevican could be one such component of these net-like matrix specialisations.

ACKNOWLEDGMENTS

We thank Dr. K. Langnäse and Dr. K.-H. Smalla for many helpful suggestions and discussions throughout this work, and K. Hartung for expert technical assistance.

This work was supported by the BMBF (GSF) and the Fonds der Chemischen Industrie.

REFERENCES

1. Herndon, M.E., and Lander, A.D., 1990, *Neuron* **4**: 949–961.
2. Margolis, R.U., and Margolis, R.K., 1994, *Methods in Enzymology*, Vol. 245 (J.M. Abelson and M.I.Simon, eds.), Academic Press, Pasadena, PP. 105–126.
3. Doege, K., Sasaki, M., Horigan, E., Hassel, J.R., and Yamada, Y., 1987, *J. Biol. Chem.* **262**: 17757–17767.
4. Zimmermann, D.R., and Ruoslahti, E., 1989. *EMBO J.* **8**:2975–2981.
5. Rauch, U., Karthikeyan, L., Maurel, P., Margolis, R.U., and Margolis, R.K., 1992, *J. Biol. Chem.* **267**: 19536–19547.
6. Yamada, H., Watanabe, K., Shimonaka, M., and Yamaguchi, Y., 1994, *J. Biol. Chem.* **269**: 10119–10126.
7. Jaworski, D.M., Kelly, G.M., and Hockfield, S., 1994, *J. Cell Biol.* **125**:495–509.
8. Seidenbecher, C.I., Richter, K., Rauch, U., Fässler, R., Garner, C.C., and Gundelfinger, E.D., 1995, *J. Biol. Chem.* **270**: 27206–27212.
9. Yamada, H., Watanabe, K., Shimonaka, M., Yamasaki, M., and Yamaguchi, Y., 1995, *Biochem. Biophys. Res. Comm.* **216**: 957–963.
10. Bignami, A., Hosley, M., and Dahl, D., 1993, *Anat. Embryol.* **188**: 419–433.
11. Brown, D.A., 1992, *Trends Cell Biol.* **2**:338–343.
12. Meyer-Puttlitz, B., Milev, P., Junker, E., Zimmer, I., Margolis, R.U., and Margolis, R.K., 1995, *J. Neurochem.* **65**: 2327–2337.
13. Hockfield,S., Kalb, R.G., Zaremba, S., and Fryer, H., 1990, *Cold Spring Harbor Symp. Quant. Biol.* **LV**:505–514.
14. Celio, M.R., and Blümcke, I., 1994, *Brain Res. Rev.* **19**: 128–145.
15. Brückner, G., Brauer, K., Härtig, W., Wollf, J.R., Rickmann, M.J., Derouiche, A., Delpech, B., Girard, N., Oertel, W.H., and Reichenbach, A., 1993, *Glia* **8**: 183–200.

ENHANCEMENT OF HIPPOCAMPAL LONG-TERM POTENTIATION *IN VITRO* BY FUCOSYL-CARBOHYDRATES

H. Matthies, Jr.,[1] S. Staak,[2] K. H. Smalla,[2] and M. Krug[1]

[1]Institute of Pharmacology and Toxicology
University "Otto von Guericke"
Magdeburg
[2]Federal Institute for Neurobiology
Magdeburg, Germany

1. INTRODUCTION

Long-term potentiation (LTP) is an elementary model for use-dependent synaptic strengthening and widely accepted for studies of basic mechanisms underlying learning and memory formation (1). Similar to learning of several behavioral tasks in mammals and birds LTP consists of different phases (2). Induction of LTP in hippocampal CA1 region depends on the activation of NMDA-receptors and subsequent intracellular effectors, e.g.phospholipases, protein kinases, phosphatases and proteases (3) while the maintenance of LTP requires newly synthesized proteins (4).

Although the molecular mechanisms of synaptic remodelling accompanying LTP (5) are not fully understood there are several indications for an involvement of fucosylated glycoconjugates in the late nondecremental phase of LTP. Particularly, after induction of LTP in hippocampal slices an enhanced incorporation of ^{3}H-fucose into glycoproteins has been observed (6). Furthermore, 2-deoxy-D-galactose which prevents the formation of $\alpha(1-2)$ glycosidic linkages of fucose to galactose residues in glycoproteins has an amnesic effect in different learning tasks (7,8) and suppresses the maintenance of LTP (9). Therefore, a particular role of fucose containing carbohydrate structures in neuronal reorganization processes underlying the maintenance of LTP can be assumed.

In a previous study of LTP *in vivo* in order to imitate an increased amount of fucoglycoepitopes on cell surfaces L-fucose or 2'-fucosyllactose were applied intrahippocampally prior to tetanization. 2'-Fucosyllactose or L-fucose caused a significant increase of the POP-spike amplitude lasting for more than 48 hours (10). In *in vivo*-experiments with chronically implanted electrodes, the field excitatory postsynaptic potential (fEPSP) cannot be recorded at the locus of its generation; therefore it is difficult to discuss fEPSP-data from such experiments. Therefore, extending our previous experiments, the influence of

2'-fucosyllactose compared to 3-fucosyllactose on hippocampal CA1-LTP was tested in an *in vitro* approach in which in addition to population-spike amplitude (POP-spike) the fEPSP-slope was recorded.

2. METHODS

Slice preparation from rat hippocampus and extracellular recordings from the CA1 stratum radiatum were done as described earlier (11). Briefly, 400 μm thick slices of the right hippocampus were placed in a submerge chamber containing artificial cerebrospinal fluid (ACSF) immediately after the preparation. The ACSF contained the following compounds (in mM): NaCl, 124; KCl, 4.9; KH_2PO_4, 1.2; $MgSO_4$, 1.3; $CaCl_2$, 2.5; $NaHCO_3$, 25.6; D-glucose, 10; (pH 7.4). ACSF was saturated with 95% O_2/5% CO_2. POP-spike and fEPSPs were evoked by stimulating the Schaffer collateral-commissural pathway. Extracellular field potential recordings were obtained from the pyramidal cell layer (POP-spike) and the basal dendritic layer of CA1 pyramidal cells (fEPSP). The POP-spike was evaluated by measuring the initial rising slope (mV/ms) between the onset of the response and its negative peak in recordings from the stratum radiatum. Stable recordings were obtained over a 30 min baseline period. LTP was induced by three 100 Hz-stimulus trains, each containing 50 pulses at double pulse width with a 2 min intervall between each train. Thereafter, potentials were measured at 5–10 min intervals during the experiment to analyse test responses. Fucosyllactose derivatives were purchased from Oxford GlycoSystems. Each of the oligosaccharides was applied to incubation medium 60 min before LTP induction (final concentration 0.2 mM) and washed out 5 min after the last tetanus. Statistical evaluation was carried out with the two-tailed Mann-Whitney U-test.

3. RESULTS

High frequency stimulation of hippocampal slices according to our stimulation protocol induced long-term potentiation which was stable for more than 3 hours. No significant alterations of the input/output curves induced by the oligosaccharides were detected (data not shown). The effects of 2'-fucosyllactose and 3-fucosyllactose on LTP compared to potentiation in control experiments are presented in Fig. 1.

Bath application of 2'-fucosyllactose led to an increase of the POP-spike amplitude of about 400% (Fig. 1A) compared to corresponding increases of about 250% in control LTP. Moreover, within the first two hours after induction of LTP significantly higher fEPSPs are recorded when the slices were treated with 2'-fucosyllactose (Fig. 1B). No differences in POP-spike amplitude and fEPSPs between control LTP and 3-fucosyllactose treated slices were observed (Fig. 1C and 1D).

4. DISCUSSION

Bath applications of 2'-fucosyllactose caused a dramatic increase in population spike amplitude and the field excitatory postsynaptic potentials compared to untreated tetanized slices. The corresponding isomer 3-fucosyllactose did not elicit such an improvement of the POP-spike amplitude or fEPSP slope. Furthermore, the size of control potentials and the input/output curves were not significantly altered by the trisaccharides.

Enhancement of Hippocampal Potentiation

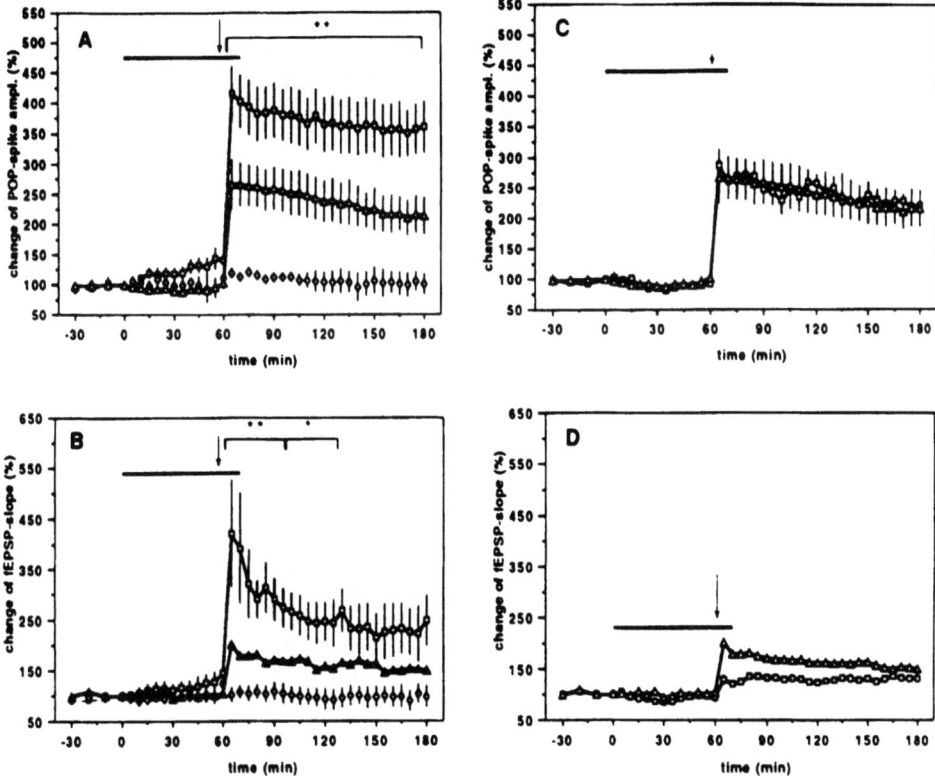

Figure 1. Influence of 2'-fucosyllactose (A,B) and 3-fucosyllactose (C,D) on POP-spike (A,C) and fEPSP (B,D) in CA1-LTP. The bars indicate the time of oligosaccharide application, the arrow indicates the time point of high frequency stimulation to induce LTP. Symbols in A and B: triangle for control-LTP, square for 2'-fucosyllactose-treatment/LTP, rhombus for 2'-fucosyllactose-treatment/control input; Symbols in C and D: triangles for control-LTP, squares for 3-fucosyllactose-treatment. Asterisks indicate significant deviations between LTP under 2'-fucosyllactose and LTP control (** $P<0.02$, * $P<0.05$; Mann-Whitney U-test, n =10). Error bars are S.E.M.

Thus, extending our previous observations (10) about the LTP prolonging effect of 2'-fucosyllactose in behaving animals the investigation of CA1-LTP in hippocampal slices revealed that in addition to the augmentation of the POP-spike amplitude 2'-fucosyllactose also enhanced fEPSP potentiation. This points to changes in synaptic transmission due to the action of 2'-fucosyllactose but not the corresponding isomer 3-fucosyllactose, thus, excluding unspecific osmotic effects etc. At present there is no clear explanation for both effects of 2'-fucosyllactose-treatment, the strong initial increase of POP-spike resp. fEPSP slope and the maintenance of these elevated levels in the late, nondecremental phase of LTP. However, glycoepitopes of cell recognition molecules have been shown to be involved and regulated during brain developement (12,13) and in reorganization processes (12,14). Furthermore, functional synaptic plasticity, i.e. LTP, can be influenced by the interaction of oligomannosidic carbohydrates with NCAM (15). It is tempting to speculate that 2'-fucosyllactose interacts in a similar manner with cell adhesion processes followed by alterations in signal transduction. On the other hand, most neurotransmitter receptors and ion channels are highly glycosylated. At least, 2'-fucosyllactose might interfere with agonist binding or desensitization. For instance, interaction of carbohydrate-binding lectins with the glycan chains of kainate receptors leads to remarkable changes in desensitization and ion permeability (16). It could be speculated that a direct interaction of

2'-fucosyllactose with receptors or cell recognition molecules is responsible for the effects during induction of LTP. For the alterations observed in maintenance of LTP an effect on glycoprotein processing has to be taken into consideration. Investigations are in progress to identify the "target" proteins and intracellular events affected by 2'-fucosyllactose.

ACKNOWLEDGMENTS

This research has been supported by the Federal Government of Germany (BMBF 07NBL06).

REFERENCES

1. Bliss, T.V.P., and Collingridge, G.L., 1993, *Nature* 361: 31–39.
2. Matthies, H., Frey, U., Reymann, K., Krug, M., Jork, R., and Schroeder, H., 1990, *Exitatory Amino Acids and Neuronal Plasticity* (Ben-Ari, ed.), Plenum Press, New York, pp. 359–368.
3. Reymann, K.G., and Staak, S., 1994, *Proteinkinase C in the CNS - focus on neuronal plasticity* (P.L. Canonico, U. Scapagnini, F. Pamparana and A. Routtenberg eds.) Masson, Milano, Parigi, Barcellona, pp. 31–56.
4. Frey, U., Krug, M., Reymann, K.G., and Matthies, H., 1988, *Brain Res.* 452: 301–313.
5. Hosokawa, T., Rusakov, D.A., Bliss, T.V.P., and Fine, A., 1995, *J. Neurosci.* 15: 5560–5573.
6. Angenstein, F., Matthies, H., Staeck, S., Reymann, K.G., and Staak, S., 1992, *Neurochem. Int.* 21: 403–408.
7. Rose, S.P.R., and Jork, R., 1987, *Behav. Neural Biol.* 48: 246–258.
8. Jork, R., Grecksch, G., and Matthies, H., 1986, *Pharm. Biochem. Behav.* 25: 1137–1144.
9. Krug, M., Jork, R., Reymann, K., Wagner, M., and Matthies H., 1991, *Brain Res.* 540: 237–242.
10. Krug, M., Wagner, M., Staak, S., and Smalla, K.-H., 1994, *Brain Res.* 643: 130–135.
11. Reymann, K.G., Malisch, R., Schulzeck, K., Brödemann, R., Ott, T., and Matthies, H., 1985, *Brain Res. Bull.* 155: 249–255.
12. Schachner, M., and Martini, R., 1995, *TINS* 18: 183–191.
13. Mai, K. J., and Schönlau, C, 1992, *Histochem. J.* 24: 878–889.
14. Jork, R., Potter, J., Bullock, S., Grecksch, G., Matthies, H., and Rose, S.P.R., 1989, *Neurosci. Res. Comm.* 5: 105–110.
15. Lüthi, A., Laurent, J.-P., Figurov, A., Muller, D., and Schachner, M., 1994, *Nature* 372: 777–779.
16. Huettner, J.E., 1990, *Neuron* 5: 255–266.

GLYCOCONJUGATES AND NUCLEAR MEMBRANE LECTIN FROM RAT BRAIN CELL NUCLEI

N. G. Aleksidze, R. G. Akhalkatsi, and T. Bolotashvili

Tbilisi State University
Department of Biochemistry and Biotechnology
Tbilisi, Republic of Georgia

1. INTRODUCTION

At present, many scientists are interested in the nucleo-cytoplasmic interrelation, but unfortunately many questions relating to this problem are still unsolved and await an answer. After the discovery of lectin-binding glycoconjugates and carbohydrate specific lectins, extensive publications have appeared in literature on the role of glycoconjugates and lectins in the regulation of DNA-, and RNA-polymerases, protein synthesis, and nucleo-cytoplasmic exchanges through nuclear pores etc (1–3). It is necessary to emphasize that transcriptional factors (4), some of the nuclear proteins (5,6), poly(A)-polymerase, DNA polymerase, RNA polymerase II and many biologically active moietes are O-glycosylated proteins containing monosaccharides: glucose, galactose, mannose, fuccose, N-acetyl-D-glucosamine, N-acetyl-D-galactosamine etc., easily accessible for some carbohydrate-specific plant lectins (3,7,8).

It should be borne in mind that most of the experiments have been mainly carried out on the lungs, liver, other organs and tissues. An attempt has been made to investigate nuclear glycoconjugates and specific lectins of rat brain cells.

2. MATERIALS AND METHODS

For all experiments male white Wistar linear rats, weighing 100–120g were used and the brain cell nuclear suspension was prepared by the method of Chauveau (9). The nuclear protein fractions were prepared as described earlier (10). The pellet, containing purified nuclei, was resuspended in PBS-A (20mM KH_2PO_4, 0.9% NaCl, 5mM phenylmethylsul-

---

From a colloquium organized by Pam Fredman (Mölndal, Sweden) and Guido Tettamanti (Milano, Italy). (Supported by ARIN, Italy.)

fonylfluoride [PMSF], pH 5.0) in correlation 1 to 3 (v/v), then homogenised and centrifuged for 20 min at 16000g. The supernatant, protein fraction soluble in PBS-A (PF-1), was dialyzed against PBS (20mM KH_2PO_4, 0.9% NaCl, pH 7.4). The pellet, crude nuclear membrane fraction was washed and to the sediment PBS-B (40 mM KH_2PO_4, 0.9% NaCl, 5mM PMSF, 0.1% Triton X-100, pH 7.4), (v/v 1:3) was added. Then it was homogenised and incubated for extraction for 30 min at 4°C. After the centrifugation (16000g/10 min) the supernatant, protein fraction extracted by 0.1% Triton X-100 was marked as PF-2. To the pellet PBS-C (20mM KH_2PO_4, 5mM PMSF, 1.5 mM dithiothreitol, 0.5% Triton X-100, pH 5.0) was added, after which it was homogenized and kept for extraction for one hour at 4°C. The sample was centrifuged (16000g/20 min) and the supernatant (PF-3) was dialyzed against PBS and just as PF-1 and PF-2 it was kept in the refrigerator before it was used. The protein concentration was determined by the method of Lowry et al., (11). Sodium dodecyl sulphate polyacrylamide gel-electrophoresis was carried out according to the method of Laemmli (12) using 5–12% acrylamide gradient on slabs. Gels were calibrated by molecular weight standard protein kit (Sigma). The hemagglutination activity was determined by serial 2-fold dilution of protein fraction in microtiter U-plate with trypsin-treated rabbit erythrocytes (1). The lectin binding ability of glycoconjugates was determined by hapten-inhibitory technique in the agglutination system after the co-preincubation of lectins and glycoconjugates in PBS (13). For technical convenience in all the experiments a plant lectins kit was used containing: *Sambucus nigra* agglutinin (SNA), *Arachis hypogaea* agglutinin (PNA), *Ricinus communi* agglutinin (RCA), Soybean agglutinin (SBA), Concanavalin A (Con A), *Pisum Sativum* agglutinin (PSL), Wheat germ agglutinin (WGA), *Solanum tuberosum* agglutinin (STA), purchased from "Diagnostikum Lvov".

3. RESULTS

The biochemistry, as well as the function of nuclear glycoconjugates and lectins of brain cells are poorly documented. Consequently, first of all, rat brain cell nuclear glycoconjugates, extracted by PBS-A and PBS-B were prepared as described in methods. After the dialysis against the PBS and gel filtration of both protein fractions, PF-1 and PF-2, on Protein PAK 300 SW by high pressure liquid chromatography (Watters) six protein fractions were identified in PBS-A and five in PBS-B, with a wide range of molecular weight from 10 to 180 kDa. None of them showed hemagglutination activity, but some of them selectively displayed plant lectin-binding capacity. It should be pointed out, that proteins, with molecular weights 104 and 38 kDa, extracted by PBS-A, displayed high binding ability to galactose/lactose specific lectin SNA and to D-mannose, D-glucose and N-acetyl-D-glucosamine specific lectin Con A. Substantial evidence was derived from the hapten-inhibitory technique. In particular, it was shown, that co-preincubation of lectins with PF-1 and PF-2 completely blocked hemagglutination of trypsin-treated erythrocytes. In subsequent experiments neither of the proteins extracted by PBS-A and PBS-B had any influence on WGA and STA hemagglutination activity. On the other hand, the hemagglutination activity of PSL specific to D-mannose> D-glucose and N-acetyl-D-glucosamine was slightly inhibited with 104 and 38 kDa molecular weight proteins soluble in PBS-A, but it was inhibited much more strongly by 52 and 18 kDa proteins extracted from rat brain cell nuclear membranes with PBS-B, containing noniogenic detergent Triton X-100 (14).

In current experiments with identification of Con A binding glycoconjugates, gel electrophoresis and following blotting was carried out (15). As lectins have a specific affinity for the sugar moietes of glycoconjugates, radioactive ^{125}I-Con A binding proteins, soluble in PBS-A and PBS-B were controled by the incubation of filters in buffer containing ^{125}I-Con A and its haptens such as methylglucoside and methylmannoside in the concentrations blocking Con A active centers (final concentration 0.2M). According to these experimental results (16) four ^{125}I-Con A binding glycoconjugates soluble in PBS-A with molecular weights 180, 175, 125 and 63 kDa have been discovered, among which the content of protein with molecular weight 180 kDa was the highest. On the other hand, ten ^{125}I-Con A positive band in protein fraction, extracted by 0.1% Triton X-100 solution (PBS-B) from brain cells nuclear membrane has been displayed. On the electrophoregramme three glycoconjugates with molecular weights 180, 70 and 63 kDa were intensively stained, the rest of the proteins with molecular weights 175, 150, 125, 120, 107, 95 and 57 kDa were only slightly stained, like minor proteins. Most recently Gerace et al., (17) showed that Con A binding glycoprotein 180 kDa is localized at or near the nuclear pore in liver cells. Well argumented evidence has been presented by Davis and Blobel (18) on the participation of proteins 63–65 kDa in structural organization of liver nuclear pore. It is well documented that 63–65 kDa protein has a high affinity to WGA and by its presence the nuclear active transport of proteins is completely blocked (19).

It should be noted that in our experiments WGA displayed a slight binding capacity to the protein fraction extracted by PBS-B (10), but completely loses it in relation to separated bands (14). Moreover, the separated brain cell nuclei had no affinity to WGA either, but were agglutinated by Con A (13).

Proceeding form the above-mentioned we assume the existence of Con A like lectin in brain cell nuclei and undertake further studies for the separation of lectin-like proteins. In special experiments the trypsin-treated, glutaraldehyde fixed erythrocyte column was used for the affinity chromatography. One ml of protein (4mg of protein) PF-3 fraction was added to the column (2 × 7cm) and after the saturation of erythrocytes, the column was washed until no further protein (A=280) was leaving it. Specifically bound lectin-like proteins were then eluted with 0.25M Glycine-HCl buffer (pH 3.0) as described earlier (20). The collected protein (pH 3 fraction) was dialyzed against PBS, sequentially rechromatographed on Protein Pak 60, Protein Pak 300 SW and was tested on hemagglutination activity. As was shown protein band with molecular weight 12–17 kDa had hemagglutination activity and it was blocked by N-acetyl-D-glucosamine, hence, it was designated as N-acetyl-D-glucosamine specific nuclear membrane lectin (NML). Simultaneously NML activity was blocked by glycoconjugates extracted with PBS-A and PBS-B. The relation constant of lectin to PBS-A and PBS-B protein fraction concentration was 6.6×10^{-2} and 1.6×10^{-2} accordingly. It seems that brain cell nuclear membrane glycoconjugates and carioplasmic soluble proteins are characterised by a high affinity to NML(20). On the basis of the current data and the above mentioned results it is concluded, that glycoconjugates with molecular weights 104 and 38 kDa (PF-1) as well as protein fraction PF-2 and NML lectin may play an important role in the nucleo-cytoplasmic relationship, particularly in regulating the enzymic action and in nuclear pore activity. The experiments aimed at the realization of this hypothesis are in progress.

REFERENCES

1. Lutsyk, M.D., Panasjuk, E.N., and Lutsyk, A.D., 1981, *The Lectins*, Lvov.
2. Hubert, J., Seve, A.P., Facy, P., and Monsigny R., 1989, *Cell Differentiation and Development*, 27: 69–81.
3. Hubert, J., and Seve, A.P. 1994, *Lectins: Biology, Biochemistry, Clinical Biochemistry*, Volume 10 (Van Drieshe, E., Fisher, J., Beekmans, S., and Big-Hansen, T.C., eds), Hellerup, (Denmark), PP. 220–226.
4. Jackson, S.P., and Tjian, R., 1988, *Cell*, 58: 125–133.

5. Reeves, R., Chang, D., and Chung, S.C., 1981, *Proc. Natl. Acad. Sci. USA*, **78**: 6704–6708.
6. Reeves, R., and Chang,D., 1983, *J. Biol. Chem.*, **257**: 679–687.
7. Kurl, R.N., Holmes, S.C., Verney, E., and Sidransky, S., 1988, *Biochemistry*, **27**: 8974–8980.
8. Bhattacharya, P., and Basu, S., 1982, *Proc. Natl. Acad. Sci. USA*, **79**: 1488–1491.
9. Chauveau, J., Moule, J., and Rouiller, Ch., 1956, *Exp.Cell Research*, **11**: 317–321.
10. Akhalkatsi, R., Bolotashvili, T., Alexidze, G., and Aleksidze, N., 1996, *Bull. Acad. Sci. Georgia,* **153**: 277–279.
11. Lowry, O.H., Rosebrough, H.J., Farr, A.L., and Randell, R.J., 1951, *J. Biol. Chem.*, **193**: 265–285.
12. Laemmli, U.K.,1970, *Nature (London)*, **227**: 680–685.
13. Akhalkatsi, R., Bolotashvili, T., and Aleksidze, N., 1996, *Bull. Acad. Sci. Georgia*, **153**: 109–112.
14. Akhalkatsi, R., Bolotashvili, T., Alexidze, G., and Aleksidze, N., 1996, *Bull. Acad. Sci. Georgia*, **153**: in press.
15. Burnette, W.N., 1981, *Anal. Biochemistry*, **112**: 195–203.
16. Akhalkatsi, R., Bolotashvili, T., Solomonia, R., and Aleksidze, N., 1996, *Bull. Acad. Sci. Georgia*, **154**: in press.
17. Gerace,L., Ottariano, J., and Kondor-Koch, C., 1982, *J. Cell Biol.*, **95**: 826–837.
18. Davis, L.I., and Blobel, G., 1986, *Cell*, **45**: 699–709.
19. Finlay, D.R., Newmeyer, D.D., Price, T.M., and Forbes, D.J., 1987, *J. Cell Biol.*, **104**: 189–200.
20. Akhalkatsi, R., Bolotashvili, T., and Aleksidze, N., 1996, *Bull. Acad. Sci. Georgia*, **154**: in press.

Section 33: Expression and Metabolism of Glycoconjugates

151

GLYCOSYLTRANSFERASE ACTIVITIES IN 15 HUMAN MENINGIOMAS

E. Sottocornola, I. Colombo, S. Rapelli, and B. Berra

Institute of General Physiology and Biochemistry
School of Pharmacy, Milan, Italy

1. INTRODUCTION

Since the biological significance of glycolipid changes associated with intracellular adhesion, recognition and growth has been demonstrated (1), extensive analysis has been carried out on different tissue samples. Particularly, gangliosides have been reported to be modified in experimental as well as in human tumors and the occurrence of tumor associated gangliosides has been reported for several tumor types (2, 3). An alteration of the activity of the enzymes involved in the metabolism of these molecules (Figure 1) has been suggested as a mechanism for their different expression (4).

The aim of our study was to characterize the glycosphingolipid pattern in 15 human meningiomas and to investigate any possible correlation between these profiles and the glycosyltransferase activities.

Meningiomas are tumors of mesenchymal origin and are considered to be histologically benign neoplasms, but they may display, to a certain extent, atypical or anaplastic features (5). Meningiomas, except for the anaplastic forms, do not invade brain tissue from which they can be separated without any contamination (6). However meningiomas can be contaminated with non-tumors cells, primarily derived from the vascular system, stroma and infiltrative lymphoid cells. The proportion of contaminants, when lower then the 25% of the tumor mass, as in our case, is unable to affect the ganglioside phenotype detected (7).

2. MATERIALS AND METHODS

2.1. Tissues

The 15 tissue specimens were kindly provided by Dr. R. Campanella (Neurosurgery Institute, Ospedale Maggiore di Milano, Milan, Italy); 7 were males and 8 females. The

---

From a colloquium organized by Pam Fredman (Mölndal, Sweden) and Guido Tettamanti (Milano, Italy). (Supported by ARIN, Italy.)

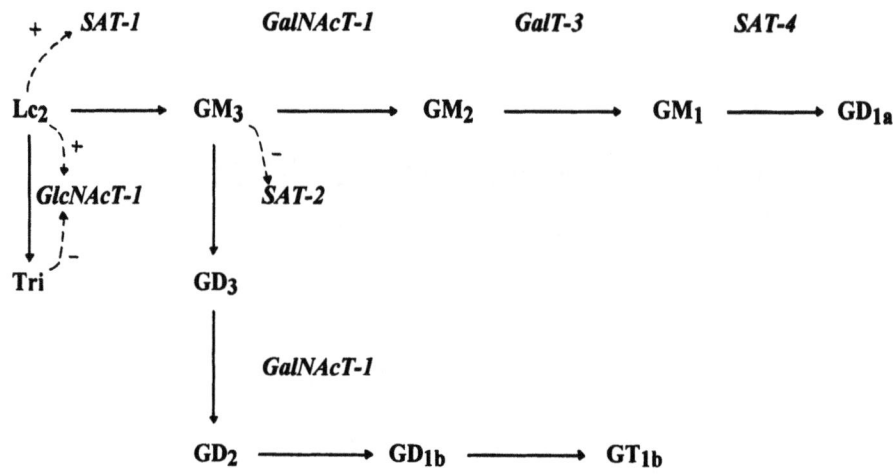

Figure 1. Effect of glycolipid concentration and ratio on SAT-1, SAT-2 and GlcNAcT-1 activities.

tumors were classified according to the WHO histological typing of tumors of the central nervous system: meningotheliomatous (9), fibrous (3), fibroblastic (2) and atypic (1).The age of the patients ranges from 32 to 79 years. Meningioma samples were stored at −20°C after surgery, transported in dry ice to our department and here stored at −80°C until use.

2.2. Extraction and Identification of Glycolipids from Tissues

Gangliosides and neutral glycolipids are extracted and purified according to Berra et al. (8). Lipid bound sialic acid content is determined according to Svennerholm (9). Quantitative analysis is performed by HPTLC (HPTLC Precoated Plates Silica Gel 60, MERCK, Darmstadt, Germany) and scanning densitometry (CAMAG, TLC Scanner II, Muttenz, Switzerland connected to a Shimadzu C-R 3A Chromatopak System, Tokyo, Japan).

2.3. Enzyme Preparation and Assay

The enzyme preparation and assay were performed according to Gornati et al. (10). All the steps during enzyme preparation are performed between 0° and 5°C. Briefly, 100–200 mg of tissue samples were homogenized in 2 ml of SHEM buffer (0.32 M sucrose, 20 mM Hepes pH 7.0, 1 mM EDTA, 0.1% mercaptoethanol) and then centrifuged at 100,000 × g for 1 hour. The pellet is resuspended in 0.5 ml of cacodylate HCl buffer (20 mM sodium cacodylate, 1 mM EDTA, 5% glycerol, 0.1% mercaptoethanol, 0.15 mM NaCl, 25 mM $MgCl_2$, 5 mM benzamidine), pH 6.4. The standard conditions used for the assays are summarized in Table 1.

The total protein amount was measured according to the Lowry method (11).

2.4. Reagents

All reagents used are of analytical grade (BDH Laboratory Supplies Poole, England).
Acceptors for glycosyltransferase assays (GM_3, GD_3, Lc_2, GM_1, αGM_1) were from Inalco S.p.a. (Milan, Italy). Radioactive donors for glycosyltransferases, CMP-[^{14}C]NeuAc, UDP-[^{14}C]GalNAc and UDP-[^{14}C]GlcNAc were purchased from Dupont NEN Products

Table 1. Standard conditions for the glycosyltransferase assays

SYNTHASE	SAT-1	SAT-2	SAT-4	GalNAcT-1	GlcNAcT-1
Acceptor	Lac-Cer	GM3	GM1 or aGM1	GM3 or GD3	Lac-Cer
mM	0,4	0,4	0,4	0,4	0,4
Donor	CMP-[^{14}C]NeuAc	CMP-[^{14}C]NeuAc	CMP-[^{14}C]NeuAc	UDP-[^{14}C]GalNAc	UDP-[^{14}C]GlcNAc
mM	0,4	0,4	0,4	0,4	0,4
cpm/nmol	3500	3500	3500	5000	3500
Metal ion	MgCl2	MgCl2	MgCl2	MnCl2	MnCl2
mM	5	5	5	10	25
Detergent CF54 mg	50	50	100	200	50
Protein, mg	100	100	100	100	100
Cacodylate buffer pH 6,4 mM	200	200	200		
Hepes buffer pH 7,6-8 mM				200	200
Reactive volume ml	50	50	50	50	50
Reactive time, h	1	1	1	1	1

(Boston, MA, USA) and nonradioactive donors, CMP-NeuAc, UDP-GalNAc and UDP-GlcNAc were from Sigma Chemical Co. (St. Louis, MO, USA).

3. RESULTS AND DISCUSSION

In Table 2 we report the glycosphingolipid composition and the total lipid bound sialic acid content of the 15 meningiomas analysed. According to a previous paper (3) meningiomas are divided in three groups: "GM_3 rich" (10) which comprises tumors where the GD_3 content is less than 25% of the GM_3 content; "GD_3 rich" (4), which includes tumors where the GD_3 content is greater than 60% of the GM_3 content. One tumor, in which the GD_3 content is from 30 to 60% of the GM_3 content, is classified as "GM_3-GD_3 rich" (1). We found no correlation between this classification and the histopathological features and localization of the tumors and the age and sex of the patients.

Although the amount of total sialic acid is fairly different, GM_3, GD_3 and GD_{1a} are the main ganglioside species and are present in all the samples. Only in few tissues, and in lower amounts, we found other gangliosides, such as GM_2, GM_1, GD_2, GD_{1b} and GT_{1b}. This agrees with many previous data that underline a shift of the ganglioside profile from polysialylated to mono or disialylated species (3, 8).

The enzymes and their activity values are reported in Table 3. The picture outlined is quite heterogeneous and complex. However, very interesting observations can be obtained comparing Table 2 and 3: the activity of Glucose-N-Acetyl Transferase-1 (GlcNAcT-1), that catalyzes the synthesis of Trihexosylceramide (Tri) from Lactosylceramide (LacCer), decreases when LacCer/Tri ratio increases. Particularly, the enzyme activity rises following an increase of the LacCer content; the opposing effect is caused by the increase in the Tri content. Reciprocal contents of LacCer and Tri seem to modulate as activator and inhibitor, respectively, the activity of GlcNAcT-1.

Table 2. Sialic acid content and glycosphingolipids composition of the meningiomas

Samples	sialic acid	Lac-Cer	Tri	GM$_3$	GM$_2$	GM$_1$	GD$_3$	GD$_{1a}$	GD$_2$	GD$_{1b}$	GT$_{1b}$
GM$_3$ rich											
1	59	32,00	15,00	201,94			5,00	14,46			
A	94	71,00	36,00	353,19			41,30	14,37			
B	111	66,00	48,00	384,59			13,83	17,69			
C	53	43,00	26,00	158,05			6,69	29,14			
D	55	55,00	18,00	182,27			5,69	16,94			
F	73	16,89	4,44	216,08	3,02	6,79	15,10	17,67	3,17		
G	29	73,64	3,06	91,18	3,28	4,55	2,50	4,27			
L	37	85,42	2,78	93,74	6,09	21,61	5,18	11,14			
P	65	73,57	3,02	227,68		7,84	7,01	6,52			
R	44	11,11	2,81	141,18			11,72	8,25			
GD$_3$ rich											
E	35	21,00	12,00	10,57			43,34	3,42	3,37	12,86	20,00
H	73	8,33	4,55	84,96		8,15	106,14	7,76	3,51		
N	44	10,00	0,55	5,74	2,71	13,13	22,64	6,30	10,67	63,07	10,04
Q	35		0,16	3,86			20,44	9,42	13,34	31,14	18,19
GM$_3$/GD$_3$ rich											
M	91	110,19	14,44	59,96	3,02	15,66	30,64	11,90	3,24		2,09

The values indicate the ng total lipid bound sialic acid and the ng glycosphingolipid species/mg tissue

Sialic Acid Transferase-1 (SAT-1) activity presents the opposite behaviour: an increase of the LacCer/GM$_3$ (substrate and product of the reaction, respectively) ratio seems to be related with the raise of the enzyme activity and this positive trend is sustained by increasing LacCer amount. Under the same concentrations of endogenous GM$_3$ a lower availability of LacCer (that is to say, lower LacCer/GM$_3$ ratio values) is followed by a weak SAT-1 activity, whereas an higher availability of LacCer (higher LacCer/GM$_3$ ratio

Table 3. Glycosyltransferase activities in the meningiomas

Samples	SAT-1	SAT-2	SAT-4 (aGM$_1$)	SAT-4 (GM$_1$)	GalNAcT-1 (GM$_3$)	GalNAcT-1 (GD$_3$)	GlcNAcT-1
GM$_3$ rich							
1	310,6	51,9	198,1	48,8	33,7		
A	221,5		36,2				10,8
B	153,4		62,1		28,2		19,4
C	184,6		146,2			n.d.	7,7
D	255,8	87,5	339,2	35,8	22,5	20,0	
F	13,0	4,0	4,0	28,0		22,0	
G	406,0		150,0		10,0		
L	289,0		44,0				51,0
P	320,0						12,0
R	47,0	8,0	39,0	9,0	469,0		
GD$_3$ rich							
E	16,2	19,1	38,6	14,9	52,3		3,7
H	18,0	7,0	44,0		4,0	13,0	
N	489,0	60,0	409,0		16,0	4,0	17,0
Q	704,0	148,0	261,0	9,0	27,0	92,0	61,0
GM$_3$/GD$_3$ rich							
M	221,0		213,0		5,0	4,0	4,0

Glycosyltransferase activities are expressed as pmol/mg tissue*h values) increases the SAT-1 activity. The LacCer availability seems to be the main factor involved in the regulation of SAT-1.

Weaker levels of Sialic Acid Transferase-2 (SAT-2) activity, the enzyme that sinthetizes GD_3 starting from GM_3, are observed for increasing GM_3/GD_3 ratio values. Particularly, the higher amounts of GM_3 seems to be responsible for the descending trend of SAT-2 activity. Increasing GM_3 contents drastically reduce the enzyme activity. GD_3 concentrations appear to have no effect on the same enzyme. We tried to exemplify this picture in Figure 1.

In conclusion, our data suggest that some glycolipids species might have a possible role, as activator or inhibitor, in the regulation of the enzyme involved in their metabolism.

Unfortunately we could not go deeper into the study of the relationships between the glycosphingolipids pattern and the glycosyltransferase activities by analysing the mRNA of the 15 meningiomas. Probably due to inadequate storage of the tissue samples after surgery, the total RNA, extracted from the tumors with standard methods (12), was degraded and useless for analysis.

ACKNOWLEDGMENTS

We would like to aknowledge Dr. R. Campanella for supplying the 15 human meningioma specimens.

This work was supported partially by MURST and Ragione Lombardia 1441 grants.

REFERENCES

1. Hakomori, S., 1993, *Biochem. Soc. Transaction* 21: 583–595.
2. Hakomori, S., 1985, *Cancer Res.* 45: 2405–2414.
3. Shinoura, N., Dohi, T., Kondo, T., Yoshioka, M., Takakura, K. and Oshima, M., 1992, *Neurosurgery* 31: 541–549.
4. van Echten, G. and Sandhoff, K., 1993, *J.Biol.Chem.* 268: 5341–5344.
5. Lekanne Deprez, R.H., Riegman, P.H., van Brunen, E., Warringa, U.L., Groen, N.A., Stefanko, S.Z., Koper, J.W., Avezaat, C.J.J., Mulder, P.G.H., Zwarthoff, E.C. and Hagemeijer, A., 1995, *J. Neuropath. Exp. Neurol.* 54: 224–235.
6. Davidsson, P., Fredman, P., Collins, P., von Holst, H., Mansson, J.E. and Svennerholm, L., 1989, *J. Neurochem.* 53: 705–709.
7. Wikstrand, C.J., Fredman, P., McLendon, R.R., Svennerholm, L. and Bigner, D.D., 1994, *Molec. Chem. Neuropath.* 21: 129–138.
8. Berra, B., Papi, L., Bigozzi, U., Serino, D., Morichi, R., Mennonna, P., Rapelli, S., Cogliati, T. and Montali, E., 1991, *Int. J. Cancer* 47: 329–333.
9. Svennerholm, L. and Fredman, P., 1980, *Biochem. Biophys. Acta* 617: 97–109.
10. Gornati, R., Basu, S., Montorfano, G. and Berra, B., 1995, *Cancer Biochem. Biophys.* 15: 11–10.
11. Lowry, O.H., Rosebrogh, N.J., Farr, A.L. and Randall, R.J., 1951, *J.Biol.Chem.* 193: 265–275.
12. Chomczynski, P. and Sacchi, N., 1987, *Analyt. Biochem.* 162: 156–159.

ADMINISTRATION OF GANGLIOSIDES CHANGES THE PROPERTIES OF ADENYLATE CYCLASE SYSTEM IN THE SENSORIMOTOR AND LIMBIC STRUCTURES OF RAT BRAIN

N. Nalivaeva, S. Plesneva, U. Chekulaeva, N. Dubrovskaya, and I. Zhuravin

Institute of Evolutionary Physiology and Biochemistry
RAS
St. Petersburg, Russia

1. INTRODUCTION

cAMP is a universal secondary messenger participating in the transduction of a great number of extracellular regulatory signals. It is e.g. induced by neurotransmitters, neuromodulators, neurohormones. (7) It appears that a number of physiological actions of certain neurotransmitters absolutely require alterations in cAMP level. The level of intracellular cAMP to a large extent depends on the activity of an enzyme adenylate cyclase. In nervous tissue adenylate cyclase is present in all cell types including neurons, glia and endothelial cells and is particularly enriched in synaptic membrane fractions. Neurotransmitter regulated changes in adenylate cyclase activity require the specific binding of the neurotransmitters to its membrane receptors (2). The binding of neurotransmitters to certain receptors (e.g. β-adrenergic, type 1 dopaminergic, type 2 histaminergic) results in the stimulation of adenylate cyclase, while binding to other receptors (e.g. m2 and m4 muscarinic, α_2-adrenergic, type 2 dopaminergic) results in the inhibition of the enzyme activity. In fact, the same neurotransmitter can have a different effect on cAMP metabolism in different cells and brain structures depending upon the number and ratio of receptor subtypes present for that particular neurotransmitter (2,4).

As was shown in our previous study (13,14) different brain structures demonstrate different levels of AC activity which to a large extent also depends on the physiological state of the organism. Thus we have found that the activity of adenylate cyclase decreases in the sensorimotor and limbic structures of rats learned to perform an instrumental reflex

---

From a colloquium organized by Pam Fredman (Mölndal, Sweden) and Guido Tettamanti (Milano, Italy). (Supported by ARIN, Italy).

(reached by pushing). The decrease in the level of adenylate cyclase activity was found to be dependent on the ability of rats to learn the instrumental reflex. I.p. injections of gangliosides to rats were shown to facilitate learning and memory of the reflex (12) as well as to decrease the threshold of the sensitivity of the striatal neurons to the effect of agonists of cholinergic system (15). The reasons and mechanisms of the alteration of adenylate cyclase activity during the formation of the instrumental reflex as well as the role of gangliosides in these processes are far from clear.

The aim of the present work was to study the effect of i.p. ganglioside injections on the activity and regulatory properties of adenylate cyclase system in the sensorimotor and limbic structures of the rat brain participating in the control and realization of motor behavior.

2. MATERIAL AND METHODS

2.1. Gangliosides

Gangliosides were extracted from bovine brain available at a slaughterhouse by the method of Suzuki (10). The content of N-acetylneuraminic acid in lyophilized total bovine brain ganglioside powder was 28%.

2.2. Animals

The experiments were performed on adult Wistar male rats weighing 150–200 g. Animals were housed, 4 individuals per cage, under vivarium conditions of constant temperature, normal 12h light-dark cycle, and freely available water. Gangliosides (30 mg/kg of body weight) were i.p. injected into rats in 0.2 ml of saline solution 6 times during two weeks. Control rats (n=12) were injected with the saline solution only. Ganglioside injected animals were taken for analysis 2 weeks (n=6) and 4 weeks (n=6) after the beginning of injections.

2.3. Isolation of P_2 Fractions

Rats of control and experimental groups were decapitated and the cortex, striatum, amygdala and hippocamp were cut from both brain hemispheres. All procedures were performed at 4°C. The tissue fragments were homogenized in 0.32 M sucrose–0.04 M tris-HCL (pH 7.4) and centrifuged at 1,000 g for 10 min. The supernatants were then centrifuged at 12,500 g for 20 min, the pellet being regarded as P_2 membrane fraction. The pellets obtained finally were suspended in 0.04 M tris-HCL (pH 7.4). The aliquots of P_2 fractions from Cx, Str, Am and Hip of experimental and control rats were used for AC assay.

2.4. AC Assay

The activity of adenylate cyclase was analyzed by the method of Salomon et al (9), in the presence of an inhibitor of phosphodiesterases IBMX in incubation media (50 μl) containing 50 mM tris-HCl (pH 7.5), 5 mM $MgCl_2$, 1 mM cAMP, 20 mM creatine phosphate, 1 mg/ml creatinphosphokinase, 0.1 mM ATP, α-[^{32}P]ATP (0.5–1μCi) for 10 min at 37°C. Reaction was initiated by addition of an aliquot of P_2 fraction of studied brain structure (10–15 μg protein). The reaction products were separated by chromatography on neutral aluminum oxide according to the method of White (11). Regulatory properties of adenylate cyclase were studied in the presence of such agents as Gpp[NH]p (10^{-6} M), forskolin (10^{-6} M) and NaF (10^{-2} M) in the incubation media.

3. RESULTS AND DISCUSSION

In the absence of any effectors we have found that limbic structures of rat brain (amygdala and hippocamp) demonstrate significantly higher level of basal activity of adenylate cyclase than sensorimotor structures (cortex and striatum) (Table 1). I.p. injections of gangliosides led to a prolonged decrease in the level of basal adenylate cyclase activity in all studied structures. Even 4 weeks after ganglioside injections the activity of the enzyme remained low in all brain structures.

Analyzing the effect of non-hormonal agents on the activity of adenylate cyclase in the absence of gangliosides we have revealed differences in the regulatory properties of the enzyme between studied brain structures (Figure, control). Thus, in the cortex and striatum adenylate cyclase was regulated by Gpp[NH]p, forskolin and NaF in the traditional manner (2,5) demonstrating an increase in its activity in the presence of the effectors. In limbic structures, on the other hand, these effectors, in general, had no stimulatory influence on the activity of the enzyme while in hippocamp a stimulatory effect of NaF was observed.

I.p. injections of gangliosides do not significantly change the character of the effects of non-hormonal agents on the activity of adenylate cyclase in the cortex (2 weeks) and striatum (2 and 4 weeks) while leading to an alteration in the response of the enzyme to the effects of Gpp[NH]p and forskolin in the cortex 4 weeks after ganglioside injections (Figure 1, 2 weeks, 4 weeks). In the amygdala and hippocamp as well as in the striatum of the injected rats the enzyme becomes more sensitive to the stimulating effect of NaF. Moreover, in the limbic structures gangliosides also initiate the stimulating effect of forskolin which was not observed in control animals. In the hippocamp ganglioside injections were also found to potentiate the effect of Gpp[NH]p on adenylate cyclase activity. In general the data obtained demonstrate that ganglioside injections lead mainly to modulation of the properties of NaF- and Gpp[NH]p-binding sites of G-protein coupled to adenylate cyclase which results in the increase in the catalytic properties of the enzyme. A slight increase in the effect of forskolin on adenylate cyclase activity justifies the suggestion that ganglioside injections affect the low-affinity interaction of the agent with the catalytic subunit of the enzyme (1).

The data obtained give additional evidence that neuritogenic and neurotrophic effects of gangliosides as well as their modulatory effects on functional characteristics and

Table 1. Activity of adenylate cyclase (pmoles cAMP/mg protein/min) in sensorimotor and limbic brain structures of control and injected with gangliosides rats. Each value is the mean±SEM

Brain structures	Control	2 weeks	4 weeks
Cortex	95.47±4.11	55.37±2.93	87.89±3.35
Striatum	52.92±2.51	28.66±1.12	42.49±3.18
Amygdala	172.16±11,15	72.94±5.41	82.79±5.97
Hippocamp	124.92±10.76	64.34±3.55	60.37±4.87

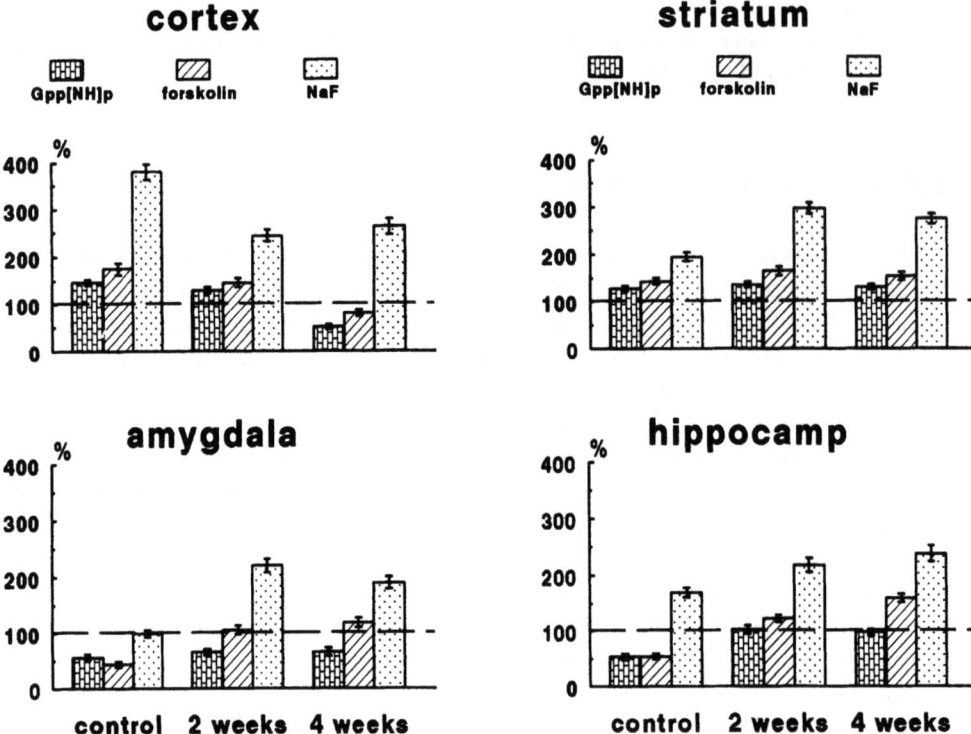

Figure 1. Effect of Gpp[NH]p, forskolin and NaF on basal adenylate cyclase activity in the sensorimotor and limbic structures of control and ganglioside injected rats (2 and 4 weeks after injections). Each value is the mean±SEM. Basal activity of the enzyme in every case is taken as 100% (dotted line).

integrative properties (6,8,12) of different brain structures might be related to the ability of these glycosphingolipids to modify regulatory properties of cAMP producing system.

ACKNOWLEDGMENT

Supported by the RBRF grant No. 96–04–50748.

REFERENCES

1. Birnbaumer, L., Abramowids, J., and Brown, A.M., 1990, *Biochim. Biophys. Acta* **1031**: 163–224.
2. Bonnet, K., 1982, *Handbook on Neurochemistry*, Volume 1 (A. Lajtha, ed.), Plenum Press, New York, pp. 257–280.
3. Dokas, L.A., and Ting, S.-M., 1993, *Neurobiology of aging* **14**: 65–72.
4. Graybiel, A.M., and Ragsdale, C.W., 1983, *Chemical Neuroanatomy* (P.C.Empson, ed.) Raven Press, New York, pp. 427–504.
5. Lemmer, B., Carlebach, R., Stiller, M., Ohm, T.G., and Nitsch, R., 1991, *Brain Res.* **565**: 225–230.
6. Partington, C.R., and Daly, J.W., 1979, *Mol. Pharmacol.* **15**: 484–491.
7. Robison, J.A., Butcher, R.W., and Sutherland, E.W., 1971, *Cyclic AMP*, Academic Press, New York.
8. Sabel, B.A., Slavin, M.D., and Stein, D.G., 1984, *Science* **225**: 340–342.
9. Salomon, Y., Londos, C., and Rodbell, M. 1974, *Anal. Biochem.* **58**: 541–548.
10. Suzuki, K., 1965, *J. Neurochem.* **12**: 629–638.

11. White, A.A., 1974, *Methods Enzymol.* **380**: 41–46.
12. Zhuravin, I.A., Nalivaeva, N.N., and Dubrovskaya, N.M., 1993, *Zhurnal Vysshei nervnoj deyatelnosti imeni I.P.Pavlova* (in Russian) **43**: 1129–1136.
13. Zhuravin, I.A., Nalivaeva, N.N., Plesneva, S.A., Dubrovskaya, N.M., Chekulaeva, U.B., and Klementiev, B.I., 1995, *Neurosci and Behav Physiol.* **25**: 117–121.
14. Zhuravin, I.A., Nalivaeva, N.N., Plesneva, S.A., Dubrovskaya, N.M., Chekulaeva, U.B., and Klementyev, B.I., 1995, *Sechenov Physiological Journal* **81**: 40–47.
15. Zhuravin, I., Nalivaeva N., and Dubrovskaya, N., 1995, *European J.Neurosci.* [Suppl.] **8**: 170.

ORGAN-SPECIFIC EXPRESSION OF MEMBRANE LIPIDS, ESPECIALLY OF GANGLIOSIDES, DEPENDING ON THE STATE OF THERMAL ADAPTATION

Hinrich Rahmann and Ute Balshüsemann

University of Hohenheim
Zoological Institute, Stuttgart, Germany

1. INTRODUCTION

As can be followed from various compensatory responses especially of ectothermic (cold-blooded) vertebrates on behavioral, electrophysiological, morphological and metabolic level, the ability of these animals to adapt to fluctuations in their temperature environment is mainly based upon changes within their nervous systems, in which the synapses were found to be the most sensitive substructures of the whole organism. On this basis it has to be emphasized that major functional properties of biological membranes, as for instance fluidity, surface charge, conductivity, depend on the composition of membranous lipids, which, however, were shown to vary considerably not only among the different organs, but also with the state of thermal adaptation of the organism.

With regard to this among the different classes of lipids which compose the membrane to its majority (phospholipids, cholesterol, neutral glycosphingolipids) there is one minor but very peculiar class of negatively charged glycosphingolipids, which over the very last decades has come more and more into the center of scientific interest. These gangliosides called lipids constitute on the one side only about 1% of all lipids in general of all body organs, however, on the other side in the nerve fiber terminals (synapses) they are being accumulated up to 10%. In the brain of ectothermic vertebrates the functional headgroup of gangliosides, which is composed of different numbers (1 to 6) and of O- or N-acetylated and negatively charged sialic acids, was found to vary dramatically with changing temperature according to the general rule "the lower the temperature, the higher the polarity (degree of sialylation) of brain gangliosides" (1).

This kind of temperature-dependent changes in the polarity of neuronal gangliosides can be correlated with alterations in the ultrastructural organization of the synaptic termi-

---

From a colloquium organized by Pam Fredman (Mölndal, Sweden) and Guido Tettamanti (Milano, Italy). (Supported by ARIN, Italy).

nals, with accumulations of glycogen in the synaptoplasm and especially with changes in the content of extracellular calcium within the synaptic cleft (2). In this context a functional temperature-dependent model on Ca^{2+}-ganglioside interactions had been developed explaining the phenomenon that in ectothermic vertebrates the neuronal processes of conduction and especially of transmission of impulses in synapses can occur in a similar way as in warm-blooded vertebrates (birds, mammals) although the ambient temperature had changed in the range of some 20 to 30 degrees, thus enabling homeoviscosity of the lipid micro-environment domain of functional membrane proteins (3).

Since this kind of adaptive molecular mechanism (changes in the degree of sialylation of gangliosides and of extracellular calcium in the synaptic cleft) had been demonstrated only for the brain, the present study was carried out to investigate comparatively whether or not a similar temperature-dependent sialylation reactivity might occur also in other tissues or whether or not also other molecular adaptations exist in the various other membranous lipids.

Therefore different molecular reactivities of membranous phospholipids, cholesterol, neutral glycolipids, and of sialo-glycosphingolipids (gangliosides) of brain, lateral line nerves, muscle and liver from warm (22°C) versus cold (4°C) acclimated crucian carps, and in addition from some antarctic icefish (± 0°C) versus tropic fish (28°C) had been investigated.

2. MATERIALS AND METHODS

Eurythermic european crucian carps (*Carassius carassius*) were held under standardized laboratory conditions (12h/12h, day/night-cycle; full diet), and were acclimated for 6 weeks to 4°C and 20°C, respectively. The fish were killed by ice-water and decapitation.

Tissues (brain, lateral line nerve, muscle and liver) were homogenized and successively extracted with 10 volumes of varying composition of chloroform and methanol (1:2, 1:1 and 2:1) (4). The combined extracts were partitioned with chloroform/methanol/water (4:2:1) and lower phases were washed (4:1:1). Phases were dried; gangliosides in the upper phase were separated from phospholipids by DEAE-Sephadex column chromatography. Lower phases were separated by Silica column chromatography in neutral lipids with chloroform, neutral glycosphingolipids with chloroform/methanol (3:1) and in phospholipids with methanol.

Phospholipids were quantified by the method of Fiske-Subarow and cholesterol by means of an enzymatic kit (Sigma No. 352). Neutral glycosphingolipids were determined by the content of hexose (5) and ganglioside-quantification was done by the method of Svennerholm (6). Individual lipids were determined by high-performance thin-layer chromatography. The following solvent systems were used: methylacetate/n-propanol/chloroform/methanol/0,25% $CaCl_2$ (30:25:25:4) for phospholipids; chloroform/methanol/12 mM $MgCl_2$/33% NH_3 (61:38:6,6:2,7; first run) and chloroform/methanol/12 mM $MgCl_2$ (55:40:9; second run) for gangliosides. Lipids were visualized with phosphomolybdic acid (phospholipids) or with resorcinol/HCl (gangliosides). The fatty acid analysis was performed by gas chromatography. Extracted lipids were hydrolyzed with 1N methanolic HCl (20 min.; 80°C), dried and separated with chloroform/propanol/water (8:4:3). Fatty acids were derivatized with DMF and DMA (10 min.; 60°C). Gas chromatography was carried out with a DB 1701 column (temperature gradients: 60–80°C with 4°C/min.; 80–220°C with 2°C/min.; F/D-injector 260°C; carrier gas: H_2).

3. RESULTS AND DISCUSSION

Total lipid content differed considerably among the investigated organs (Fig.1): The content of cholesterol and of neutral glycosphingolipids was remarkably higher in neuronal tissues than in muscle and liver. The relative proportion of gangliosides in all organs was less than 1%. Varying environmental temperature caused different changes in the lipid content and composition in the various fish organs: While the total lipid content in brain and muscle appeared unaffected by temperature, a decrease of total lipids in lateral line nerve and liver of cold acclimated crucian carps was found. Previous investigations of various authors indicated similar results (7,8). The relative composition of total lipids was influenced in all investigated organs in a different way: While in muscle and liver, a decrease of phospholipids was found, in brain (4,7 to 5,3 μmol NeuAc/g dry weight) and lateral line nerve (2,2 to 2,7 μmol NeuAc/g dry weight) the content of gangliosides increased significantly in cold adapted fishes (Fig.2).

Additionally to these quantitative changes lowered temperatures altered the composition of ganglioside, which switched to more polar fractions in brain as had been followed from the quantitative evaluation of HPTLC-chromatograms (Fig.2). These data, according to which the brain of cold-adapted fishes contains more tri- and polysialylated gangliosides than that of warm adapted ones confirm previous investigations (9,10,11,12). In contrast to this however the headgroups of gangliosides in lateral line nerve, muscle and liver were not influenced by temperature changes. Comparing additionally the ganglioside composition of various organs of closely related perciforme fishes, which are genetically adapted to extreme temperature habitats of the earth like tropics, moderate climate and Antarctica the same trends became obvious (data not shown here).

Contrary to preliminary investigations the ratio of saturated versus unsaturated fatty acids of gangliosides remained unchanged in all examined organs. In contrast to these sialo-glycosphingolipids, in neutral glycosphingolipids fatty acids from brain and muscle, but not from liver reacted sensitively concerning the degree of saturation, but not concerning the chain length of their fatty acids.

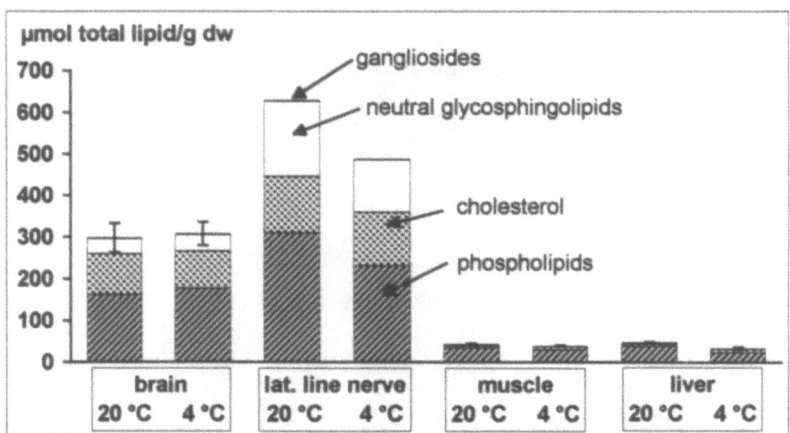

Figure 1. Influence of temperature changes on lipid content and composition of brain lateral line nerve, muscle and liver of crucian carps. In case of lateral line nerve there was only one pooled sample containing the nerves of 18 fishes.

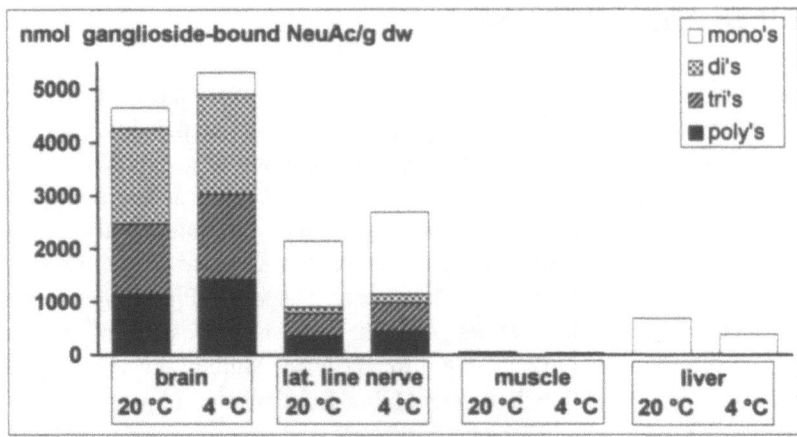

Figure 2. Influence of temperature changes on ganglioside content and composition of brain, lateral line nerve, muscle and liver of crucian carps. Ganglioside-composition was evaluated by densitometric quantification of HPTLC-chromatograms and summarized in fractions containing 1 (= mono's) to 6 (= poly's) sialic acid residues. In case of lateral line nerve there was only one pooled sample containing the nerves of 18 fishes.

The data speak in favour of the idea, that the individual cellular functions of the various organs are being guaranteed by specific adaptive molecular mechanisms.

4. CONCLUSIONS

In membranes of all investigated tissues temperature changes induce adaptive molecular processes on the basis of fatty acid adaptation (degree of saturation) within the lipid bilayer. In excitable membranes (esp. brain cells), however, additional molecular adaptive processes occur on the basis of alterations of the polar head group of complex sialo-glycosphingolipids (gangliosides) thus enabling a homeoviscosity of the micro-environment of functional proteins, although the environmental temperature has changed.

ACKNOWLEDGMENTS

We want to thank Dr. H. Menges and Dr. Opper from the institute for physiological chemistry (Prof. Dr. Wesemann, University of Marburg) for fatty acid analysis.

REFERENCES

1. Rahmann, H., 1992, *Curr. Aspects Neurosci.* **4**: 87–125.
2. Rahmann, H., and Körtje, K.H., 1993, *J. Hirnforsch.* **34**: 351–359.
3. Rahmann, H., 1995, *Behav. Brain Res.* **66**: 105–116.
4. Svennerholm, L., and Fredman, P., 1980, *Biochim. Biophys. Acta* **617**: 97–109.
5. Svennerholm, L., 1956, *J. Neurochem.* **1**: 42–53.
6. Svennerholm, L., 1957, *Biochim. Biophys. Acta* **24**: 604–611.
7. Hazel, J.R., 1979, *Am. J. Physiol.* **236**: R91-R101.
8. Rafael, J., and Braunbeck, T., 1988, *Fish Physiol. Biochem.* **5**: 9–19.
9. Rahmann, H., Hilbig, R., Probst, W., and Mühleisen, M., 1983, *J. therm. Biol.* **8**: 107–109.
10. Breer, H., and Rahmann, H., 1976, *J. therm. Biol.* **1**: 233–235.
11. Hilbig, R., Rahmann, H., and Rösner, H., 1979, *J. therm. Biol.* **4**: 29–34.
12. Becker, K., and Rahmann, H., 1995, *Comp. Biochem. Physiol.* **111B**: 299–310.

N-LINKED OLIGOSACCHARIDE EXPRESSION IN WHOLE TISSUE AND SPECIFIC GLYCOPROTEINS OF RODENT CNS

D. R. Wing, Y-J. Chen, R. A. C. Clark, R. A. Dwek, and S. E. Zamze

Glycobiology Institute
Department of Biochemistry
University of Oxford
Oxford, United Kingdom

1. INTRODUCTION

There is increasing evidence that the carbohydrates of brain-tissue glycoproteins play key recognition roles in the CNS. In particular, oligosaccharides expressed on the cell surface and extracellular matrix may participate in cell-cell and cell-substratum interactions. This is exemplified by (i) the developmentally-regulated α-2,8-polysialylation of N-CAM, which impedes the homophilic binding of the cell adhesion molecule (1), and (ii) the L2/HNK-1 carbohydrate epitope which is expressed on a variety of neural glycoproteins, all of which show cell-adhesive properties (2). Not all glycoforms of a particular glycoprotein simultaneously express L2/HNK-1 carbohydrate and the epitope itself may be developmentally regulated independent of the protein backbone (3). Expression of this carbohydrate therefore occurs at particular times and sites during development.

A knowledge of oligosaccharide structures is clearly essential if the molecular mechanisms underlying recognition are to be understood. This has been illustrated outside the nervous system by the selectin family recognition of the sialyl Lewis x carbohydrate determinant, sugar-lectin interactions that are shown to be involved in leukocyte migration in inflammation and injury (4).

In parallel studies, protein glycan expression has been studied in both whole brain tissue from the rodent (rat) and in diverse neural glycoproteins — namely tenascin-R (from mouse), found in the extracellular matrix (5), and gp110 and gp116 (from rat), glycoproteins of the post-synaptic membrane (6).

Tenascin-R is located exclusively in the CNS, has 13 potential N-glycosylation sites and is L2/HNK-1 reactive. *In vitro* it is known to modulate interactions between cells and

---

From a colloquium organized by Pam Fredman (Mölndal, Sweden) and Guido Tettamanti (Milano, Italy). (Supported by ARIN, Italy.)

their surroundings. In contrast, glycoproteins of gp 110, gp 116 (and gp 180) are highly concentrated in and probably uniquely associated with the post-synaptic apparatus where events are mediated following neurotransmitter binding.

2. METHODS

Tenascin-R was immunoaffinity purified from adult mouse brain in the laboratory of Prof. M. Schachner (see acknowledgements). Gps 110 and 116 were purified as Con A binding glycoproteins of synaptic junctions from forebrains of 5 - 6 week old Wistar rats in the laboratory of Prof. J. Gurd (see acknowledgements, (6)). Whole brain tissue was obtained from adult Wistar rats, and was lyophilised and de-fatted (7).

Oligosaccharides were released from the de-salted glycoprotein preparations — ranging in quantity from several hundred µg for the synaptic gps to several g for the tissue preparation - by anhydrous hydrazinolysis (8). The glycans were fluorescently labelled with 2-aminobenzamide (2-AB) by reductive amination (9) and characterised by Bio-Gel P4 gel filtration chromatography (8), weak anion exchange (WAX) HPLC (9), normal phase HPLC (10), chemical and enzymatic neutralisation (7), exoglycosidase sequencing (10) and matrix-assisted laser-desorption ionisation mass spectrometry (MALDI-MS) (11).

3. RESULTS

Bio-Gel P4 gel filtration chromatography indicated that approximately 70% of rat brain tissue glycans released by hydrazinolysis were acidic (voided) and 30% were neutral (retarded). The anionic glycans were a complex mixture of sialylated, sulphated and

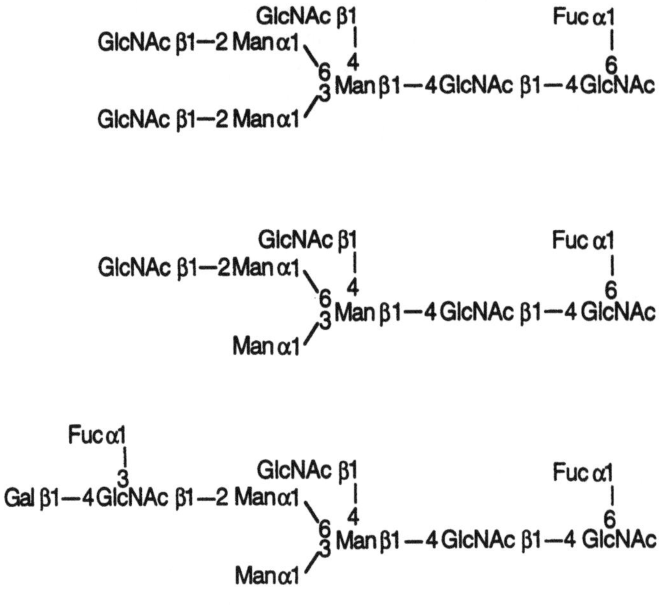

Figure 1. Structures of selected complex, N-linked neutral, oligosaccharides from brain tissue, showing characteristic features of core fucosylation, outer-arm fucosylation, a bisecting N-acetyl glucosamine and arm truncation. The glycans shown were major constituents of the neutral oligosaccharides of Tenascin-R. Fuc: fucose. Gal: galactose. GlcNAc: N-acetylglucosamine. Man: mannose.

uronic acid containing species as indicated by sequential neutralisation procedures, involving *Arthrobacter ureafaciens* sialidase (to desialylate), alkaline phosphatase (to dephosphate), methanolysis (to desulphate) and base treatment (to de-esterify residual carboxyl groups, as found on glucuronic acid). Structural features included the presence of the sialyl Lewis x carbohydrate determinant on a number of glycans, also, from a competitive ELISA study, several L2/HNK-1 reactive glycan fractions were found, suggesting the presence of terminal 3-sulphated glucuronic acid epitopes or similar structures.

The neutral glycans from the preparation of whole brain tissue were characterised by the presence of the oligomannosidic family, and by complex structures possessing outer-arm fucosylation, core fucosylation, a bisecting GlcNAc and arm truncation, as illustrated in Fig 1.

The neutral glycans of Tenascin-R, approximately 50% of the total, were dominated by oligosaccharides showing the complex structures described above for the whole tissue. Most members of the oligomannosidic family, however, were absent. In contrast, the glycosylation of gp110 and gp116 was mainly neutral (> 70%) in both cases, dominated by oligomannosidic glycans.

4. DISCUSSION

The conditions of hydrazinolysis used in this study were optimal for the release of N-linked glycans. Present evidence suggests that the library so obtained from brain tissue itself would mainly contain oligosaccharides expressed on cell surfaces and the extracellular matrix. The complexity of the anionic glycans indicated the potential in nervous tissue for future analytical and functional studies. This study substantiated in adult rat brain, the presence of two types of carbohydrate structure important in interactions involving molecular recognition, namely the sialyl Lewis x determinant and the L2 /HNK-1 carbohydrate epitope.

Structural details obtained for the neutral oligosaccharides revealed several characteristic features for neural tissue, and, by comparison with the mouse (7), little, if any, effect of rodent species. Expression of oligomannosidic glycans was relatively abundant. Their distribution, however, was shown to depend on the glycoprotein. Thus tenascin-R, as an extracellular matrix component, was restricted in its 'high mannose' content, whereas these structures were the major glycans of the post-synaptic glycoproteins. Other cell-surface expressed glycoproteins of the CNS known to carry oligomannosidic glycans include Thy-1 (12) and L1 (13). In the latter case, the glycans have been shown to be involved in cis-heterophilic interactions with N-CAM.

The characteristic features of the complex glycans observed in rodent brain tissue in the present study were also observed as major structures of the neutral glycans of tenascin-R. It is clear that fucosylation, both in α 1.6 linkage on the reducing terminal GlcNAc and in α 1,3 linkage on outer-arm GlcNAcs (to give the Lewis x structure), are prominent. The Lewis x carbohydrate, being closely related to its sialylated counterpart, may play a determining role in carbohydrate-protein interactions. The possibility also exists that core fucosylation influences the overall display of the oligosaccharide from the protein backbone and therefore of carbohydrates involved in recognition processes.

The degree of outer-arm truncation is of particular interest. In some relatively abundant structures, the extent of truncation resulted in the mannose residue being the non-reducing terminal monosaccharide (see Fig. 1). The notable feature was that this degree of truncation occurred on the 3-arm. In the classical N-linked processing pathway it is known that the 3-arm of the 'Man-5' structure is first extended by a GlcNAc in β 1–2 linkage through the action of N-acetylglucosaminyltransferase I (GnT-I). This step is a key one in that all other processing steps are consequent upon it. If these pathways are followed in brain tissue, and in the processing of the glycans of tenascin-R, it follows that an N-acetylglucosaminidase (β-hexosaminidase) must have been active in causing the observed truncation. The functional significance of this proposed pathway is not known at present.

In view of the preponderance of structures possessing a bisecting GlcNAc, it is also clear that the CNS has an active GnT-III or that the glycan substrates for this enzyme are particularly amenable. Again, the functional consequences of this structural feature are unknown, although recent preliminary evidence indicated a reduction in EGF binding to its receptor in a human glioma cell line when GnT-III was overexpressed (14). Thus changes in the signalling of a growth-factor receptor of brain origin may be modulated by changing the glycosylation of the receptor.

ACKNOWLEDGMENTS

Tenascin-R was supplied by Dr P. Pesheva from the laboratory of Prof. M. Schachner (Zurich). Gp 110 and 116 were supplied by Prof. J. Gurd (Toronto).

REFERENCES

1. Acheson, A., Sunshine, J.L., and Rutishauser, U., 1991, *J. Cell Biol.* **114**: 143–153.
2. Schachner, M., Antonicek, H., Fahrig, T., Faissner, A., Fischer, G., Kunemund, V., Martini, R., Meyer, A., Persohn, E., Pollerberg, E., Probstmeier, R., Sadoul, K., Sadoul, R., Seilheimer, B., and Thor, G., 1990, *Morphoregulatory Molecules* (G.M.Edelman, B.A.Cunningham, and J.P.Thiery, eds.), Wiley, New York, PP. 443–468.
3. Yoshihara, Y., Oka, S., Watanabe, Y., and Mori, K., 1991, *J.Cell Biol.* **115**: 731–744.
4. Lasky, L.A., 1992, *Science* **258**: 964–969.
5. Morganti, M.C., Taylor, J., Pesheva, P., and Schachner, M., 1990, *Expl. Neurol.* **109**: 98–110.
6. Gurd, J.W., 1989, *Neurobiology of Glycoconjugates,* (R.U. Margolis, and R.K. Margolis, eds.), Plenum Press, New York, PP. 219–242.
7. Wing, D.R., Rademacher, T.W., Field, M.C., Dwek, R.A., Schmitz, B., Thor, G., and Schachner, M., 1992, *Glycoconjugate J.* **9**: 293–301.
8. Ashford, D., Dwek, R.A., Welply, J.K., Amatayakul, S., Homans, S.W., Lis, H., Taylor, G.N., and Rademacher, T.W., 1987, *Eur. J.Biochem.* **166**: 311–320.
9. Guile, G.R., Wong, S.Y.C., and Dwek, R.A., 1994, *Anal. Biochem.* **222**: 231–235.
10. Guile, G.R., Rudd, P.M., Wing, D.R., Prime, S.B., and Dwek, R.A., 1996, *Anal. Biochem..* (in press).
11. Harvey, D.J., Rudd, P.M., Bateman, R.H., Bordoli, R.S., Howes, K., Hoyes, J.B., and Vickers, R.G., 1994, *Organic Mass Spectrom.* **29**: 753–765.
12. Parekh, R.B., Tse, A.G.C., Dwek, R.A., Williams, A.F., and Rademacher, T.W., 1987, *EMBO J.* **6**: 1233–1244.
13. Horstkorte, R., Schachner, M., Magyar, J.P., Vorherr, T., and Schmitz, B., 1993, *J. Cell Biol.* **121**: 1409–1421.
14. Rebbaa, A., Yamamoto, H., Moskal, J.R., and Bremer, E.G., 1996, Internat. Symp. '*Molecular and Cell Biology of Glycoconjugate Expression*', Rigi Kaltbad, Switzerland, Abstr.

ANTI-GANGLIOSIDE ANTIBODIES AND MOLECULAR MECHANISM OF DEVELOPMENT OF GUILLAIN-BARRE SYNDROME

Takao Taki, Nobuhiro Yuki, and Shizuo Handa

Department of Biochemistry
Tokyo Medical and Dental University
Tokyo, Japan

1. INTRODUCTION

Guillain-Barre syndrome (GBS) is characterized as an acute symmetrically progressive inflammatory neuropathy. Plasma exchange elicits a beneficial response (1). Since antibodies against gangliosides are commonly detected in patients with GBS, gangliosides are considered to be the target antigens of anti-neural antibodies. There is a close association between GBS and antecedent infection with a Gram-negative bacterium, *Campylobacter jejuni*, which causes acute gastroenteritis in human. The bacterium which is frequently isolated from GBS patients was serotyped as Penner type 19 (PEN 19). Sera from GBS patients after *C. jejuni* infection contain antibody against GM1 in the acute phase of illness (2,3). On the other hand, papers concerning that the carbohydrate structures of lipopolysaccharides (LPS) of *C. jejuni* show striking similarities with those of gangliosides have been accumulated (4–6). On the basis of this back ground, we examined the presence of molecular mimicry between carbohydrate structure of LPS of *C. jejuni* (PEN 19) and that of GM1 ganglioside to clarify the molecular mechanism which produces anti-GM1 antibodies as well as the pathogenic significance of the bacterium in eliciting GBS (7).

2. EXPERIMENTAL

LPS was extracted from *C. jejuni* (PEN 19) isolated from a GBS patient, by using the hot phenol-water technique. The LPS-containing fraction was was treated with 0.1 N

From a colloquium organized by Pam Fredman (Mölndal, Sweden) and Guido Tettamanti (Milano, Italy). (Supported by ARIN, Italy.)

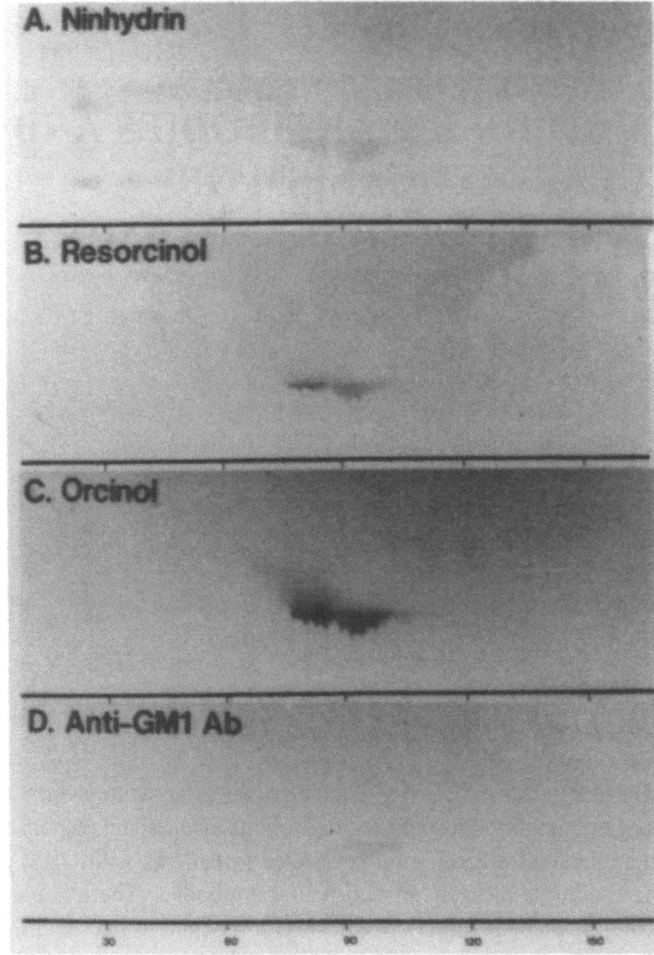

Figure 1. Chromatographic profile of the LPS. Fractions eluted from the Iatrobeads column were subjected to TLC and made visible by spraying with nihydrine (A), orcinol (B), resorcinol (C) and by the binding of cholera toxin (D).

NaOH at 56°C for 48 h. After dialysis against water and lyophilization, the crude LPS was dissolved in a solvent mixture of n-propanol-water-25% NH_4OH (75:15:10, v/v).

The LPS fraction was applied to a column packed with Iatrobeads (6RS-8060; Dia-Iatron, Tokyo) and eluted with a gradient of n-propanol-water-25% NH_4OH (75:15:10–60:40:10, v/v). Each fraction eluted from the column was monitored by TLC (Fig. 1). Fractions (94–104) showed homogenous bands when they were stained with nihydrine, orcinol, and Dittmer reagent and by binding of the peroxidase-conjugated cholera toxin, which specifically binds the GM1. These fractions were collected and used for further analysis. Gas chromatography/mass spectrometry analysis of the LPS fraction showed the presence of sugar components such as Gal, Glc, GalNAc, NeuAc and Hep, KDO as well as the fatty acid components, 3 hydroxy myristic acid and palmitic acid. Ethanolamine was detected by the amino acid analysis.

Figure 2 shows the NMR spectrum of the purified LPS. Anomeric proton resonances a, b, c and d, suggested the presence of βGal, βGal, βGalNAc and αHep, respectively. The

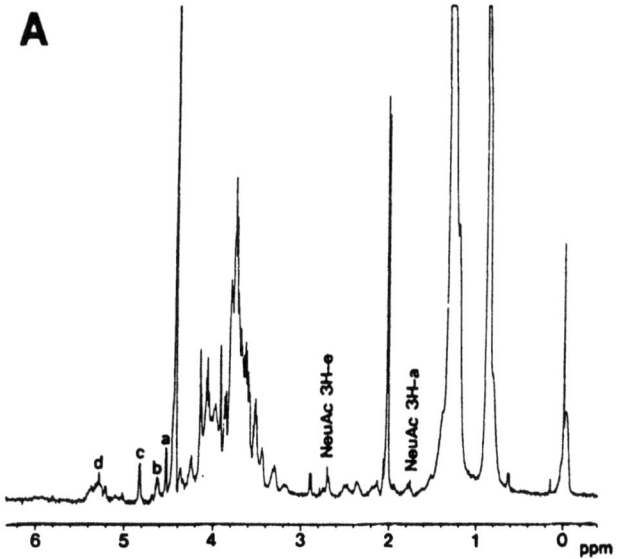

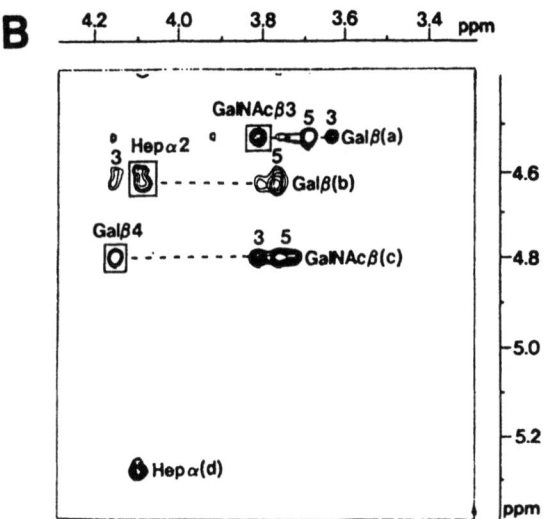

Figure 2. Proton NMR spectra of the LPS sample. A, One-dimensional proton NMR spectrum; B, two-dimensional NOESY spectrum.

equatorial and axial protons of ^{3}H of NeuAc are detected as indicated in the figure. To determine the linkage between sugars, two dimensional(2D) NMR analysis was done (Fig. 2b). The 2D NMR spectrum indicates the presence of βGal(a) bound at 3 position of βGalNAc, another βGal (b) linking at the 2 position of αHep and βGalNAc linking at the 4 position of βGal. From these analyses, the oligosaccharide structure, Galβ(a)1-3GalNAcβ(c)1-4(NeuAcα2-3)Galβ(b)1-2Hepα(d) was confirmed. This terminal oligosaccharide structure of LPS is identical to the tetrasaccharide of GM1 ganglioside.

LPS oligosaccharide: Galβ1-3GalNAcβ1-4Galβ1-2Hepα-LPS-core
                                                    < >
                                                     3
                                                     |
                                               NeuAcα2

GM1 ganglioside: Galβ1-3GalNAcβ1-4Galβ1-4Glcβ1-1'Cer
                                                 < >
                                                  3
                                                  |
                                           NeuAcα2

GM1 ganglioside is rich in the membrane of the motor nerve terminal which lacks the blood nerve barrier, and to which anti-ganglioside antibodies easily gain access. This is the first study to demonstrate the existence of molecular mimicry between nerve tissue components and infection agent that elicit GBS(7).

Next, we investigated the molecular mechanism of development of Fisher syndrome (FS) which is generally considered a variant form of GBS. FS is defined as a monophasic illness of acute onset characterized by external ophthalmoplegia, cerebellar ataxia, and absence of tendon reflexes. FS very frequently is associated with occurrence of autoantibody against GQ1b (8,9). We could isolate *C. jejuni* from two patients with FS, and both of the serotypes were determined PEN 2. To clarify the pathogenesis of FS and to examine the molecular mimicry theory, we studied whether the GQ1b epitope is expressed in the LPS of *C. jejuni* (PEN 2)(10). For this experiment, LPSs extracted from *C. jejuni* strains, CF 93-5, CF 93-6, CF 91-71 and CF 90-26 were used. CF 93-5 and CF 93-6 are isolated from two different patients with FS, respectively. CF 91-71 (PEN 2) was from enteritis patient with no neurological disorder. CF 90-26 (PEN 19) was isolated from a GBS patient. The LPSs were extracted and used for binding study using two different anti-GQ1b monoclonal antibodies, 7F5 and CMR 13. The result showed that both anti-GQ1b antibodies reacted with the LPSs from PEN 2 (CF 93-5, CF 93-6 and CF 91-71) but not with the PEN 19 isolated from the GBS patient. This result indicates an antigenic similarity between GQ1b ganglioside and the LPSs of *C. jejuni* isolated from FS patients. In human nerve tissues, anti-GQ1b antibody strongly stains the paranodal regions of the extramedullary portion of the oculomotor, trochlear and abducens nerve and weakly stains the deep cerebellar nuclei, the gray matter in the brainstem and spinal cord, and some large dosal ganglion cells (11). These facts suggest that GQ1b ganglioside is a target antigen in FS.

3. CONCLUDING REMARKS

Through these studies, we could demonstrate the molecular mimicry between structures of gangliosides of nerve tissues and LPS-oligosaccharides of *C. jejuni* isolated from patients either with GBS and FS. Since LPS is an adjuvant agent which activates immune system, GM1- or GQ1b-oligosaccharide bearing LPSs found in *C. jejuni* should have potent antigenic property to produce antibodies against these oligosaccharides. A particular immunogenetic background in the host, in addition to the antigenic components of the infectious agent, may be necessary for FS and GBS to develop.

REFERENCES

1. Ropper, A.H., 1992, *N. EngJ. Med.* **326**: 1130–1136.
2. Nobile-Orazio, E., Carpo, M., Meucci, N., Grassi, M.P., Captiani, E., Sciacco, M., Mangoni, A., and Scarlato, G.J., 1992, *J. Neurol. Sci.* **109**: 200–206.
3. Yuki, N., Yoshino, H., Sato, S., and Miyatake, T., 1990, *Neurology* **40**: 1900–1902.
4. Aspinall, G.O., McDonald, A.G., Raju, T.S., Pang, H., Kurjanczyk, L.A., Penner, J.L., and Moran, A.P., 1993, *Eur. J. Biochem.* **213**: 1029–1037.
5. Aspinall, G.O., McDonald, A.G., Raju, T.S., Pang, H., Kurjanczyk, L.A., Penner, J.L., and Moran, A.P., 1993, *Eur. J. Bichem.* **213**: 1017–1027.
6. Yuki, N., Handa, S., Taki, T., Kasama, T., Takahashi, M., Saito, K., and Miyatake, T., 1992, *Biomed. Res.* **13**: 451–453.
7. Yuki, N., Taki, T., Inagaki, F., Kasama, T., Takahashi, M., Saito, K., Handa, S., and Miyatake, T., 1993, *J. Exp. Med.* **178**: 1771–1775.
8. Chiba, A., Kusunoki, S., Shimizu, T., Kanazawa, I., 1992, *Ann. Neurol.* **31**: 677–679.
9. Yuki, N., Sato, S., Tsuji, S., Ohsawa, T., and Miyatake, T., 1993, *Neurology* **43**: 414–417
10. Yuki, N., Taki, T., Takahashi, M., Saito, K., Yoshino, H., Tai, T., Handa, S., and Miyatake, T., 1994, *Ann. Neurol.* **36**: 791–793
11. Chiba, A., Kusunoki, S., Obata, H Machinami, R., and Kanazawa, I., 1993, *Neurology* **43**: 1911–1917.

NEUROPROTECTIVE ACTIONS OF TAURINE

Pirjo Saransaari[1] and Simo S. Oja[1,2]

[1]Tampere Brain Research Center
University of Tampere Medical School
Tampere
[2]Department of Clinical Physiology
Tampere University Hospital
Tampere, Finland

1. INTRODUCTION

Taurine (2-aminoethanesulphonic acid) is a component of bile salts and a ubiquitous constituent of all animal cells. It is also one of the most abundant free amino acids in the brain of many mammals, the concentration even exceeding that of glutamate during ontogenic development (1). This simple sulphonic acid has been shown to be an essential nutrient for cats, and probably also for primates, in particular during development (2). It acts as an osmoregulator in marine animals (3) and has also been considered to function in the same role in the brains of terrestrial species (4). On the other hand, taurine induces hyperpolarisation and inhibits firing of central neurons (1,5,6).

Taurine protects neural cells from excitotoxicity induced by excitatory amino acids in the hippocampus (7) and cerebellum (8). It also forestalls harmful metabolic cascades evoked by ischemia and hypoxia (9). Moreover, taurine has long been known to alleviate symptoms in epilepsy (10). The mechanisms of these neuroprotective effects are not known but may be related to neuromodulatory, osmoregulatory, antioxidant and calcium ion regulatory effects of taurine.

2. TAURINE AND NEURAL CELL DAMAGE

The conditions which are known to cause neural cell damage inhibit the uptake and enhance the release of taurine in brain tissue (11). Metabolic poisons, hypoglycemic, hypoxic and ischemic conditions all distort the oxidative metabolism needed for active taurine transport across cell membranes. The saturable, sodium-dependent transport sys-

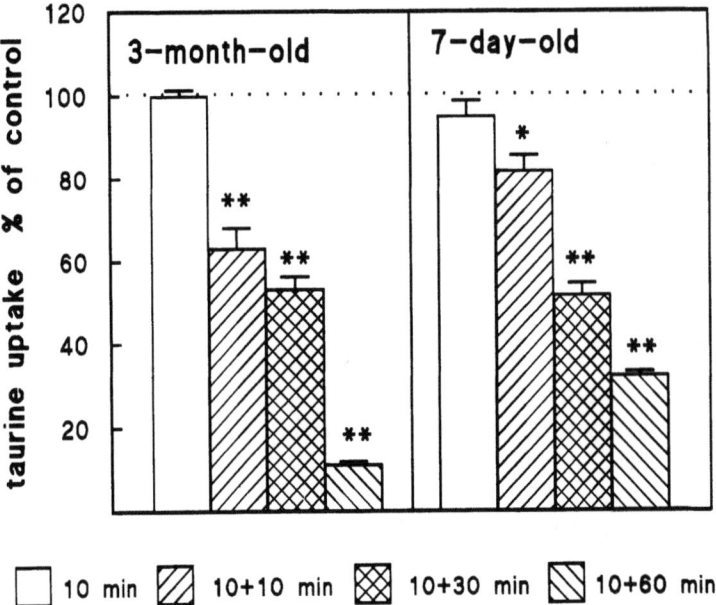

Figure 1. Taurine uptake by cerebral cortical synaptosomal preparations from 3-month-old and 7-day-old mice in ischemic conditions with increasing duration (glucose omitted, under N_2). The results are percentages of control uptake at 10 μM taurine in normal conditions. Number of independent experiments 4–8. Significance of differences: *p>0.05, **p>0.01.

tems of taurine comprise both high- and low-affinity components in neuronal and glial cell membranes (1,12). In ischemic conditions nonsaturable diffusion first increases in both immature and mature mouse hippocampus (11). However, a longer exposure to ischemia gradually reduces the uptake, the inhibition being more pronounced in the adults (Fig 1).

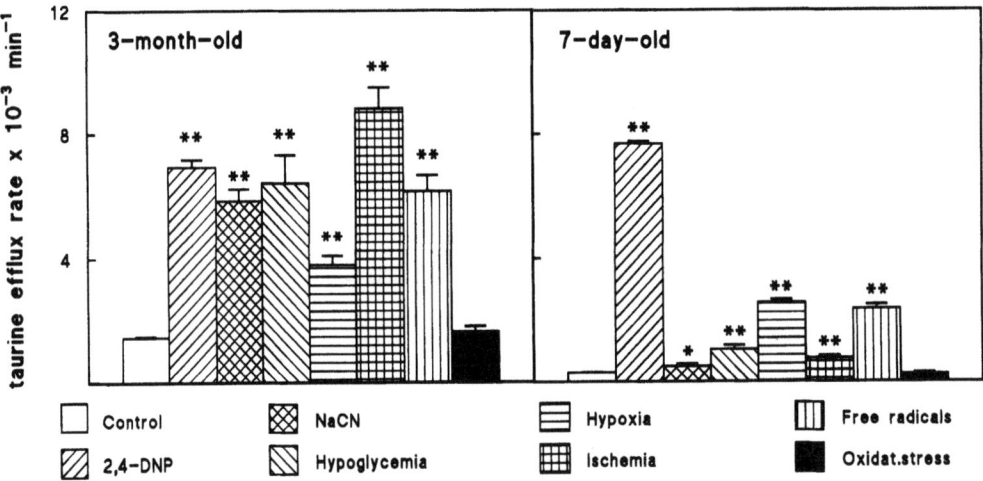

Figure 2. Release of taurine from hippocampal slices from 3-month-old and 7-day-old mice in cell-damaging conditions (see ref. 11). The results are efflux rate constants ± SEM. Number of independent experiments 4–8. Significance of differences: *p>0.05, **p>0.01.

The membrane structures become apparently more and more disrupted, first allowing leakage of taurine molecules and later impeding the functions of carrier proteins. The uptake systems of taurine seem to be more resistant to perturbation of oxidative metabolism in the immature than mature brain. For instance, the experimental conditions inducing free radical production and oxidative stress affect taurine uptake only in the adults (11).

The release of taurine is greatly potentiated by tissue-damaging conditions in both adult and developing hippocampus (Fig. 2).

The large increase in the release in hypoxic and ischemic conditions could partly be due to the reversed operation of cell membrane carriers (11). Taurine interacts with glutamatergic neurotransmission in the brain, which interaction is potentially of great functional significance. Hypoxic and ischemic conditions cause a large increase in the release of excitatory amino acids which activate glutamate receptors, particularly those of the N-methyl-D-aspartate (NMDA) type, increase the intracellular concentration of Ca^{2+} and trigger a long-lasting potentiation of NMDA receptor-gated currents (13). The glutamate agonists evoke taurine release from brain tissue, particularly in the immature brain (14). For example, an activation of NMDA receptors triggers taurine release in the hippocampus (15,16), in the immature hippocampus in particular (17). We have shown for the first time that the release evoked by NMDA in developing mice is even greater in hypoxic than normal conditions, as well as the release evoked by the kainate and 2-amino-3-hydroxy-5-methyl-4-isoxazolepropionare (AMPA) types of glutamate receptors. Moreover, taurine can also attenuate neural cell swelling (4) induced by activation of glutamate receptors (14) and hypoxia (18). Taurine released simultaneously could counteract by several mechanisms the harmful effects of excessive activation of glutamate receptors eventually leading to neuronal death. It attenuates the depolarizing effects of excitatory amino acids by increasing membrane chloride conductance (19). Taurine also inhibits Ca^{2+} influx in adult and developing brain tissue, antagonizes Ca^{2+} efflux evoked by potassium stimulation (20) and inhibits Ca^{2+} influx in the posthypoxic neuronal tissue (15).

Taurine seems thus to play an important role in the maintenance of homeostasis in the brain during impending hyperexcitation. The neural cell-damaging conditions, including ischemia and exposure to free radicals, appear on the one hand to inhibit taurine uptake, but on the other hand to greatly increase the release. The consequent increase in the extracellular taurine levels may be an important part of endogenous protective mechanisms, which is particularly efficient in the immature brain. Indeed, taurine-containing neurons in the hippocampus are more resistant than other neurons to damage induced by ischemia (21,22).

3. TAURINE AS AN ANTICONVULSANT

Taurine ameliorates epileptic symptoms in experimental animals and human patients (10) but only about one third of epileptics exhibit a clearly positive response to taurine medication, as was also the case in our own study on intractable cases of epilepsy in children (23). Taurine penetrates into the brain only very slowly in vivo (24). The molecule is rather lipophobic and the active uptake systems are not very effective (25). On the other hand, an excess of taurine in plasma is readily excreted into urine (26). The brain taurine is thus only marginally elevated after oral or systemic parenteral administration. We have therefore undertaken to develop novel lipophilic taurine derivatives which could more readily penetrate brain tissue and be useful as epileptic drugs. A few derivatives appeared fairly effective in rodent seizure models (27,28) but the one of them, taltrimide, that was subjected to clinical trials failed to ameliorate symptoms in intractable epileptic patients (29). Like the parent compound, taurine, taltrimide also affects GABAergic neurotransmission (30) which interactions may underlie its failure in trials on human epilepsy. In our opinion there are nevertheless unexplored possibilities of modifying the taurine molecule in the search for novel anticonvulsants and neuroprotectants.

REFERENCES

1. Oja, S.S., and Kontro, P., 1983, *Handbook of Neurochemistry*, Volume 3 (A. Lajtha, ed.), Plenum Press, New York, PP. 501–533.
2. Sturman, J.A., 1993, *Physiol. Rev.* **73**: 119–147.
3. Simpson, J.W., Allen, K., and Awapara, J., 1959, *Biol. Bull.* **117**: 371–381.
4. Walz, W., and Allen, A.F., 1987, *Exp. Brain Res.* **68**: 290–298.
5. Oja, S.S., Lähdesmäki, P., and Kontro, P., 1977, *Prog. Pharmacol.* **1/3**: 1–119.
6. Saransaari, P., and Oja, S.S., 1992, *Prog. Neurobiol.* **38**: 455–482.
7. French, E.D., Vezzani, A., Whetsell, W.O., Jr., and Schwarcz, R., 1986, *Adv. Exp. Med. Biol.* **203**: 349–362.
8. Trenkner, E., 1990, *Adv. Exp. Med. Biol.* **268**: 239–244.
9. Schurr, A., Tseng, M.T., West, C.A., and Rigor, B.M., 1987, *Life Sci.* **40**: 2059–2066.
10. Oja, S.S., and Kontro, P., 1983, *Acta Neurol. Scand.* **67**, Suppl. 93: 5–20.
11. Saransaari, P., and Oja, S.S., 1996, *Taurine: basic and clinical aspects* (R.J. Huxtable, J. Azuma, M. Namagawa, K. Kuriyama, and A. Baba, eds.), Plenum Press, New York, in press.
12. Huxtable, R.J., 1992, *Physiol. Rev.* **72**: 101–163.
13. Szatkowski, M., and Attwell, D., 1994, *Trends Neurosci.* **17**: 359–365.
14. Saransaari, P., and Oja, S.S., 1991, *Neuroscience* **45**: 451–459.
15. Lehmann, A., Hagberg, H., Nyström, B., Sandberg, M., and Hamberger, A., 1985, *Prog. Clin. Biol. Res.* **179**: 289–311.
16. Magnusson, K. R., Koerner, J.F., Larson, A.A., Smullin, D.H., Skilling, S.R., and Beitz, A.J., 1991, *Brain Res.* **549**: 1–8.
17. Saransaari, P., and Oja, S.S., 1994, *Adv. Med. Exp. Biol.* **359**: 279–287.
18. Laakso, M.-L., and Oja, S.S., 1976, *Acta Physiol. Scand.* **97**: 486–494.
19. Oja, S.S., Korpi, E.R., and Saransaari, P., 1990, *Neurochem. Res.* **15**: 797–804.
20. Kontro, P., and Oja, S.S., 1988, *Int. J. Neurosci.* **38**: 103–109.
21. Wu, J.-Y., Johansen, F.F., Lin, C.-T., and Liu, J.-W., 1987, *Adv. Exp. Med. Biol.* **217**: 265–274.
22. Matsumoto, K., Ueda, S., Hashimoto, T., and Kuriyama, K., 1991, *Brain Res.* **543**: 236–242.
23. Airaksinen, E.M., Oja, S.S., Marnela, K.-M., Leino, E., and Pääkkönen, L., 1980, *Prog. Clin. Biol. Res.* **39**: 157–166.
24. Oja, S.S., Lehtinen, I., and Lähdesmäki, P., 1976, *Q. J. Exp. Physiol.* **61**: 133–143.
25. Lähdesmäki, P., and Oja, S.S., 1973, *J. Neurochem.* **20**: 1400–1417.
26. Chesney, R.W., Gusowski, N., and Dabbagh, S., 1985, *J. Clin. Invest.* **76**: 2213–2221.
27. Lindén, I.-B., Gothóni, G., Kontro, P., and Oja, S.S., 1983, *Neurochem. Int.* **5**: 319–324.
28. Oja, S.S., Kontro, P., Lindén, I.-B., and Gothóni, G. 1983, *Eur. J. Pharmacol.* **87**: 191–198.
29. Airaksinen, E.M., Koivisto, K., Keränen, T., Pitkänen, A., Riekkinen, P.J., Oja, S.S., Partanen, J., Tokola, O., Gothóni, G., Neuvonen, P.J., and Airaksinen, M.M., 1987, *Epilepsy Res.* **1**: 308–311.
30. Kontro, P., and Oja, S.S., 1987, *Neuropharmacology* **26**: 19–23.

THE VOLUME RESPONSES OF BRAIN CELLS DURING OSMOTIC STRESS

How Important Is Taurine?

R. O. Law

Department of Cell Physiology and Pharmacology
University of Leicester
PO Box 138
Leicester, LE1 9HN
United Kingdom

1. INTRODUCTION

The sulphonic amino acid taurine is distributed in high concentrations throughout the animal kingdom, and as Huxtable has reasoned in a recent comprehensive review (1) "One is not tumbling into the abyss of teleology in thinking that a compound conserved so strongly and present in such high amounts is exhibiting functions that are advantageous to the life forms containing it". Ample evidence suggests that one of these advantages is the provision of osmoprotection that enables brain cells to maintain their volume when challenged by variations in extracellular osmolality (for review see (2–4)). In the clinical setting chronic anisosmotic states are typically asymptomatic, and this is in part due to adaptive variations in cellular contents of taurine and certain other low molecular weight organic solutes, collectively termed organic osmolytes. These are lost from cells during hyponatraemia, thus countering the tendency of cells to swell by osmotic uptake of water, whereas in hyperosmotic conditions shrinkage due to water loss is minimised by cellular accumulation of osmolytes.

The purpose of this chapter is to provide a brief overview of taurine's participation in anisosmotic brain cell — and hence, by extension, whole brain — volume regulation, and to suggest that taurine may play a "special role" in cellular adaptation to osmotic perturbation.

From a colloquium organized by Jan Albrecht (Warsaw, Poland) and Arne Schousboe (Copenhagen, Denmark). (Supported by Red Bull GmbH, Austria).

2. CELLULAR RESPONSES TO ACUTE OSMOTIC STRESS

2.1. Response to Hyposmotic Shock

There have been numerous studies on the release of taurine and other amino acids from cultured neurons and glia, or cells in incubated brain slices, subjected to sudden hyposmotic shock. In cultured cells the fractional release of taurine, which is believed to occur via diffusional pathways, correlates with the severity of hyposmotic stress and the extent of initial cell swelling, and has a time course corresponding to the rate of cell volume adjustment (regulatory volume decrease) (5–7). Taurine-defficient rat cerebellar astrocytes show impairment of hyposmotic volume regulation (8). Taurine release in vitro can also be evoked by high external K^+ (5,7,9), but since the required levels of K^+ greatly exceed those that would be present under even the most extreme pathophysiological conditions in vivo, this aspect will not be considered here in detail. Taurine responses to hyperosmolality in vitro have been less extensively studied, but blockage of the diffusional leakage pathway, with increased Vmax for Na^+-dependent uptake, has been reported in cultured astrocytes (10), in which prolonged exposure to hyperosmotic medium also results in enhanced de novo taurine biosynthesis (11).

2.2. Response to Non-Osmotic Stimuli

Taurine release from brain cells has also been studied under a variety of conditions that are not directly associated with osmotic disturbance. Intracranial microdialysis and microperfusion have been extensively used to monitor increases in extracellular taurine (presumed to reflect cellular leakage) that are related to brain swelling in certain pathophysiological states (for references see (12)). Rapid astrocytic swelling, with taurine efflux, occurs in CNS trauma (as reviewed in (13)). Loss of cellular taurine in association with cell swelling occurs in hepatic encephalopathy and hyperammonaemia (14,15) and during subjection of cultured astrocytes to isotonic ethanol (16). It has also been demonstrated during perfusion of rat hippocampus with isosmotic solutions of weak organic acids, possibly due to the operation of an osmotically sensitive mechanism dependent on activation of Na/H exchange (17).

3. TAURINE IN WHOLE BRAIN

3.1. Brain Taurine under Artificial Circumstances

Although studies in cultured preparations afford information about the responses of individual cells to anisosmotic stress, these inevitably differ in several important respects from whole tissue in situ, notably in the absence of an intact circulation and blood brain barrier. Incubation or superfusion media represent a large extracellular compartment, effectively precluding re-uptake of effluxed solutes, an event which may occur in the tortuous interstitium of intact tissue (possibly including slices). Moreover, cultures (though not slices) represent a single cell type population; the potential relevance of this in terms of neuronal-glial exchange of taurine is considered briefly in Section 5. There are striking differences in the regional distribution of taurine in rat brain (18) although any relation of this to local volume-regulatory capacity is unknown.

Table 1. Increases in levels of animo acids in brains of weanling mice during chronic (4 day) hyperatraemic dehydration. Values are mmol/kg (mean ± s.e.m.). Data from Thurston et al. (1983) J. Neurochem. **40**: 240–245 (ref. (18)), with permission

	Control (n=3)	Hypernatraemic (n=4)
Alanine	0.52 ± 0.01	0.53 ± 0.03
Aspartate	3.50 ± 0.09	5.46 ± 0.20
GABA	2.08 ± 0.10	2.82 ± 0.26
Glutamate	10.86 ± 0.23	13.35 ± 0.07
Glutamine	3.69 ± 0.26	7.05 ± 0.28
Glycine	1.38 ± 0.02	2.19 ± 0.17
Taurine	13.59 ± 0.49	20.97 ± 0.78
Total amino acids	35.62 ± 0.66	52.37 ± 1.87

3.2. Brain Taurine Levels during Chronic Hypernatraemia

Thurston and her co-workers (19) measured levels of taurine and other major amino acids in the brains of weanling mice undergoing 4-day hypernatraemic dehydration. Relevant data are shown in Table 1. Although in percentage terms the increase in cerebral taurine (~50%) is smaller than that of certain less abundant amino acids, its relative preponderance meant that taurine made the major contribution to the overall increase in the size of the amino acid pool. Levels of taurine also increase markedly (though from a relatively low baseline — see section 4) in chronically hypernatraemic adult rats (20). There is evidence for increased Na-dependent uptake of taurine in synaptosomes isolated from chronically hypernatraemic dehydrated rats (21), and it has been estimated that taurine constitutes nearly 50% of the adaptable intracellular osmolal pool in hypernatraemic kittens (22). Dehydration unaccompanied by hypernatraemia had no effect of synaptosomal taurine uptake (21) or on the accumulation of amino acids and other osmolytes (methylamines and polyhydric alcohols) in whole rat brain: the latter accumulate only when plasma Na is increased (23).

3.3. Brain Taurine Levels during Chronic Hyponatraemia

Cerebral taurine levels decrease during chronic hyponatraemia. Levels of taurine fell to 1/3 control values during 4-day hyponatraemia in young mice, whereas those of other amino acids decreased by approximately 2-fold (24). Relevant data are shown in Table 2.

Table 2. Decreases in levels of animo acids in brains of weanling mice during chronic (4 day) hyponatraemia. Values are mmol/kg (mean ± s.e.m.). Data from Thurston et al. (1987) *Metab. Brain Res.* **2**: 223–241 (ref. (23)), with permission

	Control (n=8)	Hyponatraemic (n=10)
Aspartate	2.46 ± 0.09	1.07 ± 0.08
GABA	2.22 ± 0.07	1.44 ± 0.07
Glutamate	9.00 ± 0.29	5.08 ± 0.30
Glutamine	8.63 ± 0.17	3.87 ± 0.20
Glycine	1.42 ± 0.08	0.63 ± 0.05
Taurine	12.7 ± 0.06	4.18 ± 0.59
Total amino acids	36.7 ± 0.6	16.1 ± 0.7

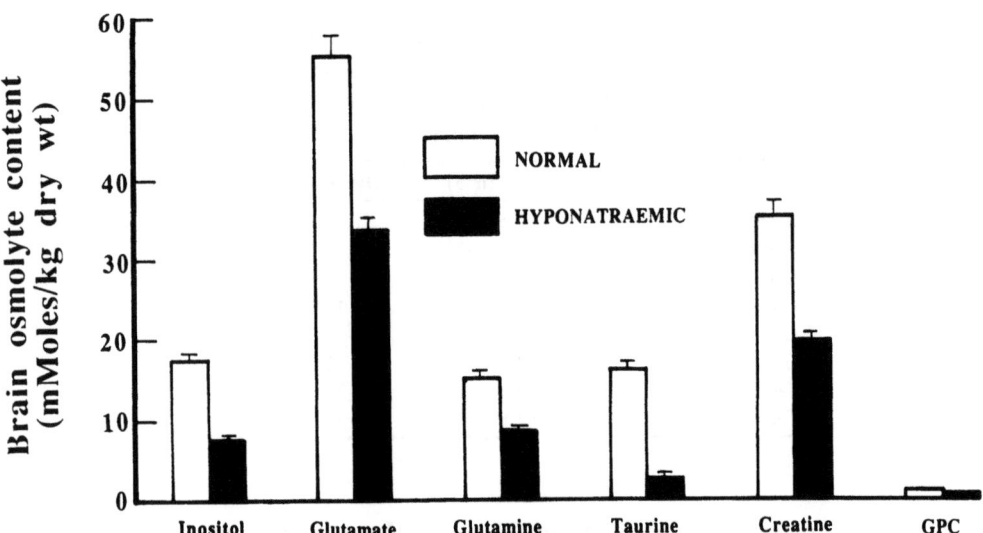

Figure 1. Brain contents of organic osmolytes in normal rats (n=15) and rats undergoing chronic (14 day) hyponatraemia (n=21). GPC, glycerophosphorylcholine. Redrawn form Verbalis & Gullans (1991) *Brain Res.* **567**: 274–282 (ref. (27)), by kind permission of Elsevier Science NL, Sara Burgerhartstraat 25, 1055 KV Amsterdam, The Netherlands.

Taurine also shows the greatest proportional decrease in the brains of acutely hyponatraemic adult rats (25). The mild hyposmolality of pregnancy in rats leads to a decrease in cerebral taurine significantly greater than that of total amino nitrogen (26). Wade and colleagues microperfused the piriform cortex of rats with fluids of decreasing osmolality (NaCl content) and showed that a fall from 305 to 280 mosmol/kg led to significant increases of taurine in the interstitial fluid (12). Further step-wise decreases in osmolality led to graded, reversible increases in fluid taurine, although it has been suggested (27) that some of this might have influxed from blood rather than reflect solely cellular loss. In contrast, extracellular levels of other amino acids only increased significantly when perfusate osmolality was reduced to an unphysiologically low level.

All these findings strongly suggest that taurine plays a special role in the responses of brain to hyponatraemia. This view is supported by a study in which Verbalis and Gullans (28) induced hyponatraemia in rats for up to 14 days by continuous infusion of DDAVP. During the initial 2 days there were comparable decreases in brain contents of taurine and other osmolytes. Similar findings have been reported during severe 24 hour hyponatraemia (29). But whereas levels of inositol and other major amino compounds (creatine, glutamine and glutamic acid) stabilised after 2 days, those of taurine continue to fall until at 14 days they were 1/7 of control levels, while those of other solutes had fallen to only ~50% (see fig. 1).

An interesting recent study has drawn attention to the presence of plentiful water channels (aquaporin 1 and 4) in the taurine-rich cells found in brain and other tissues that are characterised by high water-flux capacity, suggesting that there may be a functional correlation between these two properties (30).

4. CEREBRAL TAURINE IN RELATION TO AGE AND SPECIES

Taurine plays an important role in post-natal development of the nervous system (31, 32) but cerebral taurine levels decrease during maturation. This has been clearly demonstrated in humans and rhesus monkeys (33), and although there appear to have been no other detailed studies of cerebral taurine in relation to age, inspection of relevant literature clearly confirms that levels are lower in mature that post-natal laboratory animals.

Marked differences have been reported in the rates of K^+-stimulated endogenous taurine release from cortical and cerebellar slices in adult and developing mice, release being significantly more pronounced in the latter (9, 34). The relevance of these findings to cell volume regulatory capacity is uncertain, but K^+-stimulated taurine release from cultured cerebellar granule cells is known to be associated with cell swelling (35) and taurine efflux is likely to be a particularly effective volume-regulatory mechanism in immature brain.

Van Gelder (36) has drawn attention to the correlation between cerebral taurine content and glucose consumption in different animals. Both are 4–5 x greater in the mouse than in dogs or humans. These findings are consistent with the notion that taurine affords protection against the osmotic effects of intracellularly generated water. Moreover, there is a close association between brain water and taurine contents in laboratory rodents (37) although it is not clear whether this relates directly to osmoprotective cell volume regulation.

5. FUTURE PERSPECTIVES

There are two issues in relation to cerebral taurine transport that appear to merit particular attention at the present time. Firstly, it has been convincingly demonstrated by Ottersen and his co-workers (38, 39) that acute hyposmotic stress results in reversible glial sequestration of taurine that has been lost from neurons . While this is consistent with a specific role for glia in brain volume regulation and control of cerebral microenvironment, it inevitably poses serious questions regarding the functional relevance of those in vitro experiments in which hyposmotically-stimulated acceleration of taurine has been demonstrated (see Section 2).

Secondly, there appear to be significant differences between the rates of taurine release from acutely osmotically stressed cultured cells or cells in brain slices prepared from normonatraemic animals, and those from cells in animals that have been subjected to prolonged anisosmotic stress. Taurine efflux is retarded, with modest cell shrinkage, in hyperosmotically incubated cortical slices from normal rats (40). But in slices from chronically hypernatraemic rats, cell volumes are accurately maintained at normal values and taurine efflux is accelerated (41). This pattern of efflux persists for at least 5 hours following restoration of normonatraemia. Current studies in the author's laboratory (R.O. Law, unpublished results) indicate that chronic hyponatraemia is associated with decreased cellular taurine efflux, in contrast to the enhancement that occurs in slices from normonatraemic controls. As with chronic hypernatraemia, cell volumes are accurately maintained. Depression of efflux continues for at least 24 hours after normalisation of plasma Na. In addition, cells in slices from hyponatraemic rats display paradoxical swelling when acutely subjected to isosmotic media, which may reflect a transient imbalance between decrease taurine efflux and enhanced Na-dependent cellular uptake.

ACKNOWLEDGMENT

Work in the author's laboratory is supported by the Wellcome Trust. The second sentence of the "Introduction" is quoted with permission of The American Physiological Society (from ref. 1).

REFERENCES

1. Huxtable, R.J., 1992, *Physiol. Rev.*, **72**: 101–163.
2. Gullans, S.R., and Verbalis, J.G., 1993, *Annu. Rev. Med.*, **44**: 289–301.
3. Law, R.O., 1991, *Comp. Biochem. Physiol.*, **99A**: 263–277.
4. Law, R.O., 1994, *J. Exp. Zool.*, **268**: 90–96.
5. Kimelberg, H.K., Goderie, S.K., Higman, S., Pang, S., and Wanieswki, R.A., 1990, *J. Neurosci*, **10**: 1583–1591.
6. Pasantes Morales, H., and Schousboe, A., 1988, *J. Neurosci. Res.*, **20**: 505–509.
7. Schousboe, A., and Pasantes-Morales, H., 1992, *Can. J. Physiol. Pharmacol.*, **70**: S356-S361.
8. Moran, J., Maar, T.E., and Pasantes-Morales, H., 1994, *Neurochem. Res.*, **19**: 415–420.
9. Oja, S.S., and Kontro, P., 1989, *J. Neurochem.*, **52**: 1018–1024.
10. Sanchez-Olea, R., Moran, J., and Pasantes-Morales, H., 1992, *J. Neurosci. Res.*, **32**: 86–92.
11. Beetsch, J.W., and Olson, J.E., 1996, *Biochim. Biophys. Acta*, **1290**: 141–148.
12. Wade, J.V., Olson, J.P., Samson, F.E., Nelson, S.R., and Pazdernik, T.L., 1988, *J. Neurochem.*, **51**: 740–745.
13. Kimelberg, H.K., 1992, *J. Neurotrauma*, **9** Suppl. 1: S71-S81.
14. Hilgier, W., and Olson, J.E., 1994, *J. Neurochem.*, **62**: 197–204.
15. Hilgier, W., Olson, J.E., and Albrecht, J., 1996, *J. Neurosci. Res.*, **45**: 69–74.
16. Kimelberg, H.K., Cheema, M., O'Connor, E.R., Tong, H., Goderie, S.K., and Rossman, P.A., 1993, *J. Neurochem.*, **60**: 1682–1689.
17. Solis, J.M., Herranz, A.S., Herreras, O., Menendez, N., and Martin del Rio, 1990, *J. Neurosci. Res.*, **26**: 159–167.
18. Banay-Schwartz, M., Lajtha, A., and Palkovits, 1989, *Neurochem. Res.*, **14**: 563–570.
19. Thurston, J.H., Hauhart, R.E., and Schulz, D.W., 1983, *J. Neurochem.*, **40**: 240–245.
20. Lien, Y.-H. H., Shapiro, J.I., and Chan, L., 1990, *J. Clin. Invest.*, **85**: 1427–1435.
21. Trachtman, H., Futterweit, S., and del Pizzo, R., 1992, *Pediatr. Res.*, **32**: 118–124.
22. Trachtman, H., Barbour, R., Sturman, J.A., and Finberg, L., 1988, *Pediatr. Res.*, **23**: 35–39.
23. Heilig, C.W., Stromski, M.E., Blumenfeld, J.D., Lee. J.P., and Gullans, S.R., 1989, *Am. J. Physiol.*, **257**: F1108-F1116.
24. Thurston, J.H., Hauhart, R.E., and Nelson, J.S., 1987, *Metab. Brain Dis.*, **2**: 223–241.
25. Lien, Y.-H. H., Shapiro, J.I., and Chan, L., 1991, *J. Clin. Invest.*, **88**: 303–309.
26. Law, R.O., 1989, *J. Neurochem.*, **53**: 300–302.
27. Lehmann, A., Charlstrom, C., Nagelhus, E.A., and Ottersen, O.P., 1991, *J. Neurochem.*, **56**: 690–697.
28. Verbalis, J.G., and Gullans, S.R., 1991, *Brain Res.*, **567**: 274–282.
29. Sterns, R.H., Baer, J., Ebersoll, S., Thomas, D., Lohr, J.W., and Kamm, D.E., 1993, *Am. J. Physiol.*, **264**: F833-F836.
30. Amiry-Moghaddam, M., Nagelhus, E.A., Nielsen, S., and Ottersen, O.P., 1996, *Kidney Int.*, **49**: 928.
31. Chesney, R.W., 1987, *Pediatr. Res.*, **22**: 755–759.
32. Sturman, J.A., 1993, *Physiol. Rev.*, **73**: 119–147.
33. Sturman, J.A., and Gaull, G.E., 1976, *Taurine*, (R. Huxtable and A. Barbeau, eds), Raven Press, New York, PP. 73–84.
34. Oja, S.S., and Saransaari, P., 1995, *Neurochem. Int.*, **27**: 313–318.
35. Schousboe, A., Moran, J., and Pasantes-Morales, H., 1990, *J. Neurosci. Res.*, **27**: 71–77.
36. Van Gelder, N.M., 1989, *Neurochem. Res.*, **14**: 495–497.
37. Puza, M., Sundell, K., Lazarewicz, J.W., and Lehmann, A., 1991, *Brain Res.*, **548**: 267–272.
38. Nagelhus, E.A., Lehmann, A., and Ottersen, O.P., 1993, *Neuroscience*, **54**: 615–631.
39. Nagelhus, E.A., Amiry-Maghaddam, M., Lehmann, A., and Ottersen, O.P., *Taurine in Health and Disease* (R. Huxtable and D.V. Michalk, eds), Plenum Press, New York, PP. 325–334.
40. Law, R.O., 1994, *Biochim. Biophys. Acta.*, **1221**: 21–28.
41. Law, R.O., 1995, *Neurosci. Lett.*, **185**: 56–59.

KAINIC ACID–STIMULATED RELEASE OF TAURINE FROM THE RAT SUBSTANTIA NIGRA *IN VIVO*

L. Bianchi,[1] R. Zamfirova,[1] M. Nerini,[1] J. P. Bolam,[2] and L. Della Corte[1]

[1]Dipartimento di Farmacologia Preclinica e Clinica M. Aiazzi Mancini
Universitá di Firenze
Viale G.B. Morgagni 65
50134 Firenze, Italy
[2]MRC Anatomical Neuropharmacology Unit
Department of Pharmacology
Oxford, United Kingdom

1. INTRODUCTION

The substantia nigra pars reticulata (SNr) is one of the major output nuclei of the basal ganglia, a group of sub-cortical nuclei that are involved in the control of movement and various mnemonic and cognitive functions. It contains high levels of the inhibitory amino acid transmitter, GABA, which is localised in afferent terminals derived from the striatum and the globus pallidus and it is also the transmitter of the output neurons which project to the superior colliculus, the thalamus and the brain stem and possess local axon collaterals (1, 2). The substantia nigra also receives excitatory (glutamatergic) inputs derived from the subthalamic nucleus (3), the cortex (4) and possibly the mesopontine tegmentum (5). *In vivo* microdialysis studies in the freely moving rats have demonstrated that stimulation of excitatory amino acid receptors, which are presumably associated with the excitatory inputs, by perfusion with kainic acid (KA) induces a DNQX-sensitive release of GABA (6). At least some of this released GABA may be derived from the terminals of striatonigral neurons. Since taurine has also been proposed to be present in striatonigral neurons and has been shown to be released in the substantia nigra in response to stimulation of striatonigral neurons (7) we have carried out experiments to determine whether endogenous taurine is released in response to excitatory amino acid receptor stimulation in a manner similar to that of GABA.

From a colloquium organized by Jan Albrecht (Warsaw, Poland) and Arne Schousboe (Copenhagen, Denmark). (Supported by Red Bull GmbH, Austria).

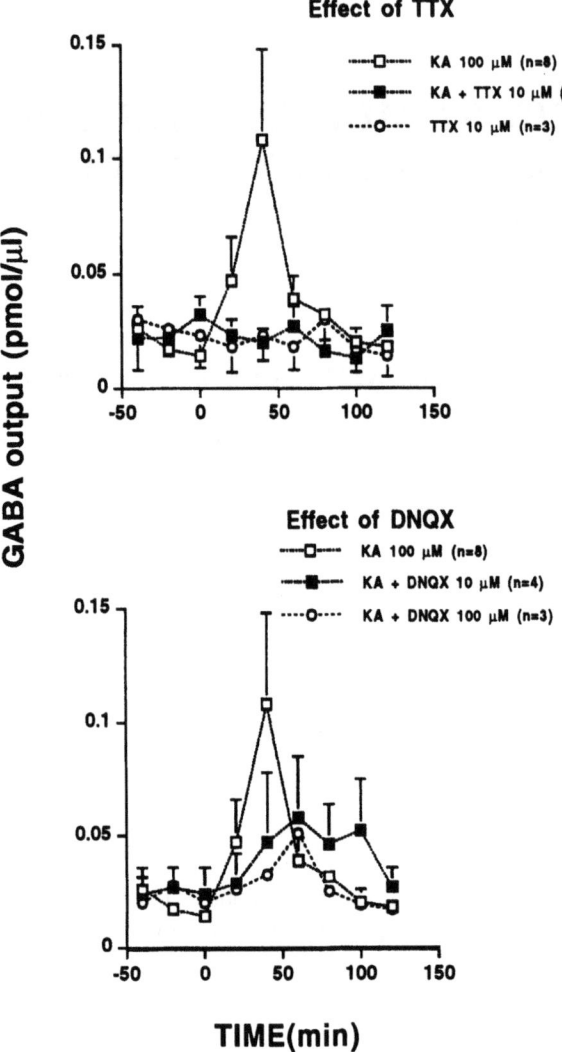

Figure 1. The effect of the local application of kainic acid (100 μM for 20 min) on the release of endogenous GABA (left panels) and taurine (right panels) from the substantia nigra. The upper panels show the effects of 10 μM tetrodotoxin administered throughout the experiment. The lower panels show the effect of 10 and 100 μM DNQX throughout the experiment. Data are expressed as the mean ± standard error of the mean of pmol of amino acid/μl of perfusate for the number of experiments indicated by n.

2. MATERIALS AND METHODS

The experiments were performed on male Sprague-Dawley rats. Single cannula microdialysis probes were implanted vertically into the right SNr (2 mm exposed tip), as previously described (6). Twenty four hours later, the probes, of the now freely moving rats, were perfused with artificial cerebrospinal fluid at a rate of 1.2 μl/min. After a perfusion period of 2 h the SNr was then perfused with 100 μM kainic acid (20 min), alone or in the presence of 10 or 100 μM 6,7-dinitroquinoxaline-2,3-dione (DNQX) or 10 μM tetrodo-

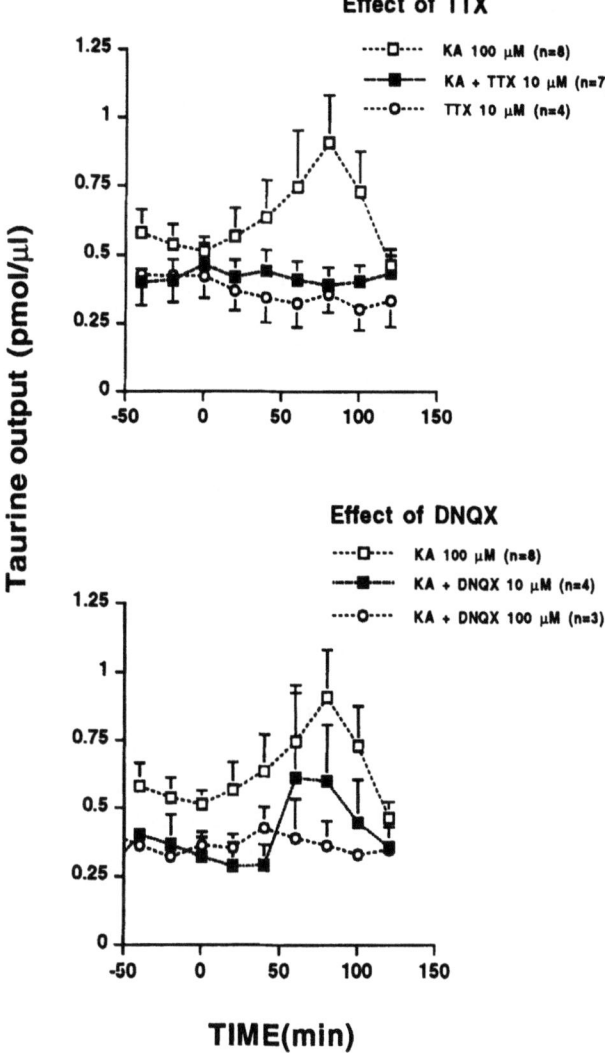

Figure 1. *(Continued)*

toxin (TTX). Fractions (20 min) were collected for 2 h before, and up to 3 h after the KA perfusion. Following derivatisation and separation (hplc), the amino acids were detected fluorimetrically. The levels of GABA and taurine were expressed as pmol of amino acid/μl of perfusate. Original values were compared by analysis of variance (6).

3. RESULTS

Basal release of GABA and taurine in the SNr was 0.022±0.003 (n=29) and 0.467±0.049 (n=30) pmol\μl perfusate, respectively. Perfusion of the SNr with 100 μM KA stimulated the release of both GABA and taurine (Figure 1). The peak release of GABA and taurine were 771% and 177% of basal levels ($p<0.05$, MANOVA), respectively. The local enhanced release of GABA in the substantia nigra was inhibited when DNQX (10 µM) was co-administered with the kainic acid (242%), whereas that of taurine was unaffected (191%) (Figure 1). Co-administration of 100 µM DNQX, however, completely blocked the KA-stimulated release of taurine in the substantia nigra (118%), without further affecting the release of GABA (255%). The KA-enhanced release of GABA and taurine was also completely blocked by 10 µM TTX (103% and 108%, respectively).

4. DISCUSSION

The main finding of the present study is that endogenous GABA and taurine are released from the substantia nigra in response to perfusion with kainic acid. This release is due mainly to the stimulation of excitatory amino acid receptors as the enhanced release was blocked, in the case of taurine and markedly attenuated in the case of GABA, by DNQX. The data are consistent with the presence of excitatory amino acids in some of the afferents of the SN (see Introduction for references) and the presence of non-NMDA excitatory amino acid receptors (8,9). The failure to completely abolish the enhanced release of GABA in response to kainic acid, implies that at least part of the release is independent of AMPA/kainate receptors.

The stimulated, but not basal release of both of the amino acids was abolished in the presence of tetrodotoxin indicating a dependence on fast sodium channels and thus axon potential propagation. This suggests that the enhanced release in response to kainic acid is not an effect directly on GABA- or taurine-containing axon terminals but is more likely an effect at the level of neuronal cell bodies or dendrites. Thus the release of GABA is likely to be derived from the local axon collaterals of GABA neurons in the nigra (1) or due to an indirect action of the kainic acid on GABA-containing terminals mediated by a class of neuron containing a different transmitter e.g. the dopaminergic neurons. Since there is no evidence for taurine-containing neurons in the substantia nigra then the latter mechanism is the more likely to apply to the release of taurine. If we thus assume that the DNQX- and the TTX-dependence of the taurine release indicates a neuronal origin of the stimulated release, taurine is likely to have been derived from afferent axon terminals. These data thus provide additional support for the possible association of taurine with the striatonigral pathway (7,10). The data do not exclude the possibility that at least part of the released taurine and GABA is derived from non-neuronal sources.

ACKNOWLEDGMENT

Supported by the EU (BIOMED grant BMH1-CT94–1402) & grants from CNR and MURST (IT). R. Zamfirova was supported by TEMPUS S-JEP-07424–95 and grant number 558/95 from the National Fund "Scientific Research" Bulgaria.

REFERENCES

1. Deniau, J.M., Kitai, S.T., Donoghue, J.P., and Grofova, I., 1982, *Exp. Brain Res.* **47**: 105–113.
2. Bevan, M.D., Bolam, J.P., and Crossman, A.R., 1994, *Eur. J. Neurosci.* In press.
3. Rinvik, E., and Ottersen, O.P., 1993, *J. Chem. Neuroanat.* **6**: 19–30.
4. Naito, A., and Kita, H., 1994, *Brain Res.* **637**: 317–322.
5. Lavoie, B., and Parent, A., 1994, *J. Comp. Neurol.* **344**: 232–241.

6. Bianchi, L., Sharp, T., Bolam, J.P., and Della Corte, L., 1994, *NeuroReport* **5**: 1233–1236.
7. Della Corte, L., Bolam, J.P., Clarke, D.J., Parry, D., and Smith, A.D., 1990, *Eur. J. Neurosci.* **2**: 50–61.
8. Bernard, V., Gardiol, A., Faucheux, B., Bloch, B., Agid, Y., and Hirsch, E. C., 1996, *J. Comp. Neurol.* **368**: 553–568.
9. Martin, L.J., Blackstone, C.D., Levey, A.I., Huganir, R.L., and Price, D.L., 1993, *Neuroscience* **53**: 327–358.
10. Clarke, D.J., Smith, A.D., and Bolam, J.P., 1983 *Brain. Res.* **289**: 342–348.

KAINIC ACID–STIMULATED RELEASE OF TAURINE AND GABA FROM THE RAT GLOBUS PALLIDUS AND SUBSTANTIA NIGRA *IN VIVO*

L. Bianchi,[1] P. Bartolini,[1] A. Colivicchi,[1] J. P. Bolam,[2] and L. Della Corte[1]

[1]Dipartimento di Farmacologia Preclinica e Clinica M. Aiazzi Mancini
Universitá di Firenze
Viale G.B. Morgagni 65, 50134 Firenze, Italy
[2]MRC Anatomical Neuropharmacology Unit
Department of Pharmacology
Oxford, United Kingdom

1. INTRODUCTION

The basal ganglia are a group of subcortical nuclei intimately involved in the control of movement. The major input to the basal ganglia arises in the cortex and is carried by the corticostriatal projection. The cortical information is processed in the striatum and then transmitted to the output nuclei of the basal ganglia, the entopeduncular nucleus and the substantia nigra pars reticulata (SNr), by either the direct or the indirect pathways. The direct pathway consists of a direct projection from the striatum to the entopeduncular nucleus or SNr, the indirect pathways consist of a striatal projection to the globus pallidus and thence to the output nuclei or to the output nuclei via the subthalamic nucleus. Striatal projection neurons, neurons of the globus pallidus, SNr and entopeduncular nucleus are GABAergic whereas those of the subthalamic nucleus use glutamate. Activation of the direct pathway leads, by virtue of its GABAergic nature, to inhibition of neurons in the entopeduncular nucleus and SNr whereas activation of the indirect pathways produces the opposite effect i.e. disinhibition (1–3) of neurons in the entopeduncular nucleus and SNr. Neurones identified as medium-size densely spiny striatonigral neurones have also been shown to accumulate exogenous taurine and release it at their terminals in the SN (4).

In vivo microdialysis studies using dual probes have shown that neurons of the direct pathway i.e. striatonigral neurons release endogenous GABA and taurine in response to excitatory amino acid receptor stimulation. Thus the administration of kainic acid to the neostriatum elicits the local release (presumably from local axon collaterals) and at the

From a colloquium organized by Jan Albrecht (Warsaw, Poland) and Arne Schousboe (Copenhagen, Denmark). (Supported by Red Bull GmbH, Austria).

same time, distal release in the ipsilateral SN, of both amino acids (5, 6). The object of the present experiment was to examine the release of endogenous taurine and GABA from the indirect pathway neurons of the pallidonigral system.

2. MATERIALS AND METHODS

The experiments were performed on male Sprague-Dawley rats. Single cannula microdialysis probes were implanted vertically into the right globus pallidus (3 mm exposed tip) and the ipsilateral SNr (2 mm exposed tip), as previously described (5). Twenty four hours later, the probes, of the now freely moving rats, were perfused with artificial cerebrospinal fluid at a rate of 1.2 µl/min. After a perfusion period of 2 h the globus pallidus was perfused with 100 µM kainic acid, alone or in the presence of 100 µM 6,7-dinitroquinoxaline-2,3-dione (DNQX). Fractions of the perfusate (20 min) were collected for the 2 h period before the kainic acid perfusion and for up to 3 h afterwards. The amino acids in the perfusate fractions were derivatised, separated by hplc and the orthophtalaldehyde derivatives were detected fluorimetrically (6). The levels of GABA and taurine were expressed as pmol of amino acid/µl of perfusate. Original values were compared by analysis of variance (5).

3. RESULTS

Basal release of GABA and taurine was 0.019 ± 0.002 (n=11) and 0.516 ± 0.047 pmol/µl perfusate (n=11), respectively, in the globus pallidus and 0.016 ± 0.003 (n=8) and 0.594 ± 0.082 (n=10), respectively, in the SNr. The application of kainic acid to globus pallidus significantly ($p<0.05$, MANOVA) enhanced the release of both GABA and taurine locally in the globus pallidus (at peak 712% and 411% of basal release, respectively) (Figure 1). The peak release of GABA (666%; $p<0.05$, MANOVA) but not taurine (190%) was also significantly enhanced in the distal probe in the SNr (Figure 1). The local enhanced release of both GABA and taurine in the globus pallidus was inhibited when DNQX (100 µM) was co-administered with the kainic acid (205% and 139%, respectively) (Figure 1). Furthermore the distal release of GABA in the substantia nigra was also blocked by administration of DNQX to the globus pallidus (249%).

4. DISCUSSION

The present experiments were designed to test the possibility that the amino acids, GABA and taurine, are released from the 'indirect' limb of the pathways of information flow through the basal ganglia in a manner analogous to the release in the 'direct' pathway. The main finding is that indeed for GABA, the stimulation of excitatory amino acids in the globus pallidus leads to the local release of GABA, presumably from the local collaterals of pallidal neurons (7), and from the terminals of pallidal neurons in the substantia nigra (8,9). The observation that the stimulated release is sensitive to DNQX implies that it is a receptor mediated phenomenon being mediated through the AMPA subclass of glutamate receptors. Furthermore, the stimulation of these receptors in the globus pallidus led to DNQX-sensitive release of GABA in the substantia nigra which is several hundred microns (or several millimetres) caudal to the globus pallidus. This implies a release from

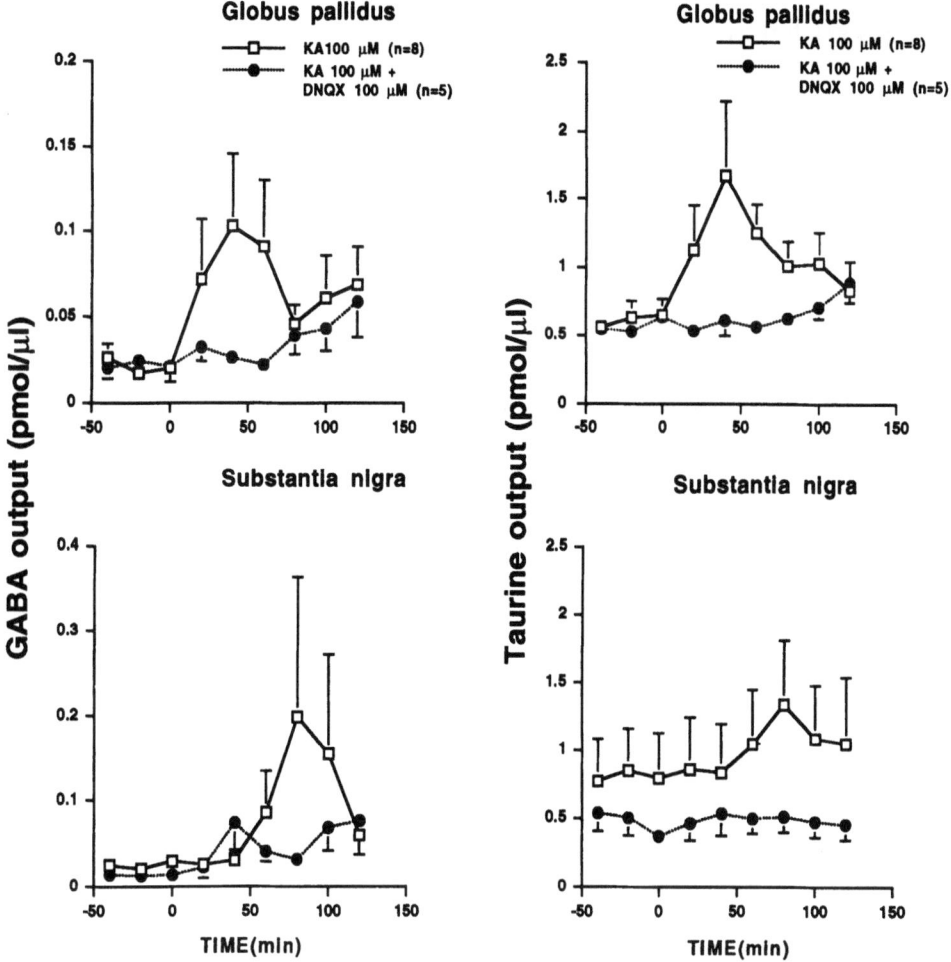

Figure 1. The time courses of GABA (left panels) and taurine (right panels) output, expressed as pmol/μl of perfusate, monitored locally in the globus pallidus (upper panels) and distally in the substantia nigra reticulata (lower panels), following intrapallidal application of 100 μM KA (collection time every 20 min.), are shown.

the axon terminals of pallidonigral neurons. This is thus analogous to the release of GABA from the local axon collaterals in the striatum and distally in the substantia nigra following the stimulation of excitatory amino acid receptors in the striatum (5,10). It must be remembered, however, that the amino acid receptors in the striatum are associated with the excitatory input from the cortex, whereas the receptors in the globus pallidus are mainly associated with an excitatory feedback system from the subthalamic nucleus. In contrast to the release of GABA the release of taurine in the pallidonigral system had a different profile. The stimulation of excitatory amino acid receptors in the globus pallidus led to a significant increase in the release taurine locally, within the globus pallidus, that was sensitive to blockade of AMPA receptors but did not affect the distal release in the substantia nigra. These observations imply therefore that although there is a receptor-dependent release in the globus pallidus this is probably not related to pallidonigral neurons as there were no changes in the substantia nigra. These findings contrast those made in the striatonigral system where the local stimulated release of taurine is independent of DNQX-sensitive receptor stimulation but occurs distally in the substantia nigra (11).

ACKNOWLEDGMENTS

This work was supported by the European Community (BMH1 CT94-1402), CNR-Bilateral Project (95.00833 CT04), MURST-60%, Roma, Italia and the MRC, UK.

REFERENCES

1. Albin, R.L., Young, A.B., and Penney, J.B., 1989, *TINS* **12**: 366–375.
2. De Long, M.R., 1990, TINS **13**: 281–285.
3. Gerfen, C.R., 1992, TINS **15**: 133–139.
4. Della Corte, L., Bolam, J.P., Clarke, D.J., Parry, D.M., and Smith, A.D., 1990, *Eur. J. Neurosci.* **2**: 50–61.
5. Bianchi, L., Sharp, T., Bolam, J.P., and Della Corte, L., 1994, *NeuroReport* **5**: 1233–1236.
6. Bianchi, L., Galeffi, F., Bolam, J.P., and Della Corte, L., 1995, *Eur J. Neurosci.*, Suppl. **8**: 206.
7. Kita, H. and Kitai, S.T., 1994, *Brain Res.* **636**: 308–319
8. Smith, Y. and Bolam, J.P., 1989, *Brain Res.* **493**: 160–167.
9. Smith, Y. and Bolam, J.P., 1991, *Neuroscience* **44**: 45–73.
10. Morari, M., O'Connor, W.T., Ungerstedt, U., Bianchi, C., and Fuxe, K., 1996, *Neuroscience* **72**: 89–97.
11. Bianchi, L., Bolam, J. P., Galeffi, F., Frosini, M., Palmi, M., Sgaragli, G. P., and Della Corte, L., 1996, *Taurine: basic and clinical aspects* (R. Huxtable, J. Azuma, M. Nakagawa, K. Kuriyama, and A. Baba, eds), Plenum Publishing Corporation. In press.

DIFFERENT SPECIFICITIES FOR TAURINE ANALOGUES AND THEIR TARGET SITES IN BRAIN

M. Marangolo,[1] D. Zisterer,[1] D. C. Williams,[1] K. F. Tipton,[1] H. B. F. Dixon,[2] and L. Della Corte[3]

[1]Department of Biochemistry
 Trinity College, Dublin, Ireland
[2]Department of Biochemistry
 Cambridge University, United Kingdom
[3]Dipartimento di Farmacologia
 Università di Firenze, Italia

1. INTRODUCTION

The amino acid taurine is present in high concentrations in the brain but its functions are far from clear. It has been variously suggested to be an osmoregulator, an antioxidant, a membrane-stabiliser, a modulator of trophic factors and/or calcium ions and a neurotransmitter/neuromodulator (for review see 1). Taurine has been shown to bind to $GABA_A$ receptors, as a competitive inhibitor of the binding of the receptor agonist [^{3}H]-muscimol (2). However, it appears that taurine is not transported to any appreciable extent by neurone or astrocyte GABA carriers but shares a common carrier with β-alanine (3).

As an approach to studying its functions in the CNS, we are examining the interactions of a number of taurine analogues (Fig. 1) with receptors and transport systems.

2. MATERIALS AND METHODS

C6 glioma cells, originally cloned from a rat astrocytoma (4), were routinely grown as monolayer cultures in 75 cm^2 flasks containing Dulbecco's Modified Eagle's Medium, with 10% (v/v) foetal calf serum, glutamate (2 mM) and gentamycin (100 mg/ml) at 37°C. Cells were re-seeded by trypsinization twice weekly. Cells used in the uptake assay were grown in 24-well plates, seeded initially at 2 x 10^5 cells/well in 1 ml of growth medium.

From a colloquium organized by Jan Albrecht (Warsaw, Poland) and Arne Schousboe (Copenhagen, Denmark). (Supported by Red Bull GmbH, Austria).

Table 1. Taurine analogues and other compounds used in this work

Taurine	$H_2NCH_2CH_2SO_3H$
***N*-Methyltaurine**	$CH_3NHCH_2CH_2SO_3H$
***N,N*-Dimethyltaurine**	$(CH_3)_2NCH_2CH_2SO_3H$
***N,N,N*-Trimethyltaurine**	$(CH_3)_3N^+CH_2CH_2SO_3H$
2-Aminoethylphosphonic acid	$H_2NCH_2CH_2PO_3H$
2-Aminoethylarsonic acid	$H_2NCH_2CH_2AsO_3H_2$
3-Aminopropylarsonic acid	$H_2NCH_2CH_2CH_2AsO_3H_2$
D,L-3-Arsonoalanine	$H_2NCH(COOH)CH_2AsO_3H_2$
Isethionic acid	$HOCH_2CH_2SO_3H$
GES (Guanidinoethane sulfonic acid)	$H_2N C(=NH) NHCH_2CH_2SO_3H$
Pyridine-3-sulfonic acid	pyridine-SO$_3$H
Piperidine-3-sulfonic acid	piperidine-SO$_3$H
TAG (6-Aminomethyl-3-methyl-4H-1,2,4-benzothiadiazine-1,1-dioxide hydrochloride)	benzothiadiazine dioxide structure
β-Alanine	$H_2NCH_2CH_2COOH$
GABA (γ-Aminobutyric acid)	$H_2NCH_2CH_2CH_2COOH$
Carnitine	$(CH_3)_3N^+CH_2CH(OH)CH_2COOH$
Choline	$(CH_3)_3N^+CH_2CH_2OH$
HEPES	piperazine with $CH_2CH_2SO_3H$ and CH_2CH_2OH

Prior to measuring [^{3}H]taurine uptake, cells were washed (1 x 1 ml) with Krebs-Ringer-HEPES (KRH) assay buffer, pH 7.4, 37°C (120 mM NaCl, 4.7 mM KCl, 2.2 mM CaCl$_2$, 1.2 mM MgSO$_4$, 1.2 mM KH2PO4 and 10 mM HEPES, 5 mM glucose was added immediately prior to use). When Na$^+$ was omitted, NaCl was replaced with either LiCl or choline chloride (120 mM). Following this wash, 0.5 ml of identical warmed medium, containing [^{3}H]taurine (specific radioactivity 9.6 Ci/mmol), was added to each well. Taurine analogues, when added, were present in the assay buffer.

After incubation at 37°C (incubation time was varied up to a maximum of 40 min depending on the experiment) the assay buffer was removed and uptake was terminated by washing the cells with ice-cold KRH buffer (2 x 1 ml), followed by solubilization of the cells with 0.5 ml of 1% (w/v) sodium dodecyl sulphate (SDS). Each sample was transferred to a plastic scintillation vial, to which 10 ml of aqueous-based scintillation fluid (ECO-Scint, National Diagnostic) was added, and the radioactivity was measured.

Synaptic membranes were prepared by a modification of the method of Gordon-Weeks (5). For the binding assay, synaptic membranes were treated with Triton X-100 by incubation for 30 min at 37°C with about 35 ml of 50 mM Tris-citrate buffer, pH 7.1, containing 0.05% Triton X-100. After centrifugation at 40,000 g for 20 min the pellet was resuspended in 35 ml of 50 mM Tris-citrate buffer, pH 7.1, and the incubation, centrifugation and resuspension procedure was repeated three times before the pellet was resuspended in the same buffer to give a final protein concentration of 1 mg/ml. Samples containing 0.1–0.5 mg of protein were incubated, in triplicate, at 4°C for 30 min in 1 ml of 50 mM Tris-citrate buffer, pH 7.1, containing appropriate concentrations of [^{3}H]muscimol (specific radioactivity 14.9 Ci/mmol), in the presence or absence of various taurine analogues (100μl of 1 mM stock solutions). After a 30 min incubation, the reaction was terminated by centrifugation at 8,800 x g in an Eppendorf minifuge (in a cold room at 4°C). The supernatants were decanted and the pellets were washed rapidly but superficially with 1 ml of 50 mM Tris-citrate buffer, pH 7.1. The pellet was then solubilized with 8 M urea (100 ml). 1 ml of aqueous-based scintillation fluid (ECO-Scint) was added, and the contents were thoroughly mixed. The minifuge tubes were placed in plastic scintillation vials and the radioactivity was assayed by liquid scintillation counting.

The taurine analogues studied are shown in Table 1. They were either obtained commercially or synthesized in the laboratory of H.B.F. Dixon. Binding and transport data were analysed by non-linear regression.

3. RESULTS AND DISCUSSION

[^{3}H]Taurine uptake into C6 glioma was found to be time-dependent and to require the presence of Na$^+$ ions. It was saturable, with a K_m = 30 ± 11 μM and V_{max} = 22 ± 2 pmol/2×10^5 cells/min being obtained by non-linear regression analysis. Uptake was not inhibited by HEPES at concentrations up to 10 mM. GES and β-alanine strongly inhibited [^{3}H]taurine uptake into C6 glioma cells (concentrations of 1 mM reducing the rate of transport of 20 μM taurine to 30 ± 2.5 and 21.6 ± 1.6 % of control values, respectively). N,N-dimethyltaurine (DMT), N,N,N-trimethyltaurine (TMT), 2-aminoethylphosphonic acid, 2-aminoethylarsonic acid, 3-aminopropylarsonic acid, D,L-3-arsonoalanine, isethionic acid, pyridine-3-sulphonic acid, piperidine-3-sulphonic acid, TAG, carnitine and choline did not inhibit glial [^{3}H]taurine uptake significantly.

N-Methyltaurine (MT) and GABA at concentrations of 1 mM reduced the rate of transport of 20 μM taurine to 61.6 ± 1.2 and 53.0 ± 4.2% of control values, respectively.

Initial rate studies at varying concentrations of [^{3}H]taurine and a series of fixed concentrations of N-Methyltaurine and GABA (in the concentration ranges 0–25 mM and 0–50 mM, respectively) inhibited glial [^{3}H]taurine uptake competitively. However, the dependence of the apparent K_m/V_{max} upon the competitor concentration was hyperbolic in both cases. Kinetic behaviour of this type would suggest either the presence of more than one transport system or interaction of these inhibitors at a site that is distinct from that for taurine uptake decreasing the affinity for taurine, rather than preventing its binding (partially competitive inhibition). Comparative studies on the interactions of these and other analogues with synaptosomal taurine uptake are in progress.

Taurine was found to inhibit the binding of [^{3}H]muscimol to the Triton X-100-washed membranes from rat brain, in agreement with the results of Bureau & Olsen (2), who had previously reported this compound to be a competitive inhibitor of muscimol binding to the GABA$_A$ receptor. MT, DMT and TMT also inhibited [^{3}H]muscimol binding to the high-affinity GABA$_A$ receptors in rat brain in a fully competitive manner with K_i values of 310 ± 37 μM, 270 ± 33.7 μM and 312 ± 37 μM, respectively. Preliminary studies have shown that [^{3}H]muscimol binding to the high-affinity GABA$_A$ receptors of rat brain (Triton X100-washed membranes) is also significantly inhibited by taurine, TAG and GES.

The different specificities of taurine analogues for GABA$_A$ receptors and for the glial-cell taurine transporter should result in useful tools for investigating the factors contributing to the function(s) of taurine in the central nervous system.

ACKNOWLEDGMENT

The support of the European Community BIOMED programme (Contract No. BMH1-CT94-1402) is gratefully acknowledged.

REFERENCES

1. Huxtable, R.J., 1992, *Physiol. Rev.* **72**: 101–160.
2. Bureau, M.H., and Olsen, R.W., 1991, *Eur. J. Pharmacol.* (Mol. Pharmacol. Section) **207**: 9–16.
3. Larsson, O.M., Griffiths, R., Allen, I.C., and Schousboe, A., 1986, *J. Neurochem.* **47**: 426–432.
4. Benda, P., Lightbody, J., Sato, G., Levine, L., and Sweet, W., 1968, *Science* **161**: 370–371.
5. Gordon-Weeks, P.R., 1987, *Neurochemistry*: a Practical Approach, (A. J. Turner, and H. S. Bachelard, eds.), IRL Press, Oxford, PP. 1–26.

THE EFFECT OF TAURINE AND SALINE ON NEUROACTIVE AMINO ACID LEVELS IN THE MOUSE BRAIN

M. Mijanovic,[1] K. Valjevac,[1] and Lj. Jozanc[2]

[1]Institute of Pharmacology and Toxicology
Fac. of Medicine, Univ. of Sarajevo
[2]Out-patient Clinic Vrazova
Sarajevo, Bosnia and Herzegovina

1. INTRODUCTION

Neurochemical and neurophysiological studies have focused attention on taurine as one of the major constituent of the free amino acid pool in mammalian neural tissue (1). Taurine is one of those enigmatic substances about which so much, and so little is known. This substance of high chemical stability and low metabolic reactivity (2) is present in relatively large amounts in the brain (3), heart (2), sperm (4), cultured glial cells (5) and skin (6). Even though the presence of taurine in the organism of mammals has been studied a lot over the period of the last few years, the role of this widely spread amino acid has not been proved yet. The entire list of effects is ascribed to this amino acid, from neurotransmitter (3,7) to osmoregulator (8), and the attention is focused on brain, in which taurine is present in high level. High concentrations of taurine in the brain led to the investigation of the role of this amino acid in the normal brain, as well as in mental and neurological disorders. The aim of this paper was to determine the influence of *i.v.* taurine on neuroactive amino acid levels in the mouse brain.

2. MATERIAL AND METHODS

Albino mice of both sexes (Veterinarska stanica, Zagreb) weighing 25 ± 3 g were used in all experiments. Mice were randomly divided into groups of 9 animals and kept under the same standard conditions. Food and water were *ad libitum*. Taurine (50 mg/kg) was administered intravenously to experimental animals. Changes of neuroactive amino

---

From a colloquium organized by Jan Albrecht (Warsaw, Poland) and Arne Schousboe (Copenhagen, Denmark). (Supported by Red Bull GmbH, Austria).

acid levels in whole mouse brain were determined by the TLC method (9), 15, 20, 30, 40 and 120 minutes after the injection. Control animals were injected intravenously with an equivalent amount of physiological saline solution (90 mg/kg) and sacrificed after the same time intervals as the experimental animals. The results were expressed in μmol/g brain tissue, as the mean ± standard error of the mean. Student t-test was used to determine significance. The obtained results were confirmed by high performance liquid chromatography (HPLC) technique. Standard samples of glutamic acid and GABA, then phenol, ninhydrin, ethanol and TLC-aluminium silica-gel 60/kieselguhr F_{254} pre-coated sheets (20 × 20 cm) were all obtained from "Merck", Darmstadt, Germany; taurine was from "Sigma Chemical Co.", St. Louis; ammonia, acetone and sodium chloride were from "Kemika", Zagreb, Croatia. All chemicals were analytical grade. The absorbance values were measured with a "Perkin Elmer" 124-UV/VIS spectrophotometer.

3. RESULTS AND DISCUSSION

As shown in Table 1, the concentrations of glutamic acid, taurine and GABA in the mouse brain, 20 minutes after i.v. taurine injection, were significantly increased ($p<0,001$) by 101%, 99% and 47%, respectively. In the period of 30 minutes after taurine injection, significant ($p<0,01$) decreases of glutamic acid and taurine by 22.5 and 37% respectively as well as a notable increase of GABA level by 32% were noticed. In 40 minutes a significant decrease ($p<0,001$) of glutamic acid level was registered and, at the same time, taurine and GABA did not differ from the control level. The data suggest that 20 minutes is the main time point when taurine exhibits its major effect. At that time point the levels of all investigated amino acids were significantly increased.

Surprisingly the levels of the investigated amino acids were markedly affected by the injection of saline itself. So, both the taurine and GABA levels decreased as compared with normal mouse brain. The results taken together indicate that *i.v.* taurine in dose of 50 mg/kg changes the levels of neuroactive amino acids, in the way that restores the balance

Table 1. The effect of i.v. injection of taurine (50 mg/kg) on the levels of glutamic acid, taurine and GABA in mouse brain compared to the control (μmol/g wet weight of brain)

		Glutamic acid	Taurine	GABA
Intact mouse brain		11,08 ± 0,08 (6)	8,33 ± 0,12 (6)	2,64 ± 0,17 (6)
15 min.	Saline	8,56 ± 0,75 (6)	3,40 ± 0,17 (6)	2,86 ± 0,16 (6)
15 min.	Tau 50 mg/kg	8,71 ± 1,07 (6)	3,73 ± 0,23 (6)	2,70 ± 0,17 (6)
20 min.	Saline	8,10 ± 0,25 (6)	4,48 ± 0,18 (6)	1,73 ± 0,04 (6)
20 min.	Tau 50 mg/kg	16,27 ± 0,39*** (6)	8,90 ± 0,16*** (6)	2,54 ± 0,09*** (6)
30 min.	Saline	12,67 ± 0,67 (5)	4,57 ± 0,04 (5)	2,18 ± 0,11 (5)
30 min.	Tau 50 mg/kg	9,82 ± 0,31** (6)	2,88 ± 0,10*** (6)	2,88 ± 0,09*** (6)
40 min.	Saline	17,87 ± 0,21 (6)	6,24 ± 0,25 (6)	2,75 ± 0,03 (6)
40 min.	Tau 50 mg/kg	12,13 ± 0,85*** (6)	6,22 ± 0,28 (6)	2,38 ± 0,22 (6)
120 min.	Saline	14,59 ± 0,48 (5)	6,10 ± 0,30 (6)	2,70 ± 0,03 (6)
120 min.	Tau 50 mg/kg	13,37 ± 0,25 (6)	7,70 ± 0,04*** (6)	2,57 ± 0,02 (6)

* $p < 0,05$
** $p < 0,01$
*** $p < 0,001$

between one excitatory (Glu) and one inhibitory (GABA) amino acid, consistent with the suggested nontransmitter, osmoregulatory function of taurine (8, 10–15).

4. CONCLUSIONS

Intravenous application of taurine modified the level of all investigated amino acids, compared to the control. At the same time, intravenous injection of saline caused significant changes of estimated amino acid levels compared to the intact mouse brain.

REFERENCES

1. Lombardini, J.B., 1977, *J. Neurochem.* **29**: 305–312.
2. Huxtable, R.J., Laird, H.E., and Lippincott, S.E., 1979, *J. Pharmacol. Exp. Ther.* **211**: 465–471.
3. Van Gelder, N.M., 1972, *Brain Res.* **47**: 157–165.
4. Holmes, R.P., Goodman, H.O., Shihabi, Z.K., and Jarow, J.P., 1992, *J. Andrology* **13**: 289–292.
5. Martin, L.D., Madelian, V., and Shain, W., 1989, *J. Neurosci. Res.* **23**: 191–197.
6. Watanabe, H., Watanabe, M., Jo, N., Kiyokane, K., and Shimada, M., 1995, *Cell. Molecul. Biol.* **41**: 49–55.
7. Davison, A.N., and Kaczmarek, L.N., 1972, *Nature* **234**: 107–108.
8. Jones, D.P., Miller, L.A., and Chesney, R.W., 1993, *Am. J. Phisiol.* **265**: F 137–145.
9. Mijanovic, M., and Gaon, I.D., 1897, *Farmacevtski Vestnik* **38**: 247–249.
10. Mijanovic, M., and Gaon, I.D., 1987, *Period. Biol.* **89**: 325–326.
11. Usami, S., and Ottersen, O.P., 1995, *Brain Res.* **676**: 277–284.
12. Goldstein, L., and Davis, M., 1994, *Am. J. Physiol.* **267**: R426–431.
13. Decavel, C., and Haton, G.I., 1995, *J. Compar. Neurol.* **354**: 13–26.
14. Nakada, T., Hida, K., and Kwee, I.L., 1992, *Neurosci. Res.* **15**: 115–123.
15. Shupliakov, D., Brodin, L., Srinivasan, M., Grillner, S., Cullheim, S., Storm Mathisen, J., and Ottersen, O.P., 1994, *J. Compar. Neurol.* **347**: 301–311.

NITRIC OXIDE IN THE CENTRAL NERVOUS SYSTEM OF RATS SUFFERING FROM RABIES VIRUS INFECTION OR EXPERIMENTAL ALLERGIC ENCEPHALOMYELITIS

A.-M. Van Dam,[1] S. R. Ruuls,[2] C. Marquette,[3] J. Bauer,[2] C. J. A. de Groot,[4] H. Tsiang,[3] C. D. Dijkstra,[2] and F. J. H. Tilders[1]

[1]Department of Pharmacology
Medical Faculty
Research Institute Neurosciences
Vrije Universiteit
[2]Department of Cell Biology and Immunology
[3]Rabies Unit
Institut Pasteur
Paris, France
[4]Department Pathology
Amsterdam, The Netherlands

1. INTRODUCTION

The interest in nitric oxide (NO) as an intercellular messenger in the brain has grown considerably over the last few years. Different types of enzymes involved in NO production have been found in the central nervous system. Constitutive forms of NO-synthase (cNOS) are present in neurons and brain endothelium and their activity is calcium- and calmodulin-dependent. In glial cells an inducible form of NO-synthase (iNOS) has been detected which activity is not controled by calcium.

The physiological functions of brain-derived NO have been a focus of intense investigation. It has been reported to control cerebral blood flow (1), to mediate long term potentiation in hippocampal neurons (2,3) and to increase cGMP levels in the cerebellum by activation of guanylate cyclase (4). Under neuropathological conditions e.g. brain ischemia, NO has been considered as a mediator of neuronal cell death (5,6). However, during inflammatory processes in the brain, the role of NO is less well defined. Recently,

---

From a workshop organized by Paul J.L.M. Strijbos (London, United Kingdom) and Jan De Vente (Maastricht, The Netherlands).

it has been shown that mRNA for iNOS is upregulated in the brain during several viral infections including herpes simplex, borna disease and rabies, and also after the induction of experimental allergic encephalomyelitis (EAE) (7). Here we will briefly describe the cellular localization of iNOS and the formation of nitrite in the central nervous system (CNS) of rats suffering from rabies virus infection or EAE.

2. RABIES VIRUS INFECTION

After infection of animals or man with rabies virus, various severe neurological symptoms occur during the course of this lethal disease. The most prominent symptoms are paralysis and alternate phases of anxiety, aggression and mental confusion (8,9). They are probably due to impairment of neuronal functioning as measured by electrical recordings of brain activity (10). Malfunctioning of the serotonergic system has been suggested to underly some of the clinical symptoms. For instance, a decrease in serotonergic receptor number in the cortex, striatum and hypothalamus has been reported (11). Neuropathologically, an encephalitis with immune cells infiltrating the brain, neuronal cell death and glial cell proliferation can be observed.

By using a rat model we investigated whether NO is produced in the brains of rabies virus infected animals. This model of rabies virus infection is routinely used at the rabies unit of the Institute Pasteur in Paris. Adult rats were injected i.m. in the masseter muscle with an inoculum consisting of 20% brain homogenate in phosphate-buffered saline prepared from Swiss neonatal mice (day 5–7) that have been inoculated i.c.v. with rabies virus (strain: CVS IP 15). Control rats received an i.m. injection with phosphate-buffered saline. Five days after infection when the rats are in the terminal stage of the disease, the brains of the rats were fixed by intracardial perfusion with 4% paraformaldehyde in 0.1 M phosphate buffer. After a 2 h postfixation period, the brains were sectioned on a vibratome (50 µm). Immunocytochemical procedures for iNOS and GSA-I-B4 isolectin (macrophage/microglia marker) staining were performed according to previous descriptions (12). iNOS positive cells were absent in the brains of control animals whereas massive amounts of iNOS positive cells were found throughout the brain of rabies virus infected rats. These cells represent activated microglia or infiltrated macrophages as determined by GSA-I-B4 isolectin staining. A large number of the iNOS positive cells were preferentially localized near bloodvessels and are therefore likely to be infiltrated blood-derived macrophages. The overall expression of iNOS in macrophages in the brain suggests that significant amounts of NO are produced which may play a role in the development of neurological symptoms and may mediate neuronal cell death. Further studies using iNOS inhibitors may contribute to the understanding of the role of NO in the brains of rats suffering rabies virus infection.

3. EXPERIMENTAL ALLERGIC ENCEPHALOMYELITIS

EAE is an experimentally induced inflammatory, autoimmune disease of the central nervous system (CNS) and serves as an animal model to study demyelinating diseases of the CNS in general and multiple sclerosis (MS) in particular (13).

EAE is induced by a subcutaneous injection of brain homogenate or with myelin basic protein (MBP), emulsified in complete Freund's adjuvant in susceptible animals (active immunization). Activated T-cells isolated from the spleen and lymph nodes of such immunized animals, can transfer the disease to naive recipients (AT-EAE). During the course of disease, the animals suffer from weight loss and general malaise preceeding the neurological deficits such as reduced tail tonus or complete paralysis of the tail. This is followed by paresis or complete paralysis of the hind limbs and sometimes paralysis up to the diaphragm. Neuropathologically, the animals show perivascular infiltrations of T-cells and monocytes/ macrophages forming lesions mainly localized in the spinal cord and cerebellum.

In the model used in our experiments, Lewis rats were immunized in the hind footpad with homogenates of guinea pig spinal cord emulsified in complete Freund's adjuvant. Control rats received an injection with complete Freund's adjuvant only. Before, during and after the expression of neurological symptoms, the rats have been transcardially perfused with 4% paraformaldehyde in 0.1 M phosphate buffer. After a 2 h postfixation, the cerebellum and spinal cord of the rats have been sectioned on a vibratome (50 µm; cerebellum) or cryostat (8 µm; spinal cord). Immunocytochemical procedures for iNOS, ED1 (macrophage marker) and NADPH-diaphorase staining were performed according to previous descriptions (12; Ruuls et al., submitted). In the brains of control rats, no iNOS positive cells can be localized. At the early onset of clinical symptoms (day 10 after immunization) a small number of iNOS positive cells can be found near or in lesion sites that are morphologically similar to ED1 positive macrophages and not to astrocytes present at these sites. At day 12–15 when clinical symptoms are overt, the total number of iNOS positive cells is increased in the spinal cord and cerebellum. At the same time high numbers of ED1 positive macrophages are localized in the lesion sites. At 24 days after immunization, all rats had recovered from clinical symptoms. Still, in the spinal cord large cellular infiltrates are present that contain ED1 positive macrophages and activated microglial cells. A very small percentage of these infiltrates contained iNOS positive cells. For ultimate cell identification we used a double-labeling technique of NADPH-diaphorase (representing cNOS and iNOS activity) which stains cells blue combined with ED1 that stains macrophages brown. ED1 positive macrophages and few microglial cells localized near lesions also demonstrate NADPH-diaphorase activity indicating that infiltrated macrophages and occasional microglial cells produce NO in the brains of rats suffering from EAE (Table 1).

In addition we studied the production of NO by brain macrophages isolated at different stages of the disease. Nitrite, the stable derivative of NO has been measured according to Ding (14) in the culture medium of cells isolated from brains and spinal cords (15) that have been cultured for 48 h. Cytospin preparations of such cell isolations stained with ED1 reveal that these cultures consist of more than 90% pure macrophages. Nitrite production showed a progressive increase per 10^5 cells at 0–10–12 days after active immunization. Cells collected 17 and 21 days after immunization showed a similar nitrite production as found at day 12. Thus, although we found a decrease in iNOS immunoreactivity in macrophages localized in lesions in situ during the recovery phase of EAE (day 24), this is not reflected in decreased nitrite production by macrophages isolated at this stage of the disease. This indicates that during the recovery phase of EAE mechanisms are activated in vivo that down regulate the expression of iNOS and production of NO by brain macrophages which is absent in vitro.

3.1. Is Nitric Oxide Production during EAE Good or Bad?

It has been demonstrated that oligodendrocyte cytotoxicity in vitro is mediated by microglial-derived NO (16). This strongly supports the idea that NO may contribute to demyelinating processes during EAE/MS resulting in clinical symptoms such as paralysis.

Various studies have now tried to elucidate the role of NO in the development of clinical symptoms during EAE. Although acute models of EAE have been used when loss of oligodendrocytes is scarce, some studies demonstrated an attenuation of the severity of clinical symptoms when mice or rats were treated with the NOS inhibitor aminoguanidine from the day they were transfered with activated T-cells to induce AT-EAE (17,18). In contrast, other studies showed no attenuation of clinical symptoms or even an aggravation of symptoms when rats were treated with the NOS inhibitors L-NAME or L-NMMA from day 7 after active immunization with myelin basic protein (MBP) (19,20). These apparantly conflicting results with regard to the role of NO in the development of clinical symptoms during EAE may be carried back to differences in EAE models and treatment protocols used by the various research groups. Moreover, NO is certainly involved in pathological processes but further studies are indispensable.

3.2. Regulatory Mechanism of Nitric Oxide Production during EAE

We have demonstrated that there is a local production of iNOS in lesion sites in the cerebellum and spinal cord during EAE which increases when clinical symptoms are present and decreases during the recovery phase of the disease. The fact that iNOS is hardly expressed further away from the lesion sites and is attenuated in the recovery phase of EAE suggests that a (local) NO down-regulatory mechanism is present in the central nervous system. In this way, the brain may be prevented from excessive production of NO that may result in too much damage to the brain.

Various immunological studies have reported that NO production in peripheral macrophages is inhibited by regulatory cytokines including IL-4, IL-10 and transforming growth factor β (TGFβ) (21). From these cytokines, only the presence of TGFβ has been demonstrated in the CNS so far. For instance, in post-mortem material of MS patients and in brains of rats suffering from EAE, TGFβ expression is detected in brain lesions (22,23). In addition it has been reported that astrocytes produce TGFβ in vitro (24). Moreover, when microglial cells are stimulated with endotoxin to produce NO, coculture of these cells with astrocytes strongly reduces iNOS expression and NO production (25). This reduction in NO production can be reversed by TGFβ antibodies (Vincent et al., submitted). Therefore, we put forward the hypothesis that locally produced TGFβ down-regulates the expression of iNOS in the brains of rats during EAE.

In support of this hypothesis it has been demonstrated that administration of recombinant TGFβ results in a reduction of clinical symptoms during EAE whereas administration of TGFβ antibodies aggravates the symptoms (26,27). However, whether TGFβ attenuates clinical symptoms of EAE by down-regulating NO production in the brain is still unknown and needs to be investigated.

4. MULTIPLE SCLEROSIS

The animal model EAE has been found to mimick the neuro(patho)logical features of MS to a certain extent. In the brains of rats suffering from EAE, we demonstrated an increased expression of iNOS and nitrite production during the course of disease. If NO plays a significant role in the disease course of EAE and MS, an elevated expression of iNOS in brain tissue of MS patients may be expected. So far, the expression of NADPH-diaphorase in reactive astrocytes in active demyelinating lesions has been reported (28,29). By using an antibody directed to human iNOS, we could localize iNOS positive cells in

Table 1. Production and cellular characterization of NO-related molecules in the central nervous system of rats suffering from rabies virus infection or EAE and in brains of MS patients

	Rabies Virus Infection	Experimental Allergic Encephalomyelitis	Multiple Sclerosis
iNOS immunoreactivity	+	+	+
nitrite production	ND	+	ND
in macrophages	+	+	+
in astrocytes	-	-	-
NADPH-diaphorase activity	ND	+	+
in macrophages		+	+
in astrocytes		-	+

ND = not determined
+ = present
- = absent

active brain lesions of MS patients. By comparing the morphology of the iNOS positive cells to cells stained for CD68 (KP1; macrophage marker) in adjacent sections, we conclude that in the brains of MS patients iNOS is produced by macrophages and not by reactive astrocytes (de Groot et al., submitted). In addition, iNOS mRNA was also recently demonstrated to be localized in macrophages in the brain of MS patients (30). Moreover, the NADPH-diaphorase staining in reactive astrocytes probably reflects cNOS activity of which the relevance for NO production under these conditions remains to be established (Table 1).

5. CONCLUSIONS

In summary, during rabies virus infection, EAE and MS, we demonstrated the induction of iNOS in infiltrated macrophages and nitrite production by these cells, reflecting the production of NO. This NO may contribute to the neuronal dysfunctioning that is typical for the respective diseases (Table 1). In particular, the role of NO during EAE needs further investigation because it has been demonstrated to act as a protective and as a damaging molecule. Furthermore, we propose that a NO-downregulatory mechnism is present in the brain, resulting in the decrease in NO production in the recovery phase of EAE and protecting the brain from excessive damage by NO.

REFERENCES

1. Tanaka, K., Gotoh, F., Gomi, S., Takashima, S., Mihara, B., Shirai, T., Nogawa, S., and Nagata, E. 1991, *Neurosc. Lett.* **127**: 129–132.
2. Schuman, E.M. and Madison, D.V. 1991, *Science* **254**: 1503–1506.
3. Bon, C., Bohme, G.A., Doble, A., Stutzmann, J-M., and Blanchard, J-C. 1992, *Eur. J. Neurosci.* **4**: 420–424.

4. Garthwaite, J., Charles, S.L., and Chess-Williams, R. 1988, *Nature* **336**: 385–388.
5. Nowicki, J.P., Duval, D., Poignet, H., and Scatton, B. 1991, *Eur. J. Pharmacol.* **204**: 339–340.
6. Garthwaite, J. 1991, *Trends NeuroSci.* **14**: 60–67.
7. Koprowski, H., Zheng, Y.M., Heber-Katz, E., Fraser, N., Rorke, L., Fu, Z.F., Hanlon, C., and Dietzschold, B. 1993, *Proc. Natl. Acad. Sci. U.S.A.* **90**: 3024–3027.
8. Tsiang, H. 1993, *Adv Virus Res* **42**: 375–411.
9. Murphy, F.A. 1977, *Arch. Virol.* **54**: 279–297.
10. Gourmelon, P., Briet, D., Court, L., and Tsiang, H. 1986, *Brain Res.* **398**: 128–140.
11. Ceccaldi, P-E., Fillion, M-P., Ermine, A., Tsiang, H., and Fillion, G. 1993, *Eur. J. Pharmacol.* **245**: 129–138.
12. Van Dam, A-M., Bauer, J., Man-A-Hing, W.K.H., Marquette, C., Tilders, F.J.H., and Berkenbosch, F. 1995, *J. Neurosci. Res.* **40**: 251–260.
13. Raine, C.S. 1984, *Lab. Invest.* **50**: 608–635.
14. Ding, A.H., Nathan, C.F., and Stuehr, D.J. 1988, *J. Immunol.* **141**: 2407–2412.
15. de Groot, C.J.A., Sminia, T., and Dijkstra, C.D. 1989, *Immunobiology* **179**: 314–327.
16. Merrill, J.E., Ignarro, L.J., Sherman, M.P., Melinek, J., and Lane, T.E. 1993, *J Immunol* **151**: 2132–2141.
17. Cross, A.H., Misko, T.P., Lin, R.F., Hickey, W.F., Trotter, J.L., and Tilton, R.G. 1994, *J. Clin. Invest.* **93**: 2684–2690.
18. Zhao, W., Tilton, R.G., Corbett, J.A., McDaniel, M.L., Misko, T.P., Williamson, J.R., Cross, A.H., and Hickey, W.F. 1996, *J. Neuroimmunol.* **64**: 123–133.
19. Zielasek, J., Jung, S., Gold, R., Liew, F.Y., Toyka, K.V., and Hartung, H.P. 1995, *J. Neuroimmunol.* **58**: 81–88.
20. Ruuls, S.R., Van Der Linden, S., Sontrop, K., Huitinga, I., and Dijkstra, C.D. 1996, *Clin. Exp. Immunol.* **103**: 467–476.
21. Oswald, I.P., Gazzinelli, R.T., Sher, A., and James, S.L. 1992, *J. Immunol.* **148**: 3578–3582.
22. Woodroofe, M.N. and Cuzner, M.L. 1993, *Cytokine* **5**: 583–588.
23. Issazadeh, S., Mustafa, M., Ljungdahl, A., Hojeberg, B., Dagerlind, A., Elde, R., and Olsson, T. 1995, *J. Neurosci. Res.* **40**: 579–590.
24. Da Cunha, A. and Vitkovic, L. 1992, *J. Neuroimmunol.* **36**: 157–169.
25. Vincent, V.A.M., Van Dam, A-M., Persoons, J.H.A., Schotanus, K., Steinbusch, H.W.M., Schoffelmeer, A.N.M., and Berkenbosch, F. 1996, *GLIA* **17**: 94–102.
26. Santambrogio, L., Hochwald, G.M., Saxena, B., Leu, C-H., Martz, J.E., Carlino, J.A., Ruddle, N.H., Palladino, M.A., Gold, L.I., and Thorbecke, G.J. 1993, *J. Immunol.* **151**: 1116–1127.
27. Johns, L.D. and Sriram, S. 1993, *J. Neuroimmunol.* **47**: 1–8.
28. Bo, L., Dawson, T.M., Wesselingh, S., Mork, S., Choi, S., Kong, P.A., Hanley, D., and Trapp, B.D. 1994, *Ann. Neurol.* **36**: 778–786.
29. Brosnan, C.F., Battistini, L., Raine, C.S., Dickson, D.W., Casadevall, A., and Lee, S.C. 1994, *Dev. Neurosci.* **16**: 152–161.
30. Bagasra, O., Michaels, F.H., Zheng, Y.M., Bobroski, L.E., Spitsin, S.V., Fu, Z.F., Tawadros, R., and Koprowski, H. 1995, *Proc Natl. Acad. Sci U.S.A* **92**: 12041–12045.

THE EFFECT OF GLUTAMATE, ANTIOXIDANTS, AND GANGLIOSIDES ON Na^+,K^+-ATPASE ACTIVITY IN RAT BRAIN CORTEX SYNAPTOSOMES

N. F. Avrova, V. A. Tyurin, I. O. Zakharova, T. V. Sokolova, and Y. Y. Tyurina

Institute of Evolutionary Physiology and Biochemistry
Russian Academy of Sciences
St.-Petersburg, Russia

1. INTRODUCTION

Excitotoxicity mediated by glutamate is considered to contribute significantly to neuronal damage and death in various neurodegenerative conditions including ischemic injury. The decrease of Na^+,K^+-ATPase activity found to take place in ischemic and other neurodegenerative conditions leads to defective mechanism of repolarization and removal of potential-dependent Mg^{2+} block of NMDA receptors (1). It may result from ATP depletion or from the effect of glutamate leading to increased phosphorylation of this enzyme by protein kinase C (PKC) which causes its inhibition (1–2). According to the recent data the increase in PKC content and translocation to membrane fraction is accompanied by quick loss of its activity at early stages of ischemia (10–15 min) or glutamate application. In some models in fact activation was shown of Na^+,K^+-ATPase as a result of modulation of the degree of its phosphorylation by glutamate (2, 3). Another possibility of modulation of activity is inactivation of the enzyme owing to the increased formation of reactive oxygen species (ROS), which oxidize SH groups of the enzymes and cause accumulation of lipid peroxidation products. Taking into account the above-mentioned data one of the main mechanisms of the decrease of Na^+,K^+-ATPase activity in ischemic conditions and by glutamate may be its inactivation by ROS.

The aim of the present work was to see if the application of glutamate to rat brain cortex synaptosomes causes a significant decrease of Na^+,K^+-ATPase activity and to show the modulation of this effect by antioxidant alpha-tocopherol, enzyme of antioxidative defence — superoxide dismutase (SOD), and gangliosides which inhibit free radical reactions (4, 5), as we proved earlier on.

From a workshop organized by Paul J.L.M. Strijbos (London, United Kingdom) and Jan De Vente (Maastricht, The Netherlands).

2. MATERIALS AND METHODS

Synaptosomes were isolated from brain cortex of Wistar rats (6); they were shown to retain their integrity by electron microscopic and fluorescent methods using potential sensitive probe dis-C_3-(5). They were incubated in tris-HCl-buffered saline (4), 1 mM glutamate was applied for 15 minutes. Preincubation of synaptosomes with 10^{-11}–10^{-7} M of ganglioside GM1 was performed for 1 hour, with 10 U/ml of SOD and 10^{-4} M of alpha-tocopherol — for 15 minutes.

Na^+,K^+-ATPase activity was measured by continuously monitoring the oxidation of NADH at 340 nm with linked enzyme system (7); reaction was started by addition of 2 mM ATP. The spectra were registered using spectrophotometer Specord 40 (Germany). Na^+,K^+-ATPase activity was estimated as the difference between the rate of the reaction in the presence and absence of 0,05% strophantin G (7). The statistical significance of differences was evaluated by Student's t test.

3. RESULTS AND DISCUSSION

The application of glutamate to rat brain cortex synaptosomes was found to result in the significant decrease of Na^+,K^+-ATPase activity (table), the mean ± S.E.M. of 10 experiments was 70.9 ± 3.2 % of the controls ($p<0,05$ by Student's t test or $p<0.01$ by paired Student's t test). The control values taken for 100% constituted 14.7 ± 0.9 μmol NADH/mg protein/hour. The degree of inhibition varied from 10 to 60 % in individual experiments. In the presence of antagonist of NMDA receptor MK-801 this effect was not revealed. The inhibition of Na^+,K^+-ATPase activity appears to be the consequence of the enchancement of the ROS formation and of the decrease of SH group content of proteins in synaptosomes caused by glutamate shown by us to take place under the experimental

Table 1. The effect of glutamate and various concentrations of GM1 on the Na^+,K^+-ATPase activity in rat brain cortex synaptosomes

Sample	GM1 concentration	Na^+,K^+-ATPase activity % of control values	Number of experiments
Control	-	100 %	10
Glutamate	-	70.9±3.2 *	10
Glutamate	10^{-8} M	95.8±6.6 **	10
Glutamate	$4·10^{-8}$ M	102.3±6.4 **	8
Glutamate	10^{-7} M	98.1±7.0 **	10

M±s.e.m. Synaptosomes were preincubated during 1 hour with various concentrations of GM1 or without it, then glutamate was added for 15 minutes. *Significant difference from control values by paired Student's t test ($p<0.01$). **Significant difference from data on the effect of glutamate alone by Student's t test ($p<0.01$).

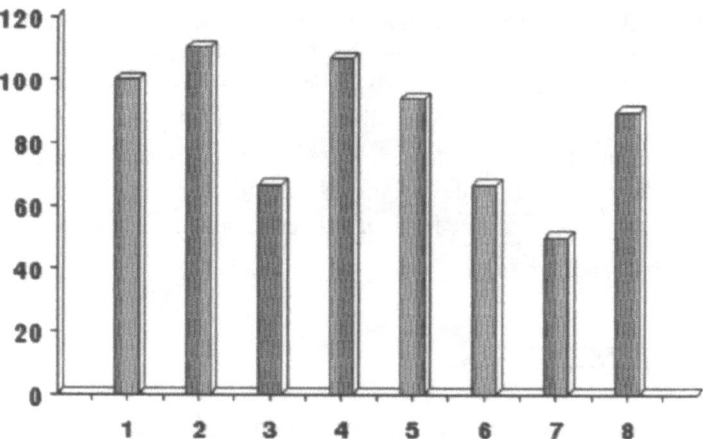

Figure 1. The effect of glutamate, alpha-tocopherol and superoxide dismutase on the Na^+,K^+-ATPase activity in rat brain cortex synaptosomes. (Results of one typical experiment from 3 experiments carried out in duplicate). 1 — control, 2 — SOD (10 U/ml), 3 — 1 mM glu, 4 — 1 mM glu + SOD (10U/ml), 5 — 1 mM glu + 10^{-4} M alpha-tocopherol, 6 — 0.5 mM NADPH, 7 — 1 mM glu + 0.5 mM NADPH, 8 — 1 mM glu + 0.5 mM NADPH + 10^{-4} alpha-tocopherol.

conditions used (in preparation). Glutamate also significantly diminished the reduced glutathione and ascorbate content in brain cortex homogenate and increased MDA content in cultured cerebellar granule cells.

Preincubation of synaptosomes with 10^{-8}, $4 \cdot 10^{-8}$, 10^{-7} M of ganglioside GM1 was shown to prevent the partial inactivation of Na^+,K^+-ATPase by glutamate, the differences between the effect of glutamate alone and in the presence of GM1 being statistically significant ($p<0.01$) (see Table 1). This effect of GM1 was dose-dependent in the range from 10^{-11} to 10^{-8} M. The activity of Na^+,K^+-ATPase after preincubation with 10^{-8}–10^{-7} M GM1 in the presence of glutamate did not differ from control values ($p>0.05$ in each case). It appears that the effect of GM1 is not the consequence of the activation of the enzyme because in the conditions of experiments no significant difference was observed in its activity in the presence of various concentrations of GM1 and in the absence of glutamate as compared to controls (not shown).

The suggestion that preincubation with GM1 prevents the inactivation of enzyme caused by oxidation of SH groups is based on our previous findings. According to our data gangliosides GM1, GT1b, GD1a etc. inhibit the free radical reactions induced in synaptosomes or in vivo, prevent oxidative destruction of enzymes, receptors and lipid fatty acids. Their effect is not observed in liposomes made of lipids isolated from synaptosomal membranes or in denatured synaptosomes or in the presence of polymyxin B (inhibitor of lipid-dependent protein kinases) and appears to be mediated by enzymes (4, 5). All this was confirmed in our recent experiments. Gangliosides were shown to prevent the generation of ROS and the decrease of protein SH groups in synaptosomes caused by glutamate.

Furthermore, in the present studies it was shown that the inactivation of Na^+,K^+-ATPase by glutamate may be prevented by alpha-tocopherol and SOD. Thus, in three experiments glutamate was found to diminish the enzymatic activity in synaptosomes to 53.3 ± 1.5% of the controls, but in the case of the preincubation of synaptosomes with SOD prior to glutamate application the activity of the enzyme constituted 105.1 ± 8.1% of the control values ($p<0.05$, n=3). In Figure 1 the results of a typical experiment are given. It is shown that in the presence of NADPH (cofactor of NO synthase) the decrease of Na^+,K^+-ATPase activity is even more pronounced, alpha-tocopherol normalizes to a great extent the activity of the enzyme inactivated by glutamate alone or by combined effect of NADPH and glutamate.

In conclusion, the data provide evidence that inhibitory effect of glutamate on Na^+,K^+-ATPase activity in rat brain synaptosomes is mediated by activation of free radical reactions. The results obtained appear to contribute to the understanding of the mechanism of the protective effect of gangliosides against glutamate neurotoxicity shown by Manev and co-authors (8) and were confirmed by our studies in which we used cultured rat cerebellar granule cells (in preparation).

ACKNOWLEDGMENT

Supported by the Grant from Russian Foundation for Fundamental Research N 94-04-11425.

REFERENCES

1. Zeevalk, G., Nicklas, W.J., 1992, *J. Neurochem.*, **59**: 1308–1314.
2. Marcaida, G., Kosenko, E., Minana, M.-D., Grisolia, S., Felipo, V., 1996, *J. Neurochem.*, **66**: 99–104.
3. Cardell M., Wieloch T., 1993, *J. Neurochem.*, **61**: 1308–1314.
4. Tyurin, V.A., Tyurina, Y.Y., Avrova, N.F., 1992, *Neurochem. Int.*, **20**: 401–407.
5. Tyurina, Y.Y., Tyurin, V.A., Avrova, N.F., 1993, *Mol. @Chem. Neuropathol.*, **19**: 205–217.
6. Hajos, F., 1975, *Brain Res.*, **93**: 485–489.
7. Leon, A., Facci, L., Toffano, G., Sonnino, S., Tettamanti, G., 1981, *J. Neurochem.*, **19**: 350–357.
8. Manev, H., Favaron, M., Siman, R., Guidotti, A., Costa, E., 1991, *J. Neurochem.*, **57**: 1288–1295.

NITRIC OXIDE LEVEL DRAMATICALLY INCREASES IN THE RAT BRAIN DURING EPILEPTIFORM SEIZURES

V. Bashkatova,[1] A. Vanin,[2] V. Mikoyan,[2] G. Vitskova,[1] V. Narkevich,[1] and K. Rayevsky[1]

[1]Institute of Pharmacology
Russian Academy of Medical Sciences
Moscow, Russia
[2]Institute of Chemical Physics
Russian Academy of Sciences
Moscow, Russia

1. INTRODUCTION

Nitric oxide (NO), a highly reactive, short-lived radical, is believed to be a transneuronal messenger in the central nervoussystem. NO is generated by the enzyme NO-synthase, which was shown to occur widely in the central nervous system. NO has been implicated in several brain disorders such as Alzheimer's and Parkinson's deseases, stroke, trauma, epilepsy, etc. Under these pathological conditions excessive release of excitatory amino acid glutamate results in increase of NO content in the brain to a toxic level. Epilepsy is one of the diseases in which excitatory amino acids are implicated and NO could be an important pathogenetic factor in the mechanisms that underlie seizure induction, propagation and progression. However, acute effects of neuronal NO overflow on brain functions are still a matter of controversy. NO-synthase, the enzyme responsible for NO synthesis in mammalian brain from L-arginine, can be blocked by arginine analogues such as L-NNA. These compounds can be used as a tool to investigate the role of nitric oxide in various cellular mechanisms. Recent reports claim NO to be either pro- or anticonvulsant (1,2). The aim of the present study was to prove whether model convulsive states of different origin in rats may involve any alteration in NO content and in content of secondary products of lipid peroxidation in the brain cortex.

From a workshop organized by Paul J.L.M. Strijbos (London, United Kingdom) and Jan De Vente (Maastricht, The Netherlands).

2. MATERIALS AND METHODS

Male Wistar rats weighing 200–220 g were used. Various epilepsy models were used such as maximal electroshock seizure (MES, 150mA, 0,2 sec), convulsions induced by administration of thiosemicarbazide (TSC), 30 mg/kg, penthylenetetrazol (PTZ), 120 mg/kg (both subcutaneously, s.c.) and N-methyl-D,L-aspartate (NMDLA), 28 ug (intracerebroventricularly, i.c.v.). N-nitro-L-arginine (L-NNA) was dissolved in saline and injected intraperitoneally (i.p.) in doses of 10 or 250 mg/kg. All chemicals were from Sigma. L-NNA was administered 30 min prior to convulsive agent application. Immediately after injection of convulsants, the rats were placed in plexiglas boxes and were observed for a period of 1–2 hours. The convulsive response appears 2–3 min after injection of PTZ or NMDLA and 40–60 min after TSC . Animals were sacrificed by decapitation 30 min after the treatment with NO scavengers, the brains were quickly removed, frontal cortex was dissected and frozen for biochemical analysis. Determination of NO content in the rat brain structures was performed as follows (3): diethyldithiocarbamate (DETC), 500 mg/kg i.p. and mixture of FeSO4, 37.5 mg/kg and sodium citrate 165 mg/kg s.c. were injected. DETC penetrated the blood-brain barrier and formed a complex with intracellular non-haem ion. This complex trapped NO in the tissue, thereby generating a paramagnetic mononitrosyl-iron DETC complex. The EPR spectra were recorded at 77 K on a Brucker EPR 300E spectrometer at a frequency of 9.33 kHz, hf-modulation frequency 0.5 mT, microwave power 20 mW, and time constant 0.05 s. The concentration of NO trapped was calculated from the intensity of the third hyper-fine splitting line of the resonance at $g = 2.035$ (= perpendicular) of the NO-Fe-(DETC) complex (Fig.1).

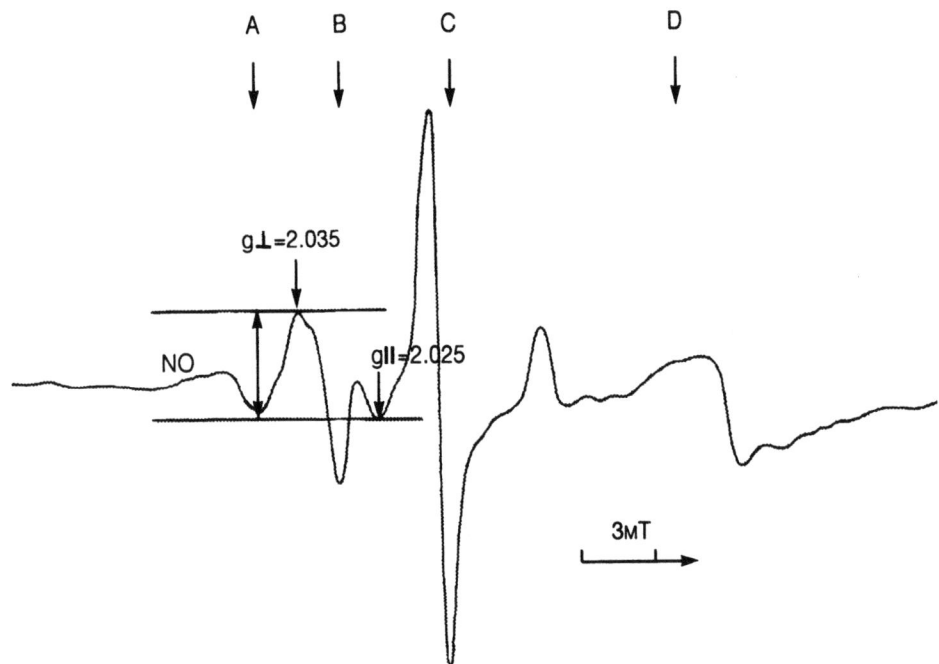

Figure 1. EPR signals obtained from the brain cortex of control rat. "NO" (arrow) indicates the size of NO-Fe-DETC signal.

Determination of Lipid Peroxidation processes in brain tissue was performed by measuring thiobarbituric acid reactive substances — TBARS (4). Briefly, 10 percent (w/v) tissue homogenate was mixed with sodium dodecyl sulfate, acetate buffer (pH 3,5) and aqueous solution of thiobarbituric acid. This mixture was heated and kept at 950°C for 60 min. The red pigment produced was then extracted with n-butanol-pyridine mixture and estimated at 532 nm on account of the absorbance. All results are expressed as means + s.e.m. For statistical analysis, the Mann-Whitney U-test for non-parametric data was used. Student's t-test was used for the other comparisons. Difference was considered to be significant for $P<0.05$.

3. RESULTS

In our experimental conditions MES was manifested by typical tonic hindlimb extension followed by clonic convulsions. In chemical models of seizures the convulsive response consisted of first a twitch, then short-lasting episodes of clonic seizures (false clonic convulsions), continuous clonic seizures with wild running and ended with falling of the animal (genuine clonic convulsions) and tonic seizures. Injections with scavenger substanceswere well tolerated by the rats and produced no apparent behavioral changes. Two doses of L-NNA (10 and 250 mg/kg) were tested on the seizures induced by PTZ, in experiments with MES 250 mg/kg of L-NNA was the only dose used. L-NNA (10 mg/kg) was effective as it significantly reduced the latency of the first convulsive twitch (1.2 ± 0.7 min), however, it failed to effect the latency of tonic seizures. A similar but more pronounced effect was observed after injection of L-NNA (250 mg/kg). In contrast to the lower dose used, this dose significantly prolonged the latency of the onset of each behavioral sign. Nearly 4-fold increase in the brain NO level (Fig.2) and 2-fold elevation of TBARS content (data are not shown) were found at the peak of the tonic extension phase of the MES in comparison to control. NO was entirely absent when L-NNA (250 mg/kg), an inhibitor of NO synthase was injected into animals one hour prior to MES application. However the rats pretreated with L-NNA did manifast clonic seizures but there were no tonic ones.

The similar increase in brain NO level was found under the seizures induced by TSC, PTZ or NMDLA (Fig.3).

Lipid Peroxidation processes measured as the content of TBARS from the cerebral cortex of rats after MES was found to have increased to 138% v/s control. In rats injected with PTZ, TSC and NMDLA corresponding level of LPO products appeared to have increased up to 82, 67 and 200% in comparison with control. Pretreatment of rats with NO-synthase inhibitor L-NNA (250 mg/kg) failed to affect the increased TBARS level.

4. DISCUSSION

Our results show that convulsive model states of different origin in rats, i.e. produced by electrical superstimulation (MES) or by injection of several convulsive agents that interact with different neuronal targets are followed by the significant enhancement of lipid peroxidation processes and a dramatic increase in NO generation in the brain. This phenomenon described earlier for convulsions in rats treated with kainic acid (5) might be attributed to an activation of the enzyme NO-synthase (NOS). It is our hypothesis that a burst-like increase in the brain NO content following immediately after convulsive seizure

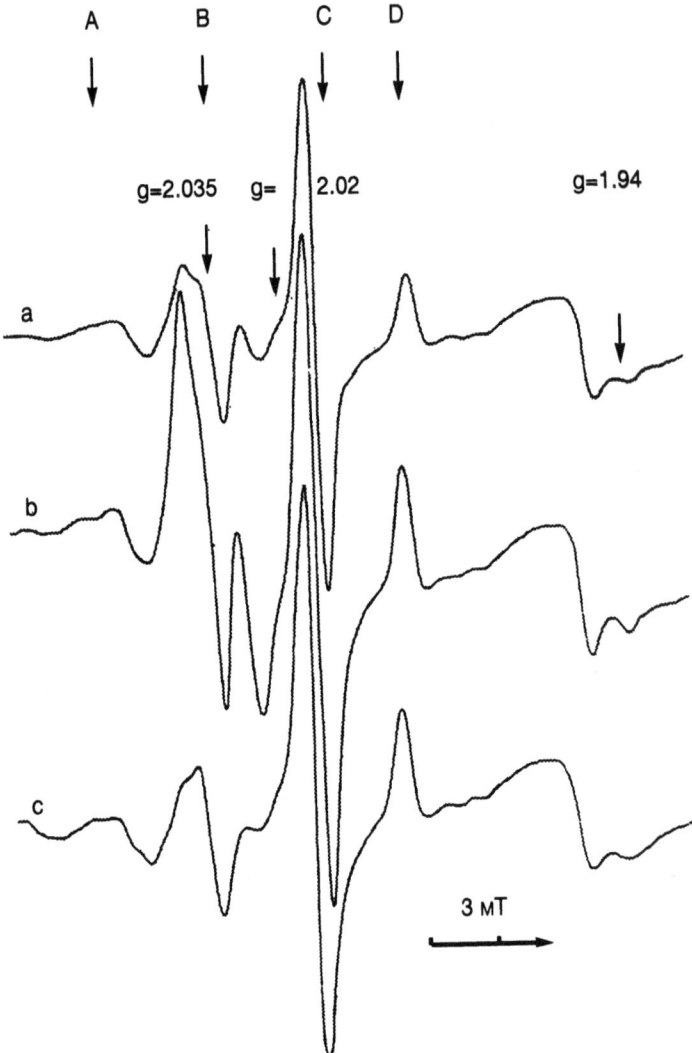

Figure 2. Typical EPR spectra obtained from brain cortex of rats pretreated with scavenger substances 30 min before decapitation: control (a); MES (b); L-NNA one hour prior MES (c). EPR conditions were the same as those in Fig.1.

is initiated by an overstimulation of ionotropic glutamate receptor subtypes, i.e. NMDA or AMPA/kainate, and due to the activation of a particular type of NOS, e.g. type I (neuronal NOS), type II (inducible NOS) or type III (endothelial NOS). It is known that seizure followed by a substantial increase in lipid peroxidation in brain tissue might be diminished by substances with antioxidant properties. From our results it could be speculated that NO formation plays an important role in the regulation of seizures. It is still controversial whether NO is functioning as pro- or anticonvulsant. NO may exert protective activity on the early stages of epileptogenesis and fail to affect further phases of seizures. The main results of this study, i.e. the rapid burst-like increase in NO content in the rat brain cortex observed during or immediately after the convulsive seizure support recently published data (5,6). These results are in agreement with the hypothesis on the trigger role of NO in

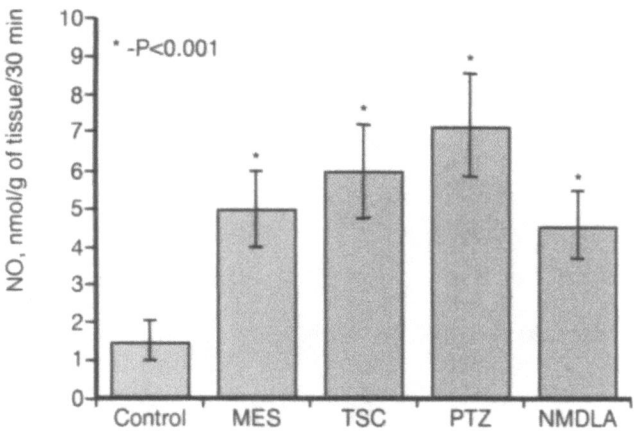

Figure 3. The enhancement of NO formation in brain cortex of rats with convulsive seizures of different origin. Each column represents NO level in saline control, maximal electroshock seizure (MES), thiosemicarbazide (TSC), penthylenetetrazol (PTZ), N-methyl-D,L-aspartate (NMDLA).

pathophysiology of convulsive disorders. The mechanism of this phenomenon is currently under investigation.

ACKNOWLEDGMENTS

This work was supported by the Russian Fund for Fundamental Research (RFFI, project code 95-0411861a) and INTAS (project code 94-500).

REFERENCES

1. De Sarro, J., Di Paola, E.D., De Sarro, A., Vidal, M.J., 1993, *Eur. J. Pharmacol.* **230**: 151–158.
2. Maggio, R., Fumagally, F., Donati, E. et al., 1995, *Brain. Res.* **679**: 184–187.
3. Mikoyan, V.D., Kubrina, L.N., Vanin, A.F., 1994, *Biophisica* (Russian), **39**: 915–918.
4. Ohkava, H., Ohishi, N., Yagi, K., 1979, *Analyt. Biochem.* **95**: 351–358.
5. Mulsch, A., Busse, R., Mordvintcev, P., Vanin, A., Nielsen, E., Scheel-Kruger, J., Olesen, S.P., 1994, *NeuroReport.* **5**: 2325–2328.
6. Bashkatova, V.G., Mikoyan, V.D., Kosacheva, E.S., Kubrina, L.N., Vanin, A.F., Rayevsky, K.S., 1996, *Dokl. Acad. Nauk* (Proc. Russ. Acad. Sci.). **348**: 119–121.

Section 35: Free Radicals, NO, and Brain Pathology

RESISTIVITY OF LEECH RETZIUS NERVE CELLS TO LONG-LASTING OXIDANT

Zorica Jovanovic and B. B. Beleslin

Department of Pathological Physiology
Faculty of Medicine
Belgrade, Yugoslavia

1. INTRODUCTION

Numerous experimental and clinical observations suggest that reactive oxygen species (ROS) play a significant role in the primary and secondary processes of tissue injuries (1, 4, 6). Nervous systems are especially sensitive to ROS for several reasons (9). The neuronal membrane lipids are particularly enriched in cholesterol, polyunsaturated fatty and ascorbic acid. On the other hand, the brain is poor in catalytic activity and has moderate amounts of glutation peroxidase and superoxyde dismutase. Lipid peroxydation, the undesired oxidative modification of the polyunsaturated fatty acyl chain is a major contributor to membrane damage in cells (5).The idea that oxygen radicals are important mediators of brain damage related to ischemia and reperfusion is not new (7). Recently, it has been demonstrated that free radicals are generated during both ischemic and reperfusion periods in the rat brain and their levels increase during the onset of reperfusion (13). The results obtained with sinoatrial heart nodes of rabbits showed that exposure to *tetra*-butylhydroxide leads to an increase of inward rectifying potassium current which activates a negative potential and may maintain the resting potential of cells (10). On the other hand, under the whole cell patch clamp, exposure of guinea pig ventricular myocytes to oxygen radical did not affect inward rectifier potassium current, suggesting that cell membrane was relatively intact (3). The present study therefore, was carried out to investigate the effect of chemical mixtures of long lasting oxidant, hydrogen peroxide and ferrous chloride which produce free radicals, on resting membrane potential in leech Retzius nerve cell. Our data support the view that leech ganglia have an efficient defence system against reactive oxygen species.

From a workshop organized by Paul J.L.M. Strijbos (London, United Kingdom) and Jan De Vente (Maastricht, The Netherlands).

2. MATERIAL AND METHODS

Our experiments were performed on Retzius nerve cells in the abdominal ganglia of horse leeches (*Haemopis sanguisuga*). The method of dissection was identical to that described previously (2).

The resting potential was recorded intracellularly with glass microelectrodes filled with 3 M KCl having resistance of about 10 MΩ in leech saline. Usually the microelectrode was connected through Ringer bridge and Ag-AgCl electrode *via* a negative capacitance high input resistant amplifier to a pen recorder. With respect to the other chloride silver electrode in Ringer solution: it communicates with test solution in the bath by way of glass tube filled with 3 M KCl in agar. Microeletrodes with a tip potential less than 5 mV in artificial solution, were selected for use. The ganglia were bathed in leech Ringer containing (mM/l): 115 NaCl, 4 KCl, 2 $CaCl_2$, 1.2 Na_2HPO_4, 0.3 NaH_2PO_4 (pH 7.2). Experiments were performed on leech Retzius nerve cells. Changes in leech Ringer did not by themselves induce changes in resting potential and firing rate of spike potentials. Chemical mixtures of hydrogen peroxide and ferrous chloride, or each chemical separately, were added to Ringer. The bath volume was 2 ml and the solution changes were completed within 30 s. All experiments were done at room temperature (20–24° C).

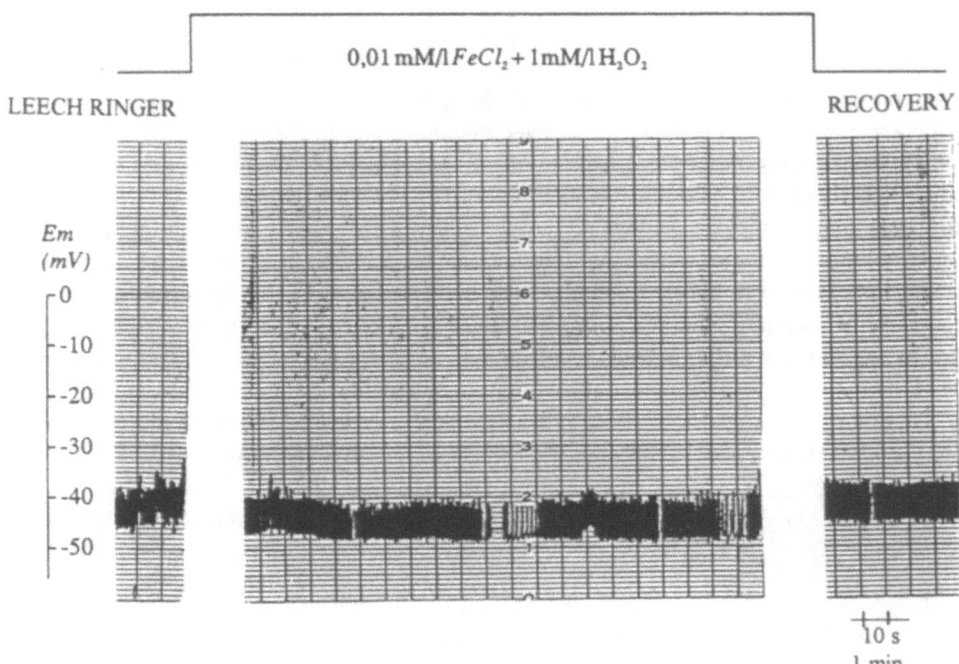

Figure 1. Pen recording of an experiment showing the effect of 1 mM/l hydrogen peroxide and 0.01 mM/l $FeCl_2$ on resting membrane potential of Retzius nerve cells. The experiment was started with the ganglion bathed in leech Ringer solution for 15 minutes.

3. RESULTS

Transition metal ions are important in the production of radical species. In the presence of hydroxide peroxide and a transient metal such as iron, the extremely reactive hydroxyl radical may be generated (12). Figure 1 illustrates one example in which Retzius nerve cells have been washed out with a mixture of 1 mM/l hydroxide peroxide and 0.01 mM/l $FeCl_2$.

As can be seen there were only some insignificant changes in resting membrane potential. Similar effect has been obtained in control experiments after adding only 0.01 mM/l $FeCl_2$.

4. DISCUSSION

The data presented in this study clearly demonstrate that hydrogen peroxide treatment with ferrous chloride did not significantly change the resting membrane potential of leech Retzius nerve cells. The present results suggest that leech ganglia have an efficient system against oxidative stress.

The negative effect of hydrogen peroxide and ferrous chloride is little surprising since transition metals act as catalysts in the oxidative deterioration of biological macromolecules. A variety of studies have demonstrated that the toxicity associated with these metals may lie in the ability of iron chelates to catalyse the formation of reactive oxygen species. Evidence, indicates that chelated ion acts as a catalyst for the Fenton reaction, facilitating the conversion of superoxide anion and hydrogen peroxide to hydroxyl radical, a species frequently proposed to start lipid peroxydation (12).

In our case it is difficult to suppose that peroxidation of lipid membrane is involved, since it was not possible to obtain significant changes in resting membrane potential. This is something unexpected since hydrogen peroxide at 1 mM/l concentration results in plasma membrane depolarisation, and increases plasma membrane blebbing, cell swelling and membrane permeability (11). On the other hand peroxidation of membrane lipids has demonstrated to affect various transmembrane processes such as receptor activation and formation of intracellular second messenger and calcium homeostasis (14).

There are several explanations why leech Retzius nerve cells should be resistant to reactive oxygen species. The simplest could be that leech Retzius nerve cells have a low concentration of polyunsaturated fatty acids, which are very sensitive to radical injury. This possibility is unlikely since neuronal membrane are rich in lipids. Another explanation could be that they have an efficient scavenging enzyme system which reacts rapidly with reactive oxygen species. According to biochemical studies, hydroxyl radical that is generated from hydrogen peroxide by Fenton reaction easily removes hydrogen atoms from methylene carbon of an unsaturated fatty acid of membrane phospholipids (8). Peroxidation of lipid that inactivates membrane-associated enzymatic protein, increases membrane permeability. This metabolic and physico-chemical alteration of cell membrane would produce intracellular Ca^{2+} overload. However, since we have insignificant changes in resting potential with hydrogen peroxide, it is reasonable to suppose that lipid membrane peroxidation, which could induce permeability changes did not occur. On the other hand since changes in the resting membrane potential were not significant it could further be expected that hydrogen peroxide was decomposed by a number of enzymatic and non-enzymatic antioxidant systems in order to protect other cellular components against harmful effects of these reactive oxygen metabolites.

In conclusion it is obvious that leech ganglia are a good model for studying protective mechanisms against reactive oxygen species.

5. SUMMARY

The effect of the long lasting oxidant, hydrogen peroxide, on leech Retzius nerve cell was investigated. It was found that a mixture of 1 mM/l hydrogen peroxide and 0.02 mM/l $FeCl_2$ did not significantly change the resting potential in leech Retzius nerve cells. A similar effect has been obtained in control experiments after adding only 0.01 mm/l $FeCl_2$. It has been concluded that leech ganglia are good models for studying protective mechanisms against reactive oxygen species.

ACKNOWLEDGMENT

This work has been supported by Ministry of Science and Technology of Serbia.

REFERENCES

1. Ames, B.N., Shigenaga, M.K. and Hagen, T.M., 1993, *Proc. Natl. Acad. Sci.,* **90**, 7915–7922.
2. Beleslin, B. and Mihailovic, Lj., 1967, *Iugosl. Physiol. Phramacol. Acta*, **3**: 85–86.
3. Cerbai, E., Ambrosio, G., Porciatti, F., Chiariello, M., Giotti, A. and Mugelli, A., 1991, *Circulation*, **84**: 1773–1782.
4. Del Maestro, R., 1991, *The Humana Press Inc.* pp. 25–51.
5. Dix, T. and Aikens, J.H., 1993, *Chem. Res. Toxicol.* **6,** 2–18.
6. Faden, A.I., 1987, *Clin. Neuropharmacol.* **10**: 193–204.
7. Flamm, E.S., Demopoulos, H.B., Seligman, M.L., Posner, R.G. and Ransohoff, J., 1978, *Stroke*, **9**: 445–447.
8. Hess, M.L. and Manson, N.H., 1994, *J. Mol. Cell. Cardiol.* **16**: 969–986.
9. Kostos, H.A. and Povilshock, J.T., 1986, *CNS Trauma*, **3**: 257–263.
10. Sato, N., Nishimura, M., Tanaka, H., Homma, N. and Watanabe, Y., 1989, *Br. J. Pharmacol.*, **98**: 721–723.
11. Scott, J.A., Fishman, A.J. Homey, J., Fallon, J.Y., Khan, B.A., Peto, C.A. and Rabito, C.A., 1989, *J. Free Rad. Biol. Med.* **6**: 361–367.
12. Stohs, S.J. and Bagchi, D., 1995, *Free Radic. Biol. Med.* **18**: 321–336.
13. Zini, I., Tomasi, A., Grimaldi, R., Vannini, V. and Agnati, F.F., 1992, *Neurosci. Let.* **138**: 279–282.
14. Van der Vliet, A. and Bast, A., 1992, *Chem. Biol. Interactions*, **85**: 95–116.

EFFECTS OF NITRIC OXIDE ON THE CATECHOLAMINE RELEASE FROM CULTURED BOVINE ADRENAL CHROMAFFIN CELLS

Ken Lee,[1,2] Vladimír Dolezal,[2] and Georg Hertting[2]

[1]Department of Physiology
Gifu University School of Medicine
Gifu, Japan
[2]Institute of Pharmacology
University of Freiburg
Freiburg im Breisgau, Germany

1. INTRODUCTION

Nitric oxide (NO), which has recently been shown to subserve an extraordinary diverse range of physiological functions in vivo, was first recognised as an endothelial derived relaxing factor in blood-vessels (1, 2), and has been identified recently as an important second messenger or neurotransmitter in both the central and peripheral nervous systems (2, 3). NO is apparently formed under the catalysis of NO synthase (NOS) from L-arginine via oxidation of one of the guanidino groups of arginine with the stoichiometric formation of L-citrulline (4). The NOS protein is at least three different isoforms have so far been identified, and the form present in neural and endothelial tissue is constitutively expressed whereas that in macrophage is inducibly expressed. Neural NOS has been identified in the central nervous system such as in the cerebellum and in the supraoptic nuclei of the hypothalamus, and in the peripheral nervous system, such as in adrenal medulla. NO is considered to play an important physiological role even in these tissues. However, few investigations have been performed on the effects of NO on catecholamine (CA) release from adrenal chromaffin cells, so that the authentic physiological significance of NO on the cells is still unknown.

In the present study, to clarify the effects of NO on the CA release from the chromaffin cells, we have examined a possible effects of NO generating compounds, such as sodium nitroprusside (SNP) and L-arginine, on the CA release using bovine adrenal chromaffin cells co-cultured with non-chromaffin cells of adrenal medulla.

From a workshop organized by Paul J.L.M. Strijbos (London, United Kingdom) and Jan De Vente (Maastricht, The Netherlands).

2. MATERIALS AND METHODS

2.1. Isolation and Culture of Chromaffin Cells

Bovine adrenal chromaffin cells were isolated, purified and cultured as described previously (5, 6, 7). In some experiments, suspended chromaffin cells obtained after density gradient centrifugation were further purified by differential plating (8). The cells were maintained at 37°C in a humid atmosphere of 5% CO_2 in air.

2.2. Measurement of Catecholamine Release

The cultured cells were preincubated at 37°C for 10 min in fresh Krebs-HEPES buffer and were then stimulated by various concentrations of SNP or L-arginine. In the experiments on the CA release in Ca^{2+}-free medium EGTA (5 mM) was substituted for Ca^{2+}. Cas which were released into the medium and that remained in the cells after incubation were determined using HPLC with electrochemical detection.

2.3. Permeabilization of Adrenal Chromaffin Cells by Digitonin and Ca^{2+}-Induced CA Release from the Permeabilized Cells

Permeabilization of the cells by a detergent digitonin was carried out as described previously (9). After permeabilization, the cells were incubated for 10 min at 37°C in Krebs-HEPES buffer containing no digitonin and various concentrations of $CaCl_2$. CA release from the digitonin-treated cells was determined as described above.

2.4. Measurement of Cytosolic Free Calcium Concentration

Cytosolic free calcium concentration ($[Ca^{2+}]_i$) was measured using a fluorescent dye fura-2 as described previously (6). The cells preloaded with fura-2 were suspended in Krebs-HEPES buffer and the fura-2 fluorescence of these cells was measured in dual excitation wavelength Ca^{2+} spectrofluorometer, with excitation at 340 nm and 380 nm and emission at 500 nm. $[Ca^{2+}]_i$ was calculated from the ratio of fluorescence at the two excitation wavelengths (10).

3. RESULTS AND DISCUSSION

Fig. 1 shows the effects of NO generating compounds on basal noradrenaline (NA) release from cultured bovine adrenal chromaffin cells prepared with or without purification by differential plating. The basal release of NA was constant up to 10 min from either preparation. Incubation of chromaffin cells which had been purified by differential plating, in medium containing 50 µM SNP or 100 µM L-arginine did not influence basal NA release (Fig. 1A). In contrast, NA release was significantly enhanced time-dependently by any of these compounds (Fig. 1B) in cells prepared without differential plating. Upon incubation with SNP or L-arginine the evoked release of NA reached plateau after 6 min. Similar results were obtained for adrenaline release (data not shown). To examine a possible cell damage caused by NO in cultured cells, experiments of trypan blue exclusion were performed. This experiment showed no significant difference from the cells prepared

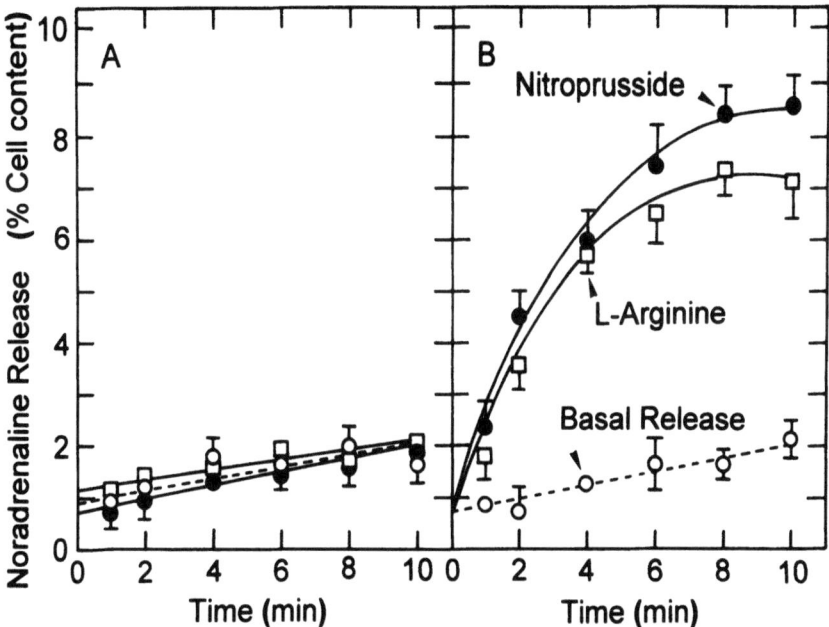

Figure 1. Effects of NO generating compounds on NA release from cultured bovine adrenal chromaffin cells prepared with (A) or without (B) differential plating. O, control; ●, 50 μM SNP; □, 100 μM L-arginine.

with or without differential plating purification (cellular viability between 90 and 95% in both groups). Previous studies have shown that NO generating compounds such as SNP (donor of exogenous NO) and L-arginine (substrate of endogenous NOS) do not affect the basal CA release. Our results obtained using purified chromaffin cells in culture (more than 90% of chromaffin cells) are in concert with this conclusion (Fig. 1A). In contrast,

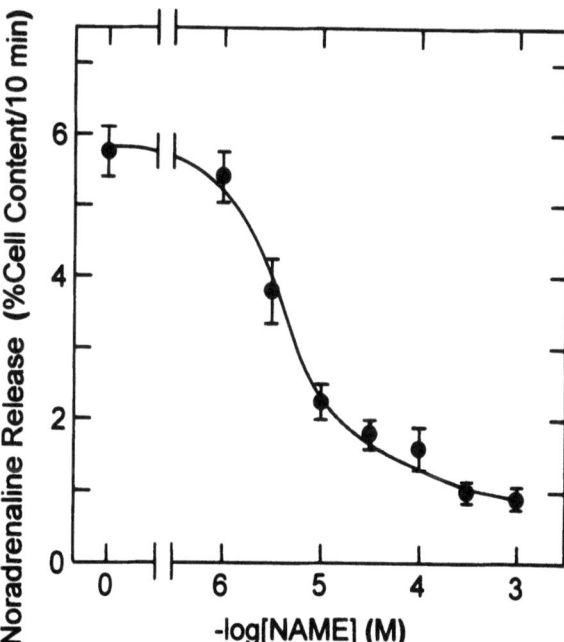

Figure 2. Inhibition of L-arginine-induced NA release by an arginine derivative. The cells were incubated with 100 μM L-arginine for 10 min in the absence or presence of indicated concentrations of NAME, and NA released was determined.

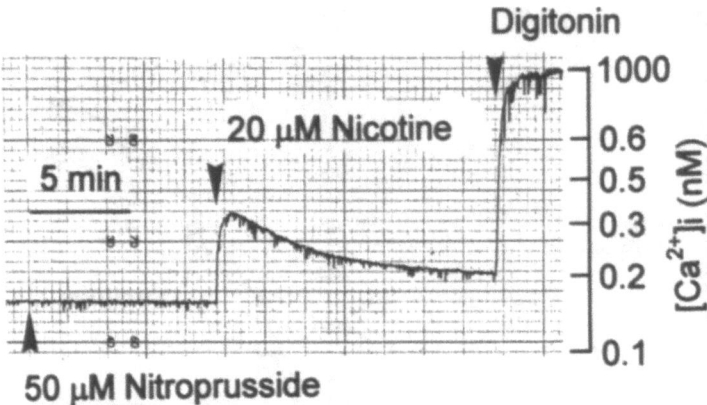

Figure 3. Lack of the effect of NO generating compounds on $[Ca^{2+}]_i$.

we have demonstrated that chromaffin cells co-cultured with non-chromaffin cells of adrenal medulla (about 70% of chromaffin cells in culture) do respond to SNP and L-arginine by the increase of CA release (Fig. 1B). The releasing effect of these compounds was not due to non-specific tissue damage because cellular viability was the same in controls and in the cells treated with different NO donors. In addition, the increase of $[Ca^{2+}]_i$ evoked by stimulation of nicotinic receptors was preserved in the presence of both SNP (Fig. 3) and L-arginine. These results indicate that the reported failure to observe the enhancement of CA release by NO donors (11) may be due to the use of cultures of pure chromaffin cells. Furthermore, our results obtained using the stimulation with L-arginine indicate that the NOS is present rather in non-chromaffin cell of adrenal medulla than in chromaffin cells and that NO formed exerts its effect on chromaffin cells.

Fig. 2 shows the inhibition of L-arginine-induced NA release by an arginine-derivative. L-Arginine-induced NA release was inhibited by N^G-nitro-L-arginine (NAME) in a concentration-dependent manner, and the release of NA was blocked by the compound at a concentration of 50 µM. Similar results were obtained for adrenaline release (data not shown). Since NAME is a specific inhibitior of the constitutive form of NOS, these result suggest that NOS in the cells corresponds to the constitutive type.

Fig. 3 shows typical tracings illustrating lack of the effects of NO donors on $[Ca^{2+}]_i$ in suspended adrenal chromaffin cells. Addition of SNP (Fig. 3) and L-arginine (data not shown) had no effect on $[Ca^{2+}]_i$. The cells clearly responded to the stimulation by nicotine and even after the treatment with these compounds $[Ca^{2+}]_i$ rapidly increased by 0.584 ± 0.089 µM (n = 10). From these results, NO generating compounds-induced CA release appeared not to be prerequisite for the elevation of $[Ca^{2+}]_i$. When we determined the effects of these compounds on CA release in calcium-free medium, the release evoked by SNP and L-arginine was fully extracellular calcium-dependent (data not shown). The reason for the discrepancy of these two results is not clear. A possible explanation would be the spatial distribution of the increase of $[Ca^{2+}]_i$. If there were an increase of $[Ca^{2+}]_i$ limited to thin submembrane area it would be undetectable in the whole population of cells by the approach used. This explanation is based on recent reports on the role of $[Ca^{2+}]_i$: in exocytosis from bovine chromaffin cells which have shown that the optimal signal for triggering release does not correspond to maximum $[Ca^{2+}]_i$ attained but to the increase in submembrane region which is determined by the rate of Ca^{2+}-influx (12).

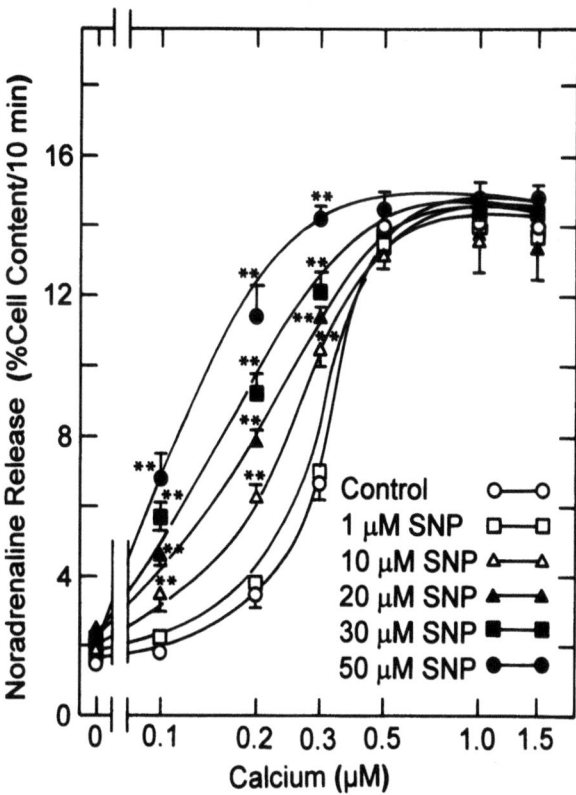

Figure 4. Effect of SNP on the concentration-response curves of NA release for Ca^{2+} in digitonin-permeabilized cells. **, $P < 0.01$; significantly different from the control value obtained without SNP.

Fig. 4 shows the effects of various concentrations of SNP on concentration-response curves of NA release for calcium in digitonin-permeabilized cells. In the digitonin-permeabilized cells, the NA release increased concentration-dependently with the increasing Ca^{2+} concentration and reached a plateau at 0.5 µM. In the presence of increasing concentrations of SNP, the concentration-response curves for Ca^{2+} were significantly shifted to the left while the maximum release response remained unchanged. At 50 µM SNP, the EC_{50} value for Ca^{2+} (79.7 ± 22.6 nM, n = 9) was about three times smaller than that observed in control (297 ± 11.4 nM, n = 9). This increase of the sensitivity of releasing machinery for Ca^{2+} results in the amplification of exocytotic release in response to a small increase of $[Ca^{2+}]_i$.

In conclusion, our data indicate that the release of CA from bovine chromaffin cells in culture is increased by substances involved in NO signalling pathway (L-arginine: precursor of endogenous NO; SNP donor of exogenous NO) only when the cells are co-cultured with non-chromaffin cells of adrenal medulla. This evoked release of CA is extracellular calcium dependent, and the presence of NO donor SNP increases the sensitivity of releasing machinery for calcium ions.

REFERENCES

1. Furchgott, R.F., 1989, *Acta Physiol. Scand.* **139**: 257–270.

2. Moncada, S., Palmer, R.M.J., and Higgs, E.A., 1991, *Pharmacol. Rev.* **43**: 109–142.
3. Dawson, T.M., Dawson, V.L., and Snyder, S.H., 1992, *Ann. Neurol.* **32**: 297–311.
4. Palmer, R.M.J., and Moncada, S., 1989, *Biochem. Biophys. Res. Commun.* **158**: 348–352.
5. Lee, K., and Sekine, A., 1993, *Naunyn-Schmiedeberg's Arch. Pharmacol.* **348**: 275–281.
6. Lee, K., Miwa, S., Koshimura, K., Hasegawa, H., Hamahata, K., and Fujiwara, M., 1990, *J. Neurochem.* **55**: 1131–1137.
7. Lee, K., Ito, A., Koshimura, K., Ohue, T., Takagi, Y., and Miwa, S., 1995, *J. Neurochem.* **64**: 874–882.
8. Waymire, J.C., Bennett, W.F., Boehme, R., Hankins, L., Waymire, K.G., and Haycock, J.W., 1983, *J. Neurosci. Methods* **7**: 329–351.
9. Dunn, L.A., and Holz, R.W., 1983, *J. Biol. Chem.* **258**: 4989–4993.
10. Grynkiewicz, G., Poenie, M., and Tsien, R.Y., 1985, *J. Biol. Chem.* **260**: 2019–2022.
11. O'Sullivan, A.J., and Burgoyne, R.D., 1990, *J. Neurochem.* **54**: 1805–1808.
12. Cheek, T.R., and Burgoyne, R.D., 1993, *J. Exp. Biol.* **184**: 183–196.

Section 35: Free Radicals, NO, and Brain Pathology

MODIFICATION BY NITRIC OXIDE OF ACETYL-CoA AND ACETYLCHOLINE METABOLISM IN NERVE TERMINALS

M. Tomaszewicz, H. Bielarczyk, A. Jankowska, and A. Szutowicz

Department of Clinical Biochemistry
Medical University of Gdańsk
Gdańsk, Poland

1. INTRODUCTION

Excessive production of NO in the brain is an important factor in pathomechanism of various encephalopathies including Alzheimer's (AD), Huntington's disease and postischemic brain injury. NO was postulated to mediate excitotoxic mechanisms caused by sustained activation of NMDA receptors taking place in the brain during ischemia (1). Deteriorative effects are thought to be caused by both NO itself as well as derived peroxynitrite free radical (1). Peroxynitrite was found to affect energy metabolism by inhibition or inactivation of glycolytic enzymes, aconitase and mitochondrial respiratory chain activities (2). It was found to stimulate acetylcholine (ACh) release in different brain preparations (3). This activation apparently increases demand of cholinergic neurons for energy and acetyl-CoA, necessary for repolarization and restoration of the transmitter pool. These needs remain however in conflict with NO-evoked inhibition of energy pathways. The inhibition of pyruvate oxidation was always found to cause the suppression of ACh synthesis (4). It has been thought to be caused by the shortage of acetyl-CoA in nerve terminal mitochondria and subsequent inhibition of its transport to synaptoplasmic compartment. Our experiments provided evidence that it may be the case (5). On the other hand, we also have shown that Ca influx to nerve ending cytoplasm might, to some degree, compensate these shortages by stimulation of direct acetyl-CoA efflux from mitochondria (6).

Therefore, we think that in the presence of NO acetyl-CoA transport to terminal cytoplasm ought to be accelerated, despite of suppression of its formation in mitochondria, to provide an adequate amounts of acetyl units for ACh resynthesis. One might suppose that it can be facilitated by inactivation of aconitase by NO which should tend to cause a backward increase of acetyl-CoA in mitochondria. To elucidate these contradic-

---

From a workshop organized by Paul J.L.M. Strijbos (London, United Kingdom) and Jan De Vente (Maastricht, The Netherlands).

tions we studied relationships between compartmetalization of acetyl-CoA and ACh in synaptosomes in the presence of sodium nitroprusside (SNP) as NO generator.

2. MATERIALS AND METHODS

Isolated rat brain synaptosomes (5) were incubated for 30 min at 37°C with continuous shaking 100 cycles/min, in medium containing in a final volume of 2.0 ml: 20 mM Na-HEPES buffer, 1.5 mM Na-phosphate buffer (final pH 7.4), 90 mM NaCl, 30 mM KCl, 2.5 mM Na-pyruvate, 2.5 mM Na-L-malate, 0.01 mM choline chloride, 0.01 mM eserine sulfate, 32 mM sucrose and 2.5–3.0 mg of synaptosomal protein. Incubation was terminated by separation medium from synaptosomes by centrifugation through the layer of silicon oil followed by deproteinization (5). Pyruvate and ACh contents were determined in supernatants, while acetyl-CoA in whole synaptosomes and its particulate subfraction (5,6). Whole brain mitochondria were incubated in high K medium as described elsewhere (6). Enzyme activities in brain subcellular fractions were determined at 37°C as described earlier (7). Samples were preincubated with SNP for 5 min at 37°C before reaction started. Data are means ± S.E.M. from 5–7 experiments done in duplicate.

3. RESULTS AND DISCUSSION

3.1. Effect of SNP on Enzyme Activities

SNP in concentrations used in metabolic experiments (section 3.2 and 3.3) strongly suppressed activity of mitochondrial aconitase (Tab. 1) with apparent IC_{50} equal to 0.06 mM

Table 1. Effect of SNP on enzyme activities in rat brain subcellular fractions

Enzyme	Fraction	Control specific activity nmol/min/mg protein	Inhibition SNP 0.2 mM	SNP 1.0 mM %
Aconitase	C	219.3 ± 12.8	78.8 ± 2.9	90.8 ± 4.0
	B_p	25.5 ± 4.0	78.7 ± 2.0	77.0 ± 7.6
Pyruvate	C	45.8 ± 13.3	21.0 ± 7.1	42.1 ± 6.2
dehydrogenase	B_p	39.0 ± 16.5	19.0 ± 8.1	42.2 ± 5.0
Oxoglutarate	C	153.3	46.7	56.4
dehydrogenase	B_p	83.3	15.0	43.0
ATP-citrate lyase	S_3	10.7	58.0	100
	B	2.1	100	100

Data are means ± S.E.M from 5 experiments; for oxoglutarate dehydrogenase and ATP-citrate lyase from 2 experiments. Abbreviations: S_3, whole brain cytoplasm; B, synaptosomes; C, whole brain mitochondria; B_p, synaptosomal particulate fraction.

(not shown). Cytosolic aconitase was much less affected. Pyruvate dehydrogenase (PDH) and oxoglutarate dehydrogenase activities were also inhibited by SNP, although with relatively smaller potency (IC_{50} above 1.0 mM). Isocitrate dehydrogenase (NADP), choline acetyltransferase, carnitine acetyltansferase were not changed by this compound.

On the contrary SNP totally abolished activity of ATP-citrate lyase (ATP-CL) in synaptosomes (Tab. 1).

These data indicate that NO may exert wide spread and probably not very specific inhibitory effects on structurally and functionally unrelated enzymes, which activities are not dependent on the presence of a Fe-S prosthetic group (8). Irrespective of the mechanism, strong inhibition of ATP-CL by NO allows to claim that its increased concentration in the brain may inhibit transfer of acetyl units to synaptoplasm by this pathway and impair lipid synthesis in glial cells and ACh formation in cholinergic neurons (9, 10).

3.2. Effect of SNP on Mitochondrial Acetyl-CoA Metabolism

The inhibition of pyruvate utilization in whole brain mitochondria by SNP appeared to be proportional to its suppressory effect on PDH activity (Tab. 1, 2). It would point out that PDH and not pyruvate transport is a rate limiting step for intramitochondrial synthesis of acetyl-CoA. SNP caused significant decrease of acetyl-CoA content in mitochondria both in the presence and in the absence of external Ca^{2+} (Tab. 2). This drop was nonproportionally greater than the inhibition of pyruvate utilization, in spite of that that simultaneous inhibition of aconitase should reduce acetyl-CoA entry into tricarboxylic acid cycle. It may indicate that NO increases a direct efflux of acetyl-CoA from mitochondria by Ca-independent mechanism. In addition, this increase of permeability seems to be pretty specific as free CoA-SH level was found to be increased in these conditions (Tab. 2). The rise of CoA-SH could result from NO or free radical-evoked hydrolysis of acyl-CoA derivatives.

3.3. Effect of SNP on Acetyl-CoA and ACh Compartmentalization in Synaptosomes

In these incubation conditions ACh synthesis and release from synaptosomes were tightly linked. ACh release triggered its synthesis. On the other hand, steady rate of ACh

Table 2. Effect of SNP on acetyl-CoA metabolism in whole rat brain mitochondria

Conditions	Pyruvate *nmol/min/mg protein*	Acetyl-CoA *pmol/mg protein*	CoA-SH *pmol/mg protein*
No Ca^{2+}			
Control	64.1 ± 4.7	30.3 ± 6.5	113 ± 31
0.2 mM SNP	48.4 ± 4.4*	16.6 ± 3.8*	175 ± 43
1.0 mM SNP	42.3 ± 2.5*	13.6 ± 5.0*	175 ± 38
0.01 mM Ca^{2+}			
Control	53.8 ± 5.4	29.4 ± 8.8	140 ± 15
0.2 mM SNP	46.4 ± 4.3	13.9 ± 3.6*	182 ± 23*

* Significantly different from respective control, $p < 0.05$.

Table 3. Effect of SNP on pyruvate utilization and acetyl-CoA contents in rat brain synaptosomes

Conditions	Pyruvate utilization	Acetyl-CoA content	
	nmol/min/mg protein	Mitochondria	Synaptoplasm
		pmol/mg protein	
No Ca^{2+}			
Control	13.5 ± 1.6	11.4 ± 1.3	27.5 ± 11.5
0.2 mM SNP	10.8 ± 1.1*	10.3 ± 2.1	30.9 ± 7.7
1.0 mM SNP	9.7 ± 1.0*	9.2 ± 1.5	36.6 ± 10.7
1mM Ca^{2+}			
Control	8.0 ± 1.2+	9.9 ± 0.9	35.1 ± 12.5
0.2 mM SNP	7.8 ± 0.9+	10.8 ± 1.5	46.6 ± 6.2*+

*Significantly different from respective control, $p < 0.05$; +significantly different from respective no Ca^{2+}, $p < 0.05$.

release over 30 min period required its adequate synthesis (9). We have shown previously (6), that the addition of Ca^{2+} to K-depolarized synaptosomes caused inhibition of pyruvate utilization but simultaneously triggered quantal ACh release and stimulated its synthesis (Tab 3, 4). We explained this apparent discrepancy by the fact that rise of cytosolic Ca^{2+} increased permeability of mitochondrial membrane to acetyl-CoA, thereby facilitating its supply to the site of ACh synthesis (5, 6). SNP inhibited pyruvate utilization in synaptosomes in Ca-free medium. It failed to do so in the presence of external Ca where pyruvate utilization was already inhibited (Tab. 3). In the absence of Ca, SNP affected significantly neither synaptosomal mitochondria nor synaptoplasmic acetyl-CoA level and Ca-independent (nonquantal) (9) ACh release/synthesis (Tab, 3, 4). In the presence of 1 mM Ca^{2+} SNP did not cause changes in acetyl-CoA contents in intrasynaptosomal mitochondira (in situ), but significantly increased (33%) acetyl-CoA level in synaptoplasm (Tab. 3) as well as quantal ACh release (78%) (Tab. 4).

These data indicate that NO may change cholinergic transmission by interference with quantal ACh release. However excessive, long term activation of ACh releaseby pathological concentrations of NO probably would require simultaneous stimulation of direct acetyl-CoA transport through the mitochondrial membrane. In these conditions it

Table 4. Effect of SNP on acetylcholine metabolism in rat brain synaptosomes

Conditons	Total ACh release		Quantal ACh release
	No Ca^{2+}	1 mM Ca^{2+}	1mM Ca^{2+}
		pmol/min/mg protein	
Control	8.9 ± 1.8	28.7 ± 1.1+	19.8 ± 3.2
0.2 mM SNP	8.8 ± 3.0	44.1 ± 3.1*+	35.3 ± 10.8*

*Significantly different from respective control, $p < 0.05$; +significantly different from respective no Ca^{2+}, $p < 0.05$.

could be the only pathway of cytoplasmic acetyl-CoA supply since ATP-CL pathway was found to be strongly suppressed by NO (Tab. 1). NO-evoked loss of acetyl-CoA from mitochondria in connection with its decreased production and increased utilization for ACh synthesis could, exacerbate existing energy deficits and, at least in part, contribute to preferential damage of cholinergic neurons in various brain pathologies. The inhibition of ATP-citrate lyase by NO could diminish synthesis of structural lipids in different brain cells, as citrate was found to be a main source of acetyl-CoA for these synthetic pathways (10). Hence, except of already known cytotoxic effects of NO (1,2) this one might be an additional cause of degeneration of brain cells.

ACKNOWLEDGMENTS

This work was supported by K.B.N. project No 6 PO4A 01310.

REFERENCES

1. Dawson, V.L., Dawson, T.M., 1996, *Proc. Soc. Exp. Biol. Med.* **211**: 33–40.
2. Lees, G.J., 1993, *Neuroscience* **54**: 287–322.
3. Ohkuma, S., Katsura, M., Guo, J.L., Hsegawa, T., Kuriyama K., 1995, *Neurisci. Lett.* **183**:151–154.
4. Gibson, G.E., Barclay, I., Blass, J.P., 1982, *Ann., New York Acad. Sci.* **378**: 382–403.
5. Bielarczyk, H., Szutowicz A., 1989, *Biochem. J.* **262**: 377–380.
6. Szutowicz, A., Bielarczyk H., Skulimowska, H., 1994, *Neurochem. Res.* **19**: 1107–1112
7. Szutowicz, A., Stêpieñ, M., Bielarczyk, H., Kabata, J., £ysiak, W., 1982, *Neurochem. Res.* **7**: 799–810.
8. Castro, L. Rodriguez, M., Radi, R., 1994, *J. Biol. Chem.* **269**: 29409–29415.
9. Szutowicz, A.,Tomaszewicz, M., Szutowicz, A., 1996, *Acta Neurobiol. Exp.* **56**: 323–339.
10. Patel, T.B., Clark, J.B., 1980, *Biochem. J.* **188**: 163–168.

ROLE OF THE PHOSPHATIDYLINOSITOL TRANSFER PROTEIN IN INTRACELLULAR MEMBRANE TRAFFIC

Gerry T. Snoek,[1] Klaas Jan de Vries,[1] Philip I. H. Bastiaens,[2] and Karel W. A. Wirtz[1]

[1]Centre for Biomembranes and Lipid Enzymology
Institute for Biomembranes
Utrecht University
Utrecht, The Netherlands
[2]Department of Molecular Biology
Max Planck Institute for Biophysical Chemistry
Göttingen, Germany

1. INTRODUCTION

Proteins which are able to transfer phospholipids between membranes *in vitro*, have been the subject of investigation for many years (1, 2). Much information about their biochemical properties is available. Since a few years the actual cellular function of these proteins is subject of intensive investigations. Brain tissue has been the major source of one of the most interesting of these proteins: the phosphatidylinositol transfer protein (PI-TP). This protein is present in all mammalian tissues examined sofar as well as in many lower species and micro-organisms (reviewed in 1, 2). The protein functions as a carrier and, in addition to PI, can also bind a PC molecule. The affinity for PC is about 16-fold lower than for PI. Because of this dual specificity, it has been proposed that *in vivo* PI-TP is able to cause net transfer of PI between membranes in exchange for PC (3). As a consequence, PI-TP could be involved in restoring the phosphatidylinositol-4,5-bisphosphate (PIP_2) levels in the plasma membrane after stimulation of PLC-mediated PIP_2 degradation. This could be accomplished by direct transfer of PI from the endoplasmic reticulum (ER) where the phospholipids are synthesized, to the site of phosphorylation in the plasma membrane. On the other hand, PI-TP could also increase PI transfer to the plasma membrane by manipulation of the PI/PC content of secretory vesicles budding from the *trans*-Golgi membranes. These possible involvements of PI-TP in PI transfer from the ER/Golgi membranes to the plasma membrane are shown in Figure 1.

From a workshop organized by Joanna B. Strosznajder (Warsaw, Poland), Jolanta Baranska (Warsaw, Poland) and Moti Liskovitch (Rehovot, Israel).

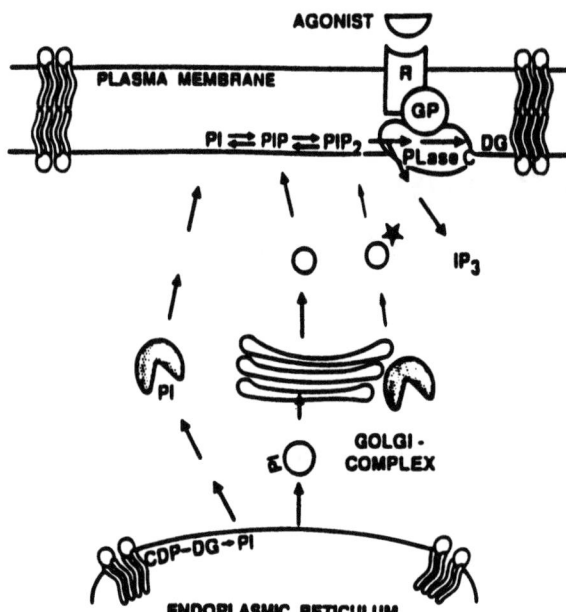

Figure 1. The proposed role of PI-TP in the transfer of PI from the endoplasmic reticulum to the plasma membrane (adapted from ref. 4).

2. CELLULAR LOCALIZATION

To obtain information about possible cellular functions, we studied the intracellular localization of PI-TP in mammalian cells by indirect immunofluorescence. It was found that PI-TP is localized in the nucleus and in the cytosol and is associated with perinuclear structures which have been shown to be Golgi membranes by double labeling experiments (5). Localization studies were performed with quiescent Swiss mouse 3T3 cells and with cells upon stimulation by various hormones/growth factors known to stimulate the PI-cycle in these cells. In stimulated cells, a relocalization of PI-TP between Golgi membranes and vesicles was observed. However, increased levels of PI-TP near or at the plasma membrane were not found under any of these conditions (6). Therefore, a function of PI-TP in the direct monomer transfer of PI to the plasma membrane seems unlikely.

3. CELLULAR FUNCTION

Information about possible functions of PI-TP were obtained from studies in which the regulatory factors of specific cellular processes were identified using semi-intact (permeabilized) cells or cell-free systems. Many cellular processes in permeabilized cells are hampered because specific cytosolic factors which are essential for these processes have leaked out of the cells. Fractionation of the cytosol can then lead to the identification of the cytosolic factors required for these processes. In permeabilized HL60 cells it was shown that PI-TP is able to reconstitute the GTPγS-dependent PLC-activity (7), the PIP_2 synthesis (8) and the ARF-, GTPγS-dependent secretion (9). In semi-intact PC12 cells it was shown that PI-TP is one of the three cytosolic proteins which are necessary for the reconstitution of the Ca^{++}-, ATP-dependent secretion of noradrenalin (10). In a cell-free system containing isolated *trans*--Golgi membranes, it was shown that PI-TP is able to stimulate the budding of secretory vesicles from these membranes (11).

4. TWO ISOFORMS; PI-TPα AND β

Assuming that in order to perform a specific task, a protein must be present at the correct site in the cell, we tried to bring our observations on the localization of PI-TP in line with possible cellular functions of the protein. This attempt was hampered by the detection of a PI-TP isoform. The cDNA encoding the isoform, PI-TPβ, was identified by its ability to rescue SEC14 yeast mutants (12). The deduced amino acid sequence is 77% identical to PI-TPα. Independently two novel phospholipid transfer proteins were identified and purified in our laboratory. From chicken liver a protein was isolated which, in addition to PI and PC, was able to transfer sphingomyelin (SM) with a high efficiency (the SM/PI-TP, 13). Furthermore, a minor protein which was shown to be cross-reactive with highly specific antibodies raised against synthetic peptides representing epitopes in PI-TPα, was purified from bovine brain. This protein was identified in Swiss mouse 3T3 cells as well (5) and was described as a 36kDa protein with an isoelectric point of 5.4 (14). Comparison of the N-terminal amino acid sequences of SM/PI-TP from chicken liver and of the 36kDa protein from bovine brain demonstrated that these proteins are identical to PI-TPβ from rat brain (12, 15).

5. CHARACTERIZATION OF PI-TPβ

PI-TPβ is not abundant in bovine brain making its purification difficult (15, 16). Because of the sequence homology with SM/PI-TP from chicken liver, we investigated whether PI-TPβ is able to transfer sphingomyelin (SM). For this purpose we used pyrene-labeled phospholipids (Figure 2B).

In the assay the protein-mediated transfer of fluorescently labeled phospholipids from quenched donor vesicles to non-quenched acceptor vesicles is measured. The increase of the fluorescent signal is a measure for the lipid transfer activity (17). The relative transfer activities of PI-TPα and β for PI, PC and SM are shown in Figure 2A. Both isoforms demonstrate comparable transfer activities towards PI and PC. In addition, PI-TPβ demonstrates a significant capacity to transfer SM in agreement with the observation on SM/PI-TP from chicken liver. This characteristic difference *in vitro* might indicate distinct cellular functions of PI-TPα and β.

To obtain evidence for this we investigated the cellular localization of both isoforms by two independent methods. The availability of antibodies which are able to discriminate between PI-TPα and β enabled us to study the localization by indirect immunofluorescence. It was shown that in exponentially growing Swiss mouse 3T3 cells, PI-TPα is mainly localized in the nucleus and in the cytosol and that PI-TPβ is mainly associated with perinuclear structures which most likely represent Golgi membranes (15).

These results were confirmed by using a different technique: the microinjection of fluorescently labeled PI-TPα and β into living cells. For this purpose, purified PI-TPα and β were labeled with the fluorescent dyes, Cy3 and Cy5, which are structurally highly similar molecules but which can be discriminated because of different excitation and emission wavelenghts. PI-TPα was labeled with Cy3 and PI-TPβ with Cy5. The covalent binding of the fluorescent dyes to the proteins was confirmed by SDS polyacrylamide gelelectrophoresis. In addition, it was shown that coupling of the dyes did not affect the *in vitro* phospholipid transfer activities of both PI-TP isoforms (16). A 1:1 mixture of CY3-PI-TPα and Cy5-PI-TPβ was microinjected into living fetal bovine heart epithelial

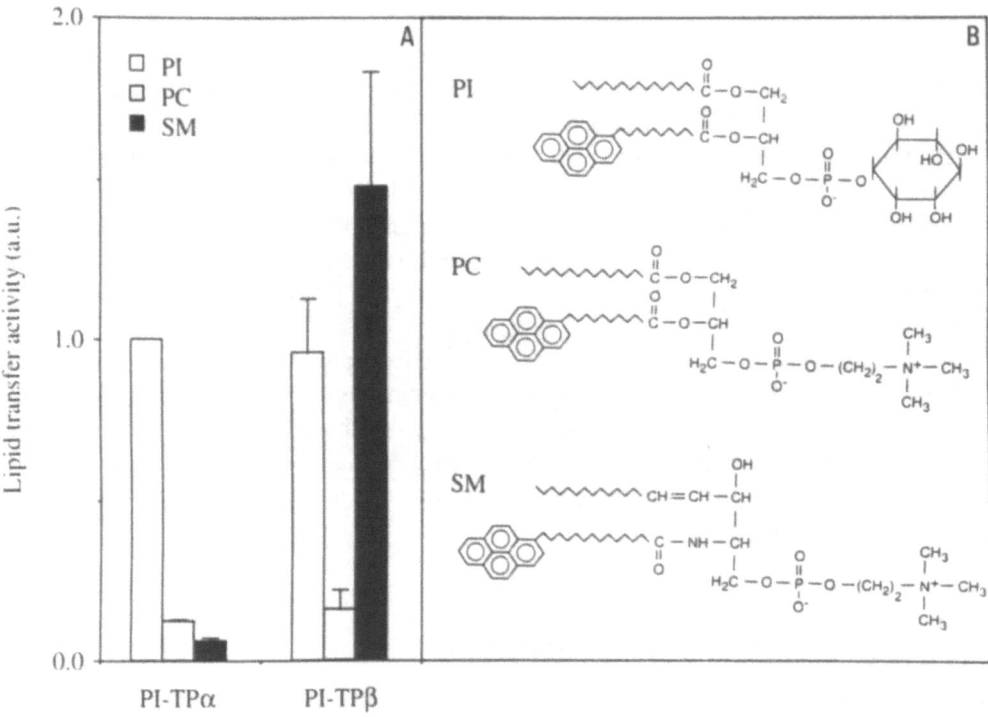

Figure 2. (A): Relative phospholipid transfer activities of PI-TPα and β. (B): Pyrene-labeled PI, -PC and -SM.

(FBHE) cells. After 30 minutes the cells were fixed and the fluorescent signals were analyzed by confocal laser scanning microscopy.

In panel A of Figure 3, the fluorescence signal of CY3-PI-TPα is analyzed. It can be observed that the highest intensity is present in the nucleus. Also some labeled protein is found in the cytosol. When, in the same optical slices, the emission signal of CY5-PI-TPβ is analyzed (Figure 3B) the most intens signal is observed in the perinuclear structures (16).

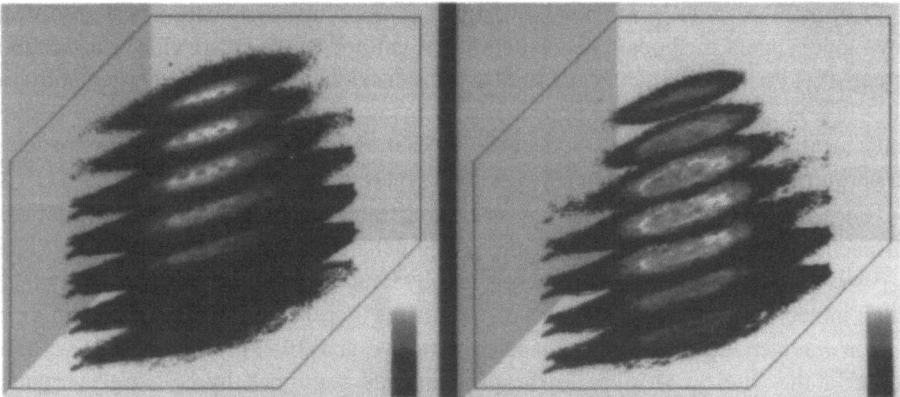

Figure 3. Cellular localization of Cy3-PI-TPα (A) and Cy5-PI-TPβ (B) in confocal optical sections of FBHE cells (adapted from ref.16).

Studies of the cellular localization of the PI-TP isoforms by i) microinjection experiments, ii) indirect immunofluorescence studies and iii) the biochemical analysis of cellular fractions of permeabilized cells have led to the same conclusion: PI-TPα is preferentially localized in the nucleus and in the cytosol and PI-TPβ is preferentially associated with perinuclear structures which most likely represent Golgi membranes (14–16).

The observation that PI-TPα and β demonstrate a distinct localization in mammalian cells, suggests that both isoforms have different cellular functions. However, when the activity of both isoforms was determined in some of the reconstitution assays described above it was found that both isoforms were equally active. This was observed for the reconstitution of PLCβ activity in permeabilized HL60 cells (15). Likewise, no difference was observed between PI-TPα and β in the stimulation of vesicle budding from isolated *trans*-Golgi membranes (11). Recently however, PI-TPβ was shown to be much more potent in restoring the GTPγS-stimulated protein secretion in permeabilized HL60 cells as compared to PI-TPα (9). The finding that PI-TPβ is preferentially associated with Golgi membranes may be of relevance to this observation.

6. CONCLUSIONS

In table 1 the properties of PI-TPα and β are summarized. The reconstitution studies seem to indicate that PI-TPα and PI-TPβ behave rather similarly. This however, appears

Table 1. Properties of PI-TPα and β isolated from bovine brain. n.d.: not detected

	PI-TPα	PI-TPβ	Reference
M_W	35 kDa	36 kDa	2, 14
Isoelectric point	5.5 - 5.7	5.4	3, 14
Occurrence	abundant in brain, pancreas	abundant in brain, liver	18, 19
intracellular localization	nucleus, cytosol	perinuclear	5, 15, 16
lipid specificity	PI>>PC	SM>PI>>PC	3, 15
Substrate for PKC	+	n.d.	6
subforms	PI-TPαI, PI-TPαII	PI-TPβI	3, 15
Stimulation of PLC activity	+	+	7
Stimulation of Ca^{++}-, ATP dependent secretion	+	n.d.	10
Stimulation of vesicle budding from Golgi membranes	+	+	11
Stimulation of GTPγS-dependent protein secretion	+	++	9

to be in contrast with the rather striking differences between these proteins, in particular the capacity of PI-TPβ to bind/transfer SM and the clearly distinct cellular localization of both isoforms.

The cellular localization of a protein provides an important indication for its possible cellular function. It should be emphasized that when investigating the cellular functions of PI-TPα and β, one should take into account that the protein has to be present at the site of action in the cell or has to be directed towards that site as a result of a stimulus. Therefore, conclusions about the cellular functions of PI-TPα and β should be based on biochemical and on localization studies. In addition, it remains to be established whether the capacity of PI-TPβ to bind and/or transfer SM is part of its cellular function.

It may be expected that in the future mammalian cells in which the ratio of PI-TPα to PI-TPβ is varied by molecular biological techniques, will provide new information on the cellular function of these two isoforms.

ACKNOWLEDGMENT

Part of this studie was carried out under the auspices of the Netherlands Foundation for Chemical Research (SON) and with financial aid from the Netherlands Organization for Scientific Research (NWO).

REFERENCES

1. Ossendorp, B.C., Snoek, G.T., and Wirtz, K.W.A., 1994, *Current Topics in Membranes* **40**: 217–259.
2. Wirtz, K.W.A., 1991, *Annu. Rev. Biochem.* **60**: 73–99.
3. Van Paridon, P.A., Gadella Jr., T.W.J., Somerharju, P.J., and Wirtz. K.W.A., 1987, *Biochim. Biophys. Acta* **903**: 68–77.
4. Helms, J.B. Thesis, Utrecht, 1991.
5. Snoek, G.T., De Wit, I.S.C., Van Mourik J.H.G., and Wirtz, K.W.A., 1992, *J. Cell. Biochem.* **49**: 339–348.
6. Snoek, G.T., Westerman, J., Wouters, F.S., and Wirtz, K.W.A., 1993, *Biochem. J.* **291**: 649–656.
7. Thomas, G.M.H., Cunningham, E., Fensome, A., Ball, A., Totty, N.F., Truong, O., Hsuan, J.J., and Cockcroft, S., 1993, *Cell* **74**: 919–928.
8. Cunningham, E., Thomas, G.M.H., Ball, A., Hiles, I., and Cockcroft, S., 1995, *Curr. Biol.* **5**: 775–783.
9. Fensome, A., Cunningham, E., Prosser, S., Khoon Tan, S., Swigart, P., Thomas, G., Hsuan, J., and Cockcroft, S., 1996, *Curr. Biol.* **6**: 731–738.
10. Hay, J.C., and Martin, T.F.J., 1993, *Nature* **366**: 572–575.
11. Ohashi, M., De Vries, K.J., Frank, R., Snoek, G.T., Bankaitis, V.A., Wirtz, K.W.A., and Huttner, W.B., 1995, *Nature* **377**: 544–547.
12. Tanaka, S., and Hosaka, K., 1994, *J. Biochem.* **115**: 981–984.
13. Westerman, J., De Vries, K.J., Somerharju, P., Timmermans-Hereijgers, J.L.P.M., Snoek, G.T., and Wirtz, K.W.A., 1995, *J. Biol. Chem.* **270**: 14263–14266.
14. De Vries, K.J., Momchilova-Pankova, A., Snoek, G.T., and Wirtz, K. W. A., 1994, *Exp. Cell Research* **215**: 109–113.
15. De Vries, K.J., Heinrichs, A.J.A. Cunningham, E., Brunink, F., Westerman, J., Somerharju, P.J., Cockcroft, S., Wirtz, K.W.A., and Snoek, G.T., 1995, *Biochem, J.* **310**: 643–649.
16. De Vries, K.J., Westerman, J., Bastiaens, P.I.H., Jovin, T.M, Wirtz, K.W.A., and Snoek, G.T., 1996, *Exp. Cell Res.* in press.
17. Van Paridon, P., Gadella, Jr., T.W.J., Somerharju, P.J., and Wirtz, K.W.A., 1988, *Biochemisty* **27**: 6208–6214.
18. Wirtz, K.W.A., Jolles, J., Westerman, J., and Neys, F., !976, *Nature* **260**: 354–355.
19. Tanaka, S., Yamashita, S., and Hosaka, K., 1995, *Biochem. Biophys. Acta* **1259**: 199–202.

Section 36: Molecular Mechanism of Lipid Mediators Action and Their Role in Neurotransmission and Signal Transduction

LIPID METABOLISM IN HUMAN NEUROBLASTOMA SK-N-BE DIFFERENTIATED WITH RETINOIC ACID

A. Petroni, N. Papini, P. La Spada, M. Blasevich, and C. Galli

Institute of Pharmacological Sciences
University of Milan
Milan, Italy

1. INTRODUCTION

Neuroblastoma cells are an important model for the study of neuron differentiation. The inhibition of cell growth and the changes to mature neurons may be induced by different agents. Retinoic acid (RA), the vitamin A metabolite, is among the most active ones. When neuroblastoma SK-N-BE is incubated with retinoic acid, cell body shape is modified, long neurites are formed and cell growth is inhibited (1,8). Retinoic acid is an activator of the peroxisome proliferator-activating receptors and RA receptor may regulate fatty acid homeostasis (4). Moreover RA affects the inositol phosphate pathway (7), a regulator of cell proliferation and differentiation, in neuroblastoma cells. Since lipids and their long chain polyunsaturated fatty acids (LC-PUFA) play an important role in neural membrane, we have investigated the changes in lipid metabolism occurring during differentiation.

2. MATERIALS AND METHODS

2.1. Materials

[^{3}H] thymidine (291Ci/mmol), and [1-^{14}C] acetic acid sodium salt (56 mCi/mmol) were purchased from Amersham International (Buckinghamshire, U.K.). All-*trans* RA and the fatty acid standard used for GLC analyses (nonadecanoic acid; 19:0) were from Sigma Chemical Co. (St. Louis, MO, U.S.A.). The materials used for tissue cultures were from Gibco (Paisley, U.S.A.). Solvents and silica gel 60H used for TLC analyses were from Merck (Darmstadt, Germany). The scintillation solvent was from Canberra Packard (Meriden, CT, U.S.A.). The capillary column for GLC analyses was a Supelco Omegawax TM 320 (30m, 0,32mm i.d.,

---

From a workshop organized by Joanna B. Strosznajder (Warsaw, Poland), Jolanta Baranska (Warsaw, Poland) and Moti Liskovitch (Rehovot, Israel).

0.25µm film thickness; Supelco, Bellefonte, PA, U.S.A.). Culture morphology examination and cell count were carried out by an Axioskop Zeiss microscope (Germany).

2.2. Cell Cultures

Human neuroblastoma SK-N-BE was donated by Dr. A. Sher (Department of Pharmacological Sciences, Milan). The cells were seeded at a concentration of 10^5 cells in 100mm-diameter dishes. Cultures were grown in RPMI 1640 medium containing 10% (vol/vol) heat-inactivated FCS, 2mM L-glutamine, 100 IU/ml of penicillin and 100 µg/ml of streptomycin, in a humidified atmosphere with 95% air and 5% CO_2 at 37°C.

Cell differentiation was obtained by treating the cells with RA 10µM the 1st and 4th days after seeding. RA was dissolved in ethanol (final concentration 0.1%) and the same amount of the vehicle was added to the control cells.

2.3. Assessment of Differentiation

Differentiation was assessed by evaluation of cell growth, neurite emission and thymidine incorporation. Cell growth was evaluated by counting cells by a phase-contrast microscope and by protein quantification according to the method of Lowry et al (5). Neurite outgrowth was determined by counting the number of cells extending a neurite longer than the cell body, expressed as a percentage of the total number of cells counted. An image analizer was connected with the phase-contrast microscope for the quantification. [^{3}H]thymidine incorporation into nucleic acids was evaluated according to the procedure of Ma et al. (6) at 3 days of cell growth.

2.4. Fatty Acids Analysis

Cell total lipids were extracted according to Folch et al (2). Lipid concentration in aliquots was determined using a microbalance (C-31:Cahn Instruments, Cerriots, CA, USA). Fatty acid methyl esters were prepared by acidic transmethylation using methanolic 3N HCl and were analyzed by GLC on a capillary column using a temperature gradient from 130 to 220°C.

2.5. Incorporation of Labeled Acetic Acid into Lipid Classes

Cells incubated without or with RA at 3 and 8 days of growth were incubated with labeled acetic acid (2.5µCi / dish at a final concentration of 11.3µM) in RPMI medium without FCS for 24h. At the end of incubation cells were washed in phosphate buffered saline and scraped off. Total lipids were extracted according to Folch et al (2). Lipid classes were separated by TLC using hexane/diethyl ether/ acetic acid (70:30:1 by volume) as developing solvent. Spots were revealed by exposure to iodine vapours, identified by the use of standards and scraped off into vials. The radioactivity was measured by a β-counter after addition of 1ml of methanol/H_2O (1:1) and 10ml of the proper scintillation fluid. Separation of cholesterol and diglyceride, which coeluted in the above solvent system, was obtained using chloroform/ methanol (98:2 by volume) as eluting solvents.

2.6. Statistical Analysis

Data are presented as means±SEM. Statistical analysis were made using Student's t test.

3. RESULTS

After about 8 days of treatment with RA $10\mu M$ SK-N-BE cells exhibited profound morphological changes, typical of neuronal differentiation, and inhibition of cell growth, when compared with untreated cells. In nondifferentiated cells there was a marked increase of cell number and protein content over time of growth unlike SK-N-BE differentiated (fig. 1B and A).

In RA-treated cells neurite outgrowth took place progressively (Fig. 1C). The incorporation of [^{3}H]thymidine in cellular nucleic acids of untreated and RA-treated cells (three samples per group) at 3 days of growth were 3,609±191 and 917±67 dpm/mg of protein, respectively.

After differentiation with RA $10\mu M$ for 8 days the levels of LC-PUFA were modified, especially there was a significant increase of 20:4 (from 7.5% to 10.8%) and 22:4 in the n-6 series (Table 1).

In the n-3 series there was an increase of 22:5. Moreover in SK-N-BE differentiated oleic acid (18:1 n-9) decreased, whereas levels of saturated fatty acids were marginally modified (data not shown).

The changes in AA content induced by RA in SK-N-BE prompted us to investigate if lipid synthesis was modified in differentiated cells compared with nondifferentiated. Then we evaluated the incorporation of labeled acetic acid into cell lipids at 3 and 8 days

Table 1. Percent levels of major n-6 and n-3 fatty acids in nondifferentiated (ND) and differentiated SK-N-BE cells at 8 days of growth

Fatty Acids		ND cells	differentiated cells	
18:2		2.3±0.4	2.0±0.4	
20:3		1.0±0.2	1.6±0.2	
20:4	n-6	7.5±0.4	10.8±0.7	a
22:4		2.1±0.3	3.3±0.3	b
22:5		0.5±0.08	0.5±0.06	
20:5		1.2±0.2	1.0±.0.06	
22:5	n-3	3.1±0.2	4.1±0.3	c
22:6		3.5±0.3	3.8±0.4	
n-6		13.8±0.8	18.2±1.2	d
n-3		7.7±0.6	8.9±2.4	

Values are the average of weight percentages (12 different experiments, in duplicate) ± S.E.M.
Significance of differences according to Student's t test: a=p<0.001; b=p<0.01; c=p<0.02; d=p<0.005

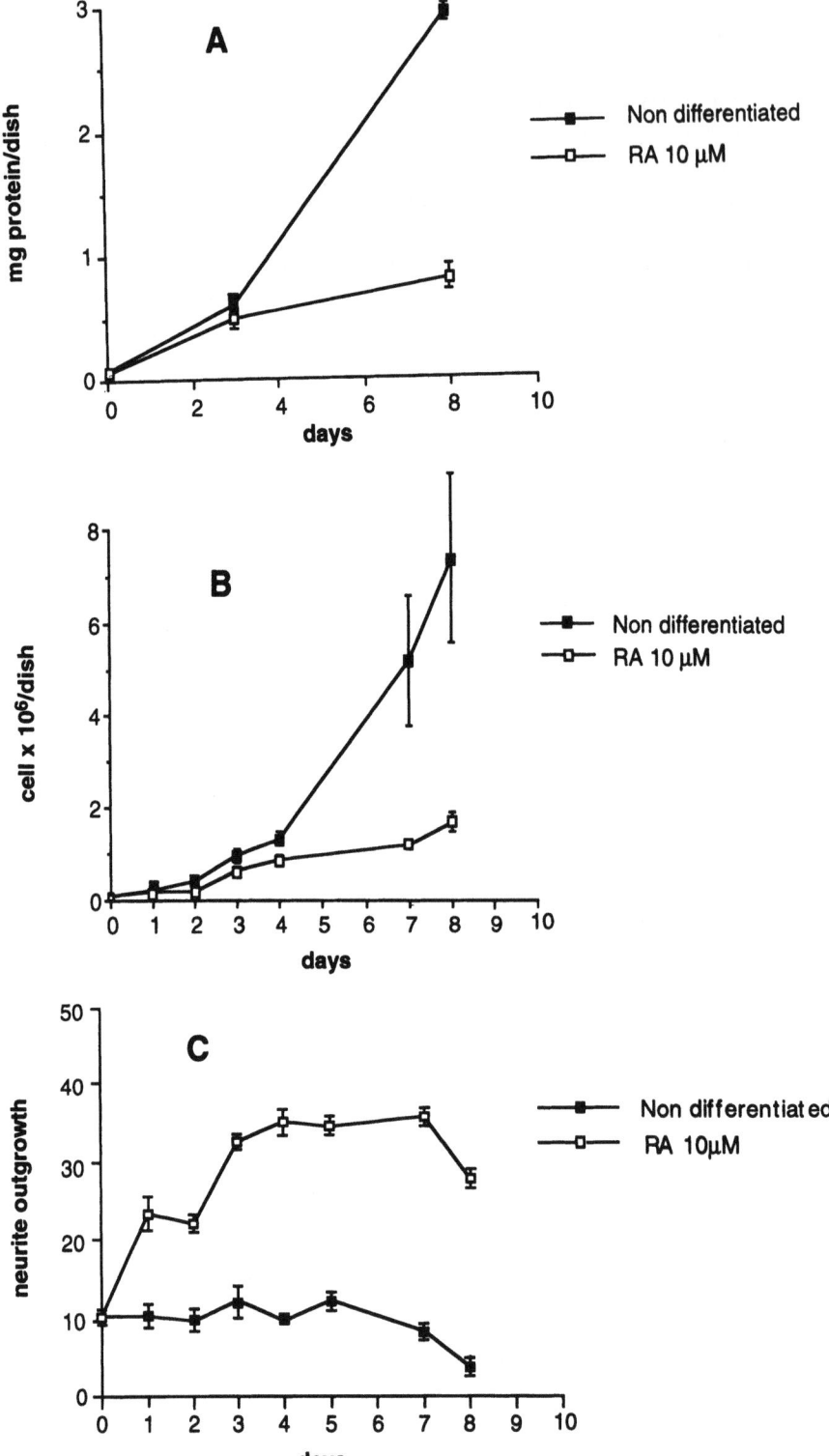

Figure 1. A: Protein content (in mg/dish) in control and RA-differentiated cells at different days of growth. B: Cell number increments. C: Neurite outgrowth expressed as a percentage of cells extending a neurite longer than the cell body. Data are average ± SEM (bars) values of determinations carried out in three experiments performed in duplicate.

Table 2. Incorporation of labeled acetate in total lipids and in individual lipid classes of SK-N-BE incubated without or with RA at 3 and 8 days of growth

	3 Days		8 Days	
	Untreated	RA-treated	Untreated	RA-treated
	cpm/µg TL		cpm/µg TL	
	A	B	C	D
Total labeling	7838±312	6835±446 ns	3132±107 a	1777±119 b, d
	% of incorporated radioactivity			
PL	69.94±0.86	69.55±0.41 ns	73.00±0.34 a	62.30±0.52 d, d
TG	7.02±0.25	9.90±0.20 d	8.90±0.66 b	21.20±0.63 d, d
FFA	0.60±0.04	1.15±0.08 c	0.55±0.03	1.08±0.18 d, ns
DAG	1.67±0.27	2.56±0.16 a	2.58±0.52	4.88±1.23 d, ns
Ch	19.23±1.08	16.16±0.42 a	13.27±0.99 c	10.98±0.92 ns, d
ChE	0.94±0.04	0.60±0.06 c	2.07±0.27 c	1.80±0.18 ns, d
Differences:		vs A	vs A	1st letter : vs C
				2nd letter : vs B

Data are the average ± S.E.M. of determinations carried out on triplicate samples obtained in two separate experiments.
Significance of differences from the indicated groups: a, $p<0.05$; b, $p<0.01$; c, $p<0.005$; d, $p<0.001$; ns, not significant.
PL, phospholipids; TG, triglycerides; FFA, free fatty acids; DAG, diacylglycerol; Ch, cholesterol; ChE, Cholesterol Esters.

of treatment without and with RA. At 3 days, total incorporation, expressed as cpm/µg of total lipids, were not different in cells grown without or with RA (Table 2).

At 8 days, there was more than 50% decline of acetate incorporation in treated cells. When compared with 3 days the reduction in the RA-treated cells was even greater. In fact at 8 days acetate incorporation in differentiated cells was about one-fourth of that in differentiated and nondifferentiated cells at 3 days, whereas nondifferentiated cells at 8 days incorporated about 50% of labeled acetate compared with cells at 3 days. The incorporation in individual lipid classes was also markedly different in cells exposed to RA versus untreated ones at 3 and 8 days (Table 2). At 3 days, in RA-treated cells, whereas incorporation into phospholipids was the same, those in cholesterol and its esters were reduced, and labeling of free fatty acids, diacylglycerol, and triglycerides was markedly increased over that in nontreated cells. At 8 days, in RA-treated versus nontreated cells there was a reduction in the incorporation in phospholipids, associated with marked accumulation into triglycerides, free fatty acids, and especiallly diacylglycerol. In untreated cells, at 8 days versus 3 days, there was an increment in the relative incorporation in triglycerides and especially cholesterol esters, whereas labeling of cholesterol was markedly reduced. In treated cells at 8 days versus 3 days, labeling of cholesterol, although not significant reduced when compared with that in nontreated cells, declined markedly.

4. DISCUSSION

Retinoic acid induced typical changes in cell morphology and cellular growth associated with differentiation and also modified PUFA metabolism. In fact in differentiated SK-N-BE the levels, especially those of arachidonic acid (20:4 n-6), were comparable with those we found in primary cultures of astroglial cells and neurons and in (rat) brain cortex (data not shown). The changes in LC-PUFA concerned mostly the n-6 series, with significant increments in levels of 20:4 and 22:4. The lack of increment in levels of LC-PUFA of the n-3 series may be related to the use of FCS, which is rich of fatty acids of n-3 series (3). This increment suggests an involvement of lipid metabolism in differentiation. Then we evaluated the de novo lipid synthesis from acetate in cells at 3 and 8 days of growth without and with RA.

During differentiation acetate incorporation into phospholipids and cholesterol was reduced. Acetate incorporation was redirected into storage lipids, and this was accompanied by accumulation of the intermediate free fatty acids and diacylglycerol. The data indicate that SK-N-BE differentiation is initially associated with enhanced synthesis of structural lipids and greater formation of LC-PUFA from their precursors for new membrane formation. When differentiation is nearly complete and LC-PUFA reach their final levels, both processes are reduced and lipid synthesis from acetate is directed toward triglycerides.

REFERENCES

1. Abemajor, E. and Sidell, N., 1989, *Environ. Health Perspect.* **80**: 3–15.
2. Folch, J., Lees, M., and Sloane Stanley, G.H., 1957, *J. Biol. Chem.* **226**: 497–509.
3. Galella, G., Marangoni, F., Risé, P., Colombo, C., Galli, G., and Galli, C., 1993, *Biochim. Biophys. Acta* **1169**: 280–290.
4. Isseman, I., Prince, R.A., Tugwood, J.D., and Green, S., 1993, *Biochimie* **75**: 251–256.
5. Lowry, O.H., Rosebrough, N.J., Farr, A.L., and Randall, R.J., 1951, *J. Biol. Chem.* **193**: 265–275.
6. Ma, Z.Q., Spreafico, E., Pollio, G., Santagati, S., Conti, E., Cattaneo, E., and Maggi, A., 1993, *Proc. Natl.Acad. Sci. USA* **90**: 3740–3744.
7. Ponzoni, M. and Lanciotti, M., 1990, *J. Neurochem.* **54**: 540–546.
8. Sidell, N., Altman, A., Haussler, M.R., and Seeger, R.C., 1983, *Exp. Cell Res.* **148**: 21–30.

THE ROLE OF Ca^{2+} AND PROTEIN KINASE C IN REGULATION OF PHOSPHATIDYLSERINE SYNTHESIS IN GLIOMA C6 CELLS

J. Barańska, M. Czarny, P. Sabała, and M. Wiktorek

Department of Cellular Biochemistry
Nencki Institute of Experimental Biology
Warsaw, Poland

1. INTRODUCTION

In living organisms there are two different pathways for the biosynthesis of phosphatidylserine (PS). The first is typical for prokaryotes and occurs in the presence of CDP-diacylglycerols and L-serine with the release of CMP. The second pathway is a base exchange reaction in which serine is directly exchanged with the base moiety of preexisting phospholipids. In general, this pathway is typical for eukaryotic organisms. In animal cells, the synthesis of PS appears to occur solely by this reaction (1). It occurs mainly in the endoplasmic reticulum (ER), is independent on metabolic energy and is characterized by a requirement for relatively high (mM) concentration of Ca^{2+} (1,2).

Besides the CTP-dependent bacterial pathway, and the base exchange reaction, a third system for PS formation has also been demonstrated in animal cells. In this system, first reported in 1958 by Hubscher et al. (3,4) incorporation of L-serine into PS is dependent on the presence of ATP and Mg^{2+}. The mechanism of this process, corroborated later by other authors, remained for a long time obscure (2). The requirement for ATP contrasted with the Ca^{2+}-dependent incorporation of serine by the base exchange reaction and suggested that this might be a different process.

Our studies (5,6) clarified this puzzling problem, indicating that the ATP-dependent PS formation is in fact the base exchange reaction, in which ATP and Mg^{2+} are utilized for Ca^{2+} accumulation inside the microsomal vesicles by Ca^{2+}-dependent, Mg^{2+}-stimulated ATPase. Due to that, at low extravesicular Ca^{2+} concentration, the concentration of Ca^{2+} in the lumen of the ER is high enough to enable the base exchange reaction. Namely, we demonstrated that at low Ca^{2+} concentrations, the stimulatory effect of ATP and Mg^{2+} on PS formation occurred in: (i) untreated rat liver microsomes, but did not occur in fully dis-

---

From a workshop organized by Joanna B. Strosznajder (Warsaw, Poland), Jolanta Baranska (Warsaw, Poland) and Moti Liskovitch (Rehovot, Israel).

rupted by a freeze-thawing and sonication procedure vesicles; (ii) was reduced by the Ca^{2+} ionophore A23187 and thapsigargin, which is a specific blocker of Ca^{2+}-ATPase; (iii) was modulated by agents that affect Ca^{2+} release by the ER receptor channels (6).

Thus, one could expect that PS synthesis would be regulated by the activity of the ER Ca^{2+}-ATPase that pumps Ca^{2+} into this structure, and by such intracellular events that are responsible for the release of Ca^{2+} from this structure. On the other hand, PS is a known activator of protein kinase C (PKC) that plays a key role in many transmembrane signalling systems in animal cells (7). Therefore, to gain better insight into molecular events in which PS takes part, in our further studies we investigated the role of Ca^{2+} and PKC in regulation of PS synthesis in intact cells, using glioma C6 line. It is worth noting that glioma C6 cells belong to the type of nonexcitable cells (8).

2. THE ROLE OF Ca^{2+} IN REGULATION OF PS SYNTHESIS IN GLIOMA C6 CELLS

In order to examine the role of Ca^{2+} in PS synthesis in glioma C6 cells we used a number of agents which are known to alter the intracellular Ca^{2+} concentration ($[Ca^{2+}]_i$), as glutamate, acetylcholine, thapsigargin, 2,5-di-tert-butylhydroquinone (DBHQ), and calcium ionophore A23187 (9,10). Glutamate and acetylcholine are neurotransmitters that interact with cell surface receptors and produce a rapid and sustained increase in the cytosolic Ca^{2+} concentration. This is due to both the release of Ca^{2+} from the ER Ca^{2+} stores caused by inositol 1,4,5-trisphosphate ($InsP_3$) and the influx of extracellular Ca^{2+} into the cell. Thapsigargin and DBHQ are specific inhibitors of the ER Ca^{2+}-ATPase. They mimic agonists by a marked increase in the level of cytosolic Ca^{2+} via discharge of intracellular Ca^{2+} stores, but without $InsP_3$ production. Calcium ionophore A23187 makes cell membrane permeabilized to Ca^{2+}.

It has been found that all of these agents result in a large decrease of PS synthesis in glioma C6 cell (9,10). Approximately 30%-40% of this process was inhibited by neurotransmitters, whereas thapsigargin and DBHQ resulted in 70% inhibition of serine incorporation into PS. Glutamate was shown to induce an inhibition of PS synthesis in a range of 1 μM–1 mM, acetylcholine — 10 nM, thapsigargin was effective at the concentration 50–150 nM, and DBHQ produced a similar effect at the concentration of 25–50 μM.

The inhibition of PS synthesis by the ionophore A23187 occurred in Ca^{2+} free medium and at low Ca^{2+} concentration (100 μM), whereas at 10 mM $CaCl_2$ it distinctly stimulated PS synthesis. Thapsigargin inhibited this process in the presence and absence of external calcium, and only at 10 mM $CaCl_2$ was this effect not so visible (Fig. 1).

The ionophore A23187 completely permeabilizes cell membranes for Ca^{2+}. Therefore, when the cells are incubated at low $CaCl_2$ (100 μM), the addition of the ionophore increases the level of Ca^{2+} in the cytosol to the same value, which is a thousand fold higher than that in the resting cell. Thus, the inhibition of PS synthesis found in these conditions can be only explained by the Ca^{2+} depletion of the ER, where high (mM) concentration of Ca^{2+} can be achieved. The same phenomenon takes place in the presence of EGTA (Ca^{2+}-free medium) (Fig. 1) On the other hand, thapsigargin is a specific inhibitor of the ER Ca^{2+}-pumping ATPase. It blocks the enzyme and Ca^{2+} leaks from the lumen of the ER stores that cannot be refilled again (8–10).

To examine if neurotransmitters and inhibitors of the ER Ca^{2+}-ATPase simultaneously with the decrease in PS synthesis indeed induce mobilization of Ca^{2+}, the effect of these agents was measured in single, intact glioma C6 cells loaded with Fura-2 applying a video

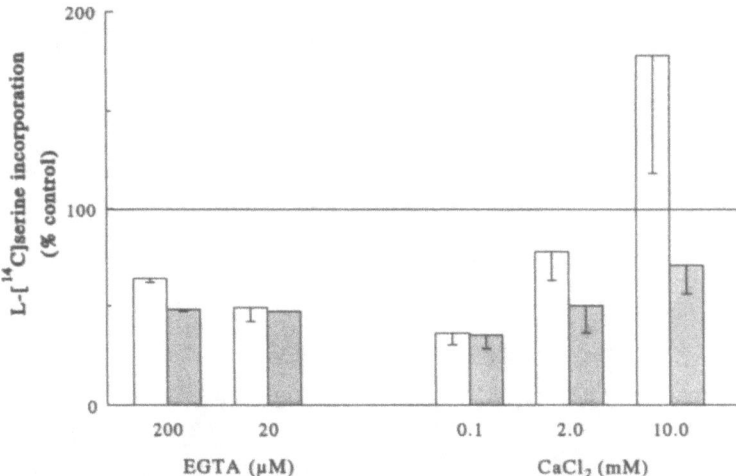

Figure 1. Effect of calcium ionophore A23187 and thapsigargin on phosphatidylserine formation in glioma C6 cells. Cells were incubated for 15 min with [^{14}C]serine in the absence (control) or presence of 5 μM A23187 (open bars) or 100 nM thapsigargin (gray bars) and at various concentration of CaCl$_2$ as indicated. The values obtained for untreated cells were taken as 100% (basal line, control) (modified from (9)).

imaging system (8,9). In the presence of extracellular Ca^{2+} all agents produced a rapid increase in [Ca^{2+}]$_i$ which consisted of two phases: a transient, initial phase associated with the depletion of intracellular stores and a sustained phase being a result of the Ca^{2+} influx to the cell (Fig. 2A and B, upper traces). In the presence of EGTA (Ca^{2+} free medium), thapsigargin and DBHQ caused a transient rise in [Ca^{2+}]$_i$ which declined again to the basal line (Fig. 2A and B, lower traces). The depletion of Ca^{2+} stores was confirmed by the observation that pretreatment of the cells with thapsigargin and DBHQ prevented appreciable Ca^{2+} release by ionomycin (Fig. 2C and D), indicating that Ca^{2+} stores were already discharged.

Thus, our data indicate that the depletion of the ER Ca^{2+} stores generated either through agonist-receptor interaction and InsP$_3$ formation, or artificially, by the action of Ca^{2+} ionophore, or by blockers of the ER Ca^{2+}-ATPase, distinctly inhibits PS synthesis. This phenomenon has been observed by us in glioma C6 cells (9,10), and by other authors in Jurkat T cells (11,12) and Ehrlich ascites tumor cells (13). These results, and those of our studies on isolated rat liver microsomes (5,6) provide a proof that serine base exchange enzyme system requires Ca^{2+} for its catalytic activity at the luminal side of the ER, and can be regulated not only by Ca^{2+}-ATPase, but also by most other signals which control cellular Ca^{2+} fluctuations.

3. TRANSMEMBRANE LOCALIZATION OF PS SYNTHESIS

An important consequence of the discovery that the depletion of the ER Ca^{2+} stores inhibits PS synthesis is a hypothesis that this process occurs at the luminal leaflet of the ER. Such hypothesis was postulated by us (6,9) and by Pelassy et al. (11), who suggested that serine base exchange enzyme system is located within or close to the intracellular Ca^{2+} stores. The luminal localization of serine base exchange enzyme was also reported by Buchanan and Kanfer (14) for rat brain microsomes treated with trypsin and phospholipases.

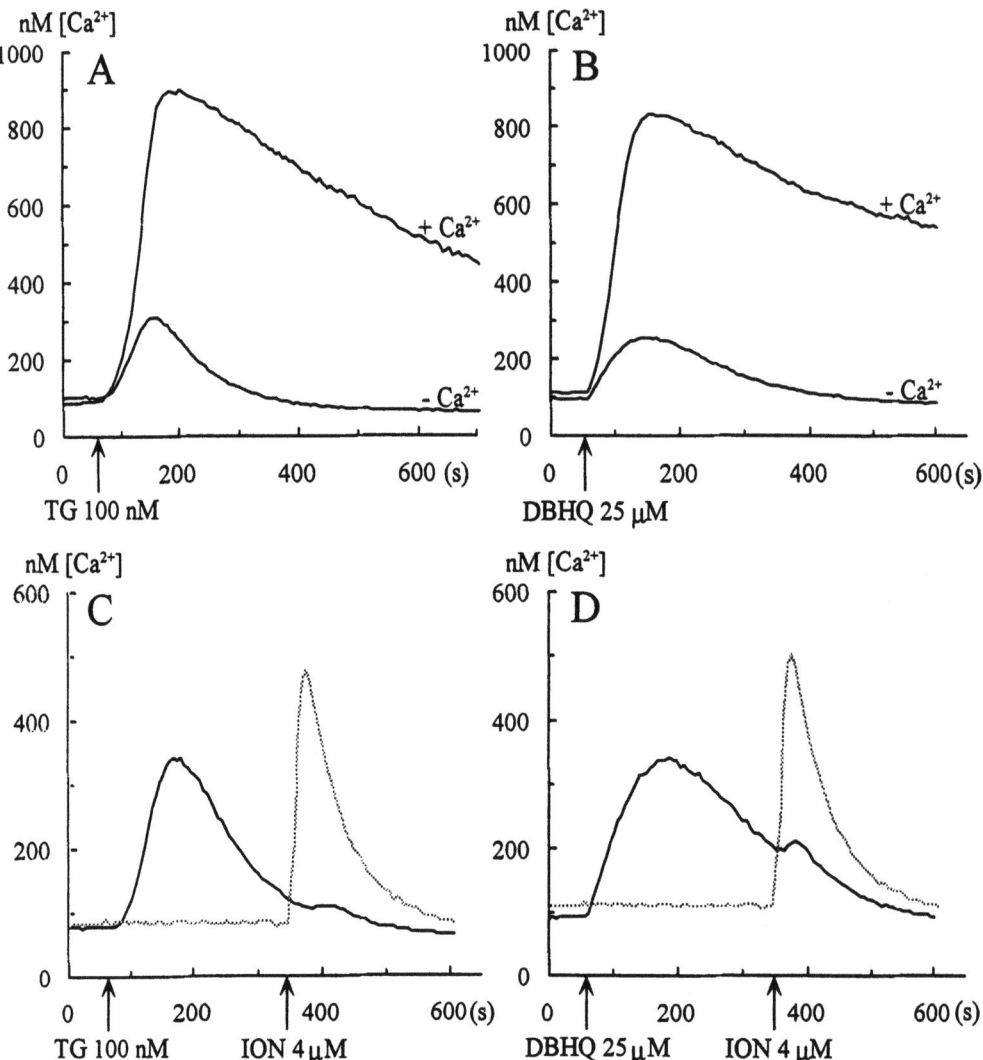

Figure 2. Changes in $[Ca^{2+}]_i$ in glioma C6 cells evoked by: A, 100 nM thapsigargin (TG); B, 25 μM DBHQ; cells treated with 2 mM $CaCl_2$ in extracellular medium (upper traces in A and B) or incubated without $CaCl_2$, in the presence of 500 μM EGTA (Ca^{2+} free medium) (lower traces in A and B). C, D, cells treated in Ca^{2+} free medium with 100 nM thapsigargin (TG) (C), or 25 μM DBHQ (D) added 6 min prior to 4 μM ionomycin (ION) (solid lines in C and D). Control, the cells treated with 4 μM ionomycin (ION) alone (dotted lines in C and D).

However, this statement remains in contrast to the other proposal. Topology experiments suggest that important domains of this enzyme are exposed on the cytoplasmic surface of hepatic microsomal vesicles (15,16). After treatment with trypsin about half of the enzyme activity was lost (16). Moreover, it has been demonstrated by Vance (17) that newly made PS is located in the cytosolic leaflet of the microsomal membrane and is preferentially translocated to mitochondria. The same observation for newly synthesized PS in rat brain microsomal vesicles were reported by Corrazi et al. (18). Besides, the authors demonstrated that also about half of serine base exchange activity was lost after treatment with reagents under non-penetrating conditions, whereas it was completely lost with reagents that penetrated vesicles. They suggested that a part of the enzyme might be present at the luminal compartment (18).

Therefore, to explain these contradictory results we propose a new model for the localization of serine base exchange enzyme. We suggest that the enzyme is a transmembrane protein containing the active site for serine accessible from the cytosolic surface of the ER and the active site for Ca^{2+} exposed at the luminal surface of the ER membrane. Newly made PS may be then located at the outer, cytosolic leaflet of the ER, being available for transfer to other membranes. PS synthesis would be then regulated by the level of Ca^{2+} within the ER lumen, as we indeed demonstrated.

4. THE ROLE OF PKC IN PS SYNTHESIS IN GLIOMA C6 CELLS

PS is a generally accepted as a very active stimulator of all isoforms of the PKC family (7). Under physiological conditions many of them are activated by diacylglycerol. In experimental model systems, PKC can be activated by phorbol esters, e.g. 12-O-tetradecanoylphorbol 13-acetate (TPA), which mimic the action of diacylglycerol. TPA is known to stimulate the metabolism of phosphatidylcholine and other phospholipids. Treatment of cells with TPA leads to increased hydrolysis of this phospholipid (19) and phosphatidylethanolamine (20). In glioma C6 cells TPA appeared to have no effect on PS degradation (21). After 4h incubation of cells prelabelled with radioactive serine, a similar 30% decrease in the amount of radioactive PS was observed in the absence and in the presence of TPA. On the other hand, 10 and 100 nM TPA inhibited serine incorporation into PS by about 30% and 60%, respectively, after 1h incubation (21) (Fig. 3). When the incubation was prolonged to 4h, this effect was not observed with 10 nM TPA, whereas 100 nM TPA still inhibited the incorporation. The inhibitory effect of TPA on PS synthesis was also reported in leukemic HL cells (22), whereas in neuroblastoma LA-N-2 cells TPA stimulated PS synthesis (23), suggesting that the effect of TPA might depend on the type of cell.

Treatment of glioma C6 cells with TPA produced, together with changes in PS synthesis, distinct changes in cellular morphology and the actin cytoskeleton (Fig. 4). Actin cytoskeleton alterations can be observed due to the selective binding and fluorescence of filamentous-actin-specific phalloidin derivative (21). The untreated, control cells exhibit an abundance of actin microfilament bundles running through much of the length of the cells (Fig. 4B). The cells exposed to TPA by 1h exhibit disruption of the actin cytoskeleton produced by both 100 nM TPA (Fig. 4E) and 10 nM TPA (Fig. 4D). In such treated cells, large areas of cytoplasm become depleted of actin bundles and ruffling of the cell periphery, and aggregates of actin staining appear. However, these TPA-induced alterations are reversible and return to the nearly normal appearance after longer treatment of cells with TPA. This reversion begins to be visible as early as after 4h incubation of cells with TPA (10 and 100 nM) and is pronounced after prolonged (18 h) 100 nM treatment (Fig. 4G).

These results show that the changes in PS formation produced by 10 nM TPA can be correlated with TPA-induced changes in the organization of the actin cytoskeleton, and can be PKC-mediated events. Actin cytoskeletal disruption is thought to be a result of a potent non-physiological activation of PKC by TPA, whereas down-regulation of PKC produced by a prolonged treatment of the cells with TPA (hours) reverse actin cytoskeletal organization (24). However, TPA used in the higher concentration (100 nM) exhibits more complex effects on cellular response. 100 nM TPA diminished PS formation even when alterations in the actin cytoskeleton were reversed. Nevertheless, the cells were still morphologically changed. The untreated, fibroblast-like glioma C6 cells (Fig. 4A), become round-up (Fig. 4C) within 1h after the addition of 100 nM TPA, whereas upon long 100

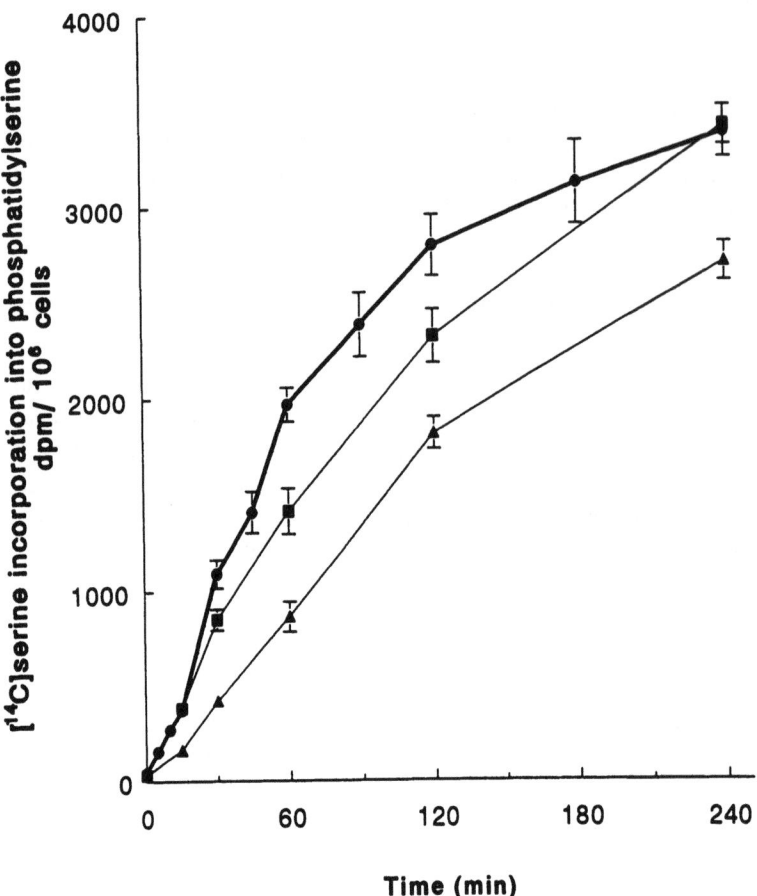

Figure 3. Time dependent effect of TPA upon the incorporation of [^{14}C]serine into phosphatidylserine. Glioma C6 cells were incubated in the absence (control,●) or presence of 10 nM TPA (■) or 100 nM TPA (▲) (21).

nM TPA treatment (18h) acquired a spindle like morphology and were slightly contracted (Fig. 4F). One can suppose that prolonged stimulation of cells with higher concentration of TPA may cause activation of Ras leading to activation of mitogen-activated protein kinase and AP-1 controlling gene expression in cascade of events initiated by PKC (25), and/or cell differentiation.

5. CONCLUSIONS

It is well known that the concentration of free Ca^{2+} in cytosol of animal cells is very low (from 0.1 μm in resting cell to 1 μM in stimulated cell), whereas Ca^{2+}-ATPase maintains a high (mM) Ca^{2+} level in the lumen of the ER (26). It is also well known that serine base exchange enzyme needs high (mM) concentration of Ca^{2+} for its activity (1). Therefore, the observation that the depletion of the ER Ca^{2+} stores strongly inhibits PS synthesis seems to be very logical. These, presently well documented data, and the hypothesis that serine base exchange enzyme system contains the active site for Ca^{2+} exposed at the luminal leaflet of the ER, suggest that PS synthesis can be regulated by intracellular events

The Role of Ca^{2+} and Protein Kinase C

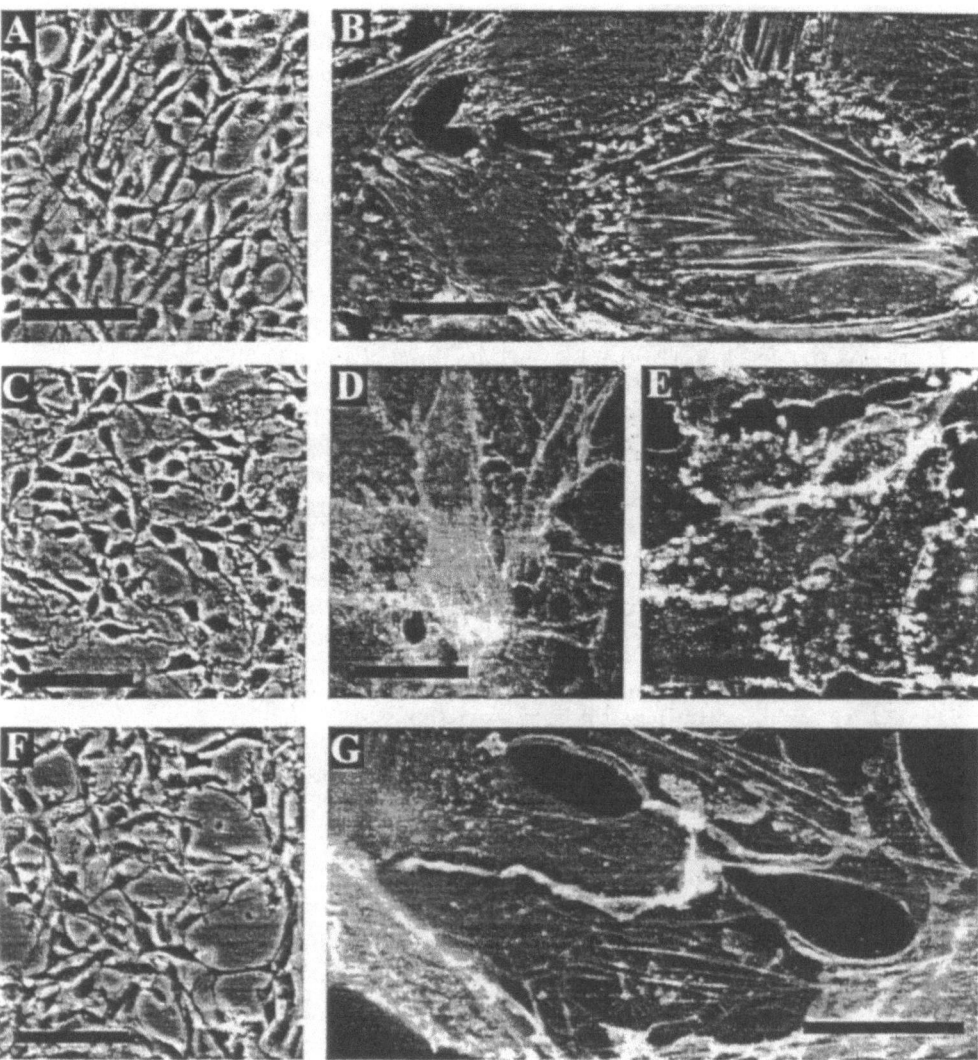

Figure 4. Effect of TPA on glioma C6 cells morphology and the actin microfilament cytoskeleton. A,C,F - morphology of glioma C6 cells: A, cultured in the absence of TPA (control cells); C, exposed to 100 nM TPA for 1h (PKC activated cells); F, exposed to 100 nM TPA for 18h (PKC down-regulated cells); B,D,E,G - fluorescence of actin cytoskeletal alterations in FITC-phalloidin labeled cells exposed to: B, control medium; D, 10 nM TPA for 1h; E, 100 nM TPA for 1h; G, 100 nM TPA for 18h. Pictures are reverse-phase light microscopy (A,C,F) and epifluorescence microscopy (B,D,E,G) photographs. Bar = 250 μm (A,C,F) and 12,5 μm (B,D,E,G) (21).

responsible for Ca^{2+} fluxes in the cell, and be under control of transmembrane signalling systems. It can be also supposed that in glioma C6 cells, PS synthesis may be regulated by the PKC activity. On the other hand, it is tempting to speculate that the production of PS may regulate the activity of PKC. If so, the decrease in PS synthesis may play a protective role against persistent and non-physiological PKC activation caused by TPA. In addition, the changes in PS synthesis can be associated with the alterations in filamentous actin cytoskeleton. Therefore, the production of PS, level of intracellular Ca^{2+}, PKC activity, and organization of the actin cytoskeleton seem to be correlated events in glioma C6 cells.

ACKNOWLEDGMENT

This work was supported by a grant from the State Committee for Scientific Research for the Nencki Institute. P.S. is the recipient of the 1996 Foundation for Polish Science fellowship programme.

REFERENCES

1. Kanfer, J.N., 1980, *Can. J. Biochem.* **58**: 1370–1380.
2. Barańska, J., 1982, *Adv. Lipid Res.* **19**: 163–184.
3. Hubscher, G., Dils, R.R., and Pover, W.F.R., 1958, *Nature* (London) **182**: 1370–1380.
4. Hubscher, G., Dils, R.R., and Pover, W.F.R., 1959, *Biochim. Biophys. Acta* **36**: 518–528.
5. Barańska, J., 1989, *FEBS Lett.* **256**: 33–37.
6. Czarny, M., and Barańska, J., 1993, *Biochem. Biophys. Res. Commun.* **194**: 577–583.
7. Nishizuka, Y., 1992, *Science* **258**: 607–614.
8. Barańska, J., Chaban, V., Czarny, M., and Sabała, P., 1995, *Cell Calcium* **17**: 207–215.
9. Czarny, M., Sabała, P., Ucieklak, A., Kaczmarek, L., and Barańska, J., 1992, *Biochem. Biophys. Res. Commun.* **186**: 1582–1587.
10. Czarny, M., and Barańska, J., 1993, *Biochem. Mol. Biol. Intern.* **5**: 967–973.
11. Pelassy, C., Breittmayer, J.P., and Aussel, C., 1992, *Biochem. J.* **288**: 785–789.
12. Breitmayer, J.P., Pellasy, C., and Aussel, C., 1996, *J. Lipid Mediators Cell Signalling* **13**: 151–161.
13. Rakowska, M., and Wojtczak, L., 1995, *Biochem. Biophys. Res. Commun.* **207**: 300–305.
14. Buchanan, A.G., and Kanfer, J.N., 1980, *J. Neurochem.* **34**: 720–725.
15. Ballas, L.M., and Bell, R.M., 1981, *Biochem. Biophys. Acta* **665**: 586–595.
16. Vance, J.E., and Vance, D.E., 1988, *J. Biol. Chem.* **263**: 5898–5909.
17. Vance, J.E., 1991, *J. Biol. Chem.* **266**: 89–97.
18. Corazzi, L., Zborowski, J., Roberti, R., Binaglia, L., and Arienti, G., 1987, *Bull. Mol. Bio. Med.* **12**: 19–31.
19. Exton, J.H., 1994, *Biochim. Biophys. Acta* **1212**: 26–42.
20. Kiss, Z., and Anderson, W.B., 1989, *J. Biol. Chem.* **264**: 1483–1487.
21. Czarny, M., Wiktorek, M., Sabała, P., Pomorski, P., and Barańska, J., 1995, *Biochem. Mol. Biol. Inter.* **36**: 659–667.
22. Kiss, Z., Deli, E., and Kun, J.F., 1987, *Biochem. J.* **248**: 649–656.
23. Singh, I.N., McCartney, D.G., Sorrentino, G., Massarelli, R., and Kanfer, J.N., 1992, *J. Neurosci. Res.* **32**: 583–592.
24. Hedberg, K.K., Bimell, G.B., Mobley, P.L., and Griffith, O.H., 1994, *J. Cell. Physiol.* **158**: 337–346.
25. Karin, M., and Smeal, T., 1992, *Trends. Biochem. Sci.* **17**: 418–422.
26. Berridge, M.J., 1993, *Nature* (London) **361**: 315–325.

THE POSSIBLE MECHANISMS INVOLVED IN NEW OPIOID AGONIST FENARIDIN AND ITS ANTAGONIST

M. I. Agadjanov and G. S. Vartanian

Department of Biochemistry
Yerevan State Medical University, Armenia

1. INTRODUCTION

The cellular mechanisms, underlying opioid action, remained to be fully determined, although there is now growing evidence that some opioid receptors may be coupled to phospholipase C, and the phosphoinositide pathway may play a part in the cellular mechanisms of opioid action (6,13,18). Simultaneously, the role of arachidonic acid and its metabolites in the cellular mechanisms of opioidergic neurotransmission is yet not defined exactly. It is generally assumed, that eicosanoids may trigger various biochemical cascades and the functional coupling of phosphoinositide specific phospholipase C (PLC) and arachidonic acid (AA) metabolites has been suggested (5, 14). Moreover, the evidence was reviewed, that different eicosanoids play a regulatory role in the release of different hormones (11), neurotransmitters and, particularly, opioid peptides (β-endorphines) (10). The role of the processes mentioned as well as their possible relationship in biochemical mechanisms of new opioid agonist fenaridin and its antagonist F remained to be determined. Opioid agonist fenaridin presents high potency of analgesic strength as compared to morphine, promedol and fentanyl and long duration as compared to fentanyl (17). Antagonist F appeared to be a "pure" opioid antagonist with high affinity to binding sites. In our earlier investigations the effect of these substances on the level of cyclic nucleotides (15) had been demonstrated as well as their influence on the activity of some glycolitic enzymes (9) and phospholipids deacylation-reacylation processes (16).

2. EXPERIMENTAL PROCEDURE

Male rats weighting 150–180 g were used. Synaptosomes were isolated by the method of Hajost (4). Phosphoinositides fraction of intact synaptosomes was labelled with [^{14}C-AA] in the presence of lysoderivatives of this phospholipids using known method (8). Recovery of PI radioactivity was ~ 70%. PI catabolism was studied in the incubation medium, containing prelabelled synaptosomes (150–200 mkg of protein), 2,5 mM $CaCL_2$, 0,32 M sucrose, 50 mM tris-HCL-bufer, 0,1 ml of studied compound (at the final concentration 10^{-5} M), at 37°C. After 5 and 10 sec. respectively the incubation was terminated and products of hydrolysis were extracted as described. Extract was analysed by TLC, and individual spots from TLC plates were analysed for [^{14}C] products. Radioactivity of samples was determined using scintillation spectrophotometer "Roshe-Bioelectronique, SL-4221" (France). Polymorphonuclear leukocytes (PMNL) were isolated from human peripheral blood, as described (1). PMNL were washed twice in the presence of human serum albumin (0,5%) and were resuspended in Dulbecco phosphate buffer. To determine the effects of compounds studied, solutions in final concentrations ($CaCL_2$ — 1,6 mM, ATP — 1 mM, ionophore A-23187 — 10 mkM, AA — 33 mkM and fenaridin (or antagonist) — 10^{-5} M), were added to a suspension of PMNL (40×10^6 cells). After 5 min the incubation was terminated by adding 1.5 volume of methanol containing PGB_2 to the medium as internal standard (50 nM/10^6 cells). The pellet was separated by centrifugation and supernatant was redissolved in water medium, where methanol concentration did not exceed 15%. Samples were applied to columns with alkylated silicagel (Sep-Pak, Waters). The water-methanol solution was evaporated to minimal volume with a fine stream of nitrogen and was extracted with chloroform. Chloroform extracts were washed twice and

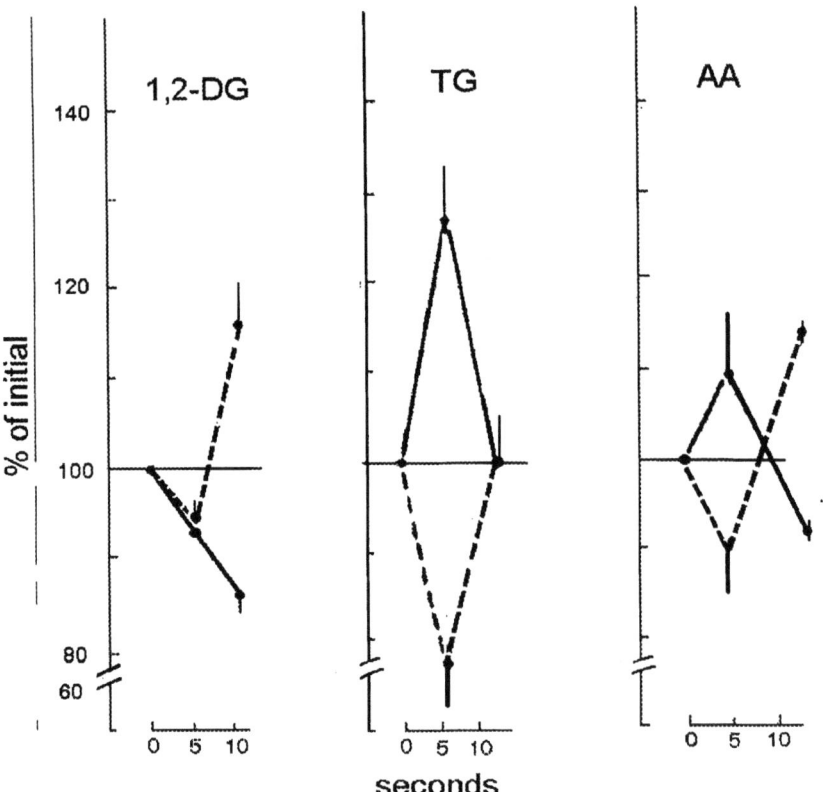

Figure 1. Effects of fenaridin and antagonist F treatment on the diacylglycerol and arachidonic acid release (% of basal) in rat brain synaptosomes, prelabelled with [^{14}C]-arachidonic acid. Results show mean ± s.e.m. from 16–20 animals. Solid line = fenaridin; dashed line = antagonist F.

evaporated to dryness and the residue was dissolved in ice cold ethanol and analysed, using HPLC (LKB, Bromma). The columns with hydrophob silicagel Sperisorb ODS-2, 5 mkM, were used as well as Lichrosorb RP18, 5 mkM, using methanol-water-acidic acid 75:25:0,1 as a solvent system. pH of medium was regulated with ammonium solution. UF-detector with changing wave-length and photodiode detector were used. Eicosanoids identification was performed, using standards. Results were treated statistically, using Student's test.

3. RESULTS AND DISCUSSION

In order to evaluate the involvement of the above processes in the biochemical mechanisms of studied agonist and antagonist's action, the possibility of PI catabolism in rat brain synaptosomes was estimated during *in vitro* action of fenaridin and its antagonist. The significant enhancement of diacyl glycerol (DG) formation has been demonstrated at already 10 sec. after antagonist administration, as well as opposite direction of agonist influence (Fig.1).

The cause of such rapidly increased availability of DG appeared to result from PLC mediated PI hydrolysis, but DG may theoretically also be derived from triacylglycerols or noninositide phospholipids. These findings suggest that opposite alterations in DG production may be due to specific -receptor coupled action via PI cycle functioning, which is in agreement with published data about antagonistic influence of proteinkinase C towards morphine-induced analgesia. In addition, significant and opposite alterations in AA mobilization have been demonstrated under the same conditions of opiate receptor blockade and facilitation (Fig.1), which may be due, on the one hand, to their opposite action on phospholipids deacylation-reacylation processes, on the other - may be connected to eicosanoids metabolism. Since eicosanoids have been shown to cause pain (2), we studied the possibility that the pharmacologic effect of mentioned substances might correlate with alterations in eicosanoid formation. However, evidence showed, that opioids modified some functions of granulocytes (12) and bidirectional communications exists between opioid and immune systems (7). The action of fenaridin and its antagonist on the level of AA transformation metabolites in human polymorphonuclear leukocytes was studied (Fig. 2,3).

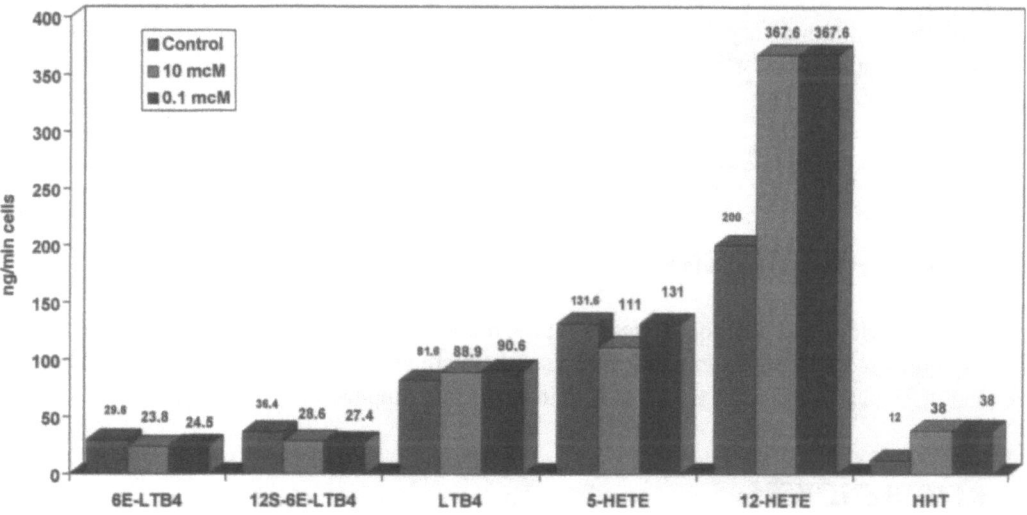

Figure 2. Effect of treatment with fenaridin on the profile of arachidonic acid transformation products in polymorphonuclear leukocytes. Results show mean from 7 determinations.

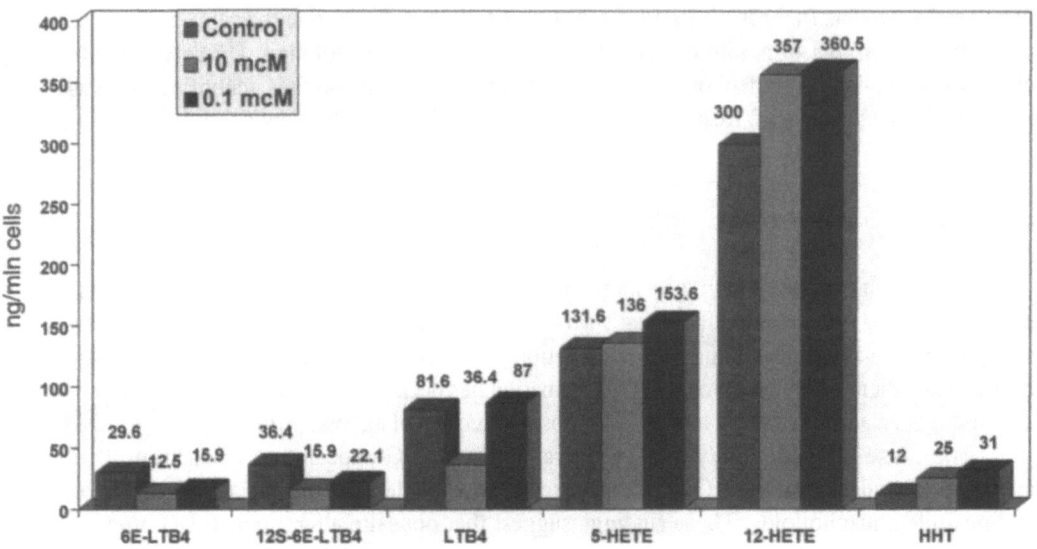

Figure 3. Effect of treatment with antagonist F on the profile of arachidonic acid transformation products in polymorphonuclear leukocytes. Results show mean from 7 determinations.

Alterations in the 12-hydroxy-5,8,10-heptadecatrienoic acid (HHT), 5-hydroeicostetraenoic acid (5-HETE), and 12-hydroeicostetraenoic acid (12-HETE) level were observed in the presence of compounds mentioned. We have found their opposite influence on the formation of hyperalgesic inflammatory agent leukotriene B_4 (LTB_4) of interest. Thus, the possibility exists, that agonist and antagonist modified the activity of leukotriene A_4 (LTA_4) hydrolase - the rate-limiting enzyme for the formation of leukotriene A_4 (LTB_4). It has been established (3), that the enzyme is catalitically bi-functional and may modulate pain and inflammation via two separate molecular pathways: formation of hyperalgesic inflammatory agent LTB_4 and degradation of endogenous opioid peptides. Moreover, studied substances influenced differently the nonenzymatic step of LTB_4 transformation to $6E-LTB_4$ and $12S-6E-LTB_4$ respectively. Thus, the effects of studied agonist and antagonist appear to be coupled to PI turnover and, may at least partly be due to alterations in arachidonic acid transformation via the lipoxygenase pathway.

ACKNOWLEDGMENT

Fenaridin and antagonist were synthesed and kindly donated by an academician of Armenian NAS, prof. Vartanian R.S. The authors are grateful to prof. A.G. Panossian and doct. U.V. Tadevosian for the great help in performed studies.

REFERENCES

1. Boyum, A., 1976, *Scand. J. Immunol.*, **5**: Suppl 5, 9.
2. Ferreira, H., Nakamura, M., 1979, *Prostaglandins* **18**: 191.

3. Griffin, K.J., Gierse, J., Krivi, G., Fitzpatrick, F.A., 1992, *Prostaglandins*, **44**: 251.
4. Hajost, F., 1975, *Brain Res.*, **93**: 485.
5. Hansson, A., Serhan, C.N., Haeggstrom, J., Ingelman-Sundberg M., Samuelsson, B, 1986, *Biochem. biophys. Res. Commun.*, **134**: 1215.
6. Jin, V., Lee, M.M., Loh, H.H., Thayer, S.A., 1994, *J.Neurosci.*, **14**: 1920; 1986, *Biochem. biophys. Res. Commun.*, **134**: 1215.
7. Jodar, L., Takahashi, M., Kaneto, H., 1994, *Yakubutsu-Seishin Kodo*, **14**: 195.
8. Manning, R., Sun, G.Y., 1983, *J. Neurochem.*, **41**: 1735.
9. Nazaryan, K.B., Vartanian, G.S., Kostanyan, A.A., Agadjanov, M.I., 1994, *J.Neurochem.*, **63**: Suppl 1, S 84.
10. Nishizaki, T., Ikegami, H., Tasaka, K., Hirota, K., Miyake, A., Tanizava, O., 1989, *Neuroemdocrinology*, **49**: 483.
11. Ojeda, S.R., Urbanski, H.F., Junier, M.-P, Capdevila, J., 1989, *Ann. N.Y. Acad. Sci.*, **559**: 192.
12. Passotti, D., Mazzonne, A., Ricevut, G., 1992, *Minerva Med.*, **83**: 433.
13. Raffa, R.B., Connelly, C.D., Martinez, R.P., 1992, *Eur. J. Pharmacol.*, **217**: 221.
14. Schaad, N.C., Magistretti, P.J., Shorderet, M., 1991, *Neurochem. Internat.*, **18**: 303.
15. Vartanian, G.S., Agadjanov, M.I., Burnazian, R.A., 1990, *Vop. Med. Chimii*, **36**: 49.
16. Vartanian, G.S., Agadjanov, M.I., Tadevosian, U.V., Batikian, T.B., 1995, *J. Neurochem.*, **65**: Suppl., S 59.
17. Vartanian, R.S., Martirosian, V.O., Vartanian, S.H., Engoian, A.P., Vlasenko, E.V., Durgarian, L.K., Azlivian, A.S., Valdman, A.V., 1989, *Chim. Pharm. J.*, **5**: 573.
18. Xu, H., Gintzler, A.R., 1992, *Proc. Natl. Acad. Sci. USA*, **89**: 1978.

Section 36: Molecular Mechanism of Lipid Mediators Action and Their Role in Neurotransmission and Signal Transduction

LIPID METABOLISM MARKERS IN MULTI-INFARCT DEMENTIA

R. Shakarishvili, S. Tabagari, G. Thakhava, and M. Topuria

Sarajishvili Institute of Neurology
Tbilisi, Georgia

1. INTRODUCTION

In recent years in the majority of economically developed countries, along with increased spread of the vascular brain diseases — on account of ageing of the population, an increase in the number of patients with vascular dementia (VD) was revealed. The study of the pathogenic characteristics of early stages of VD, and detection of distinct criteria are of great importance for a proper prevention and treatment of this disease. The study of multi-infarct dementia (MID) pathogenesis has shown that often the extent of structural alteration in the brain did not correlate with the severity of cognitive function changes (1–5).

Some authors believe that changes of the white matter of the brain are an early indication of VD. Gray matter pathology may produce secondary changes in white matter, although severe gray matter involvement is associated with only mild myelin loss. (6).

Because the pathogenic base of VD is necrobiosis of structural elements of neural tissues, efforts are made for the elucidation of specific markers of these processes in biological fluids of patients. Some authors support the idea that an increase of intermediate desintegration products of intracellular elements of neural tissues is a neurochemical indicator of multi-infarct encephalopathia and it may become the marker of the early stages of dementia (7–12). Because the changes in brain cell membrane structures represent the early microstructural correlates of VD, abnormality of membrane lipids(13) and activities of some subcellular structures, especially lysosomes, may become the biomarker of pathogenic processes underlying the base of this disease. The lysosomal system of living organizms responds to the influence of wide spectrum intrinsic and exogenous factors of which the activation rate correlates with loading and demands of oxidative metabolism. Lysosomal activation is directed to adaptive metabolic and structural reorganization in the cell, and its degree and character depend not only on the strength and duration of the pathology but also on its specificity. Much attention is paid to structural functional

---

From a workshop organized by Joanna B. Strosznajder (Warsaw, Poland), Jolanta Baranska (Warsaw, Poland) and Moti Liskovitch (Rehovot, Israel).

changes of cell structures during the investigations of various pathologies. Profoundness and duration of the membrane destructive processes in the organism determine the clinical severity of the disease. Membrane destruction, as a pathological phenomenon, is mainly cuased by involvement of lipid membranes of cells, and it leads to alterations of lipid-protein bonds, the weakening of the enzyme complex and other macromolecular membranes, as well as to a disturbance of cell metabolism.

In the presented work we investigated the functional state of some lysosomal enzymes and lipid spectrum changes in blood and cerebrospinal fluid (CSF) of patients with MID; in correlation with the severity of dementia, as well as with qualititative characteristics of brain ischemia. Patients had been selected according to NINDS AIREN criteria (14).

2. METHODS AND MATERIALS

Hachiski's Ishemia Score was also used (15). The presence and degree of dementia were identified by a complex neuropsychological evaluation of the patient's condition with respect to WHO criteria (16) and MMSE. Sixty two patients aged 50–65 (27 men and 35 women) have been examined during the investigation.

Plasma lipids were extracted by the method of Folch(17). Lipid extracts were separated then by means of thin-layer chromatography methods (18).

Activity of lysosomal glycosidases (β-galactosidase and β-glucosidase) was determined spectrophotometrically with wave-length 420 nm. Appropriate substrates — 4-nitrophenyl-β-D-galactopyranoside and 4-nitrophenyl-β-D-glucopyranoside — have been used.

3. RESULTS

Correlating with the severity of dementia our data show moderate increase of cholesterol (CHL) and lysophosphatidylcholine (LPC), decrease of phosphatidylethanolamine (PEA). In severe dementia patients decreases of cerebrosides (CRB by 15%), free fatty acids (FFA more than two fold) and activity of β-glucosidase(by more than 30%) are observed, while the activity of β-galactosidase is slightly increased as compared to the group of patients with mild dementia.

In the CSF there is a decrease of CHL(sharp), FFA, PEA(moderate) that correlates with the severity of dementia.

At the same time there is a sharp decrease of phosphatidylcholine (PCH) and CRB, in the group of patients with severe dementia.

Changes of lysosomal glycosidases (β-galactosidase and β-glucosidase) in the CSF more closely correlate with the severity of dementia than those in the blood.

As a result of the study conducted the increase of CHL (by 20%), LPC (by more than 40%), and a decrease of PCH (by 40%) was assessed in the blood of patients with moderate mono-infarct dementia as compared to the group of patients with mild multi-infarct dementia. At the same time in the group of patients with mono-infarct dementia patients the mild rise in the β-galactosidase activity and the sharp drop in β-glucosidase activity were registered in comparison with the other group. (Table 1, fig. 1).

In contrast to the blood, in the CSF of patients with mono-infarct dementia we assessed a decrease of CHL, FFA, PCH and CRB (one and half fold), but a proportionally equal increase of cholesterol esters (CHE) in comparison with multi-infarct dementia. Changes of lysosomal glycosidases activities in the CSF were similar to those in the blood (Table 2).

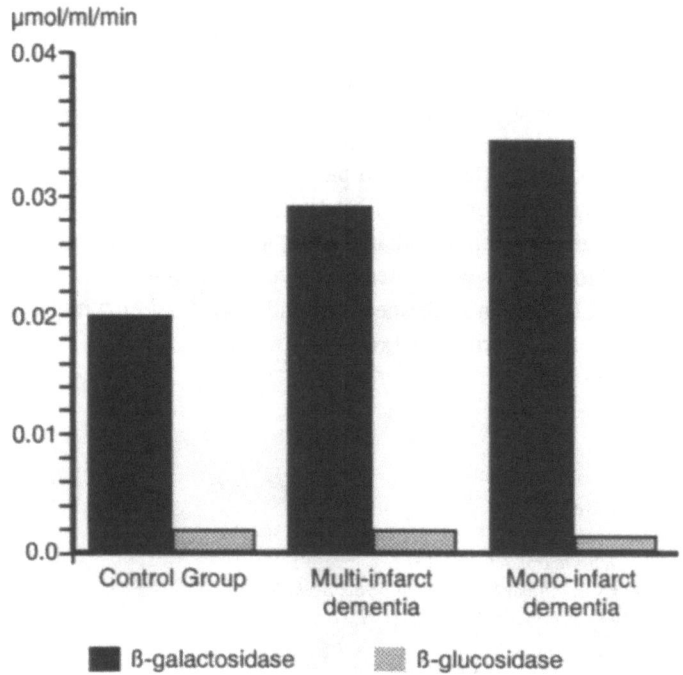

Figure 1. Changes of the activity of some lysosomal glycosidases in the blood of patients with multi- and mono-infarct dementia.

Table 1. Lipid spectrum in the blood serum of patients with mono- and multi-infarct dementia

The study group (n)	Lipid fractions (mmol/l)								
	CHL	CHE	FFA	TRG	LPC	PCH	PEA	SPH	CRB
Control group (20)	143±0.09	5.62±0.65	1.23±0.14	1.45±0.2	0.18±0.02	1.69±0.26	0.59±0.05	0.7±0.07	1.24±0.08
multi-infarct dementia (15)	2.3±0.1	6.22±0.4	1.82±0.17	1.47±0.2	0.355±0.05	1.08±0.13	0.53±0.04	0.79±0.06	1.64±0.13
mono-infarct dementia (12)	2.71±0.16	6.74±0.52	1.85±0.18	1.45±0.2	0.51±0.07	0.77±0.1	0.43±0.04	0.82±0.06	1.50±0.11
Statistical significance	$P_{1-2}<0.001$ $P_{1-3}<0.001$ $P_{2-3}<0.05$	$P_{1-2}<0.5$ $P_{1-3}<0.1$ $P_{2-3}<0.5$	$P_{1-2}<0.01$ $P_{1-3}<0.01$ $P_{2-3}<0.5$	$P_{1-2}<0.5$ $P_{1-3}<0.5$ $P_{2-3}<0.5$	$P_{1-2}<0.01$ $P_{1-3}<0.001$ $P_{2-3}<0.05$	$P_{1-2}<0.05$ $P_{1-3}<0.001$ $P_{2-3}<0.05$	$P_{1-2}<0.05$ $P_{1-3}<0.01$ $P_{2-3}<0.05$	$P_{1-2}<0.5$ $P_{1-3}<0.5$ $P_{2-3}<0.5$	$P_{1-2}<0.01$ $P_{1-3}<0.05$ $P_{2-3}<0.5$

Table 2. Lipid spectrum in CSF of patients with mono- and multi-infarct dementia

The study group (n)	Lipid fractions (mmol/l)						
	CHL	CHE	FFA	TRG	PCH	PEA	CRB
Control group (15)	0.153±0.02	0.35±0.07	0.13±0.01	0.07±0.01	0.04±0.01	0.13±0.01	0.1±0.01
multi-infarct dementia (12)	0.119±0.005	0.49±0.104	0.844±0.084	0.069±0.008	0.035±0.008	0.1±0.01	0.228±0.054
mono-infarct dementia (7)	0.075±0.003	0.738±0.13	0.438±0.049	0.063±0.006	0.023±0.001	0.104±0.01	1.144±0.23
Statistical significance	$P_{1-2}<0.02$ $P_{1-3}<0.001$ $P_{2-3}<0.001$	$P_{1-2}<0.5$ $P_{1-3}<0.1$ $P_{2-3}<0.05$	$P_{1-2}<0.001$ $P_{1-3}<0.001$ $P_{2-3}<0.001$	$P_{1-2}<0.5$ $P_{1-3}<0.5$ $P_{2-3}<0.5$	$P_{1-2}<0.2$ $P_{1-3}<0.002$ $P_{2-3}<0.05$	$P_{1-2}<0.1$ $P_{1-3}<0.005$ $P_{2-3}<0.5$	$P_{1-2}<0.02$ $P_{1-3}<0.05$ $P_{2-3}<0.05$

4. DISCUSSION

All these results may be explained by the destruction of membrane elements in white and gray matter of the brain. It has been proved that changes of the lipid spectrum and lysosomal glycosidases activity (particularly pronounced in CSF) correlates with dementia severity. We also witnessed marked changes of the lipid spectrum and activity of some lysosomal glycosidases in the group of patients with moderate dementia in comparison with mild multi-infarct dementia manifestations. Everything mentioned above points to the existence of certain biochemical changes that adequately correlate with the severity of dementia even in those cases when, according to MRT and CT findings, no changes have been registered. Changes in lipid spectrum and lysosomal enzymes activity may be regarded as markers for the severity of dementia.

REFERENCES

1. Mohr, J.P., 1982, *Stroke* **13**: 3–11.
2. Ishii, N., Nishihara, Y., Imamura, T., 1986, *Neurology* **36**: 340–344.
3. Wolfe, N., 1990, *Arch. Neurol.* **47**: 129–13.
4. Kinkel, W.H., Jacobs, L., Polanchini, I., et al, 1985, *Arch. Neurol.* **42**: 951–959.
5. Tatemichi, T.K., 1990, *Neurology* **40**: 1652–1659.
6. Brun, A., Englund, E., 1986, *Ann Neurol.* **19**: 253–262.
7. Muckle, T.J., Roy, J.K., 1985, *Lancet*, **I**: 1191–1193.
8. Tilvis, R.S., Erkinjuntti, T., Sulkava, R., Miettinen, A., 1987, *Atherosclerosis,* **65**: 237–245.
9. Fredman, P., Wallin, A., Blennow, K., Davidsson, P., Gottfries, C.G., Svennerholm, L., 1992, *Acta Neurol. Scand.,* **85**: 103–106.
10. Polischuk, I.A., Gorodkova, T.M., Chernitskaia, I.I., 1972, *S.S. Korsakov journal of Neurology and Psychiatry* **12**: 1828–1831.
11. Polischuk, 1974, *S.S. Korsakov journal of Neurology and Psychiatry* **6**: 864.
12. Drobishev, N.A., 1976, *S.S. Korsakov journal of Neurology and Psychiatry* **2**: 218.
13. Wallin, A., Gottfriess, C.G., Karlsson, I., Svennerholm, L., 1989, *Acta Neurol. Scand,* **80**: 319–325.
14. McKhann, G., Drachman, D., Folstein, M., Katzman, R., Price, D., Stadlan, E.M., 1984, *Neurology* **34**: 939–944.
15. Hacinski, V.C., Lassen, N.A., Marshall, J., 1974, *Lancet* **II**: 207–209.
16. Henderson, A.S., 1994, *Dementia WHO*, Geneva.
17. Folch, J., Lee, M., Sloane-Stanley, G.A., 1957, *J. Biol. Chem.* **226**: 497.
18. Babaskin, L.M., 1977, *S.S. Korsakov journal of Neurology and Psychiatry* **79**: 406–407.

Section 36: Molecular Mechanism of Lipid Mediators Action and Their Role in Neurotransmission and Signal Transduction

CMP-DEPENDENT DEGRADATION OF PLATELET-ACTIVATING FACTOR (PAF) BY RAT BRAIN MICROSOMES

Ermelinda Francescangeli, Serena Porcellati, and Gianfrancesco Goracci

Institute of Medical Biochemistry
University of Perugia
Perugia, Italy

1. INTRODUCTION

PAF (1-O-alkyl-2-acetyl-sn-glycero-3-phosphocholine) is a potent lipid mediator that, in the nervous tissue, participates to physiological phenomena as well as in those related to pathological events (1,2). For instance, PAF is involved in neurotransmission (3–6) and it has been recently proposed as a retrograde messenger in hippocampal long-term potentiation (7,8) and in memory formation (9). On the other hand, it has been also reported that PAF concentration increases during cerebral ischemia and convulsions (10,11) and that PAF antagonists reduce brain damage associated with these pathological conditions (12,13). Thus, it is of great importance to know of the mechanisms controlling the concentration of this lipid mediator in the nervous tissue and consequently the regulation of the enzymes involved in its synthesis and degradation.

In neural cells, PAF can be produced by two distinct metabolic routes: the *de novo* and the remodelling pathway (1,2) (Fig.1). The last step of the *de novo* pathway is catalysed by a specific phosphocholinetransferase (PAF-PCT) using 1-O-alkyl-2-acetyl-sn-glycerol (alkylacetylG) and CDPcholine as substrates (14). The limiting step for the formation of the lipid substrate in the brain is most likely the acetylation of 1-O-alkyl-sn-glycero-3-phosphate (15,16) whereas CDPcholine production is regulated at the level of CTP:phosphocholine cytidylyltransferase as shown in other cell types (17). However, it is possible that PAF-PCT might be also regulated because its activity increases following the stimulation of chick retina with neurotransmitters (3,4) and following the preincubation of rat brain microsomes with agents that might induce protein phosphorylation (18).

From a workshop organized by Joanna B. Strosznajder (Warsaw, Poland), Jolanta Baranska (Warsaw, Poland) and Moti Liskovitch (Rehovot, Israel).

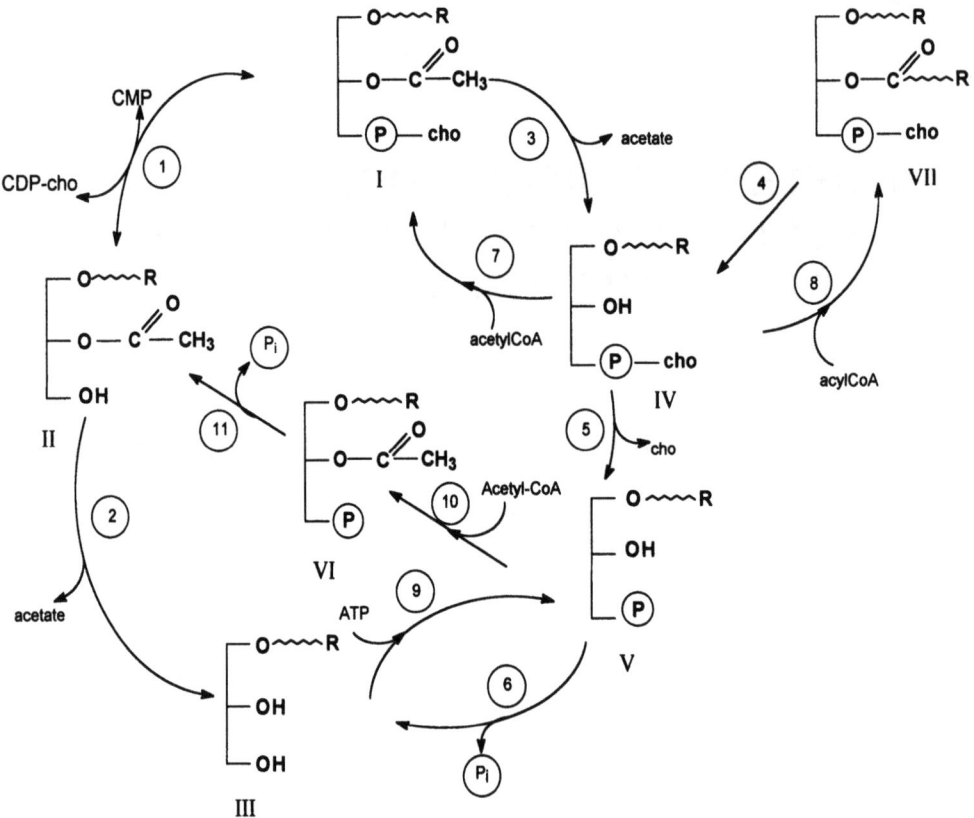

Figure 1. Metabolism of PAF and related compounds. I, 1-0-Alkyl-2-acetyl-*sn*-glycero-3-phosphocholine (PAF); II, 1-0-Alkyl-2-acetyl-*sn*-glycerol (alkylacetyl G); III, 1-0-Alkyl-*sn*-glycerol (alkyl G); IV, 1-0-Alkyl-2-lyso-*sn*-glycero-3-phosphocholine (lyso PAF); V, 1-0-Alkyl-2-lyso-*sn*-glycero-3-phosphate (alkyl GP); VI, 1-0-Alkyl-2-acetyl-*sn*-glycero-3-phosphate (alkylacetyl GP); VII, 1-0-Alkyl-2-(long-chain)acyl-*sn*-glycero-3-phosphocholine (alkylacyl GPC); 1, Phosphocholine transferase (PAF-PCT); 3, PAF acetylhydrolase; 4, Phospholipase A$_2$/CoA-independent transacylase; 5, Lysophospholipase D; 7, Lyso PAF acetyltransferase.

The second route for PAF synthesis, the remodeling pathway, utilizes an existing choline ether-phospholipid 1-O-alkyl-2-(long-chain)acyl-*sn*-glycero-3-phosphocholine (alkylacylGPC) to generate 1-O-alkyl-2-lyso-*sn*-glycero-3-phosphocholine (lysoPAF) by the direct action of a phospholipase A$_2$ or by a CoA-independent transacylase; lysoPAF is than acetylated to form PAF by a specific lysoPAF acetyltransferase (lysoPAF-AcT) (19). The presence of these enzyme activities in the nervous tissue has been documented (20–22). The limiting step for the synthesis of PAF by the remodeling pathway is very likely the production of lysoPAF but also PAF-AcT appears to be regulated by a phosphorylation-dephosphorylation mechanism (18).

The level of PAF in the nervous tissue is also controlled by PAF acetylhydrolase which converts PAF into the biologically inactive lysoPAF (23). LysoPAF can be recycled to alkylacyl-GPC or further degraded to 1-O-alkyl-*sn*-glycerol (alkylG) by a lysophospholipase D to 1-O-alkyl-2-lyso-*sn*-glycero-3-phosphate (alkylGP) which is then dephosphorylated to alkylG. The occurrence of this pathway in the microsomal fraction of rat brain has been previously reported (24).

Here we provide evidence for the presence in rat brain microsomes of a CMP-dependent pathway for the degradation of PAF that could be very likely initiated by the reversal of PAF-synthesizing phosphocholine transferase reaction. A similar mechanism has been previously described for the degradation of other cholinephosphoglycerides (25–28).

2. MATERIALS AND METHODS

2.1 Materials

1-O-[hexadecyl-1'-,2'-^{3}H(N)-2-acetyl-sn-glyceryl-3-phosphorylcholine (Spec. Rad. = 60 Ci/mmol) was from Dupont NEN (Missisauga, Ontario); 1-O-hexadecyl-2-acetyl-sn-glycero-3-phosphocholine (PAF), 1-O-hexadecyl-2-lyso-sn-glycero-3-phosphocholine (lysoPAF) and 1-O-hexadecyl-2-acetyl-sn-glycerol were from Calbiochem (La Jolla, CA); 1-O-hexadecyl-sn-glycerol, cytidine-5'-monophosphate (CMP), cytidine-5'-diphosphocholine (CDP choline), and other nucleotides were from Sigma (St. Louis, MO). Silica gel G thin-layer chromatography plates were purchased from Sigma (St. Louis, MO).

2.2 Experimental Procedure

Brain microsomes were prepared from Sprague-Dowley rats of both sexes (50–80 days old) in 0.32M sucrose, 10mM Tris-HCl buffer (pH 7.4) and 2mM EDTA, as previously described (22).

Labelled PAF (25–30 nCi; Spec. Rad.=15–20 nCi/nmol) was dissolved in 0.2% bovine serum albumin in saline and incubated in a medium containing 60mM Tris-HCl buffer (pH 8.0), 40mM $MgCl_2$, 40mM DTT, 0.05–0.1 mg of microsomal protein, in a final volume of 0.3 ml, at 37°C for 15 min (standard conditions).

Lipids were extracted by the method of Bligh and Dyer (29). Labelled phospholipids were isolated by TLC on silica gel G plates with chloroform/methanol/acetic acid/water (50:25:8:4 v/v) as developing solvent and identified by co-chromatography with authentic standards (22).

Labelled neutral lipids were isolated on silica gel G plates using the solvent system: petroleum ether (40–60°)/diethyl ether/acetic acid (60:40:1 by vol). The radioactivity was recovered together with authentic 1-O-hexadecyl-2-acetyl-sn-glycerol and 1-O-hexadecyl-sn-glycerol.

Protein concentrations were determined using bovine serum albumin as standard (30).

3. RESULTS

In order to verify whether the reaction catalysed by PAF-synthesizing phosphocholinetransferase activity could be reversed and degrade PAF to CDPcholine and alkylacetylG, rat brain microsomes were incubated with [^{3}H]PAF in the optimal conditions for enzyme activity (22).

The incubation of [^{3}H]PAF with rat brain microsomes, under standard conditions (control), produced labelled lysoPAF and alkylG (Fig.2). The presence of 2mM CMP in the incubation system (26) did not affect the radioactivity associated with lysoPAF, increased that of alkylG and induced a significative production of labelled alkylacetylG. These effects were completely abolished by the addition of CDP-choline (2mM).

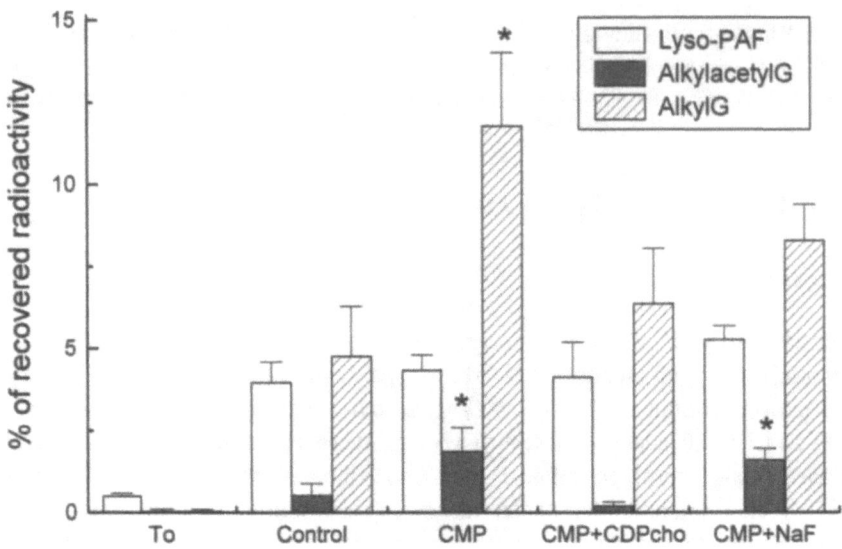

Figure 2. The effect of cytidine nucleotides on PAF degradation.

In Ehrlich ascites cells, NaF causes a 80% inhibition of the degradation of alkylacetylG by acetylhydrolase (31). In our experiments, the presence of NaF in the incubation system reduced the CMP-dependent production of radioactive alkylG but did not affect significantly neither the radioactivity of alkylacetylG nor that of lysoPAF (Fig.2).

As shown in Fig. 3, the substitution of CMP with UMP, ATP or CTP, at the same concentrations, did not significantly change the distribution of the radioactivity with respect to the control. Thus, the production of alkylacetylG and that alkylG can be considered a specific effect of CMP and attributed very likely to the reversal of PAF-synthesiz-

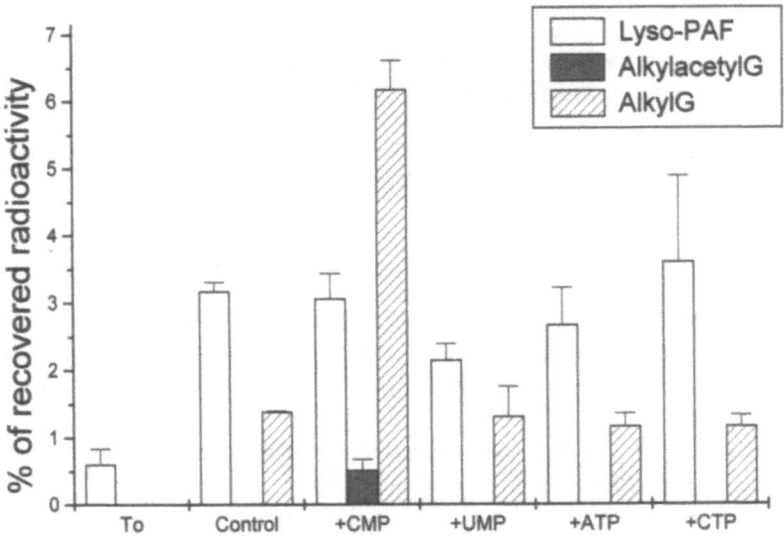

Figure 3. Effect of nucleotides on PAF degradation.

ing PCT reaction. This hypothesis is substantiated by other observations as the absolute requirement of Mg^{2+} and the inhibitory effect exerted by Ca^{2+} ions (data not shown).

4. DISCUSSION

Recent studies have postulated that the *de novo* pathway may be of particular relevance in the synthesis of PAF during brain ischemia. This hypothesis is based on the inhibition of alkylGP acetyltransferase exerted by ATP and acyl-CoA (2, 15,16), whose levels increase during such pathological situation. The decrease of the concentrations of these inhibitors allows an higher rate for the production of alkylacetylG, the lipid substrate for PAF-PCT. However, this interesting hypothesis does not account for other metabolic changes taking place in the nervous tissue at the onset of ischemia. Among other effects, the depletion of ATP increases the intracellular levels of Ca^{2+} as the consequence of membrane depolarization and release of excitatory amino acids (32). Since PAF-PCT is inhibited by Ca^{2+} ions (14), it seems unlikely that the *de novo* pathway might contribute significantly to PAF synthesis even if an higher concentration of alkylacetylG might be available to the enzyme. Furthermore, both the supply of CDPcholine, an energy requiring process, and PAF-PCT activity are reduced during ischemia (33). On the other hand, the increase of intracellular Ca^{2+} concentration activates phospholipase A_2 (34) and increases the rate of production of lysoPAF from membrane alkylacylGPC. Recently, we have demonstrated that lysoPAF acetyltransferase activity greatly increases in hippocampus during ischemia (33). This might be the consequence of its phosphorylation by protein kinase C (18) that might be also activated during ischemia (35,36). All together these observations allow us to assume that the activation of the remodelling pathway has to be mostly responsible for the increased levels of PAF during ischemia. This assumption is further supported by the results obtained by the experiments here reported because we have demonstrated that PAF can be degraded by a CMP-dependent mechanism and that CDPcholine is able to abolish this phenomenon. Since labelled alkylacetylG is produced from labelled PAF, we suggest that the reaction catalysing the last step of the *de novo* pathway for PAF synthesis is reversible and its direction is greatly dependent on the relative concentrations of CMP and CDPcholine whose ratio is linked to the energy state of the cells. This hypothesis is also based on the observations that CMP-dependent production of alkylacetylG in rat brain microsomes requires Mg^{2+} and is inhibited by Ca^{2+}. These properties fit well with the involvement of PAF-PCT in this phenomenon. It is interesting that the activity of lysoPAF-AcT is not influenced by 2mM CMP (Francescangeli and Goracci; unpublished data).

AlkylacetylG does not accumulates because is further degraded to alkylG by an acetylhydrolase sensible to NaF and this would further push backward the reaction catalysed by PAF-PCT when CDPcholine is lacking and CMP concentration increases.

In conclusion, the reversibility of PAF-PCT reaction suggest that the observed increase of PAF levels in the nervous tissue subjected to ischemia cannot be the consequence of an increased rate of the *de novo* pathway but more likely to the activation of the enzymes of the remodeling route.

ACKNOWLEDGMENTS

We thank A.Boila for technical assistance. This work was supported by a grant from Ministero della Ricerca Scientifica e Tecnologica (40%) and a contribute by CNR (950274/CT04/11513770).

REFERENCES

1. Goracci, G., 1990, *Pharmacology of cerebral ischemia*, (Krieglstein, J. and Oberpichler, H., Eds.), Wissenschaftliche Verlagsgesellshaft, Stuttgart, pp. 377–390.
2. Baker, R.R., 1995, *Neurochem. Res.* **20**: 1345–1351.
3. Bussolino, F., Gremo, F., Tetta, C., Pescarmona, G.P., and Camussi, G., 1986, *J. Biol. Chem.* **261**: 16502–16508.
4. Bussolino, F., Pescarmona, G., Camussi, G. and Gremo, F., 1988, *J. Neurochem.* **51**: 1755–1759.
5. Bussolino, F., Tessari, F., Turrini, F., Braquet, P., Camussi, G., Prosdocimi, M., and Bosia, A., 1988, *Am. J. Physiol.* **255**: (Cell physiol) 24, 1755–1759.
6. Clark, G.D., Happel, L.T., Zorumski, C.F. and Bazan, N.G., 1992, *Neuron* **9**: 1211–1216
7. Wierasko, A., Li, G., Kornecki, E., Hogan, M.V., and Ehrlich, Y.H., 1993, *Neuron* **10**: 553–557.
8. Kato, K., Clark, G.D., Bazan, N.G., and Zorunski, C.F., 1994, *Nature* **367**: 173–179.
9. Izquerdo, I., Fin, C., Schmitz, P.K., Da Silva, R.C., Jerusalinsky, D., Quillfeldt, J.A., Ferreira, M.B.G., Medina, J.H., and Bazan, N.G., 1995, *Proc. Natl. Acad. Sci. USA* **92**: 5047–5051.
10. Kumar, R., Harvey, S.A.K., Kester, M., Hanahan, D.J., and Olson, M.S., 1988, *Biochim. Biophys. Acta* **963**: 375–383.
11. Domingo, M.T., Spinnewyn, P., Chabrier, E., and Braquet, P., 1994, *Brain Res.* **640**: 268–276
12. Birkle, D.L., Kurian, P., Braquet, P., and Bazan, N.G., 1988, *J.Neurochem.* **51**: 1900–1905.
13. Panetta, T., Marcheselli, V.L., Braquet, P., Spinnewyn, B., and Bazan, N.G., 1987, *Biochem. Biophys. Res. Commun.* **149**: 580–587.
14. Francescangeli, E. and Goracci, G., 1989, *Biochem. Biophys. Res. Commun.* **161**: 107–112.
15. Baker, R.R., and Chang, H-Y., 1994, *Biochim. Biophys. Acta* **1213**: 27–33.
16. Baker, R.R., and Chang, H-Y., 1995, *J. Neurochem.* **64**: 364–370.
17. Blank, M.L., Lee, Y.J., Cress, E.A., and Snyder, F., 1988, *J.Biol.Chem.* **263**: 5456–5661.
18. Goracci, G., and Francescangeli, E., 1995, *J. Neurochem.* **64S**: 37D.
19. Snyder, F., 1995, *Biochem. J.* **305**: 689–705.
20. Woelk, H., Goracci, G., and Porcellati, G., 1974, *Hoppe-Seyler's Z. Physiol. Chem.* **355**: 75–81.
21. Blank, M.L., Smith, Z.L., Fitzgerald, V., and Snyder, F., 1995, *Biochim. Biophys. Acta* **1254**: 295–301.
22. Goracci, G., and Francescangeli, E., 1991, *Lipids* **26**: 986–991.
23. Hattori, M., Arai, H. and Inoue, K., 1993, *J. Biol. Chem.* **268**: 18748–18753.
24. Wikle, R.L., and Schremmer, J.M., 1974, *J.Biol.Chem.* **249**: 1742–1746.
25. Goracci, G., Francescangeli, E., Horrocks, L.A., and Porcellati, G., 1981, *Biochim.Biophys.Acta* **664**: 373–379
26. Goracci, G., Francescangeli, E., Horrocks, L.A., and Porcellati, G., 1986, *Biochim. Biophys. Acta* **876**: 387–391.
27. Goracci, G., Francescangeli, E., Horrocks, L.A., and Porcellati, G., 1983, *Neurochem. Res.* **8**: 971–981.
28. Goracci, G., Francescangeli, E., Floridi, A., Horrocks, L.A., and Porcellati, G., 1985, Phospholipid in the nervous system: vol.2; *Physiological roles* (Horrocks, L.A., Kanfer, J.N., and Porcellati, G., eds.), Raven Press, New York, PP 121–129.
29. Bligh, E.G., and Dyer, W., 1959, *Can. J. Biochem. Physiol.* **37**: 911–917.
30. Lowry, O.H., Rosebrough, N.J., Farr, A.L., and Randall, P.J., 1951, *J. Biol. Chem.* **193**: 265–275.
31. Blank, M.L., Smith, Z.L., Cress, E.A., and Snyder, F., 1990, *Biochim.Biophys.Acta* **1042**: 153–158.1.
32. Farooqui, A.A., Hirashima, Y., Farooqui, T. and Horrocks, L.A., 1992, *Neurochemical correlates of cerebral ischemia*, (Bazan, N.G., Braquet, P., and Ginsberg, M.D., eds) Plenum Press, New York, PP 117–138.
33. Francescangeli, E., Domanka-Janik, K., and Goracci, G., 1996, *J. Lipid Med.* in press
34. Rordorf, G., Uemura, Y., and Bonventre, J.V., 1991, *J. Neurosc.* **11**: 1829–1836.
35. Domanska-Janik, K. and Zalewska, T., 1992, *J. Neurochem.* **58**: 1432–1439.
36. Domanska-Janik, K., and Zabloka, B., 1993, *Chem. Neuropathol.* **20**: 111–123.

THE EFFECT OF ANTIDEPRESSANTS ON PHOSPHOLIPASE A2 ACTIVITY IN THE BRAIN OF RATS

Andrzej Malecki,[1] Krzysztof Kucia,[2] Irena Krupka-Matuszczyk,[2] and Henryk I. Trzeciak[1]

[1] Department of Pharmacology
Silesian Academy of Medicine
Medyków 18, 40-752 Katowice-Ligota
[2] Department of Psychiatry
Silesian Academy of Medicine
Lubliniec, Poland

1. INTRODUCTION

Phospholipase A2 (PLA2) cleaves fatty acids from the sn-2 position of phospholipids and plays an important role in phospholipid turnover of cell membranes, which in turn affects membrane fluidity and function. The products of PLA2 action are important factors in signal transduction (1). Moreover, there are suggestions that PLA2 participate in the pathogenesis of depression (2).

Pharmacological regulation of PLA2 is still poorly understood. For this reason we have examined the effects of acute treatment with different antidepressants on PLA2 activity in rat brain cortex.

2. MATERIALS AND METHODS

2.1. Animals and Drugs Administration

Experiments were performed on male Wistar rats weighing 180 ± 10g. The rats were fed standard diet as pellets and water ad libitum. Amitriptyline hydrochloride (10 mg/kg, Polfa), imipramine hydrochloride (10 mg/kg, Polfa) and mianserin hydrochloride (10 mg/kg, Sigma) were dissolved in 0.9% NaCl solution and administered in a single

---

From a workshop organized by Joanna B. Strosznajder (Warsaw, Poland), Jolanta Baranska (Warsaw, Poland) and Moti Liskovitch (Rehovot, Israel).

intraperitoneal dose in a volume of 1 ml. Control animals were given intraperitoneally the vehicle only. Each treatment group comprised 8 animals. The rats were decapitated 24 hrs after the injection of drugs.

2.2. Isolation of Plasma Membranes

The plasma membranes of rat brain cortex were isolated according to method described by Strosznajder and Strosznajder (3). The brain was rapidly removed and put on ice. Dissected cortex hemispheres were homogenized in an ice-cold Potter-Elvenhjem homogenizer in medium containing 0.32 M sucrose, 10 nM Tris-HCl buffer (pH 7,4) and 1 mM EDTA. The homogenate (10% w/v) was centrifuged for 3 min at 1100 g. The resulting supernatant was centrifuged for 10 min at 17000 g to yield a crude mitochondrial fraction (P2). Subsequently the pellet (P2) was dispersed in 1 mM Tris-HCl buffer (pH 7,4) for hypotonic shock, then vigorously vortexed and centrifuged for 20 min at 48000 g. The resulting pellet, further referred to as a brain plasma membranes, was gently resuspended in 10 mM Tris-HCl buffer (pH 7,4), immediately frozen and used for further estimation.

2.3. Phospholipase A2 Assay

The phospholipase A2 (PLA2) assay was performed according to Jelsema (4) with slight modifications (3). Briefly, the assays were performed in incubational mixtures (200 ul) containing $2,5 \times 10^4$ dpm radioactive 1-stearoyl-2-[1-^{14}C]-γ-arachidonyl phosphatidylinositol (spec. act. 48 mCi/mmol, NEN-DuPont) and 25 nmol unlabelled phospha-

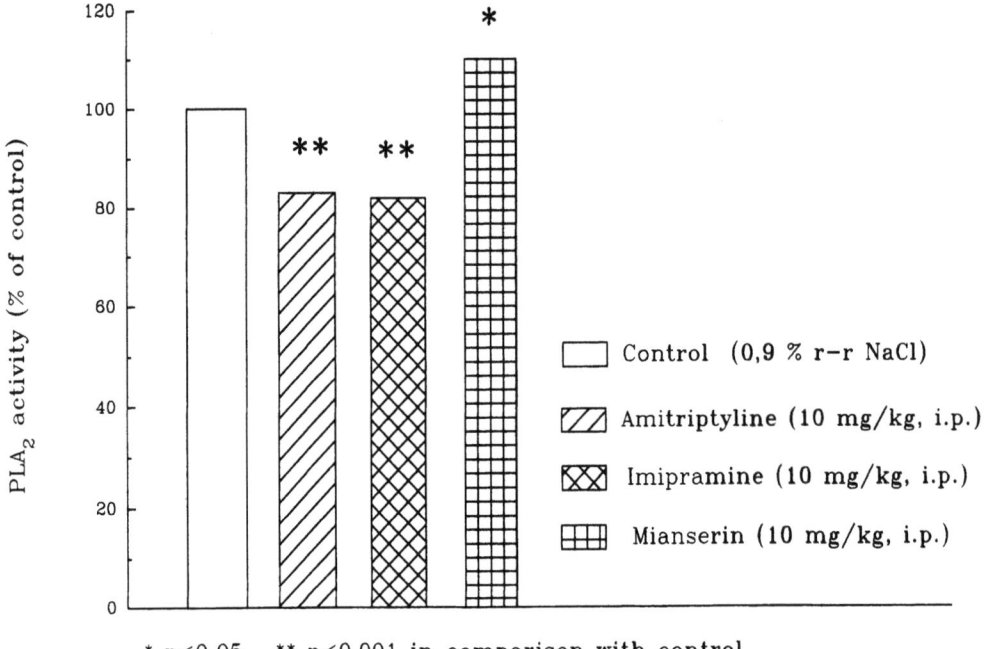

* $p<0.05$ ** $p<0.001$ in comparison with control

Figure 1. The activity of phospholipase A2 (PLA2) in rat brain plasma membranes after treatment with antidepressant drugs.

tidylinositol (Sigma), 0.01% sodium deoxycholate and 2 mM $CaCl_2$ in 10 mM Tris-HCl bufer (pH 7.8) and 200mg proteins of brain plasma membranes. The incubation was carried out at 37°C for 15 min in shaking water bath. Reaction was stopped by the addition of 3 ml Dole's reagent (isopropyl alcohol:n-heptan:1 N H_2SO_4, 40:10:1, v/v/v) and vigorous vortexing in room temperature. Following addition of 1,5 ml of n-heptane and 1 ml of H_2O, samples were vortexed and centrifuged for 10 min at 1000 g to extract the fatty acids. The enzymatically released [^{14}C]arachidonate was separated from unreacted substrate by addition of 150 mg silica gel (20–200 mesh, Fisher) to 1,5 ml aliquots of upper phase. Samples were vortexed, centrifuged (1000g for 10 min) and radioactivity was measured in Beckman LS 6000 IC scintillation counter. The value for [^{14}C]arachidonate was corrected for recovery of [^{3}H]arachidonate. PLA_2 activity was expressed as nmole of [^{14}C]arachidonate release/min/mg protein and then following as a percent of control value. Protein content was determined by the method of Lowry et al. (5) using bovine serum albumin as a standard. The Student t-test was used for statistical evaluation of data.

3. RESULTS

After amitriptyline and imipramine administration, PLA2 activity in brain plasma membrane of rats 24 hrs after the drug injection was significantly decreased 17 and 18%, respectively. On the contrary, mianserine increased PLA2 activity in brain plasma membranes by 10% in comparison to control (Fig. 1.).

4. DISCUSSION

Previously it was shown that tricyclic antidepressants (TCA) inhibit PLA2 activity in vitro (6) and in vivo (7). One of the possible reasons of the inhibition is their influence on phospholipid methylation and on the anisotropy of cortical membranes (8). On the other hand investigations performed with the freeze-fracture method showed that after long-term administration both imipramine and mianserine decrease cholesterol content in neuron membranes of rat cingulate cortex and in this way they inrease its fluidity (9). The single dose of desipramine but not mianserin decrease the beta-adrenergic response in the rat hippocampus (10). It was shown that down-regulation of beta receptors by desipramine in vitro involves PLA2 (11). Opposite to effects of other tricyclic antidepressants the action of mianserin on PLA2 activity may be explained by its action on different receptors. Acute administration of mianserin blocks alfa2-adrenergic receptors and facilitates central noradrenergic neurotransmission (12). It was proved that a single dose of mianserin, a 5HT2A and 5HT2C receptor antagonist, produces a decrease in the number of 5HT2 receptor sites in rat cerebral cortex (13). On the other hand tricyclic antidepressants cause 5HT2 receptors down-regulation in rat brains only after long-term administration (14). The stimulation of 5HT2 receptors enhance PLA2 activity in brain. This effect is presumably mediated by a specific G protein (15). Thus, the increase of PLA2 activity observed in our experiment after mianserin administration is rather unexpected.

It is believed that antidepressants may act on the level of intracellular messengers — mostly G proteins. For example, it was shown that antidepressants facilitate G-protein coupling with 5HT2 receptors in platelets of patients with depression who positively react to pharmacological treatment (16).

PLA2 activity seems to play a key role in formation and sustaining long term potentiation (LTP) in rat hippocampus which is believed to be an electrophysiological basis of synaptic plasticity and memory. Investigations with trimipraine showed that TCA inhibit LTP formation (17). The effect is most probably caused by its influence on PLA2. TCA impairs memory retention (and memory recall) both in humans and animals (18). Mianserin does not seem to have such an effect (19). Perhaps it is correlated, as demonstrated in this paper, with the influence of mianserin on PLA2 membrane activity after a single administration. Precise tricyclic antidepressants effects on PLA2 activity in rat brain need investigations with long-term administration of drugs.

REFERENCES

1. Mukherjee, A.B., Miele, L., and Pattabiraman, N., 1994, *Biochem.Pharmac.*, **48**:1–10.
2. Hibbeln, J.R., Palmer, J.W., and Davis, J.M., 1989, *Biol.Psychiatry*, **25**:945–961.
3. Strosznajder, J., Strosznajder, R., 1989, *J.Lipid Mediator.*, **1**:217–229.
4. Jelsema, C., 1987, *J.Biol.Chem.*, **262**:163–168.
5. Lowry, O., Rosebrough, N., Farr, A., Randall, R., 1951, *J.Biol.Chem.*, **193**:265–275.
6. Grabner, R., 1987, *Biochem.Pharmac.*, **36**:1063–1067.
7. Sklenovsky, A, Chmela, Z., 1978, *Activ.Nerv.Sup.(Praha)*, **20**:255–256.
8. Melzacka, M., Nocoñ, H,. 1991, *J.Pharm.Pharmacol.*, **43**:564–568.
9. Bal, A., Bird, M.M., 1991, *Brain Res.*, **550**:147–151.
10. Dijcks, F.A., Ruigt, G.S., de-Graaf, J.S., 1991, *Neuropharmacology*, **30**:1151–1158.
11. Manji, H.K., Chen, G.A., Bitran, J.A., Potter W.Z., 1991, *Psychopharmacol.Bull.*, **27**:247–253
12. Svensson, T.H., 1984, *Adv.Biochem.Psychopharmacol.*, **39**:241–248.
13. Smith, R.L., Barrett, R.J., Sanders-Bush, E., 1990, *J.Pharmacol.Exp.Ther.*, **254**:484–488.
14. Goodwin, G.M., Green, A.R., Johnson, P., 1984, *Br.J.Pharmacol.*, **83**:235–242.
15. Sanders-Bush, E., and Canton, H., 1995, *Psychopharmacology. The Fourth Generation of Progress*, (F.E. Bloom, and D.J. Kupfer,eds.), Raven Press, New York, PP. 431–441.
16. Leonard, B.E, 1994,*CNS Drugs*, **1**:285–304.
17. Bernard, J., Ohayon, M., Massicotte, G., 1994, *Psychiatry Res.*, **51**:107–114.
18. Branconnier, R.J., DeVitt, D.R., Cole, J.O., and Spera, K.F., 1981, *Neurobiol.Aging*, 3:55–59.
19. DeNoble, V.J., Schrack, L.M., Reigel, A.L., DeNoble, K.F., 1991, *Pharmacol.Biochem.Behav.*, **39**:991–996.

Section 37: The Role of Carnitine in the CNS

A BRIEF HISTORY OF CARNITINE AND ITS PRESENCE IN THE CNS

Rona R. Ramsay

School of Biological and Medical Sciences
University of St. Andrews
Irvine Building
St. Andrews, Scotland
United Kingdom

1. INTRODUCTION

Carnitine was discovered in 1905 as a significant component of muscle and its structure and optical properties were characterised by 1927 (reviewed in 1). The isomer found in all tissues across the animal and plant kingdoms is L-carnitine. Despite its structural similarity to choline, no role in neurotransmission has been found. In 1952, L-carnitine was found to be essential for the growth of the mealworm, *Tenebrio molitor*. Without carnitine, the mealworms could not use fat stores when starved. In the meantime, Friedman and Fraenkel (2) discovered that carnitine was reversibly acetylated by acetyl-CoA in the presence of muscle homogenate. Fritz (3) demonstrated that carnitine stimulated fatty acid oxidation in the liver homogenates, which led to the elucidation of the function of carnitine in transporting activated long chain fatty acids into the mitochondria for β-oxidation (reviewed in 1). This function has dominated the perception of the role of the carnitine system for nearly 40 years.

We now know that carnitine has a much wider role, that of modulating and communicating between the limited acyl-CoA pools of cellular compartments. Carnitine thus provides a mobile form of activated acyl groups connecting the intracellular CoA pools. It acts as a reservoir of these groups or as an excretory vehicle. By the modulation of the acyl-CoA pools, the system modulates energy-generating pathways. Carnitine and the carnitine acyltransferases are present in brain at levels adequate for these roles but the mechanisms of neuromodulation observed at the cellular and whole animal level remain to be elucidated.

From a workshop organized by Rona R. Ramsay (St. Andrews, United Kingdom). (Supported by Sigma Tau Ethifarm BV, Assen, The Netherlands.)

Neurochemistry, edited by Teelken and Korf
Plenum Press, New York, 1997

2. ACYL-CoA AS A KEY INTERMEDIATE

The acyl-CoA content of the cell is heterogeneous and multi-compartmentalised. For example, long chain fatty acids are activated outside the mitochondrion to provide fatty acyl-CoA for both oxidation and lipid synthesis. The fraction required for oxidation must be partitioned and transferred into the mitochondrial matrix pool. Branched chain amino acid metabolites and acetyl-CoA in excess of the TCA cycle requirements accumulate in the matrix CoA pool and must be exported to maintain free CoA for the normal function of mitochondria. The acyl-CoA pools also supply activated moieties for fatty acid elongation, cholesterol synthesis, protein acylation, lipid synthesis and turnover, etc. (reviewed in 4). Since about 90% of the cellular CoA is mitochondrial (5), the cytoplasmic pools of CoA (as yet uncharacterised), including those in the peroxisomes and endoplasmic reticulum, must be small and almost certainly predominately protein bound.

3. ACYL-CARNITINE ACTS AS A RESERVOIR OF ACTIVATED ACYL GROUPS

The carnitine pool is reversibly acylated by whatever acyl group is predominant in the local acyl-CoA pool. The proportion acylated is generally measured as short-chain (soluble) and long-chain (acid insoluble) fractions. HPLC methods for the detailed identification of all the different acyl modifications in both the CoA and carnitine pools are available (6). Except after an intense metabolic challenge, the carnitine and CoA pools are generally approximately equilibrated (7). However, the excess of carnitine present means that most of the acyl groups are on carnitine, leaving free CoA for reactions in the multiple pathways requiring it. This reservoir of activated acyl groups is immediately available when exogenous sources (either from outside the cell or from supplying pathways) fail. In the heart and insect flight muscle, acetyl-carnitine provides acetate to the TCA cycle to provide energy (8). In the erythrocyte, long-chain acyl-carnitine is the source of activated acyl groups for the membrane repair pathways when ATP is low (Figure 2) (9).

4. CARNITINE MEDIATES THE INTRACELLULAR TRANSFER OF ACYL GROUPS (ACYL TRAFFICKING)

The carnitine system (Figure 3) which transfers acyl groups to and from the CoA pools consists of a family of enzymes, the carnitine acyltransferases [EC 2.3.1.21], and the intracellular translocases (carriers) to transfer the acyl-carnitines across membranes at rates fast enough for metabolic demands. The mitochondrial system of CPT-I on the outer membrane, the carnitine translocase, and CPT-II on the inner membrane delivers fatty acyl-CoA for β-oxidation and is the best characterised system. It is now thought that similar arrangements must exist in the peroxisomes and in the endoplasmic reticulum with a malonyl-CoA-sensitive acyltransferase on the outside of the permeability barrier to CoA and an insensitive enzyme on the inside. As yet, there is no direct experimental evidence for translocases in these organelles.

The role of the system in mediating transfer of unwanted acyl groups out of mitochondria was elegantly demonstrated for propionate (10). Further, the appearance of medium chain acyl-carnitine derivatives in the urine of patients with medium-chain acyl-

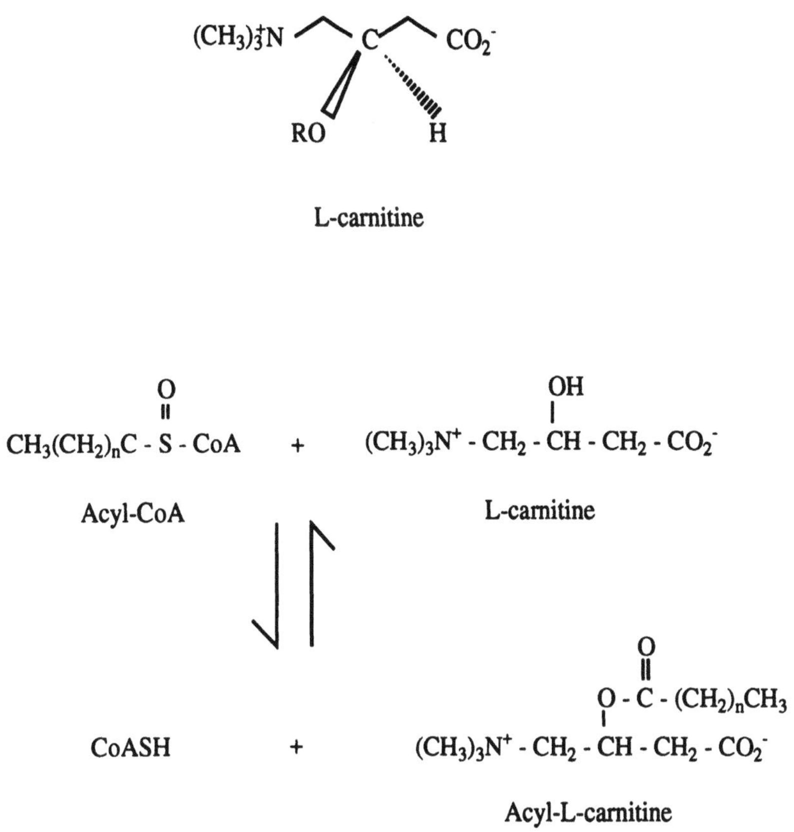

L-carnitine

Structure: (CH$_3$)$_3$N—C(RO)(H)—CO$_2^-$

Acylation reaction

$$CH_3(CH_2)_nC(=O)\text{-}S\text{-}CoA + (CH_3)_3N^+\text{-}CH_2\text{-}CH(OH)\text{-}CH_2\text{-}CO_2^-$$

Acyl-CoA + L-carnitine

⇌

$$CoASH + (CH_3)_3N^+\text{-}CH_2\text{-}CH(O\text{-}C(=O)\text{-}(CH_2)_nCH_3)\text{-}CH_2\text{-}CO_2^-$$

CoASH + Acyl-L-carnitine

Figure 1. The structure of carnitine and its reversible acylation by acyl-CoA. The equilibrium reaction is catalysed by the carnitine acyltransferase family of enzymes.

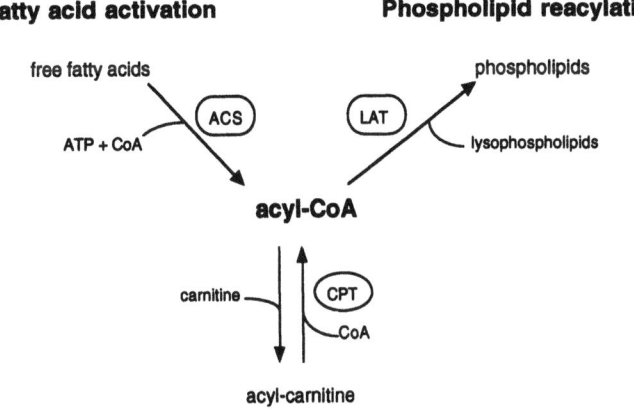

CPT-catalysed equilibrium

Figure 2. The acyl-carnitine pool is a reservoir of activated acyl groups for membrane repair. The enzymes are: ACS, acyl-CoA synthetase, CPT, carnitine palmitoyltransferase, and LAT, lysophospholipid acyltransferase.

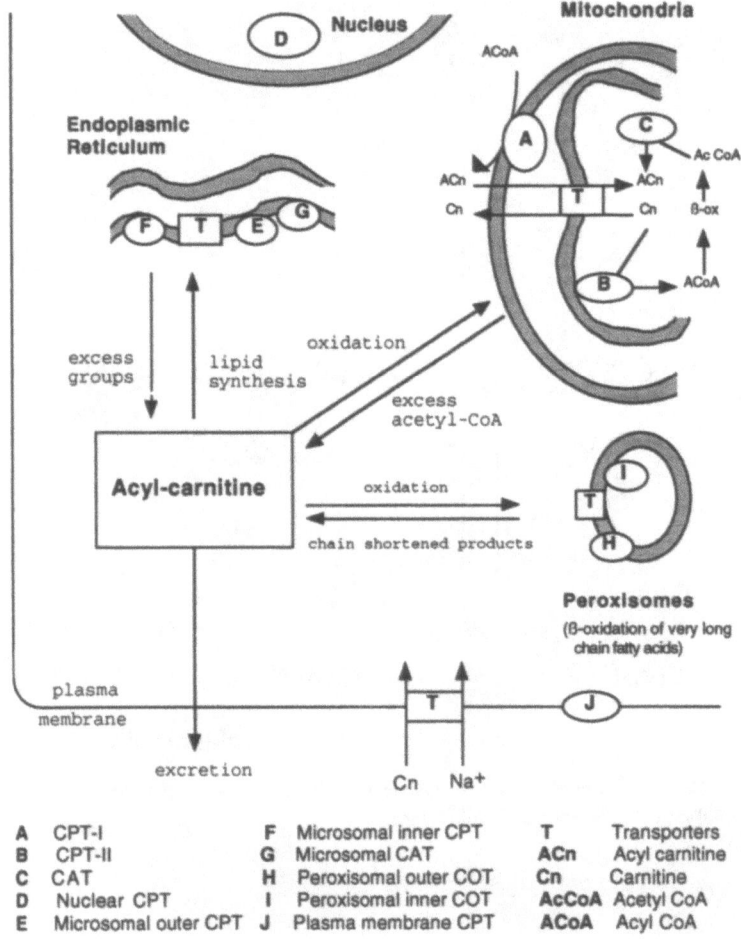

Figure 3. Carnitine-dependent enzymes and transport systems mediate the trafficking of activated acyl groups between cell compartments.

CoA dehydrogenase deficiencies illustrates that this role extends to the mediation of the excretion of excess unwanted acyl groups.

5. CARNITINE, THE CARNITINE ACYLTRANSFERASES AND THE TRANSLOCASES IN BRAIN

Carnitine is synthesised in liver (mainly) and kidney. All other tissues must accumulate it from the plasma where it is present at around 50 μM. The accumulation to the higher intracellular concentrations (see below) is driven by the sodium gradient. The uptake system has been studied in various tissues (reviewed in (1)). The brain carnitine uptake system in rat cortical slices (11), primary rat cortical cultures (12) or mouse synaptosomes (13) is sodium-dependent, saturable (with a Km value in the mM range), and inhibited by GABA but not choline. Carnitine transport in brain is further discussed by

Table 1. Carnitine levels in rat brain and other tissues. The data are taken from (17)

Tissue		Carnitine (nmol/mg non-collagen protein)	
		Free	Total
Heart		8.8	11.6
Liver		3.3	3.9
Brain			
	Hypothalamus	2.7	4.0
	Cerebellum	0.7	1.2
	Brain stem	0.7	1.4
	Hippocampus	0.9	1.3
	Basal ganglia	0.8	1.3
	Frontal cortex	0.7	1.0

Nalecz (this vol.). The reported rates of uptake are sufficient to support the reported turnover time of 220 hours (1). In vivo, transport across the blood-brain barrier is slow and also competitive with GABA (11, 14).

In contrast, carnitine binding sites in the synaptosomes were not blocked by GABA or other neurotransmitters (15). These sites formed the basis for proposing a neuro-modulatory role for carnitine but this is by no means accepted. A direct role as a neurotransmitter has been ruled out (16). The concentration of carnitine in the different brain regions varies slightly (16,17). The hypothalamus content is similar to liver, but other regions contain only about one third of that (Table 1). At a rough estimate, the intracellular content in brain averages 0.1–0.2 mM. The distribution between neurones and glia is unknown. Of the total carnitine pool, about 25% is acetylated and 10% carries long chain acyl groups (Table 1).

The levels of CAT and CPT in brain are adequate to maintain both a trafficking role and a modest rate of fatty acid oxidation (18). Developmental studies are few, but suggest that the levels of CPT are higher during neonatal period of intense lipid synthesis for myelination. Whether defects in the carnitine system might result in developmental compromise in the lipid structure is not known, but two CPT-deficient patient with brain abnormalities in addition to the common muscular problems have been reported (19,20). Only the mitochondrial enzymes CPT-I and CPT-II have been studied in detail in brain (18) and their properties are similar to those in other tissues. The CPT-I is probably of the heart type and is regulated by malonyl-CoA with a K_i of 1.5 µM (18). However, unlike the peripheral enzyme, the level and malonyl-CoA sensitivity of CPT-I in brain is not altered by starvation. Since brain does not depend on exogenous fatty acids as an energy source, this is understandable. The levels of CPT-I and CPT-II are about 3–4 fold lower than in liver (18) but are adequate to permit trafficking of long chain acyl groups between the cytoplasm and mitochondrial matrix, perhaps to oxidise unwanted (or damaged) acyl moieties.

6. FUNCTIONAL CLUES TO THE ROLE OF THE CARNITINE SYSTEM IN THE CNS

Carnitine derivatives administered to patients with low tissue levels enhance energy metabolism and the balance between glucose and fatty acid oxidation, resulting in the amelioration of muscle weakness and other symptoms. With respect to brain, acetyl-L-carnitine is thought to improve cognitive function and slow the progression of Alzheimer's disease. In seeking the mechanism underlying this action, a variety of pharmacological properties have been revealed (reviewed in 21).

In addition to a general effect on mitochondrial energy efficiency, acetyl-carnitine restores to normal the decreased mDNA transcription observed in old rats (22). It can act as an antioxidant so that rats after long-term treatment have decreased lipid damage (measured as malondialdehyde) and increased GSH levels (18).

Membrane stability which influences the permeability of the membranes and the activities of the mitochondrial energy pathway enzymes, is enhanced by acetyl-carnitine. In rat synaptic mitochondria the decreased cholesterol to phospholipid ratio is restored (23), restoring the physicochemical properties of the membranes.

The enhancement by acetyl-carnitine of cholinergic transmission may be a complex process. Certainly acetyl-carnitine could contribute to the intracellular pool of acetyl-CoA but effects on the enzymes of the pathway, on release, and on the receptors for acetylcholine have also been reported (21, 24–26).

In summary, these functional observations suggest that carnitine modulates energy production, acetate trafficking and lipid metabolism, but the mechanistic details are as yet unknown.

7. CONCLUSIONS

Carnitine and the enzymes for its function as a carrier of activated acyl groups are present in the CNS. The system functions, as in other cells, to facilitate the passage of long chain acyl groups into the mitochondria for oxidation. It buffers the intramitochondrial CoA pool and thus influences the activity of pyruvate dehydrogenase. It may also provide a reservoir of activated acyl groups for membrane repair. Additional and nervous tissue specific roles are not fully established. The clear influence of the carnitine system on energy metabolism will result in secondary alterations to all energy-dependant functions (metabotrophic effects). The study of this system in the CNS has a long way to develop. Avenues of future elucidation must include the uptake mechanism, any hormonal or signalling influences and definitive clinical studies of the pharmacological effects.

REFERENCES

1. Bremer, J., 1983, *Physiol. Rev.* **63**: 1420–1479.
2. Friedman, S. and Fraenkel, G., 1955, *Arch. Biochem. Biophys.* **59**: 491–501.
3. Fritz, I.B., 1955, *Acta Physiol. Scand.* **34**: 367–385.
4. Waku, K., 1992, *Biochim. Biophys. Acta* **1124**, 101–111.
5. Idell-Wenger, J.A., Grotyohann, L.W., and Neely, J.R., 1978, *J. Biol. Chem.* **253**: 4310–4318.
6. Pourfarzam, M., and Bartlett, K., 1991, *J. Chromatog.* **570**: 253–276.
7. Lysiak, W., Lilly, K., DiLisa, F., Toth, P.P., and Bieber, L.L., 1988, *J. Biol. Chem.* **263**: 1151–1156.
8. Childress, C.C., Sacktor, B., and Traynor, D.R., 1966, *J. Biol. Chem.* **242**: 754–760.
9. Arduini, A., Mancinelli, G., Radatti, G.L., Dottori, S., Molajoni, F., and Ramsay, R.R., 1992, *J. Biol. Chem.* **268**: 12673–12681.
10. Brass, E.P., Gandour, R.D., and Griffith, O.W., 1991, *Biochim. Biophys. Acta* **1095**: 17–22.
11. Huth, P.J., Schmidt, M.J., Hall, P.V., Fariello, R.G. and Shug, A.L., 1981, *J. Neurochem.* **36**: 715–723.
12. Vlemani, M.A., Conti, R., Spadoni, A., Rossi, S. and Arrigoni-Martelli, E., 1994, *Molec. Brain Res.* **25**: 105–112.
13. Hannuniemi, R. and Kontro, P., 1988, *Neurochem. J.* **13**: 317–323.
14. Burlina, A.P., Sershen, H., Debler, E.A. and Lajtha, A. *Neurochem. Res.* **14**: 489–493.
15. Vesci, L., Tobia, P., Corsico, N., Martelli, E.A., and Arduini, A., 1995, *J.Neurochem.* **64**: 2783–2791.

16. Shug, A.L., Schmidt, M.J., Golden, G., and Fariello, R.G., 1982, *Life Sci.*, **31**: 2869–2874.
17. Bresolin, N., Freddo, L., Vergani, L. and Angeli, C., 1982, *Exper. Neurol.*, **78**: 285–292.
18. Bird, M.I., Munday, L.A., Saggerson, E.D., and Clarke, J.B., 1985, *Biochem. J.*, **226**: 323–330.
19. North, K.N., Hoppel, C.L., Degirolami, U., Kozakewich, H.P.W., and Korson, M.S., 1995, *J. Pediat.* **127**: 414–420.
20. Shintani, S., Shiigai, T., and Sugiyama, N., 1995, *J. Neurol. Sci.*, **129**: 69–73.
21. Calvani, M., and Carta, A., 1991, *Dementia* **2**: 1–6.
22. Gadaleta, M.N., Petruzzella, V., Renis, M., Fracasso, F., and Cantatore, P., 1990, *Eur. J. Biochem.* **187**, 501–506.
23. Petruzella, V., Bagetto, L.G., Penin, F., Cafagna, F., Ruggiero, F.M., Cantatore, P., and Gadaleta, M.N., 1992, *Arch. Gerontol. Geriatr.* **14**, 131–144.
24. Ayala, C.A., 1995, *J. Neurosci. Res.* **41**, 403–408.
25. Imperato, A., Ramacci, M.T., and Angelucci, L., 1989, *Neurosci. Lett.* **107**: 251–255.
26. Janiri, L., Falcone, M., Persico, A., and Tempesta, E., 1991, *J. Neural Transm.*, **86**: 135–146.

CARNITINE AND THE MAINTENANCE OF CELL FUNCTION

Victor A. Zammit

Hannah Research Institute
Ayr, Scotland
United Kingdom

1. INTRODUCTION

Carnitine fulfills several important roles in the maintenance of cell function, all of which derive from its ability to form an ester bond, through its β-hydroxy group, with the carboxylate group of fatty acids (or, in a specialised context, branched-chain amino acids). There are two important features associated with the formation of these esters. Firstly, they are formed by the transfer of acyl chains between the respective coenzyme A esters and carnitine, thus resulting in the liberation of unesterified ("free") coenzyme A which, in turn, can then be esterified to other carboxyl group-containing acyl molecules. In view of the very limited size of the cellular pool of coenzyme A, carnitine and the carnitine acyltransferases that catalyse these reactions play a central role in determining the state of acylation of the cofactor. Secondly, the acylcarnitines so formed are substrates for specific carnitine-acylcarnitine carriers that reside within the cell plasma membrane and certain intracellular membranes. Consequently, whereas acyl-CoA esters are not permeable through membranes, acylcarnitine esters permeate very efficiently those membranes in which the carnitine-acylcarnitine carriers occur. This makes carnitine a very specialised molecule that facilitates the transport of acyl moieties across intracellular membranes while at the same time maintaining the compartmentalization of coenzyme A and acyl-CoA esters that exists in the cell, e.g. between cytosolic and mitochondrial matrix compartments.

2. THE CARNITINE ACYLTRANSFERASES

There exists a family of enzymes that catalyse the transfer of the acyl moeity from coenzyme A to carnitine. They are separate gene products with substantial similarity between their deduced primary amino acid sequences. They fall into two main classes.

From a workshop organized by Rona R. Ramsay (St. Andrews, United Kingdom). (Supported by Sigma Tau Ethifarm BV, Assen, The Netherlands).

The more numerous are those that have a greater or lesser preference for medium- or long-chain fatty acids: the octanoyl- or palmitoyl transferases, referred to as COTs and CPTs, respectively. For historical reasons, the carnitine acyltransferases of the mitochondria are the best studied, although it is now apparent that, at least in the liver, CPTs and COTs occur in the endoplasmic reticulum, peroxisomes and, possibly, the nuclear membrane (1–4). Moreover, a plasma membrane associated CPT has been described for human erythrocytes (5). Carnitine acetyltransferase appears to fall into a class of its own in that it has a much narrower substrate (chain-length) specificity and intracellular distribution. In liver mitochondria, carnitine acetyltransferase (CAT) appears to occur exclusively on the matrix side of (and loosely associated with) the inner membrane (6) but carnitine acetyltransferase activities also occur in peroxisomes and microsomes.

3. ROLES SUGGESTED FOR CARNITINE ACETYLTRANSFERASE IN BRAIN METABOLISM

3.1. Acetylcarnitine and Post-Ischaemic Recovery of the Brain

It has been shown (7) that administration of acetylcarnitine facilitates the metabolic recovery of the brain from ischaemic injury in dogs, possibly by facilitating the oxidative metabolism of pyruvate derived from glucose, which is the main substrate for the brain. The mechanisms through which acetyl carnitine could exert this effect are shown in Fig 1. Firstly, during post-ischaemic recovery, the oxidative decarboxylation of pyruvate may be impaired. Acetyl carnitine would bypass this step and deliver 2-carbon moieties to the mitochondrial matrix where, through the action of CAT, acetyl-CoA and unesterified carnitine would be generated. The intramitochondrial acetyl-CoA so generated would, in the short-term depress pyruvate oxidation further, but would provide an immediate pool of substrate for the tricarboxylic acid cycle, and thus, oxidative phosphorylation. Therefore, this would have an immediate effect on the energy status of the brain and would tend to improve the

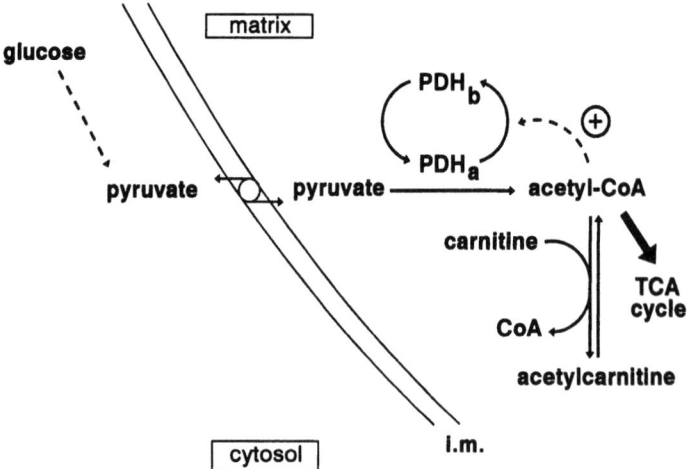

Figure 1. Mechanisms through which acetylcarnitine may improve brain energy metabolism during ischaemia and recovery in the brain.

immediate prospects of recovery from the ischaemic episode. Equally important, however, would be the longer-term effect of the delivery of additional carnitine to the brain which would subsequently lower the intramitochondrial acetyl-CoA concentration. Alhough, at first sight, this may appear to diminish the availability of substrate for the TCA cycle, in reality it would have a beneficial effect on the brain because it would de-inhibit pyruvate dehyrogenase (8), thus increasing the flux of pyruvate to acetyl-CoA. The elevated concentration of carnitine would ensure that, whereas adequate flux of acetyl-CoA into the TCA cycle can occur (because of the high affinity of citrate synthase for acetyl-CoA (9)), the concentration of intramitochondrial acetyl-CoA is maintained sufficiently low for de- inhibition of pyruvate dehydrogenase to persist. It would be anticipated that administration of unesterified carnitine by itself would give the longer-term effect, but would not enable the rapid delivery of TCA cycle substrate that may be important in the immediate post-ischaemic recovery period. This may explain why carnitine therapy by itself is not efficacious as a treatment for stroke in rats (10).

3.2. Interactions between Carnitine, GABA, and Acetylcholine

Carnitine and γ-aminobutyric acid (GABA) have been found to enhance and inhibit acetylcholine synthesis, respectively (11,12). The effect of carnitine may involve the activity of CAT in an analogous manner to that described above. The final synthesis of acetylcholine occurs in the cytosol and thus needs to involve the efflux of acetyl moyeties from the mitochondrial matrix into the cytosolic compartment. Acetylcarnitine is one molecular species through which this could be mediated. Intramitochondrial acetyl-CoA (e.g. generated from glucose or ketone bodies) is converted into acetylcarnitine through the action of CAT, followed by carrier-mediated efflux of acetylcarnitine and reconversion into acetyl-CoA in the cytosolic compartment. The same authors (12) have suggested that the inhibitory effect of γ-aminobutyric acid on acetyl-choline synthesis is mediated partly through the inhibition by GABA of CAT activity, and partly through the inhibition of carnitine uptake by the brain.

Although these mechanisms have to remain hypothetical at present, the ability of carnitine and short chain acyl esters to alleviate the effects of degenerative brain conditions such as Alzheimer's disease is well established, and similar mechanisms could operate. In addition, acetylcarnitine could be the form in which 2-carbon moyeties could be made available for biochemical pathways that occur within other intracellular compartments to which acetyl-CoA is not directly available.

4. ROLE OF CPTs AND COTs

These enzymes catalyse the transfer of medium- or long-chain fatty acyl moeities between CoA and carnitine and show overlapping chain-length specificity. It is now appreciated that several membrane-delimited systems within the cell (e.g. mitochondrial, microsomal, peroxisomal) have overt (cytosol-facing) and latent (matrix- or lumen-facing) carnitine acyltransferases that appear to be distinct gene products (1–4). This dual localization on either side of the membrane is needed to achieve the effective transfer of the acyl chain with the retention of the identity of the various acyl-CoA pools of each compartmen, when the members of the particular couple of transferases are linked through the existence within the membrane of a carnitine acylcarnitine carrier (13). The most striking characteristic of this distribution of CPTs and COTs is that the cytosolic-facing members of each pair are inhibited strongly by malonyl-CoA; reviewed in (14).

The mitochondrial system is the one that is best understood. In addition, most studies have been performed on the hepatic enzymes. The malonyl-CoA inhibition of the overt CPT of mitochondria (CPT I) was at first suggested to be primarily a mechanism for avoiding the occurrence of substantial simultaneous flux through the pathways of fatty acid oxidation and synthesis (15). However, it soon became apparent that malonyl-CoA plays an equally important role, through its inhibition of CPT I, even in tissues in which the rate of fatty acid synthesis is very low (16). Because of its central position between pyruvate (and thus glucose) metabolism and fatty acid synthesis and oxidation, malonyl-CoA has come to be recognised as a fuel sensor in such tissues as muscle and the pancreatic β-cell [17,18]. It is tempting to speculate that it fulfills a similar role in the brain.

Thus, the brain could be the first tissue in which a role for malonyl-CoA exists within cells in which the rate of lipogenesis may be high but the rate of fatty acid oxidation is low. Because of the extreme paucity of the data available on acylcarnitine metabolism in the brain and of the experimental difficulties that would be encountered in elucidating a role for malonyl-CoA, these suggestions are necessarily hypothetical at present. However, in spite of the generally held view that the brain does not oxidise fatty acids to any significant extent, such oxidation has been monitored experimentally (19). Moreover, both cell-body and synaptic mitochondria have been shown to contain malonyl-CoA-inhibitable CPT I (which also displays appreciable carnitine octanoyltransferase activity) (20). Whether this action of malonyl-CoA is related to fuel sensing still needs to be established. However, it should be appreciated that the experimental difficulties in determining malonyl-CoA concentrations in specific compartments of brain cells make this a real challenge.

5. ROLE OF CPTs IN MEMBRANE PHOSPHOLIPID TURNOVER

A newly-suggested role of CPTs is that of acting to provide a pool of long-chain acylcarnitne esters through which the turnover of membrane phospholipids, and specifically their remodelling, can occur (5). The system has been shown to operate within the erythrocyte, in which a plasma membrane associated CPT is thought to mediate this function. It is possible that in the brain such a function could be mediated by CPTs associated with other membranes. If the turnover and remodelling of brain membrane phospholipids does use the acylcarnitine route, this may explain the beneficial effects of dietary carnitine intake in certain degenerative disorders (21). For further details, the reader is referred to (22).

REFERENCES

1. Farrell, S.O., and Bieber, L.L., 1983, *Arch. Biochem. Biophys.* **222**: 123–132.
2. Chatterjee, B., Song, C.S., Kim, J.M., and Roy, A.K., 1988, *Biochemistry* **27**: 9000–9006.
3. Pande, S.V., Bhuiyan, A,K.M., and Murthy, M.S.R., 1992, *Current Concepts in Carnitine Research*, (A.L.Carter, ed.), CRC Press, Boca Raton, FL, PP. 165–178.
4. Broadway, N.M., and Saggerson, E.D., 1995. *Biochem. J.* **310**: 898–995.
5. Arduini, A., Mancinelli, G., Radatti, G., Dottori, G., Molajoni, F. & Ramsay, R.R., 1992, *J. Biol. Chem.* **267**: 12673–12679.
6. Barker, P.J., Fincham, N.J., and Hardwick, D.C., 1968, *Biochem. J.* **110**: 739–746.
7. Fiskum, G., Gamma, B., Bogaert, Y., and Rosenthal, R., 1992, *FASEB J.* **6**: A1062

8. Randle, P.J., Kerbeg, A.L., and Espinal, J., 1988, *Diabetes Metab. Rev.* **4**: 623–638.
9. Garland, P.B., Shepherd, D., Nicholls, D.G., and Ontko, J., 1968, *Adv. Enz. Regul.* **6**: 3–31.
10. Slivka, A., Silbersweig, D., and Pulsinelli, W., 1990, *Stroke* **21**: 808–811.
11. Wawrzenczyk, A., Nalecz, K.A., and Nalecz, M.J., 1994, *Biochem. Biophys. Res. Commun.* **202**: 354–359.
12. Wawrzenczyk, A., Nalecz, K.A., and Nalecz, M.J., 1995, *Biochem. Biophys. Res. Commun.* **213**: 383–388.
13. Ramsay, R.R., and Tubbs, P.K., 1976, *Eur. J. Biochem.* **69**: 299–303.
14. Zammit, V.A., 1996, *Biochem. J.* **314**: 1–14.
15. McGarry, J.D., and Foster, D.W., 1979, *J. Biol. Chem.* **254**: 8163–8168.
16. Brindle, P.J., Zammit, V.A., and Pogson, C.I., 1985, *Biochem. J.* **232**: 177–182.
17. Saha, A.K., Kurowski, T.G., and Ruderman, N.B., 1996, *Am. J. Physiol.* **269**: E283-E289.
18. Chen, S., Ogawa, A., Okneda, M., Unger, R.H., Foster, D.W., and McGarry, J.D., 1994, *Diabetes* **43**: 878–883.
19. Spitzer, J.J., and Wolf, E.H., 1971, *Am. J. Physiol.* **221**: 1426–1430.
20. Bird, M.I., and Saggerson, E.D., 1985, *Biochem. J.* **226**: 323–330.
21. Bruno, G., Scaccionoce, S., Bonamini, Patacchioli, F.R., Cesarino, E., Grassini, P., Sorrentino, E., Angelucci, L., and Lenzi, G.L., 1995, *Alzeihmer Disease and Associated Disorders* **9**: 128–131.
22. Arduini, A., Denisova, N., Virmani, A., Avrova, N., Federici, G., Arrigoni-Martelli, E. (1994) *J. Neurochem.* **62**: 1530–1536.

THE CARNITINE SYSTEM AND ACYL TRAFFICKING IN CNS

A New Task or a Biological Reappraisal?

Rita Ricciolini,[1] Maurizio Scalibastri,[1] Anna Floriana Sciarroni,[1] Secondo Dottori,[1] Menotti Calvani,[1] Lluis Lligoña-Trulla,[1] Roberto Conti,[1] and Arduino Arduini[2]

[1]Department of Biochemistry
Sigma Tau
Pomezia, Rome
[2]Department of Science
University of "G. D'Annunzio"
Pescara, Italy

1. INTRODUCTION

The carnitine system may be defined as a family of different short and long-chain acyltransferases, translocases, and their related substrates, whose common denominator is L-carnitine. The carnitine system is best known for the role played in mitochondrial metabolism, though the presence of carnitine-dependent short and long-chain acyltransferases in extra-mitochondrial compartments raises a number of intriguing question about their role. We have recently proposed that carnitine palmitoyltransferase (CPT) may be important in the pathway of phospholipid and triglyceride fatty acid turnover in neurons. The CPT action is accomplished by modulating the size and composition of the acyl-CoA pool between the activation step of the fatty acid and its transfer into complex lipids. In addition, studies on the metabolic fate of the acetate moiety of acetyl-L-carnitine revealed that the lipogenic acetyl-CoA pools present in different cellular compartments of rat brain are not necessary homogeneous. Taken together, these data suggest that the carnitine system may influence key regulatory points of lipid biosynthetic pathways.

From a workshop organized by Rona R. Ramsay (St. Andrews, United Kingdom). (Supported by Sigma Tau Ethifarm BV, Assen, The Netherlands).

2. LONG-CHAIN ACYLCARNITINE TRANSFERASE

2.1. CPT and Fatty Acid Turnover

The membrane bound long-chain acylcarnitine transferases are widely represented in organs and subcellular organelles (1, 2). The well-established role of this category of enzymes is the transport of activated long-chain fatty acids through mitochondrial membranes. In this context, CPT also plays an important regulatory function in mitochondrial fatty acid oxidation (1, 3).

The outer mitochondrial membrane CPT I catalyzes the formation of long-chain acylcarnitines that are transported across the mitochondrial inner membrane by a specific carrier, and re-converted into acyl-CoA by CPT II, which resides in the inner membrane and faces the matrix compartment. The system appears to be under tight metabolic control via the inhibition of CPT I by malonyl-CoA (the first intermediate committed to lipogenesis). CPT has also been identified in several rat brain regions, and the two isoforms of CPT localized in the outer and inner mitochondrial membrane have been studied in rat brain synaptic and non-synaptic mitochondria. However, their activity is consistently lower than that present in the liver (4). Because most regions of the brain poorly oxidize long-chain fatty acids, this has not been considered to be a major function of CPT in the brain.

Recent studies have shown that human erythrocyte CPT plays an important role in the physiological expression of membrane phospholipid fatty acids turnover, the sole possible fatty acid metabolic pathway of circulating erythrocytes (5, 6). This metabolic pathway is located in virtually all the membrane compartment of cells, and it represents an important tool to remodel the fatty acid composition of membrane phospholipids. We have also shown that neuronal cells CPT can be considered as an integral component of the pathway for membrane phospholipid and triglyceride fatty acid turnover (7). In accord with our hypothesis, Freed L.M. et al. (8) and Chang M.C.J. et al. (9) found that the treatment of a rat with an irreversible CPT inhibitor (given either orally or intravenously, respectively) caused a significant reduction of radioactive fatty acid incorporation into brain phospholipids. CPT may act as a link between the acyl-CoA synthetase step and the reacylation step of membrane lysophospholipids. In fact, the inhibition of fatty acid incorporation into phospholipids upon loss of CPT activity, was connected with the ability of this enzyme to modulate an optimal acyl-CoA/free CoA ratio throughout the reacylation process. In this respect, it has been shown that enzymes involved in the acylation of membrane phospholipids or in the de-novo synthesis of fatty acids are quite sensitive to relatively high concentrations of free CoA, which probably acts as an allosteric inhibitor toward such acyl-CoA-dependent enzymes (5, 10, 11). Although in our neuronal cell study we did not characterize which CPT isoform was responsible for the modulation of the intracellular acyl-trafficking, our findings suggest that the microsomal CPT isoform may be a good candidate. Preliminary data obtained in our laboratory also suggest that the microsomal acylation steps involved in sphingolipids synthesis are affected by the loss of CPT activity in PC12 cells (R. Ricciolini & A. Arduini, submitted). The endoplasmic reticulum is the site of a number of important lipid biosynthetic pathways (i.e., membrane phospholipid reacylation, fatty acid elongation, de-novo synthesis of sphingolipids and phospholipids), where long-chain acyl-CoAs are common intermediates. In addition, acylating reactions of both membrane complex lipids and protein may either occur on the cytoplasmic or the luminal face. Interestingly, two different liver microsomal CPT isoforms have been recently found and characterized: the overt (cytosol-facing) and latent (facing the lumen of the organelle) (12). Our proposed function about the role of extra-mitochondrial CPT fits with the overall acyl trafficking requirement of this organelle.

2.2. CPT and Membrane Phospholipid Repair

An important consequence of the potential involvement of CPT in anabolic activities in brain tissues may be its relevance in brain repair mechanisms. Free radical reactions are considered to be an important pathophysiologic determinant of a broad range of inflammatory, ischemic and neurodegenerative diseases, and brain aging (13). The CNS contains a well-integrated enzymatic and non-enzymatic antioxidant defence system, which in principle should prevent the noxious effect of a free radical attack (14). With this antioxidant network, it would seem unlikely that an oxidative challenge is capable of exerting its deleterious action on neuronal cells. However, several lines of evidence indicate that oxidative changes in various molecular components of the cell still take place. In other words, the primary antioxidant network does not seem to fully protect the cell from a free radical attack. Under these circumstances, a system capable of eventually removing and possibly repairing aberrant products of the oxidative insult would be further beneficial for cellular survival. It is becoming more and more evident that such a system, commonly regarded as the secondary antioxidant defence line, is operative towards damaged proteins, lipids, and DNA. Thus, enzymes promptly remove oxidatively damaged proteins (15), repair oxidatively damaged DNA (16), and repair peroxidized phospholipids (17). The latter repair mechanism is characterized by an enhancement of the membrane phospholipid fatty acid turnover, which removes a peroxidized polyunsatured fatty acid esterified in the phospholipid molecule (phospholipase A_2 step), and replace it with a native fatty acid (lysophospholipid acyl-CoA transferase step). It should be taken into account that the reacylation step requires an adequate supply of acyl-CoA, which in turn is generated by the ATP-dependent enzyme acyl-CoA synthase. As we have discussed in the above paragraph, the introduction of the CPT reaction in the membrane phospholipid repair pathway (6) would again guarantee a proper cycling process. In fact, given the sensitivity of CPT to the mass action ratio of the substrates, any variation of the acyl-CoA/free CoA ratio can be conpensated by this enzyme. CPT action would be then twofold: provide acyl-units at no ATP-cost and buffer the harmful elevation of free CoA. The former action becomes particularly relevant in all those physiopathological conditions characterized by a significant de-energization of the cell (hypoxia and/or ischemia).

3. CARNITINE ACETYLTRANSFERASE

3.1. Acetyl-L-Carnitine and Lipogenic Acetyl-CoA Pool

Acetyl-L-carnitine is a high energy reservoir of activated acetate moiety, which is in equilibrium with the intracellular acetyl-CoA pool. Acetyl-CoA provides acetate groups for fatty acid elongation and for fatty acid, steroid and acetylcholine synthesis. The availability of the acetate moiety from acetylcarnitine is dependent upon the activity of carnitine acetyltransferase (CAT), an enzyme which catalyzes the reversible transfer of the acetate group from carnitine to free CoA. CAT is present in synaptic and non-synaptic mitochondria, microsomes and probably also peroxisomes in all brain regions (18, 19). Its action has been often referred to buffering local changes of the acetyl-CoA/CoASH ratio in mitochondrial energy-producing process, though very little attention has been devoted to its potential action in modulating the size and availability of lipogenic acetyl-CoA pools. An important consequence of the potential involvement of acetylcarnitine in anabolic activities in brain tissues may be its relevance in the developing brain and brain repair. It is well known that during myelination there is a remarkable increase in the activity of the microsomal fatty acid elongation and desaturation system, which is also associated by an enhanced de-novo synthesis of fatty acids (20). The former system is mainly involved in desaturating and elongating essential fatty acids to 20:4ω-6 and 22:6ω-3 (21). The first step in the microsomal elongation process of fatty acids is the condensation of malonyl-CoA into a long-chain acyl-CoA. It is commonly believed that malonyl-CoA is produced by carboxylation of the cytosolic pool of acetyl-CoA, which in turn is generated through the citrate lyase reaction (22). In this context, it is clear that glucose plays a pivotal role providing the bicarbon skeleton unit for the elongation of fatty acids. However, although glucose is certainly feeding the cytosolic acetyl-CoA pool via the citrate pathway, doubts there exist as to whether the same pathway is able to feed the pool of acetyl-CoA which is used for the synthesis of the malonyl-CoA involved in the microsomal elongation process. Recent experiments conducted in our laboratory have shown that acetylcarnitine provides acetate units to the microsomal fatty acid elongation system, a metabolic compartment that does not seem to be accessible to cytosolic acetyl-CoA generated from glucose metabolism (23). Thus, the injection of uniformly radiolabelled glucose into the lateral brain ventricle of awake rats did not lead to any radioactive incorporation into brain polyunsaturated fatty acids, whereas saturated fatty acids contained the highest percent of radioactivity. Conversely, if acetylcarnitine labelled in the acetyl moiety was injected, labelling of polyunsaturated fatty acids was also observed. These data strongly suggest that the lipogenic acetyl-CoA pools present in different cellular compartments are not homogeneous, and that specific acetate-trafficking between various subcellular compartments dictate the rate of synthesis of a particular lipid component. Moreover, studies on neuronal and glial cells have revealed interesting differences between them in terms of metabolic compartmentation (24, 25, 26, 27).

4. CONCLUSION

Since the discovery in 1955 that carnitine stimulates fatty acid oxidation in liver homogenates (28), most of the research efforts have been directed to characterize the role of carnitine as a key compound in energy-producing process in mitochondria. However, the work done in our laboratory suggests that the carnitine system may be engaged in the maintenance of organelle function and/or in metabolic pathways other than those of mitochondria: we believe that the mitochondrial studies somehow biased the recognition of other role(s) for the extramitochondrial carnitine system. It is now time to propose a more general task for the carnitine system in cellular physiology, that is, to facilitate and to modulate an efficent acyl trafficking between and within subcellular organelles, irrespective of the final destination of the acyl residue. In fact, not only catabolic or biosynthetic processes may take advantage of the strategic topographic distribution of the carnitine system throughout the cell, but also signal transduction processes (protein acylation), gene expression (histone acetylation), and cytoskeleton organization (tubulin acetylation) may be affected by such system.

REFERENCES

1. Bieber, L.L., 1988, *Annu. Rev. Biochem.* **57**: 261–283.
2. McGarry, J.D., Sen, A., Brown, N.F., Esser, V., Weis, B.C., and Foster, D.W., 1992, *Current Concepts in Carnitine Research* (Carter A.L., ed) CRC Press, Boca Raton, Florida, PP. 137–151.
3. Bremer, L., 1983, *Physiol. Rev.* **63**: 1420–1480.
4. Bird, M.I., Munday, L.A., Saggerson, D., and Clark, J.B., 1985, *Biochem. J.* **226**: 323–390.
5. Arduini, A., Mancinelli, G., Radatti, G., Dottori, S., Molajoni, F., and Ramsay, R.R., 1992, *J. Biol. Chem.* **267**: 12673–12681.
6. Ramsay, R.R., and Arduini, A., 1993, *Arch. Biochem. Biophys.* **302**: 307–314.
7. Arduini, A., Denisova, N., Virmani, A., Avrova, N., Federici, G., and Arrigoni-Martelli, E., 1994, *J. Neurochem.* **62**: 1530–1538.
8. Freed, L.M., Wakabayashi, S., Bell, J.M., and Rapoport, S.I., 1994, *Brain Res.* **645**: 41–48.
9. Chang, M.C.J., Wakabayashi, S., and Bell, J.M., 1994, *Neurochem. Res.* **19**: 1217–1223.
10. Mok, A.Y., and McMurray, W.C., 1990, *Biochem. Cell. Biol.* **68**: 1380–1392.
11. Moule, S.K., Edgell, N.J., Borthwick, A.C., and Denton, R.M., 1992, *Biochem. J.* **283**: 35–38.
12. Murthy, M.S.R., and Pande, S.V., 1994, *J. Biol. Chem.* **269**: 18283–18286.
13. Hall, E.D., 1993, *Oxygen Free Radical in Tissue Damage* (Tarr, M., Samson, F., eds) Birkhäuser Press, Boston, PP. 155–173..
14. Halliwell, B., Gutteridge, J.M.C. (eds), 1989, *Free Radicals in Biology and Medicine*, Clarendon Press, Oxford.
15. Davies, K.J.A., 1988, *Cellular Antioxidant Defence Mechanisms* (Chow C.K., ed), CRC Press, Boca Raton, PP. 27–67.
16. Doetsch, P.W., Henner, W.D., Cunningham, R.P., Toney, J.H., and Helland, D.E., 1987, *Molec. Cell. Biol.* **7**: 26.
17. Arduini, A., Dottori, S., Malajoni, F., Kirk, R., and Arrigoni-Martelli, E., 1995, *The Carnitine System* (De Jong J.W., Ferrari R., eds) Kluwer Academic Publishers, The Netherlands, PP. 169–181.
18. MacCaman, R.E., McCaman, M.W., and Stafford, M.L., 1966, *J. Biol. Chem.* **241**: 930–934.
19. Bresolin, N., Freddo, L., Vergani, L., and Angelini, C., 1982, *Exp. Neurol.* **78**: 285–292.
20. Aeberhard, E., Grippo, J., and Menkes, J.H., 1966, *Pediat. Res.* **3**: 590–596.
21. Cinti, D.L., Cook, L., Nagi, M.N., and Suneja, S.K., 1992, *Prog. Lipid Res.* **31**: 1–51.
22. Srere, P.A., 1965, *Nature* **205**: 766–770.
23. Ricciolini, R., Scalibastri, M., Calvani, M., and Arduini, A., 1996, *J. Neurochem.* **66**: S65B.
24. Moore, S.A., Yoder, E., Murphy, S., Dutton, G.R., and Spector, A.A., 1991, *J. Neurochem.* **56**: 518–524.
25. Sonnewald, U., Muller, T.B., Westergraad, N., Unsgård, G., Petersen, S.B., and Schousboe, A., 1994, *Neurochem. Int.* **24**: 473–483.
26. Badar-Goffer, R.S., Bachelard, H.S., and Morris, P.G., 1990, *Biochem. J.* **266**: 133–139.
27. Hassel, B., Sonnewald, U., and Fonnum, F., 1995, *J. Neurochem.* **64**: 2773–2782.
28. Fritz I.B., 1955, *Acta Physiol. Scand.* **34**: 367–385.

REFERENCES

CARNITINE TRANSPORT AND PHYSIOLOGICAL FUNCTION IN NEURONES

Katarzyna A. Nałęcz, Agnieszka Wawrzeńczyk, Joanna Mroczkowska, Urszula Berent, Nilolai A. Lobanov, and Maciej J. Nałęcz

Nencki Institute of Experimental Biology
Polish Academy of Sciences
Warszawa, Poland

1. INTRODUCTION

Carnitine (4-N-trimethylammonium-3 hydroxybutyric acid) facilitates in the eukaryotic cell the transfer of acyl compounds, mainly long-chain fatty acids, from the cytosol to the mitochondrial matrix (1). The pathway, well characterized for peripheral tissues like liver, kidney or muscles, includes the synthesis of acylcarnitine derivatives from their acylCoA forms by the enzymes localized at the outer side of the outer mitochondrial membrane (for instance palmitoylcarnitine transferase I). Acylcarnitines are further translocated through the inner mitochondrial membrane by a carnitine carrier and subsequently, on the inner side of this membrane the acyl moieties are transferred to CoASH by a different carnitine acyl transferase (e.g. palmitoylcarnitine transferase II). The carnitine carrier responsible for the central tranlocation step catalyses an uniport of carnitine or its exchange with acylcarnitines (2). This system of carnitine-dependent transport of fatty acids into the mitochondria, the so-called "carnitine shuttle", delivers substrates for β-oxidation.

Carnitine and its derivatives accumulate in nervous tissue (3, 4), although the level of β-oxidation of fatty acids in adult brain is relatively low (5) and glucose is the main energetic substrate for an adult brain (6). Since carnitine *in vivo* is mainly synthesized in liver, it must cross the blood-brain barrier. The highest content of carnitine was detected in the hypothalamus of rat brain (3). Since the hypothalamus has access to many substances circulating in blood through fenestrations, this high accumulation may point to some limitations for carnitine in crossing the blood-brain barrier before it reaches neurones. Nevertheless, the ability of carnitine to reach all parts of the brain and to be accumulated therein to a substantial level has been well documented and thus is beyond

---

From a workshop organized by Rona R. Ramsay (St. Andrews, United Kingdom). (Supported by Sigma Tau Ethifarm BV, Assen, The Netherlands.)

question. Therefore the main purpose of the studies summarized below was an attempt to clarify the mechanism of carnitine transport through the plasma membrane of neurones and through the inner mitochondrial membrane, as well as exploration of a possible physiological functions of carnitine and its derivatives in neurones.

2. CARNITINE TRANSPORT TO NEURONES

It was known from experiments on carnitine accumulation by brain slices (7) and synaptosomal preparations (8, 9) that this process was sodium-dependent. An inhibitory effect of ouabaine, dissipating the sodium concentration gradient by inhibition of Na^+,K^+-ATPase, was also observed when accumulation of carnitine was studied with either cultured neuroblastoma NB-2a cells (10) or with isolated rat cortical neurones (11). Kinetic analysis of carnitine accumulation gave the K_m value of 123 ± 13 µM, pointing to the fact that, at physiological carnitine concentrations, the transporting system in the plasma membrane would operate below saturating conditions.

Contrary to observations on carnitine accumulation in muscles (12), there was no effect of choline and butyrobetaine on carnitine accumulation by neurones. Hemicholinium-3, a specific inhibitor of choline transporter characterized with a high affinity (13), was also without any effect on carnitine transport (10). The lack of any effect of betaines and choline excluded the importance of the trimethyl-amino group binding to the transporting protein. The inhibitory effect of serine and cysteine (10) indicated that the system transporting carnitine to neurones can be characterized by certain structural demands concerning substrate, namely the presence of a carboxylic group and a polar group in the β-position (10). This would also indicate that carnitine may be transported by a novel, not yet described, transporting system of the requirements just mentioned. On the other hand, however, a much stronger inhibitory effect of cysteine in comparison with serine could suggest an involvement of SH group(s) in the transport process. As presented in Table 1, both permeable (N-ethylmaleimide) and impermeable (mersalyl) SH-group reagents inhibited carnitine transport.

It was also observed that carnitine accumulation in synaptosomes (8) and brain slices (7, 14) was competitively inhibited by GABA. This neurotransmitter had, however, no effect on carnitine accumulation in NB-2a cells (10). All the characteristics of carnitine transport into neuroblastoma cells and the isolated cortex neurons were practically the same, indicating a presence of the same transporting proteins in the plasma membrane of both neural cell types. However, in isolated neurones from rat cerebral cortex the accumulation of carnitine was decreased by 1 mM GABA. This effect on total carnitine accumulation was correlated with a decrease of acetylcarnitine and the inhibition of carnitine

Table 1. Effect of sulphydryl reagents on the accumulation of carnitine in neural cell

Addition	Carnitine accumulation in NB-2a cells (pmol/mg protein)	Carnitine accumulation on cerebral cortical neurones (pmol/mg protein)
None	180	138
N-ethylmaleimide (1 mM)	125	80
Mersalyl (0.2 mM)	56	80

Carnitine accumulation was measured (10) after 2 min preincubation in the absence or presence of the compounds indicated. The results represent accumulation measured after 20 min.

acetyltransferase (15). This observation indicated that GABA most likely affected further steps of carnitine metabolism rather than its transport through the plasma membrane.

3. CARNITINE TRANSPORT IN BRAIN MITOCHONDRIA

There are several carrier proteins in the inner mitochondrial membrane catalysing flux of metabolites into and out of mitochondria. In the case of liver, the net-efflux of acylcarnitine from preloaded mitochondria was found much faster with longer carnitine acyl derivatives than with shorter ones (2), suggesting that the main function of carnitine carrier in this tissue is to transport long-chain fatty acids for β-oxidation. The activity of carnitine carrier was also detected in brain, although the transport activity varied with animal age, being twice higher in suckling rats than in adult animals (16). Such an observation could suggest that the carnitine carrier in brain may be involved in delivering acyl derivatives for β-oxidation, especially, if not exclusively, in suckling animals. The fact that the carrier activity decreased with a shift in brain metabolism, towards utilizing glucose as a main energetic substrate (adult animals) points to the possibility that a different function may then be swithed on (6).

The mitochondrial carnitine carrier from brain was purified and appeared to be a polypeptide with a molecular mass of 33,000 (16). After reconstitution into phosphatidylcholine vesicles, the carnitine carrier showed sensitivity to SH groups modifying reagents, N-ethylmaleimide and mersalyl. A more detailed kinetic analysis of bi-substrate initial velocities revealed that either the carnitine/carnitine or acetylcarnitine/carnitine exchange reactions followed a kinetic pattern consistent with a sequential antiport mechanism (17). Therefore, a formation of a ternary complex with the internal and external substrates molecules bound to the carrier protein before the transport reaction has been postulated (17). This is contrary to the mechanism described for the carnitine carrier isolated from rat liver mitochondria (18). The carnitine transport activity measured with the isolated protein revealed that the inhibitory effect of acylcarnitines on this reaction was much stronger with the short-chain derivatives (Fig. 1) and a linear correlation was observed between the degree of inhibition and the chain length of acylcarnitines studied.

This observation, different from the substate specificity reported for liver mitochondria, supports an earlier hypothesis that the carnitine carrier from brain mitochondria has a special physiological role, different than in other tissues. Together with an observation of a high content of the pyruvate carrier in brain mitochondria (19) it was thus proposed that the carnitine shuttle in brain would export acetyl moieties derived from glucose–pyruvate–acetylCoA pathway from mitochondria. Such a function of the carnitine carrier could

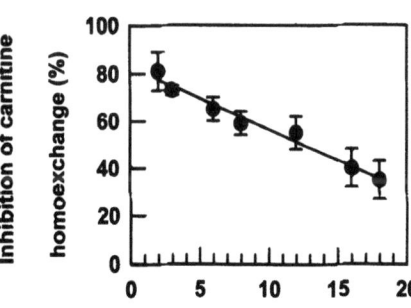

Figure 1. Inhibition of carnitine transport by acylcarnitine derivatives. The carnitine carrier was isolated from rat brain mitochondria. Proteoliposomes were preloaded with 20 mM carnitine. Transport was measured in the reconstituted system after 7 min preincubation with radioactive 0.5 mM carnitine in the presence of 3 mM acylcarnitine of the indicated amount of carbon atoms in their acyl chain. The control value of accumulation was 22.8±1.4 μmol/min per g protein. (From (17), with permission).

deliver the substrate for acetylcholine synthesis in cholinergic neurons. The possibility of carnitine influence on acetylcholine synthesis was subsequently verified.

4. EFFECT OF CARNITINE ON ACETYLCHOLINE SYNTHESIS IN NEURAL CELLS

Acetylcholine synthesis was monitored in neuroblastoma NB-2a cells and in cerebral cortex cells isolated from brains of suckling and adult animals. Acetylcholine was synthesized when glucose was administered as the precursor, no reaction was detected after addition of acetate. Carnitine was found to stimulate the synthesis of acetylcholine in a synergistic way together with choline. This effect of carnitine was observed in neuroblastoma NB-2a cells (20) and the cerebral cortical neurones isolated from adult animals (21). In neurones isolated from suckling animals administration of carnitine decreased the basic level of acetylcholine synthesis and even addition of choline did not reverse this effect (21).

The observed stimulation by carnitine of acetylcholine synthesis in cells isolated from adult brains could, at least partially, explain beneficial effects of carnitine administration to patients with neurogenerative diseases.

5. PHYSIOLOGICAL FUNCTIONS OF LONG-CHAIN ACYLCARNITINES IN NEURONES

Although β-oxidation is very low in neurones of adult brain, the isolated rat cerebral cortex cells are capable of synthesizing the long-chain acylcarnitines which can reach 15–20% of all forms of accumulated carnitine (15). Moreover, a very high accumulation of the long-chain acyl derivatives of carnitine was observed in neuroblastoma NB-2a cells. Since palmitoylcarnitine has been reported to inhibit protein kinase C (22, 23), an enzyme known to be involved in either proliferation or differentiation pathway, the effect of modified contents of palmitoylcarnitine on the activity of protein kinase C were studied in neuroblastoma NB-2a cells. Although this is a very quickly proliferating cell line, the decrease of number of cells and a concomitant decrease of incorporation of thymidine pointed to a true inhibition of proliferation under conditions when the intracellular content of palmitoylcarnitine was increased. Under the same conditions the amount of processes was observed to increase, what was correlated with the higher content of growth associated protein (B-50, GAP-43, neuromodulin). The activity of protein kinase C was measured with a peptide substrate $\{[ser^{25}]PKC\text{-}(19\text{–}31)\}$, corresponding to the pseudosubstrate domain of the protein kinase isoform α, in which alanine in position 25 was replaced by serine.

As presented in Table 2, the activity of protein kinase C was inhibited by palmitoylcarnitine in a concentration-dependent way. What has to be emphasized, that only the phorbol ester stimulated reaction was affected.

Analysis of protein phosphorylation pattern revealed that palmitoylcarnitine decreased, the phorbol-stimulated incorporation of ^{32}P to proteins of M_r 81, 50–54 and 16 kDa in the membrane fraction. It has been postulated that these could be known protein kinase C substrates present in neuronal plasma membranes, namely myristoylated alanine-rich protein kinase substrate (MARCKS), B-50 and neurogranin, and at least the identity

Table 2. Effect of palmitoylcarnitine on the activity of protein kinase C in neuroblastoma NB-2a cells

Palmitoylcarnitine μM	Protein kinase C activity (pmol ^{32}P/mg protein)	Protein kinase C activity after preincubation withphorbol ester (pmol ^{32}P/mg protein)
0	190	495
20	203	453
50	210	400
80	190	365
100	210	360
120	217	342

Cells were permeabilized with streptolysin O, 20 :M phorbol 12, 13-dibutyrate was added for 12 min preincubation. The reaction was started by the addition of synthetic peptide substrate and ATP, all other details as described in (24).

of MARCKS and B-50 has been recently confirmed by immunoprecipitation studies (data not yet published). All these observations point to palmitoylcarnitine as a natural modulator of protein kinase C and to a possible role of this carnitine derivative in stimulating nerve growth and promotion of neuronal differentiation.

6. CONCLUSIONS

All the observations indicate that carnitine is taken up by neurones through a specific transporting system. There is a functionally different carnitine carrier in brain mitochondria, involved in acetylcarnitine export to cytosol, delivering acetyl moiety for acetylcholine synthesis. The level of accumulated long-chain acylcarnitines can modulate crucial steps in signal transduction pathways in neurones. It may therefore be postulated that carnitine is a compound able to fullfil several regulatory functions in the brain that are novel to a textbook knowledge (25).

Abbreviations: CoA, CoASH — Coenzyme A moiety and free coenzyme A, respectively.

ACKNOWLEDGMENT

This work was supported by the Polish State Committee for Scientific Research Grant No 6 PO4A 065 09.

REFERENCES

1. Bremer, J., 1983, *Physiol. Rev.* **63**: 1421–1480.
2. Parvin, R., and Pande, S.V., 1979, *J. Biol. Chem.* **254**: 5423–5429.
3. Bresolin, N., Freddo, L., Vergani, L., and Angelini, C., 1982, *Expl. Neurol.* **78**: 285–292.
4. Shug, A.L., Schmidt, M.J., Golden, G.T., and Fariello, R.T., 1982, *Life Sci.* **31**: 2869–2874.
5. Warshaw, J.B., and Terry, M.L., 1976, *Dev. Biol.* **52**: 161–166.
6. Nehlig, A., and Pereira De Vasconcelos, A.P., 1993, *Prog. Neurobiol.* **40**: 163–221.
7. Huth, P.J., Schmidt, M.J., Hall, P.V., Fariego, R.G., and Shug, A.L., 1981, *J. Neurochem.* **36**: 715–723.
8. Zoccarato, F., Siliprandi, N., and Rugolo, M., 1983, *Biochim. Biophys. Acta* **734**: 381–383.

9. Hannuniemi, R., and Kontro, P., 1988, *Neurochem. Res.* **13**: 317–323.
10. Nałęcz, K.A., Korzon, D., Wawrzeńczyk, A., and Nałęcz, M.J., 1995, *Archiv. Biochem. Biophys.* **322**: 214–220.
11. Nałęcz, K.A., Wawrzeńczyk, A, and Nałęcz, M.J., 1995, *J. Neurochem.* **65** (Suppl.) S35D.
12. Rebouche, C.J., 1977, *Biochim. Biophys. Acta* **471**: 145–155.
13. Happe, H.K., and Murrin, L.C., 1993, *J. Neurochem.* **60**: 1191–1201.
14. Fariello, R.G., and Shug, A.L., 1981, *Biochem. Pharmacol.* **30**: 1012–1013.
15. Wawrzeńczyk, A., Nałęcz, K.A., and Nałęcz, M.J., 1995, *Biochem. Biophys. Res. Commun.* **213**: 383–388.
16. Kamińska, J., Nałęcz, K.A., Azzi, A., and Nałęcz, M.J., 1993, *Biochem. Mol. Biol. Int.* **29**: 999–1007.
17. Kamińska, J., Nałęcz, K.A., Nałęcz, M.J., 1995, *Acta Neurobiol. Exp.* **55**: 1–9.
18. Indiveri, C., Tonazzi, A., and Palmieri, F., 1994, *Biochim. Biophys. Acta* **1189**: 65–73.
19. Nałęcz, K.A., Kamińska, J., Nałęcz, M.J., and Azzi, A., 1992, *Archiv. Biochem. Biophys.* **297**: 162–168.
20. Wawrzeńczyk, A., Nałęcz, K.A., and Nałęcz, M.J., 1994, *Biochem. Biophys. Res. Commun.* **202**: 354–359.
21. Wawrzeńczyk, A., Nałęcz, K.A., and Nałęcz, M.J., 1995, *Neurochem. Int.* **26**: 635–641.
22. Katoh, N., Wise, B.C., and Kuo, J.F., 1983, *Biochem. J.* **209**: 189–195.
23. Katoh, N., Wrenn, R.W., Wise, B.C., Shoji, M., and Kuo, J.F., 1981, *Proc. Natl. Acad. Sci. USA* **78**: 4813–4817.
24. Alexander, D.R., Graves, J.D., Lucas, S.C., Cantrell, D.A., and Crumpton, M.J., 1990, *Biochem. J.* **268**: 303–308.
25. Nałęcz, K.A., and Nałęcz, M.J., 1996, *Acta Neurobiol. Exp.* **56**: 597–609.

CEREBRAL METABOLIC COMPARTMENTATION AS REVEALED BY ^{13}C NMR ANALYSIS OF [1-^{13}C]GLUCOSE METABOLISM

Tommaso Aureli,[1] Maria Enrica Di Cocco,[2] Caterina Puccetti,[2] Rita Ricciolini,[1] Giorgio Capuani,[1] Menotti Calvani,[1] and Filippo Conti[2]

[1]Department of Biochemistry
Sigma-Tau Research Laboratories
Pomezia, Italy
[2]Department of Chemistry
University "La Sapienza"
Rome, Italy

1. INTRODUCTION

Nuclear magnetic resonance spectroscopy (MRS) has been shown to be a powerful non invasive technique that can be used to study cerebral metabolism *in vivo* (1). ^{31}P and ^{1}H NMR spectra have yielded information on the concentration of cerebral metabolites and on their response to various pathological states (1,2).

At present, however, the interpretation of changes in ^{1}H and ^{31}P NMR spectra of the brain is complicated because MRS data are obtained from whole tissue regions that include several cell populations distributed heterogeneously. It is of basic importance to distinguish the metabolism of different cell populations and to understand how brain cell metabolism is dependent on intercellular regulations. It has been shown that glial-neuronal interactions can be studied using ^{13}C-enriched substrates and high resolution ^{13}C nuclear magnetic resonance techniques. When a ^{13}C-labeled compound (i.e. glucose or acetate) is supplied to the brain, the ^{13}C label enters several metabolites via tricarboxylic acid cycle (TCA) intermediates and amino acids at multiple carbon positions. Which carbon positions within a metabolite are to be labeled depends on the metabolic pathways travelled by the ^{13}C-enriched compounds. Hence, from the labeling pattern in the cerebral metabolites, information can be obtained about the metabolic pathways in

---

From a workshop organized by Rona R. Ramsay (St. Andrews, United Kingdom). (Supported by Sigma Tau Ethifarm BV, Assen, The Netherlands).

operation. Studies by ^{13}C MRS on cerebral metabolism have revealed that glial-neuronal interactions involves a metabolite trafficking between the two cellular compartments, i.e. a flux of glutamine from neurons to glia and a flux of glutamate into the opposite direction (3). Furthermore, studies on cultured cells have provided evidence that astrocytes are able to release TCA constituents as citrate, malate, 2-oxo-glutarate in the culture medium (4).

Recently, a pathway of pyruvate formation from TCA cycle intermediates has been suggested to occur in vivo within of neuronal compartment (5). Data from cultured cells have showed that a similar pathway of pyruvate formation occurs in astrocytes (6).

In the present study, we used ^{13}C NMR spectroscopy and [1-^{13}C]glucose to investigate some aspects of cerebral metabolic compartmentation and to follow the exchange of carbon skeletons between glia and neurons. Particular attention has been devoted to obtain experimental evidence for the activity of the pyruvate recycling pathway, and in which cellular compartment it is operating.

2. EXPERIMENTAL PROCEDURES

Male Fischer 344 rats, aged 6 months, were fasted 14–16h prior to each experiment, and injected intraperitoneally with 200 mg/kg b.w. of [1-^{13}C]glucose or unlabeled glucose dissolved in sterile isotonic saline. Control rats were injected only with saline (n=3). At 15, 30, 45 and 60 min after glucose administration, the rats (n=2 per time point) were sacrificed by the *in situ* freezing technique (7) under halothane 0.6% and N_2O/O_2 70/30% anaesthesia. The frozen brains were removed and extracted according to a modification of the Bligh-Dyer method (8), using a mixture of methanol/chloroform/water at the final proportion of 2:2:1 (v/v). One ml of blood per animal was collected and sera were extracted using the same procedure as for the brains.

^{1}H and ^{13}C NMR spectra of the aqueous phase extracts of brain and serum were obtained on a Bruker AM 500 spectrometer. An inverse-gated decoupling sequence was used for ^{13}C measurements and spectra were accumulated using a 45° pulse angle with a spectral width of 31 KHz and 32K data points. The acquisition time was 0.52 s and an additional relaxation delay of 7.48 s was used. ^{1}H NMR spectra were acquired with the following parameters: 45° pulse, 6000 Hz spectral width and 64 kwords. The acquisition time was 5.44 s and an additional delay of 10 s was used. The fractional enrichment with ^{13}C in each carbon position of a given metabolite was calculated by subtracting the natural abundance ^{13}C (1.1%) from the total ^{13}C amount in that position as determined from the ^{13}C spectra, and dividing by the concentration of that metabolite.

3. RESULTS

3.1. NMR Spectroscopy of Serum and Brain Extracts

At all time points, serum glucose levels showed no significant differences from the values obtained from the control group animals. On the other hand, after [1-^{13}C]glucose injection no significant time-dependent changes in fractional ^{13}C enrichment of C-1 in serum glucose were observed. No significant ^{13}C enrichment was observed in the other carbon positions of glucose.

3.2. NMR Spectroscopy of Serum and Brain Extracts

At different times after [1-^{13}C]glucose injection, appreciable amounts of ^{13}C label were found to be distributed in various carbon positions of amino acids and lactate. The two C-1 resonances of α and β anomers of glucose were also present at all time points.

According to the schematic representation shown in Figure 1, tricarboxylic acid intermediates and related amino acids were labeled in different carbon positions when [1-^{13}C]glucose was metabolized via the anaplerotic pathway (pyruvate carboxylase mediated) then when metabolized via the oxidative pathway (pyruvate dehydrogenase mediated). Table 1 shows the differences in the ^{13}C label distribution generated by pyruvate carboxylase (PC) or pyruvate dehydrogenase (PDH) mediated pathways for the metabolites of interest. The ^{13}C label was primarily present in C-4 of glutamate and glutamine and in C-2 of GABA (which corresponds to the C-4 position in glutamate), whereas C-2 and C-3 of glutamate and glutamine and C-3 and C-4 of GABA were more weakly labeled. A time-dependent increase in ^{13}C enrichment of C-4 in glutamate and glutamine and of C-2 in GABA was observed following ^{13}C-labeled glucose injection, consistent with the increase in [2-^{13}C] acetyl-CoA formation by pyruvate dehydrogenase. A similar trend was seen in the C-2 and C-3 positions of glutamate and glutamine and in C-3 and C-4 of GABA, as a result of ^{13}C

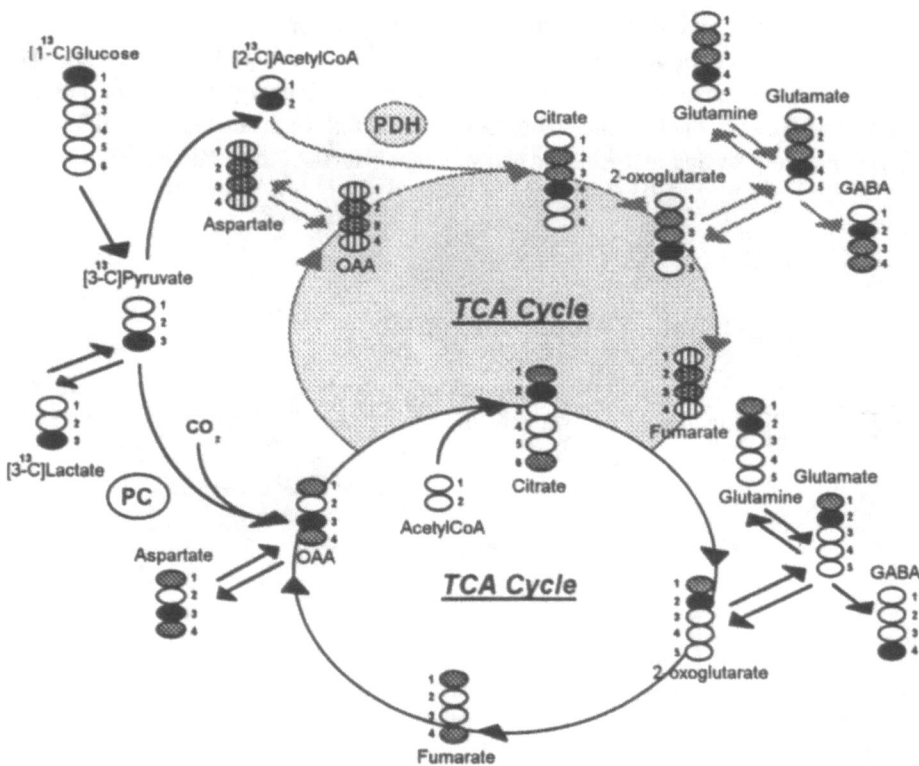

Figure 1. Metabolism of [1-^{13}C]glucose in the brain. ^{13}C Label distribution in TCA cycle intermediates via pyruvate dehydrogenase (top) and via pyruvate carboxylase (bottom). Symbols: black circles = ^{13}C label from glucose through PDH or PC reaction; grey circles = 1/2 of the original label is in this position; striped circles = 1/4 of the original label is in this position. Only two TCA cycle turns are considered.

Table 1. Fractional ^{13}C enrichments of the carbon position of various amino acids after injection of [1-^{13}C]glucose

Time (min)	Glutamate			Glutamine			GABA			Aspartate	
	C-2	C-3	C-4	C-2	C-3	C-4	C-2	C-3	C-4	C-2	C-3
15	1.30 ± 0.05	0.95 ± 0.05	4.30 ± 0.34	1.42 ± 0.37	0.73 ± 0.23	1.79 ± 0.25	3.54 ± 0.98	0.90 ± 0.12	1.06 ± 0.34	1.35 ± 0.38	1.77 ± 0.08
30	4.91 ± 0.13	4.47 ± 0.48	10.32 ± 1.34	4.33 ± 0.29	3.76 ± 0.11	5.51 ± 0.56	8.59 ± 1.06	3.65 ± 0.92	4.37 ± 1.02	5.39 ± 0.52	6.97 ± 1.41
45	6.83 ± 0.35	5.96 ± 0.36	10.09 ± 0.43	6.15 ± 0.47	4.78 ± 1.05	6.78 ± 0.07	10.42 ± 1.38	4.36 ± 0.51	4.81 ± 0.43	6.98 ± 0.81	8.70 ± 0.53
60	8.23 ± 0.56	7.54 ± 0.61	10.81 ± 1.40	8.30 ± 0.45	6.98 ± 0.16	8.33 ± 0.24	11.27 ± 1.19	6.81 ± 0.63	6.75 ± 0.96	7.44 ± 0.49	8.31 ± 0.95

Rats were injected (i.p.) with 200 mg/kg b.w. of [1-^{13}C]glucose and sacrificed after 15, 30, 45, or 60 min. Data are mean ± SD values for two rats.

label cycling through the TCA cycle. The ^{13}C enrichment of the C-2 position in glutamine and glutamate was always higher than that of the C-3 position, as was that of C-3 compared with C-2 in aspartate, thus evidencing the PC-mediated entry of ^{13}C label into TCA cycle.

Figure 2 shows ^{13}C enrichments relative to carboxyl carbons in glutamate, glutamine, GABA, and aspartate, and of the C-2 position in lactate. According to the schematic representation described in Figure 1, the C-1 position of glutamate and glutamine was labeled in the second turn of the anaplerotic pathway and in the third and subsequent turns of the oxidative pathway. Likewise, the C-1 and C-4 carbons of aspartate were labeled in the first turn of the anaplerotic pathway and in the second and subsequent turns of the oxidative pathway.

4. DISCUSSION

4.1. Cerebral Metabolic Compartmentation

The labeling pattern within the individual amino acids demonstrates the contribution of the PC- and PDH-mediated pathways to their synthesis. The anaplerotic pathway ap-

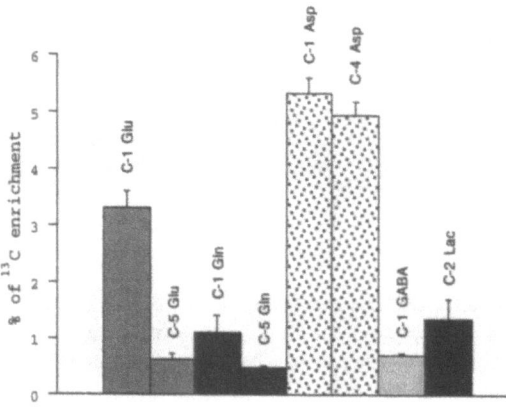

Figure 2. Fractional ^{13}C enrichment of the carboxylic positions in glutamate, glutamine, GABA, aspartate and of the C-2 position in lactate.

pears to contribute strongly to glutamine synthesis. This is expected because glutamine synthetase and pyruvate carboxylase are only localized in the glial cells (9,10). GABA synthesis has been suggested previously to take place in a compartment associated with pyruvate carboxylase activity (11). Our findings, however, do not corroborate this hypothesis, in accordance with the data by other authors (4). Nevertheless, the analysis of GABA ^{13}C labeling pattern reveals a strong similarity with that of glutamate. Furthermore, a higher ^{13}C enrichment of C-4 than of C-3 in GABA is consistent with the well-known glutamine cycle. The latter postulates that glutamine is supplied to neurons by astrocytes to serve as a metabolic precursor for replenishing GABA and glutamate transmitter pools (12). In the present study, however, GABA is labeled both through glucose metabolism in neurons and from astrocyte-derived glutamine, so that the former pathway tends to overshadow the latter. For glutamate, our data indicate a higher ^{13}C enrichment of C-2 than of C-3. A low contribution of glial metabolism to glutamate synthesis appears to be contradictory, since glutamate in astrocytes is the mandatory precursor for glutamine synthesis. It should, therefore, be highly labeled via the PC-mediated pathway. However, the ^{13}C labeling pattern of glutamate we have reported is consistent with previous hypotheses on the existence of two glutamate pools, of which the one localized in glia is notably smaller than the large neuronal pool (12). Thus, the high ^{13}C label in the C-2 position of glia-synthesized glutamate is diluted in the large neuronal pool, which is labeled via the PDH-mediated pathway. For aspartate, our data show a higher ^{13}C enrichment of C-3 compared with C-2, thereby suggesting that the anaplerotic pathway strongly contributes to aspartate synthesis. As described for glutamate, this result is consistent with a higher aspartate concentration in astrocytes than in neurons. The large glial pool associated with pyruvate carboxylation is, therefore, responsible for the high relative ^{13}C enrichment observed in the C-3 position.

4.2. Glial-Neuronal Trafficking of Metabolic Intermediates

The low yet significant ^{13}C label in the C-5 position of glutamate and glutamine, in C-1 of GABA and C-2 of lactate at 60 min after [1-^{13}C]glucose injection, was suggestive of pyruvate formation from TCA cycle intermediates and provides evidence of metabolite trafficking between the mitochondrion and the cytosol and very likely, between astrocytes and neurons. This particular ^{13}C labeling cannot be merely accounted for by the entry of the ^{13}C label in the TCA cycle either as [2-^{13}C]acetyl-CoA or [3-^{13}C]oxalacetate. These specified positions can exclusively be labeled through the entry of C-1 labeled acetyl-CoA molecules. Under our in vivo experimental conditions, we hypothesized the existence of a metabolic pathway through which one or more TCA cycle intermediates, such as malate, or oxalacetate labeled in the C-2 position, or citrate labeled in C-3, travels from mitochondrion to cytosol. Subsequently, the intermediate(s) (with citrate having first to undergo cleavage by citrate lyase) will give origin to C-2 labeled pyruvate molecules through a decarboxylating reaction (malic enzyme or phosphoenolpyruvate-carboxykinase). The pyruvate molecules, in turn, would reenter the mitochondrion and hence, the TCA cycle as [1-^{13}C]acetyl-CoA. However, an important question to be addressed is to establish the cerebral compartment where this metabolic pathway is operating. Some authors hypothesized a metabolic pathway of pyruvate recycling involving oxalacetate decarboxylation via the phosphoenolpyruvate-carboxykinase (3,5). Moreover, they suggested that this pathway was active in neurons, since the ^{13}C label was detected in C-5 of glutamate but not in C-5 of glutamine. Recent studies have suggested that a similar pathway of pyruvate synthesis through decarboxylation of a TCA cycle intermediate takes place in astrocytes

(4,6). Moreover, in a recent study on [2-^{13}C]acetate administered to fasted mice, it was observed the presence of ^{13}C label in the C-2 and C-3 positions of lactate, thereby suggesting the existence of a pyruvate synthesizing pathway from TCA cycle intermediates, which operates in the acetate-metabolizing compartment (13). Our results are consistent with the hypothesis of a pyruvate recycling pathway operating in the glial compartment. The [2-^{13}C]pyruvate so synthesized is in turn able to reenter the glial TCA cycle and to subsequently label the C-5 position of glutamine, or else it can be transferred from astrocytes to neurons as [2-^{13}C]lactate, to then enter the neuronal TCA cycle and label glutamate in C-5 and GABA in C-1.

In conclusion, the present study demonstrates that ^{13}C-NMR spectroscopy associated to the administration of ^{13}C-labeled substrates permits the investigation of cerebral metabolism. Furthermore, the analysis of labeling pattern in metabolic intermediates has allowed us to unveil new and relevant aspects of cerebral metabolic compartmentation and metabolite trafficking between astrocytes and neurons.

REFERENCES

1. Bachelard, H., and Badar-Goffer, R., 1993, *J. Neurochem.* **61**: 412–429.
2. Howe, F.A., Maxwell, R.J., Saunders, D.E., Brown, M.M., and Griffiths, J.R., 1993, *Magn. Res. Quart.* **9**: 31–59.
3. Cerdan S., Künnecke, B., and Seelig, J., 1990, *J. Biol. Chem.* **265**: 12916–12926.
4. Westergaard, N., Sonnewald, U., and Schousboe, A., 1995, *Dev. Neurosci.* **17**: 203–211.
5. Künnecke, B., Cerdan, S., and Seelig, J., 1993, *NMR Biomed.* **4**: 264–277.
6. Sonnewald, U., Westergaard, N., Petersen, S.B., Unsgård, G., and Schousboe A., 1993, *J. Neurochem.* **61**: 1179–1182.
7. Pontén, U., Ratcheson, R.A., Salford, L.G., and Siesjö, B.K., 1973, *J. Neurochem.* **21**: 1127–1138.
8. Miccheli, A., Aureli, T., Delfini, M., Di Cocco, M.E., Viola, P., Gobetto, R., and Conti, F., 1988, *Cell. Mol. Biol.* **34**: 591–603.
9. Norenberg, M.D., and Martinez-Hernandez, A., 1979, *Brain es.* **161**: 303–310.
10. Shank, R.P., Bennet, G.S., Freytag, S.O., and Campbell, G.L., 1985, *Brain Res.* **329**: 364–367.
11. Shank, R.P., and Aprison, M.H., 1988, *Glutamate and Glutamine in Mammals*, Volume II (Kvamme E. ed.), CRC Press, Boca Raton, Florida, PP.3–20.
12. Brainard, J.R., Kyner, E., and Rosenberg, G.A., 1989, *J. Neurochem.* **53**: 1285–1292.
13. Hassel, B., and Sonnewald, U., 1995, *J. Neurochem.* **65**: 2227–2234.

# AGE-DEPENDENT CHANGES OF THE *IN SITU* PROTEIN PHOSPHORYLATION IN HIPPOCAMPAL SLICES AFTER STIMULATION OF GLUTAMATE RECEPTORS

Frank Angenstein and Sabine Staak

Federal Institute for Neurobiology
Laboratory for Cellular Signalling
Magdeburg, Germany

1. INTRODUCTION

The activation of protein kinase C in the CNS is involved in the regulation of a variety of physiological processes, i.e. differentiation, neurotransmitter release, gene expression, regulation of receptors and ion channels and even more complex phenomena like long-term potentiation as well as learning and memory formation (1). Different receptors are described which are able to induce changes in the activity of the kinase after its stimulation. Among them are the metabotropic glutamate receptor subtypes mGluR1 and 5. These have been shown recently: (a) to be coupled to phosphoinositide hydrolysis, thus generating the PKC activating second messenger diacylglycerol upon stimulation, (b) to take part in the regulation of the adenylate cyclase activity and subsequently protein kinase A activation and (c) to be involved in cellular mechanisms underlying induction and/or maintenance of hippocampal long-term potentiation (2).

Using a biochemical approach we investigated the phosphorylation degree of different substrates after *in situ*-phosphorylation of hippocampal slices incubated in the presence of (1) a tumor-promoting phorbol ester (phorbol-12-myristate-13-acetate–TPA) known to activate directly the protein kinase C or (2) specific agonists of the metabotropic glutamate receptor quisqualate and trans-(1S,3R)-1-amino-1,3-cyclopentane-dicarboxylic acid (trans-ACPD).

2. MATERIAL AND METHODS

After decapitation of the animal the brain was rapidly removed, the rat hippocampus dissected out on ice and cut into 400 μm thick transverse hippocampal slices (McIlwain tissue shopper). Tissue slices were maintained in a medium containing (in mM) NaCl 134.0, KCl 6.24, $MgSO_4$ 1.3, $CaCl_2$ 2.0, $NaHCO_3$ 16 and glucose 10 (pH 7.4, 32°C). The medium was aerated with carbogen (95% O_2, 5% CO_2) throughout the experiment. After preincubation for 60 min 100 μCi (final concentration in the medium 50 μCi/ml) was added and the slices were labelled for 60 min. Then control slices were removed and the test substance added to the medium.

Experimental slices were removed at the times indicated and washed quickly in an ice-cold solution containing in mM: HEPES (pH 7.4) 20, NaF 100, $Na_4P_2O_7$ 10, EDTA 4 and 2.5% perchloric acid (PCA). They were then homogenised in 200 μl of the same buffer including 1 mM AEBSF, 10 μg/ml leupeptin and 250 μg/ml P-nitrophenyllphosphate. Each 200 μl of the homogenate was centrifuged for 15 min at 14000 g to get a supernatant of PCA-soluble proteins. The PCA-soluble protein fraction was then precipitated by 80 μl 30% TCA for 45 min at 4°C and centrifuged again. The resulting pellet was washed with 250 μl ethanol and dissolved in 50 μl sample buffer (250 mM Tris-HCl, 50 mM dithiothreitol, 3 mM EDTA, 4% SDS, 20% glycerol and 0.01% bromophenol blue, pH 8.0).Each 25 μl were used for the electrophoretic separation (5–20% SDS gel). The gel was dried and analysed by phosphoimaging on a Bio-Imaging system BAS1000 (FUJIX).

3. RESULTS AND CONCLUSION

It is known that several PKC substrate proteins (MARCKS, B50, neurogranin) share the unusual property to be soluble in 2.5% perchloric acid (PCA) (3).

If all PCA-soluble proteins were isolated from hippocampal slices after *in situ* phosphorylation and separated by polyacrylamide gel electrophoresis a distinct number of phosphoproteins was detectable in contrast to the complex pattern of phosphoproteins found in the cytosolic and membrane-bound fractions (Fig. 1). Moreover, a clear development-dependent alteration in the basal phosphorylation of those proteins was detectable by autoradiography (Fig. 1). The phosphorylation degree of some proteins (approximately 94, 46, 29 and 21 kDa) decreases age-dependent whereas for others (approximately 85, 42 and 17 kDa) an increase in the phosphorylation degree was observed. A 58 kDa protein, one of the most prominent phosphoprotein in the PCA-soluble fraction of hippocampal slices prepared from 12 day old rats disappeared in the autoradiography completely after 4 weeks, although the protein is still detectable by silver staining. The degree of protein phosphorylation was linear with respect to the protein concentration (Fig. 2). Among the 2.5% PCA-soluble poteins only the 94 kDa and a 17 kDa protein seem to be specific substrates for protein kinase C whereas other proteins may be also substrates for further kinases or were phosphorylated by other kinases as PKC. On the basis of its location after 2 dimensional separation we assume that the 94 kDa protein represents the MARCKS protein (pI about 4.5) and the 18 kDa (pI 8–8.5) protein is identical with neurogranin (Fig. 3).

Incubation of hippocampal slices of 2 or 8 week old rats in presence of the protein kinase C activating phorbol ester (phorbol-12-myristate-13-acetat) induced a strong increase in the phosphorylation degree of the 94 kDa protein to 242.0% ± 22.3% (8 week) and 667.6 ± 78.5% (2 week) and the 17 kDa protein in adult rats to 182.2% ± 17.4%. This enhanced phosphorylation could be blocked by both staurosporine and chelerythrine. In contrast to that in hippocampal slices of adult rats the addition of the metabotropic glutamate receptor agonist 20 μM t-ACPD or 0.1 mM quisqualate to the incubation medium caused only a small increase in the phosphorylation degree of the 94 and 17 kDa proteins

Age-Dependent Changes of the *in Situ* Protein Phosphorylation

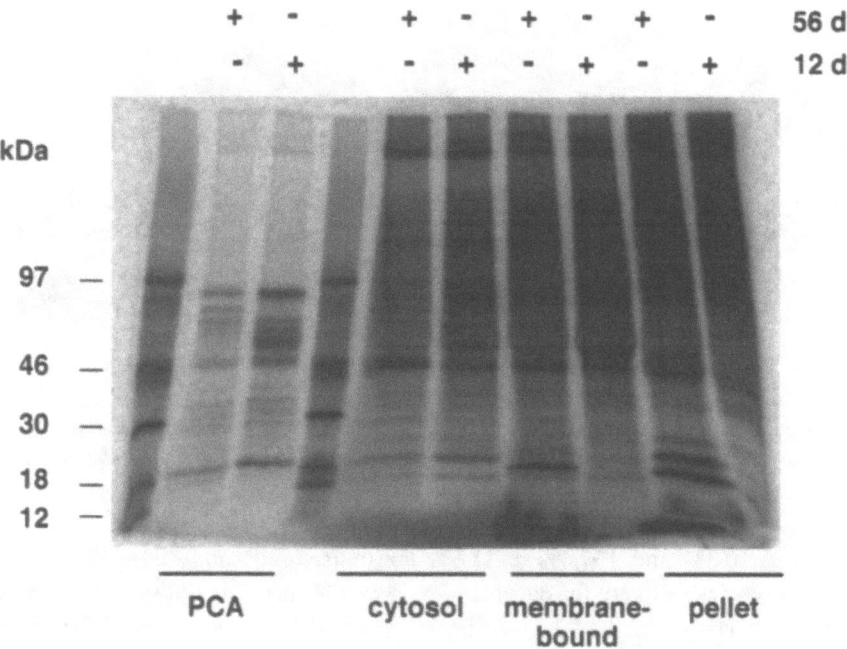

Figure 1. Basal *in-situ* phosphorylation of hippocampal slices prepared from both, 12 days and 56 days old rats. After 60 min preincubation time the slices were labelled with ^{33}P (final concentration in the medium 50 μCi/ml) for 60 min. The phosphorylation state of soluble (cytosol), Triton X100-soluble (membrane-bound) and Triton X100-insoluble proteins is shown on the right lanes. In comparison to that 2.5% perchloric acid (PCA) soluble proteins enriched from the same hippocampal slices were applied on the left lanes.

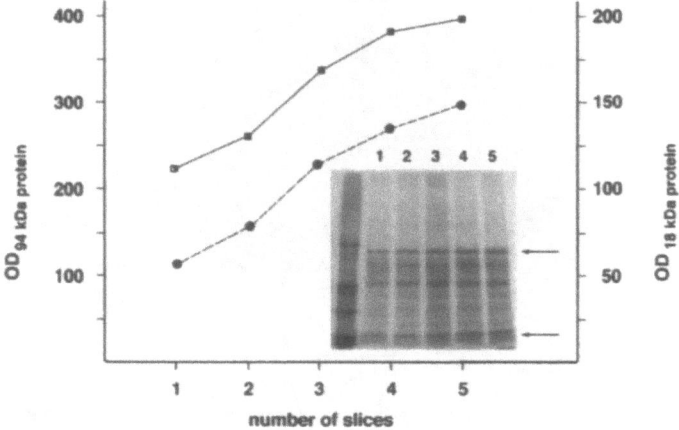

Figure 2. Autoradiography of perchloric acid soluble phosphoproteins obtained from 1 to 5 hippocampal slices.

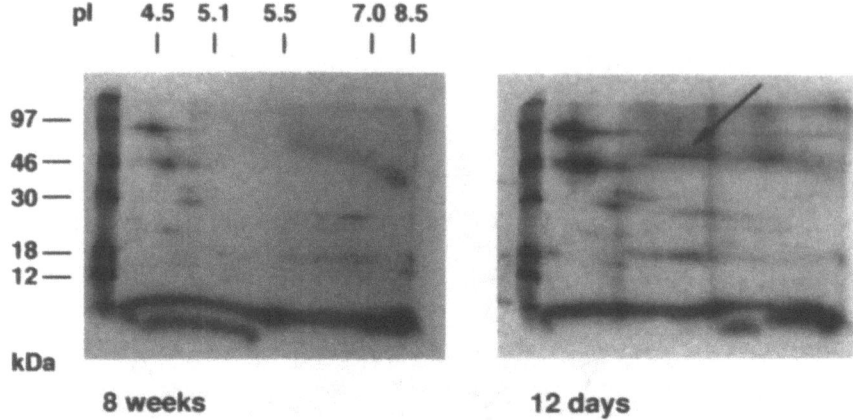

Figure 3. Autoradiography of 2.5% perchloric acid soluble proteins after two dimensional separation. Fourteen hippocampal slices from 56 day and 12 day old animals were labelled with ^{33}P (50 μCi/ml) for 60 min.

(to 131.1% ± 10.3% and 137.4% ± 11.6% for trans-ACPD and 124.4% ± 6.4% and 137.4% ± 11.6% respectively for quisqualate). However, the effect of this agonists on the *in situ* phosphorylation of the 94 kDa protein in slices from young rats were more pronounced (to 276.4% ± 55.1% by quisqualate and 184.6% ± 20.5% by trans-ACPD, Fig. 4).

The application of 0.1 mM NMDA to the incubation medium for 5 min was ineffective regarding the phosphorylation of PKC substrates. Our data demonstrate that in hippocampal slices the stimulation of mGluRs can induce an increased *in situ*-phosphorylation of different PKC proteins, particularly in slices prepared from young rats (12 day old). The activation of metabotropic glutamate receptor during high frequency stimulation have been shown recently to be critical for the maintenance of long-term potentiation (4). One possible mechanism responsible for that phenomenon might be mediated by an altered function of

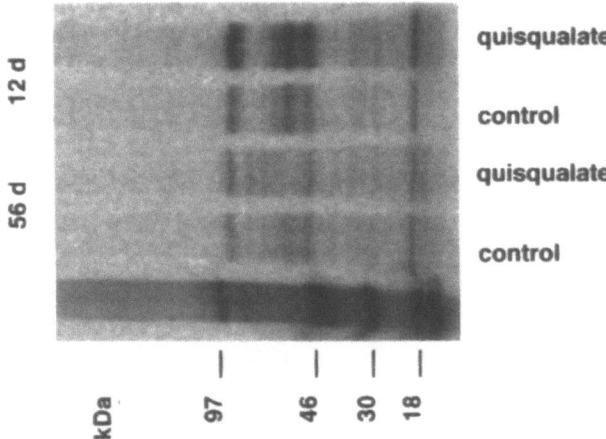

Figure 4. Effect of 0.1 mM quisqualate on the phosphorylation state of PCA-soluble proteins of hippocampal slices prepared from 8 week old rats (left lanes) or from 12 day old rats (right lanes). The slices were incubated in the same interface chamber and consequently stimulated in the same way.

the found phosphoproteins due to changes in protein kinase C dependent phosphorylation. The molecular identity of these phosphoproteins is further under investigation.

ACKNOWLEDGMENTS

The work was supported by the Bundesminister für Bildung, Wissenschaft, Forschung und Technologie (grant 07 NBL 06).

REFERENCES

1. Nishizuka, Y., 1992, *Science* **258**: 607–614.
2. Riedel, G., Casabona, G. and Reymann, K.G., 1995, *J. Neurosci.*, **15**: 87–98.
3. Baudier, J., Bronner, C., Kligmann, D. and Cole, R.D., 1989, *J. Biol. Chem.* **264**: 1824–1828.
4. Bashir, Z.I., Bortolotto, Z.A., Davies, C.H., Berretta, N., Irving, A.J., Seal, A.J.Henley, J.M., Jane, D.E., Watkins, J.C., Collingridge, G.L., 1993, *Nature* **363**: 347–350.

REGULATION OF ACTIVITY OF BRAIN PHOSPHATE ACTIVATED GLUTAMINASE ISOFORMS

R. G. Kamalyan, A. V. Gyulkhandanyan, T. D. Karapetyan, E. R. Mikaelyan, and A. G. Vardanyan

The Institute of Biochemistry
National Academy of Sciences
Republic of Armenia

1. INTRODUCTION

The glutamine hydrolysis by phosphate activated glutaminase (PAG) is a major source for the formation of the transmitter amino acid with negative effects on the neuronal activity (3, 19). It has long been known that glutamine is a better precursor than glutamate itself for GABA (19, 23). It is due to compartmentalization, neuronal differentiation and genetic regulation of enzyme controlling mechanisms.

It has been shown that the synaptosomal PAG activation by Ca^{2+} could be determined by transmitter release from nerve endings (2). Therefore, the agonists of GABA and glutamate receptors have to promote the activation of PAG by the same mechanism. In this plan we use some of possible endogenous modulators of glutamate and GABA ergic activities, ethanolamine and its O-sulfoether (EOS). The latter is the inhibitor of GABA-T as well (6,7).

There are some data with regard to the inhibition of glucose utilization by EOS in brain slices and the preferential working of glutamine and glutamate as effective energy sources (19).

Since PAG is located in the internal mitochondrial membrane (11, 21) it participates in the mitochondrial reactions, particularly in oxidative phosphorylation. It has been suggested that factors, causing an increase in liver mitochondrial matrix volume may also cause significant activation of the oxidation of fatty acids, respiration and ATP production, pyruvate carboxylation, citrulline synthesis and glutamine hydrolysis (8).

In this paper we have studied the influence of some intermediates of Krebs cycle, phosphatides components and Ca^{2+} on the glutamine hydrolysis by mitochondrial and synaptosomal fractions of rat and pig brains.

2. EXPERIMENTAL PROCEDURE

Materials: 2 month old Wistar rats were used in our experiments. In some experiments frozen pig brains were used. Reagents: Commercial ethanolamine (97%) was distilled in vacuum (12 mm mercury) and used in experiments and for EOS synthesis (13). Other reagents were obtained from Sigma Chemical Company. The mitochondrial fraction of brain was prepared as described by Brody and Bain (4). The synaptosomal fraction was obtained by sucrose gradient fractionation according to Whittaker and Barker (22).

The PAG activities of the mitochondrial and synaptosomal preparations were assayed by measuring ammonia formation after incubation with glutamate dehydrogenase for 15 min at 37°C and pH 8.0 (12). The reaction was stopped by addition of 10% TCA (final concentration 2%), followed by centrifugation (11000 g, 5 min).

Respiration of isolated mitochondria was measured polarografically with oxygen electrode. Mitochondria were isolated and suspended in a medium containing 130 mM KCl, 5 mM K_2HPO_4, 0,1 mM EGTA and 10 mM Tris-HCL, pH 7,6. At investigation of Ca^{2+} influence 10 μM $CaCl_2$ was also present in the medium and 1 μM EGTA was added instead of 0,1 mM. State 3 respiration was initiated by addition of 0,2 mM ADP. Mitochondrial protein content is 1,0–1,5 mg/ml, t = 37°C. The assay mixture contained mitochondrial or synaptosomal fraction (usually 2–3 mg protein per ml), 5 or 10 mM Hepes and other compounds indicated in the table legends. Protein was determined by the method of Lowry (15). Statistical significance was assessed by Student's test.

3. RESULTS

Table 1 demonstrates the results of the determination of ammonia formation after incubation of pig brain mitochondria with glutamine and known potential PAG effectors.

Table 1. Glutamine hydrolysis by pig brain mitochondria

Additions	Ammonia production μmole/mg of protein
P_i	4.1 ± 0.35
Succinate	-
ADP	3.0 ± 0.27
P_i + succinate	-
P_i + ADP + succinate	2.9 ± 0.26
α - oxoglutarate	1.30 ± 0.10
EOS	-
Ethanolamine	-
ADP + P_i	15.90 ± 1.20
ADP + succinate	0.70 ± 0.04
ADP + α - oxoglutarate	6.10 ± 0.45
P_i + α - oxoglutarate	7.00 ± 0.80
ADP + α - oxoglutarate + P_i	9.00 ± 0.85
P_i + EOS	7.80 ± 0.70
P_i + ethanolamine	3.90 ± 0.27
ADP + P_i + α - oxoglutarate + succinate	1.70 ± 0.03

The mitochondria were incubated for 15 min at 37°C and pH 8.0 (0.15 M Tris-HCl buffer) with 10 mM L-glutamine. To several samples were added 10 mM K_2HPO_4. succinate. α-oxoglutarate. ethanolamine and/or EOS. ADP addition was 0.2 mM. Mean of 5 experiments.

The results indicate that glutamine doesn't produce any ammonia in absence of inorganic phosphate or ADP. In our experimental conditions ADP itself didn't form ammonia, however, in combination with phosphate it stimulated hydrolysis of glutamine. Succinate which is usually regarded as a PAG activator, in our experiments inhibited ammonia production from glutamine in the presence of phosphate as well as ADP. Their combined addition relaxed the effect of the succinate. α-Oxoglutarate addition to the incubation mixture increased the ammonia formation from glutamine. This effect was protected by succinate. Basing ourselves on data about GABA generation (5, 14) we used EOS and ethanolamine in our experiments. EOS activated glutamine hydrolysis in the presence of phosphate in the incubation mixture. However, in vivo it doesn't influence the rat brain mitochondrial glutaminase (Table 2). In these experiments EOS was administered to rats (1g/kg weight) 24 hours before decapitation. In our previous experiments the maximal GABA increase was demonstrated under the same condition (10). At the same time EOS administration increases PAG activity in synaptosomal fraction by 58%. These data indicate the different functional role of two forms of glutaminase, although they both have the same mitochondrial origin. We suggest that it applied to GABA ergic nerve endings only.

In vitro (see Table 2) we observed also rat brain synaptosomal glutaminase activation by EOS as wel as Ca^{2+}. The mitochondrial enzyme was unaffected by EOS, but it reacted to Ca^{2+} in different ways depending on concentration. A concentration of 0,5 mM of Ca^{2+} inhibited glutamine hydrolysis in mitochondria, whereas 0,05 mM Ca^{2+} activated this process.

The obtained data were compared with the results of other investigations (17) and demonstrated the analogous effect of micromolar concentrations of Ca^{2+} on ATP synthesis.

The study of the respiration of brain mitochondria with succinate and glutamine as respiratory substrates has demonstrated that the micromolar Ca^{2+} concentration increases the rate of respiration in state 3.

This fact supports the opinion about the coupling of the PAG activity with the oxidative phosphorylating processes (9).

Table 3 demonstrates that after preincubation with the glutamine (10 min) and following washing of the fresh brain mitochondria part of the glutamine (about 30%) penetrates into these particles, since phosphate addition provokes ammonia production. SH-reagent N-ethylmaleimide (NEM) addition inhibits the hydrolysis of both glutamine that was added to mitochondria as well as that which penetrated into it. The differences in the ammonia production between added and penetrated glutamine are more or less related to the differences in substrate concentration. It is necessary to note that the inhibitory effect of succinate on pig PAG was not observed for the rat brain enzyme (see Table 3).

Table 2. The influence of EOS intraperitoneal administration on glutamine hydrolysis by fresh rat brain mitochondria and synaptosomes

	Ammonia production μmoles/mg protein			
	Mitochondria		Synaptomes	
Addition	Control	EOS	Control	EOS
—	2.73 ± 0.25	3.25 ± 0.28	1.20 ± 0.05	1.90 ± 0.06
P_i (10mM)	9.79 ± 0.88	10.44 ± 0.75	4.40 ± 0.28	7.20 ± 0.45
P_i + Ca^{2+} (0.5mM)	6.40 ± 0.55	10.25 ± 0.60	6.60 ± 0.35	7.00 ± 0.65
P_i + Ca^{2+} (0.05mM)	12.80 ± 0.74	12.50 ± 0.65	4.70 ± 0.30	5.20 ± 0.35

All of the used reagents (succinate, ADP, GABA, EOS) did not affect the glutamine hydrolysis in the presence of phosphate (10 mM).

4. DISCUSSION

The brain PAG is presented in two forms. Both forms have the mitochondrial origin, but they show different kinetic and regulatory properties (18). The soluble form of PAG is a fragment of the membrane bound enzyme (21). Torgner et al (21) showed that its kinetic properties in immobilization of the bearer approach those of membrane bound enzyme.

The synaptosomal PAG is also a mitochondrial enzyme, but its function, probably, directly adapted to the nerve transmission. This fact influences enzyme operation and is responsable for the differences of the PAG in cell bodies and nerve endings mitochondria. In our in vivo experiments GABA-ergic agonist and inhibitor of GABA-T EOS activated the synaptosomal PAG, but did not activate the mitochondrial enzyme. The addition of Ca^{2+} to synaptosomes in presence of phosphate activates glutamine hydrolysis, while in mitochondria the same Ca^{2+} concentrations inhibits mitochondrial PAG.

It is known that key regulatory enzymes of the mitochondrial oxidative metabolism of heart, liver and mesenteric lymphocytes (pyruvate, NADH-isocitrate and 2-oxoglutarate dehydrogenases) can be activated by micromolar concentration of Ca^{2+} (1, 16). Our data of oxygen consumption in respiratory state 3 has shown that micromolar Ca^{2+} concentrations increase the rate of glutamine and succinate oxidation by brain mitochondria by 38 and 70% respectively. Therefore, the Ca^{2+} increased simultaneously PAG activity and glutamine oxidation in brain mitochondria. We suggest that ammonia production by PAG leads to local pH changes and modulation of H^+-pump. The mechanism of Ca^{2+} action on PAG needs further elucidation. In the nerve endings its effect may be due to stimulation of transmitter release mechanism, and in cellular mitochondrial fraction it may sooner be connected with mitochondrial energetic processes.

Table 3. Glutamine hydrolysis by rat brain mitochondria

Additions	Ammonia production μmol/mg protein	Additions	Preincubation with glutamine (10 min) then washing mitochondria – Ammonia production μmol/mg protein
ADP	4.51 ± 0.22	P_i	0.93 ± 0.05
P_i	5.14 ± 0.47	N-ethylmaleimide	0.14 ± 0.01
Succinate	1.37 ± 0.09	glutamine + P_i	4.49 ± 0.37
α-oxoglutarate	2.74 ± 0.15	glutamine + P_i + N-ethylmaleimide	0.81 ± 0.05
Ethanolamine	1.06 ± 0.04		
P_i + ADP	7.40 ± 0.65		
ADP + succinate	4.11 ± 0.53		
P_i + succinate	6.17 ± 0.50		
ADP + P_i + succinate	6.17 ± 0.55		
ADP + α-oxoglutarate	7.20 ± 0.73		
P_i + α-oxoglutarate	9.05 ± 0.77		
ADP + P_i + α-oxoglutarate			
P_i + EOS	6.63 ± 0.55		
P_i + succinate + α-oxoglutarate	9.66 ± 0.75		

REFERENCES

1. Baumgarten, E., Brand, M.D., Pozzan, T., 1983, *Biochem J.* **216**: 359–367.
2. Benjamin, A.M., 1981, *Brain Res.* **208**: 363–377.
3. Bradford, H.F., Ward, H.K., Thomas, A.J., 1978, *J. Neurochem.* **30**: 1453–1459.
4. Brody, T.M., Bain J.A., 1952, *J. Biol. Chem.* **195**: 685–693.
5. Fletcher, A., Flowler, L.J., 1980, *Biochem. Pharmacol.* **29**: 1451–1454.
6. Fowler, L.J., 1973, *J. Neurochem.* **21**: 437–440.
7. Fowler, L.J., John, R.A., 1972, *Biochem. J.* **130**: 569–573.
8. Halestrap, A.P., 1989, *BBA* **973**: 355–382.
9. Kovacevic, Z., McGivan, J.D., 1983, *Physiol. Rev.* **63**:547–605.
10. Krimyan, A.Yu., Gulbayazyan, T.A., Aracelyan, L.M., Kamalyan, A.R., Kamalyan, R.G., 1992, *Neurokhimia* **11**: 65–71.
11. Kwamme, E., Torgner, I.Aa., Roberg, B., 1991, *J. Biol. Chem.* **266**: 13185–13192.
12. Kwamme, E., Tveit, B., Svenneby, G., 1970, *J. Biol. Chem.* **245**: 1871–1877.
13. Lloyd, A.G., Tudball, N., Dodgson, K.S., 1961, *BBA* **52**: 413–419.
14. Loscher, W., 1981, *J. Neurochem.* **36**: 1521–1527.
15. Lowry, O.H., Rosebrough, N.J., Farr, A.L., Randall, R.J., 1951, *J. Biol. Chem.* **193**: 263–275.
16. McCormac, J.G., Browne, H.M., Dawes, N.J., 1989, *BBA* **973**: 420–427.
17. Murphy, A.N., Kelleher, J.K., Fiskum, G., 1990, *J. Biol. Chem.* **265**: 10527–10534.
18. Nimmo, G.A., Tipton, K.F., 1979, *J. Neurochem.* **33**: 1083–1092.
19. Nobrega, J.N., Coscina, D.V., 1983, *Pharmacol. Biochem. and Biohav.* **262**: 243–252.
20. Reubi, J.C., Van den Berg, C.J., Cuenod, M., 1978, *Neurosci. Lett.* **10**: 171–174.
21. Torgner, I.A., Roberg, B., Kwamme, E., 1995, *J. Neurochem.* **65**: S158A.
22. Whittaker, V.P., Barker, L.A., 1972, *Methods of Neurochem.* **2**: 1–52.
23. Yu, A.C.H., Hertz, I., Hertz, L., 1984, *J. Neurochem.* **42**: 951–960.

Section 38: Aminoacid Neurotransmitters

EFFECT OF GABA-ERGIC SUBSTANCES ON CEREBRAL BLOOD FLOW AND INTRACELLULAR Ca^{2+} IN HYPOKINESIA

V. P. Hakopian, K. V. Melkonian, G. A. Kevorgian, and A. S. Kanayan

Department of Pharmacology
Yerevan State Medical University
Yerevan, Republic of Armenia

1. INTRODUCTION

Successful adaptation to daily physical and psychological demands (stressors) begins in the brain which coordinates the physiological response useful for coping with an acute emergency. However, when the stress response continues unabated, it can be damaging not only for the body, but also for the brain (9,17). Prolonged and repeated stress plays a significant role in the development of cerebrovascular disorders (12). Hypokinesia (HK) as well as stress factor is considered an important clinical problem favoring the premature ageing and cerebrovascular diseases (11). Consequently, the necessity of solving the problem of the cerebral hemodynamic and neurochemical reorganization's problem obtains priority value in conditions of HK (3,13).

Proceeding from the unique compensatory role of the endogenous γ-aminobutyric acid (GABA) system as the most important link of neurochemical mechanisms regulating cerebral blood circulation (15), we deemed it necessary to carry out investigations to discover the influence of GABA-ergic substances on the observed changes.

As a result of our investigations, it has been discovered, that in hypokinetic conditions significant cerebral hemodynamic and metabolic changes are developing, which promote the development of structural and functional disturbances in brain. Our results show that GABA-agonists manifest a clearly antihypoxic and cerebroprotective effect, especially, in early stages of HK.

2. MATERIALS AND METHODS

Male mature white rats were used weighing 160–180 g. HK was modeled in individual small cages (7). The experiments were carried out in both early (from the 15th till the 60th day) and late (from day 60 et seq.) periods of HK.

In order to study the parameters characterizing cerebral blood supply, changes were induced in the morpho-functional state of the capillary system of brain cortex by adenosine-triphosphate (ATP) method without injection so as to discover the intraorganic microcirculatory channels (6). Two main parameters were studied describing the functional capacity of the microcirculation of brain cortex: the mean diameter (MD) of functioning capillaries and the number of sharply narrowed capillaries (SNC) in 100 visible areas.

The modern electromicroscopic methods were used in order to carry out pathomorphological investigations in cortical and subcortical structures of the brain.

The rate of influx of the labeled [^{45}Ca]CaCl$_2$ into neurocytes of the brain and its translocation between subcellular fractions was investigated by well-known radioisotopic methods (5). Radioactivity was counted with liquid scintillation counter SL-4221 (Roche Bioelectronique Kontron, France).

During the study GABA (San Diego, California), Nootropil (piracetam) (Polfa, Starograd), bicuculline (Sigma Chemicals Co, USA) have been used. They were administered intraperitoneally for one week in doses of 2mg/kg, 20mg/kg, 0,2mg/kg, respectively.

The results of investigations were processed by methods of variation statistics.

3. RESULTS

The obtained results favoured the idea of phasing of changes. There were significant differences between data received in early and late periods of HK. As demonstrated in Fig. 1, in early HK the number of SNC sharply increases approximately 1.5 times and the MD FC, as shown in table 1, decreases 17.5% compared to control. There is a tendency

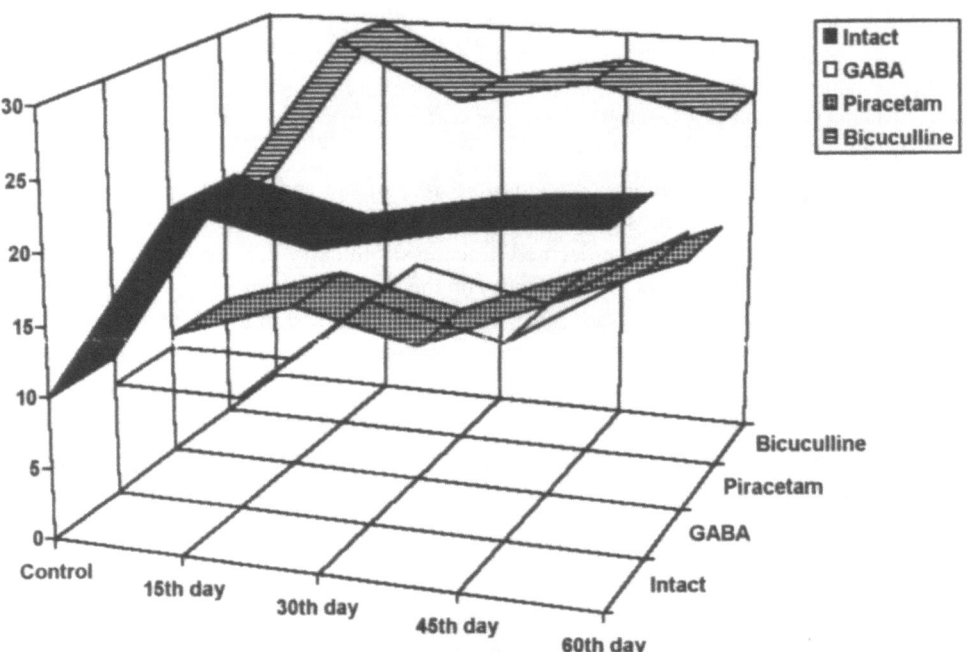

Figure 1. The number of sharply narrowed capillaries in brain cortex in 100 visible areas in conditions of hypokinesia HK and under the influence of GABA-ergic substances (M ± m), n=57.

Table 1. Middle diameter of functioning capillaries of brain cortex in conditions of hypokinesia and under the influence of GABA-ergic substances (μm) (P<0,05)

Condition	Control	Hypokinesia			
		15th day	30th day	45th day	60th day
Intact	5.7 ± 0.4	5.15 ± 0.9	5.32 ± 0.7	5.19 ± 0.8	5.096 ± 0.3
GABA	8.7 ± 1.1	7.27 ± 1.02	5.13 ± 0.83*	6.2 ± 0.91	5.17 ± 0.7*
Piracetam	7.2 ± 0.5	5.31 ± 0.9*	6.2 ± 0.94*	5.81 ± 0.8*	5.76 ± 0.87*
Bicucullin	4.7 ± 0.36	4.0 ± 0.9*	4.79 ± 0.8	4.2 ± 0.8	4.12 ± 1.0*

from the 30th day of HK to normalization of the functional capability of microcirculatory channels. Late stage HK was accompanied by sudden worsening of the functional capability of the brain cortex, which was displayed as a significant increase of the number of SAC and decrease of MD FMC.

Besides, during the experiment it was established that the morphologic picture of the capillary system changed significantly. Deformed capillaries were registered which obtained shoe-like forms with peripheral contractions and regress was registered in capillary channels.

In control conditions the vasodilating effect of GABA was more pronounced than that of piracetam. GABA and piracetam interact identically on the functional state of the brain cortex. By analyzing results of the influence of GABA and piracetam on the capillary channels of brain cortex, it was seen, that these two substances display an unidirectory influence, i.e., MDFC increases under the intraperitoneal administration of GABA and piracetam in control rats in 52.6% and 26.3% (P<0.05), and the number of SNC decreases 20% and 10%, respectively (P<0.05).

We afterwards attempted to detect the influence of the specific antagonist of GABA receptors — bicuculline — on MDFC and the number of SNC of the brain cortex. It was discovered that bicuculline in control rats leads to vasoconstriction, and therefore to an increase in the number of SNC in 100 visible areas by 17.5±2.0% and a 1.7 fold decrease in MDFC (P<0.05).

Significant changes were observed in the microcirculatory system on the 30th and 60th days of HK under the influence of GABA, piracetam and bicuculline, as shown in Fig. 1 and Table 1.

On the other hand, it has been established that HK leads to development of morphologic changes in cerebral tissue. These changes are observed from the 15th day of HK, and become more progressive during the experiment. During the experiment peripheral chromatolysis and later (on 45th day) neuronophagia were observed, which were accompanied by the appearance of glial nodules in cerebral tissue. The hyperemia of vessels and edema of cerebral membranes gradually increase. On the 45th day of HK areas of hemorrhage in cerebral membranes, brain cortex and cerebral ventricles, as well as intravascular thrombosis are detected.

In early periods of HK bicuculline was shown to posses greater cerebroprotective activity than other GABA-ergic substances. Following the administration of GABA peripheral chromatolysis was observed in neurocytes of terminal slices of cortex. The application of piracetam worsened the hemodynamic and cellular structural state during late periods of HK, which was expressed in diffuse formation of glial nodules, areas without neurocytes and vascular abnormalities. Diffuse vascular hyperemia, edema of cerebral membranes and hemorrhagic foci in cerebral tissue and ventricles were also observed. However, the use of bicuculline from the 45th day of HK displayed significantly less protective effect.

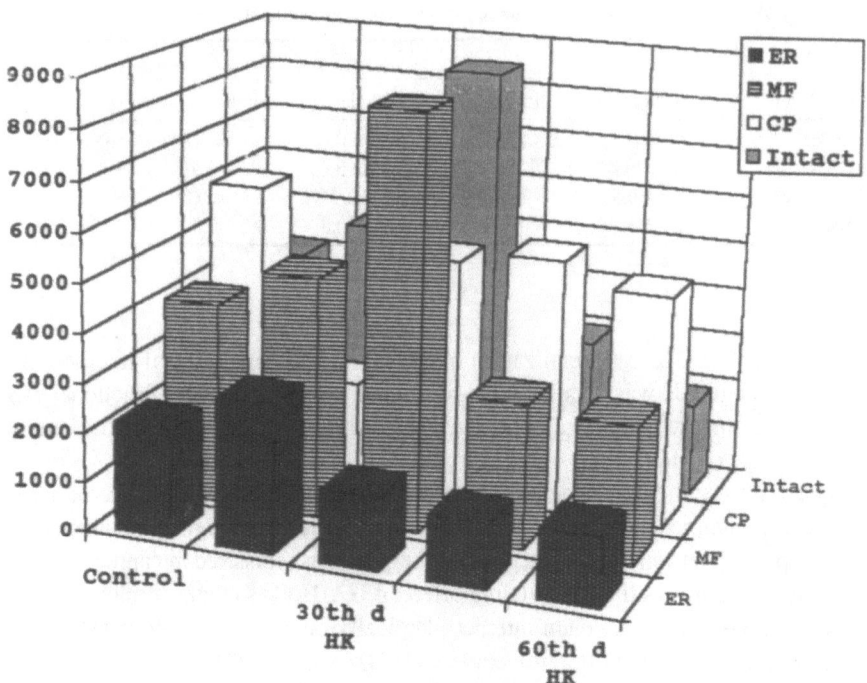

Figure 2. Translocation of $^{45}Ca^{++}$ between subcellular fractions in brain neurocytes in conditions of HK. NOTE: DPM — disintegration of the [^{45}Ca]CaCl$_2$ per minute in 1 mg fresh tissue; MF — mitochondrial fraction; CP — cytoplasm; ER — endoplasmic reticulum.

The received data testify to a significant rise of the intracellular (i.c.) Ca^{++} level in neurocytes, which during investigation increased most in the mitochondrial fraction (MF) of cells. So, brain MF sharply intensifies the Ca^{++} accumulation on 30–45 days. The Ca^{++} level tends to go down gradually during HK. The reaction of cytoplasmic fraction (CP) of the brain is expressed with initial lowering of the entrance rate in the brain to a third of its earlier value on 15th day with tendency to normalization on 45th day (Fig.2). In early periods of HK, under the influence of bicuculline more expressively decreases the level of the i.c. Ca^{++} decreases more expressively, simultaneously promoting the output of an increasing amount of labeled $^{45}Ca^{++}$ from MF.

4. DISCUSSION

These studies make it possible to conclude that locomotor hypoactivity is able to increase the incidence of hemodynamic, morphologic and metabolic damages in both cerebral blood vessels and cerebral tissue.

Furthermore, correlation has been established between the periods of HK and its damaging action on the brain. Thus, the results obtained in early HK show the probability of the increase in activity of the hypothalamic-pituitary-adrenal system as this condition acts as an extreme stress factor (7,12) which is reflected in the worsening of the functional capability of microcirculatory state in the brain, promoting the development of structural changes in cerebral tissue and neurochemical disorders. Especially, the obtained data suggest that in conditions of HK there are sudden rises of the i.c. Ca^{++}, which play a keyrole in pathogenesis of the cerebro-vascular disturbances, since it's established, that the balanced entrance of Ca^{++} into cells is considered one of the basic mechanisms of the vascular and metabolic disharmony regulation, which is in accordance with modern concepts (8,16).

The evident pathomorphological changes are involved in pathophysiologic mechanisms of hemodynamic disorders, increasing the risk of cerebral discirculation in conditions of HK (14).

It is possible that in the tendency to normalization of the investigated parameters on days 30–45th of HK, adaptive reorganization occurs in the organism, in which the GABA system takes a great part. This is expressed by increased activity of glutamate-decarboxylase due to stress interaction, which leads to an increase of GABA levels, which in turn suppresses the activity of the sympathetic system (1).

Our results show, that in conditions of early HK agonists of GABA receptors manifest a clear antihypoxic and cerebroprotective effect, leading to improvement of the microcirculatory state of the brain and promoting the outflow of the increased level of i.c. Ca^{++} from MF, thus preventing the development of subsequent brain damage. Moreover, our results may provide information on cerebral blood flow and metabolism that can be useful in the prophylaxis and treatment of the negative changes developing from HK.

Possibly, the observed effects are immediated via GABA receptors located in the wall of cerebral vessels. GABA and piracetam express vasodilating activity in conditions of HK, which leads to an increase of vascular permeability. The above mentioned effects of piracetam confirm the results (2), according to which piracetam usually leads to an increase in vasoconstrictive prostaglandins (PGE, $PGF_{2\alpha}$), which, however, don't fully reverse the vasodilatation caused by piracetam itself. Analysis of the possible mechanisms of piracetam in conditions of HK suggests that the level of GABA in cerebral tissue increases in conditions of cerebral hemodynamic abnormalities under the influence of piracetam. On the other hand, the cycling of GABA in brain tissue with formation of the pyrrolidonic ring, intensifies the vascular effect (15). It was shown, that the functional capability of the microcirculatory system of brain cortex becomes worse in conditions of hypokinesia, which is accompanied by the development of ischemia, causing acidosis. Piracetam, a cyclic analogue of GABA, contains a pyrrolidonic ring and represents a GABA prototype which is split in conditions of acid pH. Under the influence of piracetam, cyclic AMP accumulates and its level increases in hypokinetic conditions (18).

It has been discovered, that in conditions of HK piracetam prevails over GABA by vasoactive effects. This activity of piracetam in late periods of HK is not a compensatory reaction, since vasodilatation intensifies more and more, and is accompanied by increasing penetrability of the vessel wall, favoring the development of perivascular edema and hemorrhagic and necrotic nidi in the brain. These results make it worthwhile to consider the use of piracetam in clinic as a psychotropic agent even to the contraindicating of patients in prolonged confinement and together with other vasodilating agents with central influence.

Bicuculline, as a selective antagonist of GABA receptors, leads to the elimination of cerebral vasodilatation, thus preventing the development of subsequent cell damage. Simultaneously, in early HK bicuculline manifests a clear cerebroprotective effect, expressively promoting the outflow of the increased level of Ca^{++} from MF,CF and enoplasmatic rethiculum (ER) displaying a Ca^{++}-antagonisitic effect thus preventing the development of subsequent cell damage. In this view GABA and piracetam significantly yield to bicuculline, which leads to regarding GABA-ergic substances as a prospectively useful group of chemical compounds for finding new effective correctors of cerebrovascular disturbances in conditions of early HK.

The obtained results corroborate the participation of the GABA-system as an endogenous modulator in adaptive changes of cerebral hemodynamic disturbances in conditions of HK. Earlier it has been demonstrated, that the GABA system seems to participate in an inhibitory manner in the regulation of corticotropin-releasing hormone (4), which is postulated to coordinate the endocrine, metabolic and behavioral responses to stress (10).

In conclusion, the received data prove scientifically the neurogenous conception of the compensatory regulation of cerebral blood circulation, which modulates neurotransmission and intensifies metabolism in cerebral tissue. Thus, these data obtain a potential therapeutic value.

REFERENCES

1. Acosta, G.B., et al., 1990, *Gen. Pharmacol.*, **21**: 517–520.
2. Bhattacharya, S.K., et al., 1989, *Ind. J. Exp. Biol.* **27**: 261–264.
3. Brierley, E.J., Johnson, M.A., James, O.F., Turnbull, D.M., 1996, *Quart J. Med*, **89**: 251–258.
4. Calogero, A.E., 1995, *Stress. Basic mechanisms and clinical implications*, New York, **771**: 31–41.
5. Chernuch, A.M., Copteva, L.A., 1977, *J. Meditsina*, Moscow, **5**: 329–338.
6. Chilingarian, A.M., 1977, *J. Experim. & Clin. Medicine*, Yerevan, **5**: 19–28.
7. Fedorov, I.V., 1982, *Metabolism in conditions of hypodynamia*, Moscow.
8. Galoyan, A.A., Kevorkian, G.A., Voskanian, L.H., et al., 1988, *Neurochem.Res.*, **13**: 493–498.
9. Hakopian, V.P., Melkonian, K.V., Kocharian, A.G., Mirzoyan, N.R., 1995, *J. Angiology*, **37**: 89–90.
10. Koob, G.F., 1985, *Perspectives on Behavioral Medicine*, (R.B. Williams, ed) Acad. Press, New York, Ed. **2**: 39–52.
11. Mabry, Th.R., Gold, P.E., McCarty, R., 1995, *Stress. Basic mechanisms and clinical implications*, New York, **771**: 512–523.
12. Macho, et al., 1980, *Catecholamines and stress: Recent advances*, N.Y., PP. 399–408.
13. Melkonian, K.V., Hakopian, V.P., 1995, *4th. Intern.Conf. on Endothelin*, London, UK, PP 113.
14. Melkonian, K.V., Hakopian, V.P., Kanayan, A.S., 1995, *J. Pharmacological Research*, Italy.
15. Mirzoyan, S.A, Hakopian, V.P., 1985, *GABA and Cerebral Hemocirculation*, Edited by the Scientific Academy of Armenia, Yerevan, PP 17–38.
16. Siesjo Bo, K., 1994, *Can. J. Physiol. Pharmacol.*, **72**: (Suppl. 1) PP. 37.
17. Smith, M.A., Makino, Sh., Kvetnansky, R., Post, R.M., 1995, Effects of Stress on Neurotrophic Factor Expression in the Rat Brain. In: *Stress. Basic mechanisms and clinical implications*, New York, **771**: 234–240.
18. Weth, G.M.D., 1981, 1981, *2nd Intern. Symp. on Nootropic Drugs*, Mexico, PP 34–45.

Section 39: Neuropeptides

FUNCTIONAL EVALUATION OF THE BENZODIAZEPINE CHOLECYSTOKININ TYPE-B RECEPTOR ANTAGONISTS L-365,260, L-740,093, AND YM022

John Dunlop,* Neil Brammer, Non Evans, Ian Pass, and Chris Ennis

Wyeth Research (UK) Ltd
Maidenhead, Berks, United Kingdom

1. INTRODUCTION

Two identified receptor subtypes, designated type-A and type-B (1), mediate the physiological actions of the brain/gut peptide cholecystokinin (CCK). In the central nervous system the CCK-B receptor represents the predominant receptor population and a central role for CCK in mediating neurotransmission and/or neuromodulation has been proposed (2). Additionally, the CCK-B receptor has been implicated in a number of psychiatric disorders including panic attacks (3) and anxiety (4). Consequently, this receptor subtype has received considerable attention as a therapeutic target with a number of selective, high-affinity ligands now available. In particular, a number of molecules incorporating a benzodiazepine template have been developed as selective high-affinity CCK-B receptor antagonists. Examples of these are L-365,260 (5), L-740,093 (6) and YM022 (7) which represent three structurally similar (Fig. 1) members of this class.

In this study a functional evaluation of these molecules was undertaken employing a Chinese hamster ovary cell line transfected with the human CCK-B receptor gene (hCCK-B. CHO (8)). The mobilisation of intracellular Ca^{2+} in hCCK-B.CHO cells in response to the CCK-B receptor agonist CCK-4 was used as a functional assay. Additionally, recent experiments have been undertaken measuring agonist-induced extracellular acidification in hCCK-B.CHO cells with the Cytosensor™ microphysiometer (9), in order to further characterise the antagonist activity of the compounds. The results reveal a clear distinction between the antagonist activity of these structurally related molecules. Thus, whilst L-365,260 exhibits properties consistent with competitive antagonism in the Ca^{2+}-mobilisation assay, both L-740,093 and YM022 behave in a manner consistent with

---

* Present address: Wyeth-Ayerst Research, Princeton, New Jersey.

Neurochemistry, edited by Teelken and Korf
Plenum Press, New York, 1997

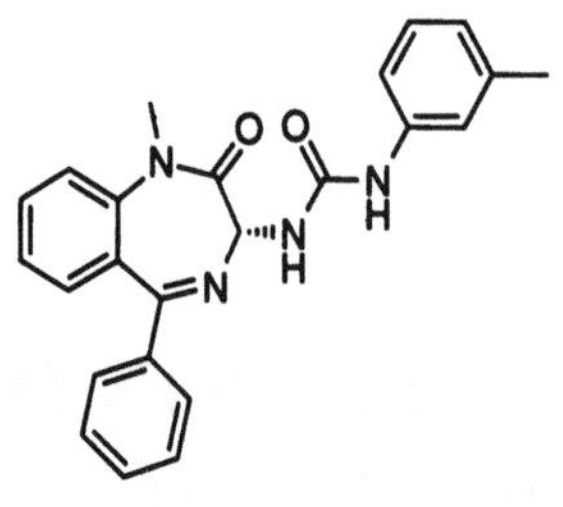

L-365,260

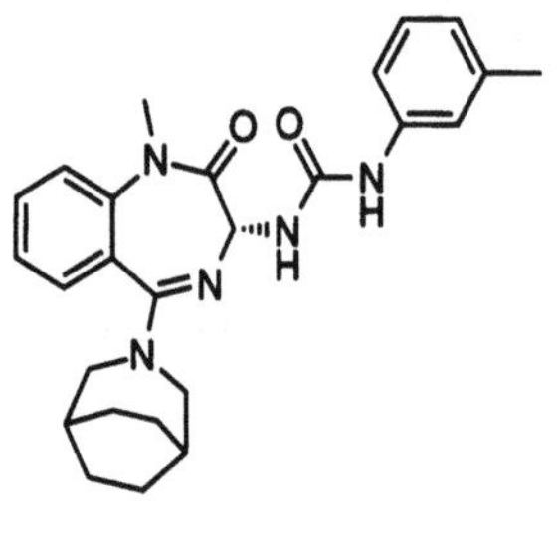

L-740,093

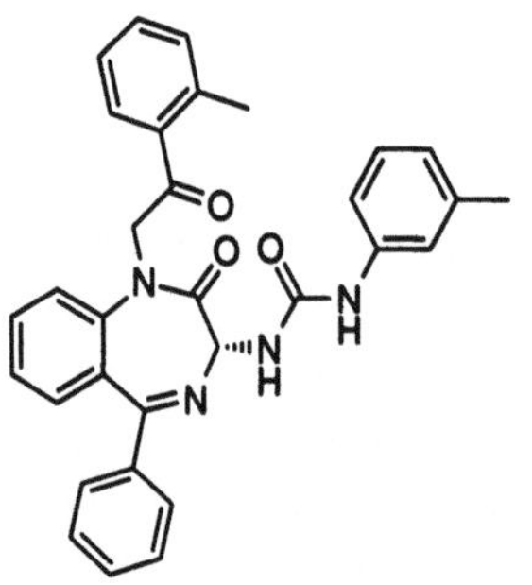

YM-022

Figure 1. Antagonist of cholecysto-kinin-B receptors.

non-competitive antagonism. The recent studies employing microphysiometry have further demonstrated the irreversible antagonist activity of YM022, thus accounting for the apparent non-competitive nature of antagonism displayed by this compound in the Ca^{2+}-mobilisation assay.

Table 1. CCK receptor binding affinities estimated from [^{125}I]-CCK-8S displacement curves. Values expressed as nM. *Taken from ref. 6

compound	CCK-B affinity	CCK-A affinity	selectivity
L-365,260	12.6	589	47-fold
L-740,093	0.49	1600*	3000-fold
YM022	0.05	316	6000-fold

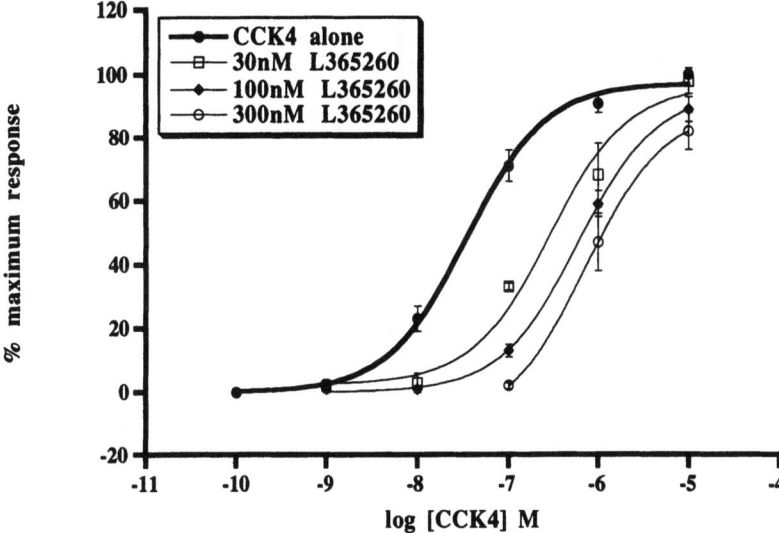

Figure 2. Effect of L-365,260 on CCK-4 stimulated Ca^{2+}-mobilisation in hCCK-B.CHO cells.

2. METHODS

Receptor binding affinities were estimated from [^{125}I]-CCK-8S displacement curves employing membranes derived from hCCK-B.CHO and hCCK-A.CHO cells. Measurements of agonist stimulated Ca^{2+}-mobilisation were performed with hCCK-B.CHO cells loaded with the Ca^{2+}-sensitive fluorescent indicator FURA-2 (8). The microphysiometer experiments were performed as described elsewhere (9).

3. RESULTS

CCK receptor binding affinities for L-365,260, L-740,093 and YM022 are presented in Table 1. Figure 2 illustrates the effect of increasing concentrations of L-365,260 on the CCK-4 stimulated Ca^{2+}-mobilisation in hCCK-B.CHO cells. In the absence of antagonist the estimated EC_{50} for CCK-4 was 36 nM, whilst in the presence of L-365,260 the following apparent EC_{50} values were estimated; 295 nM (30 nM L-365,260), 603 nM (100 nM L-365,260) and 741 nM (300 nM L-365,260).

Figure 3 illustrates the effect of increasing concentrations of L-740,093 on CCK-4 stimulated Ca^{2+}-mobilisation. In the presence of 30 nM L-740,093 the estimated EC_{50} for CCK-4 was 316 nM and the maximum response to CCK-4 was reduced to 55 ± 6 % of control. Figure 4 illustrates the effect of 30 nM YM022 on CCK-4 stimulated Ca^{2+}-mobilisation. In the presence of YM022 the estimated EC_{50} for CCK-4 was 151 nM and the maximum response to CCK-4 was reduced to 48 ± 11 % of control. The experiment presented in figure 5 demonstrates the effect of 30 nM L-365,260 (a) and 30 nM YM022 (b) on the stimulation of extracellular acidification in hCCK-B.CHO cells in response to 30 nM CCK-4.

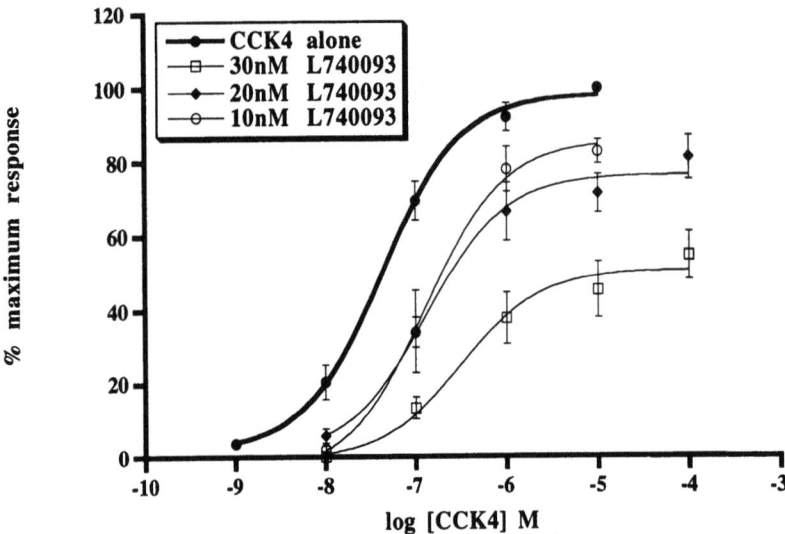

Figure 3. Effect of L-740,093 on CCK-4 stimulated Ca^{2+}-mobilisation in hCCK-B.CHO cells.

4. DISCUSSION

The CCK-B receptor binding affinities together with the rank order of affinity (YM022 > L-740,093 > L-365,260) presented in Table 1 are in good agreement with the data available in the literature (5–7). Additionally each molecule exhibits selectivity for the CCK-B receptor subtype although the degree of selectivity varies considerably.

Functional antagonism for these molecules was confirmed by their ability to antagonise the CCK-4 stimulated mobilisation of intracellular Ca^{2+} in hCCK-B.CHO cells. Using this approach the nature of the antagonist activity exhibited by these molecules was found

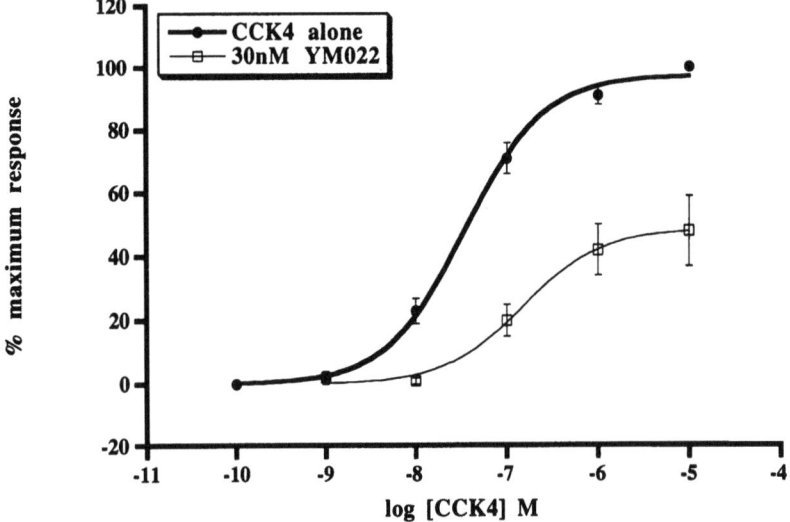

Figure 4. Effect of YM022 on CCK-4 stimulated Ca^{2+}-mobilisation in hCCK-B.CHO cells.

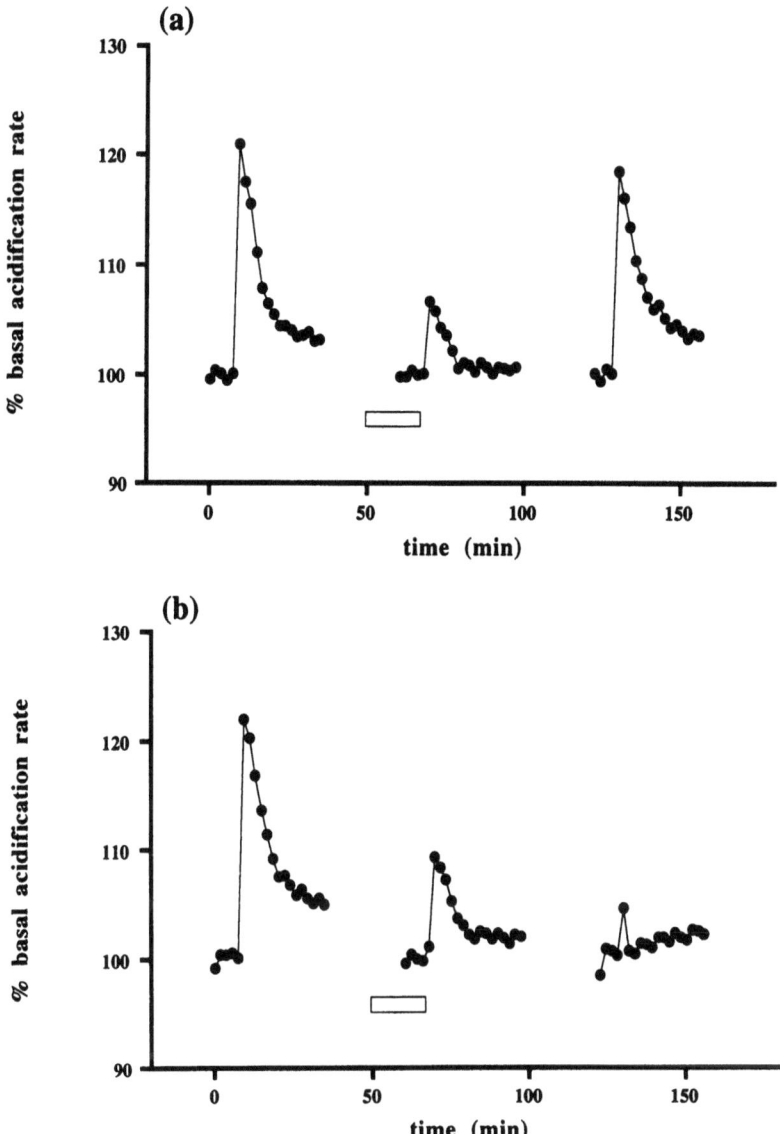

Figure 5. Effect of L-365,260 (a) and YM022 (b) on the stimulation of extracellular acidification in hCCK-B.CHO cells in response to 30 nM CCK-4. In both cases three consecutive CCK-4 stimulations were introduced for 30 s duration. The bar indicates the presence of antagonist which was present 15 min prior to, and throughout, the second stimulation.

to be different (Figs. 2–4). Thus, L-365,260 exhibits properties consistent with a competitive antagonist, producing a rightward shift in the dose response curve to CCK-4 with no reduction in the maximum effect. In contrast, both L-740,093 and YM022 behave in a manner consistent with non-competitive antagonism, reducing the maximum response to CCK-4 to 55 and 48 % control, respectively. This was accompanied by a rightward shift in the dose response curve to CCK-4, although to a lesser extent than that observed in the presence of L-365,260.

Recent experiments employing the technique of microphysiometry have demonstrated the irreversible antagonist activity of YM022 (Fig. 5). Thus, following the removal of YM022 and subsequent stimulation with CCK-4 the agonist effect was not restored. This observation would account for the non-competitive profile exhibited by YM022 in the Ca^{2+}-mobilisation assay. In contrast, the antagonist effect of L-365,260 in the microphysiometer experiment was fully reversible, consistent with its competitive antagonist profile in the Ca^{2+}-mobilisation assay. It will be of interest to evaluate the effect of L-740,093 in the microphysiometer.

Based on the structural similarity of the molecules studied here (Fig. 1) a similar pharmacological profile at the CCK-B receptor subtype might have been predicted. This is not the case and in particular the introduction of the aromatic ketone group in YM022 markedly influences the dissociation of this molecule from the receptor when compared with L-365,260 which lacks the substituent group. The structural similarity of these two compounds would further suggest a common site of action at the receptor level. Taken together with the competitive antagonism exhibited by L-365,260 this would suggest YM022 is best described as an irreversible competitive antagonist at the CCK-B receptor subtype.

REFERENCES

1. Moran, T.H., Robinson, P.H., Goldrich, M.S. and McHugh, P.R., 1986, *Brain Res.* **362**: 175–179.
2. Williams, J.A., 1982, *Biomed. Res.* **3**: 107–115.
3. Bradwejn, J., Koszycki, D., Couetoux du Terre, A. et al., 1994, *Arch. Gen. Psychiatry* **51**: 486–493.
4. Hughes, J., Boden, P., Costall, B. et al., 1990, *Proc. Natl. Acad. Sci. USA* **87**: 6728–6732.
5. Lotti, V.J. and Chang, R.S.L., 1989, *Eur. J. Pharmacol.* **162**: 273–280.
6. Patel, S., Smith, A.J., Chapman, K.L. et al., 1994, *Mol. Pharmacol.* **46**: 943–948.
7. Saita, Y., Yazawa, H., Honma, Y., Nishida, A., Miyata, K. and Honda, K., 1994, *Eur. J. Pharmacol.* **269**: 249–254.
8. Dunlop, J., Brammer, N. and Ennis, C., 1996, *Neuropeptides*: in press.
9. McConnell, H.M., Owicki, J.C., Parce, J.W. et al., 1992, *Science* **257**: 1906–1912.

Section 40: Brain Proteins

A ROLE OF THE PHOSPHOPROTEINS DYNAMIN AND RABPHILIN-3A IN NEUROTRANSMITTER RELEASE FROM SYNAPTIC VESICLES

Else Marie Fykse

Norwegian Defence Research Establishment
Division for Environmental Toxicology
Kjeller, Norway

1. INTRODUCTION

Neurotransmitters are generally accepted to be stored in synaptic vesicles in the nerve terminal (1). The release from the nerve terminal is triggered by depolarization evoked entry of Ca^{2+}. Neurotransmitter release is known to be modulated by several protein kinases including type II Ca^{2+}/calmodulin-dependent protein kinase (CaMKII), protein kinase C (PKC) and cAMP dependent protein kinase A (PKA) (2). Protein phosphorylation can regulate neurotransmitter release by several different mechanisms. One mechanism could be that protein kinases are phosphorylating ion channels and their associated proteins. Such phosphorylation would result in an increase or decrease in the level of ions like Ca^{2+}, Na^+ or K^+ in the nerve terminal. Another possibility is that proteins involved in synaptic vesicle cycling is phosphorylated and that this will modify the function of synaptic vesicles. The synapsins are a family of phosphoproteins associated with synaptic vesicles important in regulating exocytosis. They are substrates for at least four different kinases; CaMKI and II, PKA and proline directed kinase (3). GAP-43 (B-50) is another abundant phosphoprotein in the nerve terminal plasma membrane, but the role of phosphorylation of GAP-43 in exocytosis is still elusive (4). Munc-18 (n-sec 1) is isolated as a syntaxin binding protein that is essential for docking and/or fusion of synaptic vesicles with the presynaptic plasma membrane (5). Phosphorylation of Munc-18 by PKC decreases the interaction with syntaxin (6). This could be important for regulating docking and /or fusion of synaptic vesicles to the presynaptic plasma membrane.

In this paper, phosphorylation of proteins which are candidates for mediating some regulatory functions in synaptic vesicle cycling are discussed. The proteins selected as examples are dynamin and rabphilin-3A. Dynamin is a peripheral membrane protein thought to be involved in endocytosis, whereas rabphilin-3A is a synaptic vesicle protein probably involved in exocytosis.

2. DYNAMIN

Dynamin is a GTP, microtubule and phospholipid binding protein that was originally identified as a microtubule-motor protein (7). Dynamin is a member of a growing family of GTP binding proteins, and purified dynamin shows a microtubule stimulated and a phospholipid stimulated GTPase activity. In *Drosophila*, the *shibire* gene encodes a homologue of dynamin. A mutation in this gene results in a defect in endocytosis, which indicates a role for dynamin in endocytosis (8,9,10). Dynamin has been shown by sequencing to be identical to dephosphin (11). Dephosphin is a major phosphoprotein that is rapidly dephosphorylated upon depolarization of the nerve terminal (12,13). The dephosphorylation is mediated by the Ca^{2+} and the calmodulin stimulated phosphatase, calcineurin (14), whereas the rephosphorylation of dynamin occurs more slowly and it is mediated by PKC (15).

There are at least three different dynamin genes in mammals, neuronal dynamin (dynamin I), ubiquitous dynamin II and testis dynamin (16,17). The sequence similarity between dynamin I and II are 79 per cent. The most conserved region is the N-terminal part of the protein that binds GTP. The least conserved part is the C-terminal domain. This domain mediates the phospholipid binding and microtubule binding. The C-terminal is also the site for protein-protein interactions, and dynamin interacts with several proteins *in vitro*. Figure 1 shows the protein structure of dynamin. Dynamin I is mainly expressed in the central nervous system, whereas dynamin II is expressed in all tissues tested (17). Dynamin I is a candidate for participating in a neuron-specific membrane trafficking pathway, whereas dynamin II could play a role in a general membrane trafficking pathway.

Dynamin I, probably the C-terminal part, is a substrate for PKC, whereas dynamin II is not a substrate at all (17). Furthermore, the GTPase activity of dynamin I is stimulated by phosphorylation with PKC (11), which indicate that the GTPase activity of the conserved N-terminal is under the control of the less conserved C-terminal part of the molecule. Since dynamin II is not a substrate for PKC, dynamin I and II could have distinct functional roles in a neuron-specific and a general membrane trafficking pathway, respectively.

Several lines of evidence indicate a function of dynamin in an early stage of endocytosis. Transfection of the dynamin I gene, with mutations in the GTP binding domain, into mammalian cells results in a blockade of receptor mediated endocytosis (18). Furthermore, dynamin I can self assemble *in vitro* to form rings and helical stacks (19) and in the presence of the non-hydrolyzeable form of GTP, GTPγS, it has been localized to the collar around the neck of endocytotic synaptic vesicles (20). Upon depolarization and exocy-

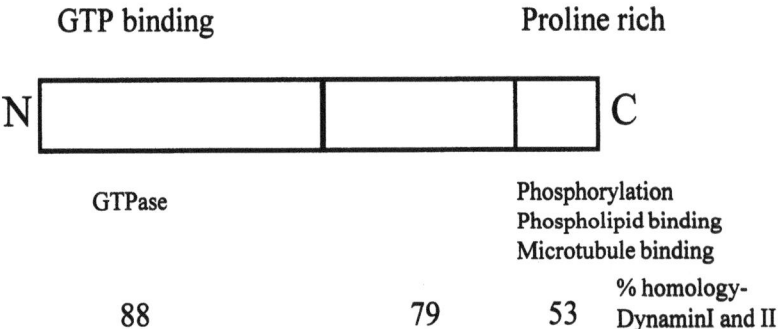

Figure 1. Protein structure of dynamin (96 kDa).

tosis of the nerve terminal, dynamin I is rapidly dephosphorylated by calcineurin, and low GTPase activity predominates (11). In resting nerve terminals, however, dynamin I is phosphorylated by PKC, and a high GTPase activity predominates (11). Such mechanisms could serve as molecular switches to reversibly inhibit or stimulate the GTPase activity of dynamin I in the nerve terminal.

3. RABPHILIN-3A

Synaptic vesicles contain a peripheral membrane protein called rabphilin-3A that binds selectively to the GTP-form of rab3A and rab3C (21,22). Rabphilin-3A is composed of two functionally different domains. These domains are the N-terminal Rab3A binding domain and the C-terminal phosphatidylserine and Ca^{2+} binding domain. In rab3A knockout mice, the level of rabphilin-3A in the nerve terminal is greatly reduced. These synapses revealed a potential defect in synaptic vesicle recycling, suggesting a potential role of rabphilin-3A in Ca^{2+} dependent synaptic vesicle recycling (23). Furthermore, over- expression of rabphilin-3A in chromaffin cells strongly enhances transmitter release, indicating a functional role for rabphilin-3A in exocytosis (24). During exocytosis, GTP-rab-3A is hydrolyzed to GDP-rab3A, and GDP-rab3A and rabphilin-3A is shown to dissociate from the vesicle membrane (25,26). Such findings are supporting an important role of rabphilin-3A in the synaptic vesicle release process.

Protein phosphorylation is known to regulate the probability of release of neurotransmitters, and rabphilin-3A is identified as a novel target for such regulatory processes (27,28,29). Rabphilin-3A is shown to be a substrate for PKA and CaMKII *in vitro*. In rat rabphilin-3A the phosphorylation sites are localized to serine^{234} and serine^{274}. The first serine is a target for PKA and CaMKII, whereas serine^{274} can be phosphorylated with CaMKII (27). These sites are located in the middle part of rabphilin-3A (see figure 2), between the N-terminal rab3A binding domain and the C-terminal Ca^{2+} binding domain. The phosphorylation of rabphilin-3A by two different kinases at closely linked sites are intriguing (27), and a role for rabphilin-3A in exocytosis would be attractive. In recombinant bovine rabphilin-3A four other phosphorylation sites are identified. These are serine^{34}, threonine^{205}, threonine^{209} and threonine^{537} (29). However, since the mouse and rat homologues of rabphilin-3A have no threonine residues corresponding to threonine^{205} and threonine^{209} of the bovine protein, these phosphorylation sites are probably of less physiological significance. Two moles of phosphate only are incorporated into one mole of bovine rabphilin-3A as well (29).

A role of rabphilin-3A phosphorylation in regulating synaptic vesicle cycling as a function of Ca^{2+} would be attractive. However, the Ca^{2+} and phospholipid binding of rabphilin-3A are not affected by phosphorylation by CaMKII, neither is the interaction of rab3A and rabphilin-3A (29). It could be speculated that rabphilin-3A complexed with GTP-rab3A may interact with a protein in the presynaptic plasma membrane. If phospho-

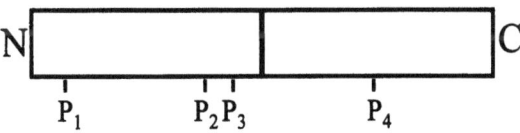

Figure 2. Protein structure of rabphilin-3A (80 kDa). The *in vitro* phosphorylation sites common for rat, mouse and bovine rabphilin-3A are indicated (28,29). P_1: Ser^{34} (bovine); P_2: Ser^{234} (rat); P_3: Ser^{274} (rat); P4: Thr^{587} (bovine).

rylation of rabphilin-3A affects its interaction with such a putative protein, phosphorylation could be very important for modulating neurotransmitter release. In this connection it is shown that rabphilin-3A binds to a protein molecule with a M_r of 115 kDa through the C-terminal domain in the presence of phosphatidylserine and Ca^{2+} (30).

The studies described in this paper are restricted to phosphorylation of rabphilin-3A *in vitro*. It is unknown if the same sites are phosphorylated *in vivo*. Currently, these questions are under investigation (31).

4. CONCLUSION

The significance of phosphorylation events in the synaptic vesicle neurotransmission in the nerve terminal is still not clear. The currently known phosphoproteins do not account for all the regulatory phenomena caused by protein kinases. Additional work will be required to identify such proteins and their role in neurotransmission. In figure 3 a diagram of dynamin I and rabphilin-3A in the synapse are shown.

Dynamin I is rapidly dephosphorylated during depolarization (a), and dephosphorylation directly decreases the rate of GTP hydrolysis, and this might be important in initiating endocytosis. Dynamin I is localized to the collar around the neck of endocytotic vesicles (20). PKC (b) phosphorylates dynamin I to complete the phosphorylation cycle, and inactivating dynamin I by increasing the GTPase activity.

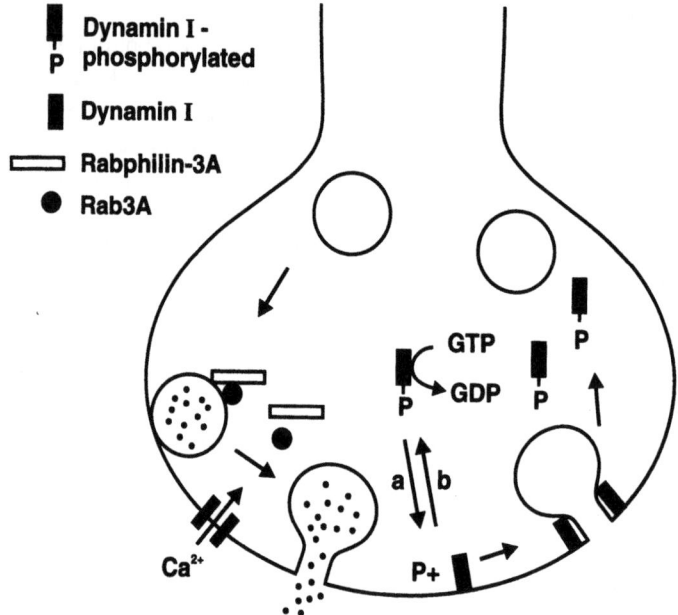

Figure 3. A diagram illustrating rabphilin-3A and dynamin I in the synaptic vesicle cycling in the nerve terminal. Rabphilin-3A binds selectively rab3A-GTP. During depolarization and Ca^{2+} influx into the nerve terminal rab3A-GTP is hydrolyzed to rab3A-GDP, and rab3A-GDP and rabphilin-3A dissociate from the synaptic vesicles (25,26). The rise in Ca^{2+} activates docked synaptic vesicles to fuse with the plasma membrane and release neurotransmitter. The role of rabphilin-3A phosphorylation remains to be seen.

REFERENCES

1. Fykse, E.M., and Fonnum, F., 1996, *Neurochem. Res.* **21**: 1059–1066.
2. Greengard, P., Valtorta, F., Czernik, A.J., and Benfenati, F., 1993, *Science,* **259**: 780–785.
3. Jahn, R., and Südhof, T., 1994, *Ann. Rev. Neurosci.* **17**: 219–246.
4. Dekker, L.V., De Graan, P.N.E., Oestreicher, A.B., Versteeg, D.H.G., Gispen, W.H., 1989, *Nature,* **342**: 74–76.
5. Hata, Y., Slaughter, C.A., Südhof, T.C., 1993, *Nature,* **366**: 347–351.
6. Fujita, Y., Sasaki, T., Fuku, K., Kotani, H., Kimura, T., Hata, Y., Südhof, T.C., Scheller, R.H., and Takai, Y., 1996, *J. Biol. Chem.* **271**: 7265–7268.
7. Shpetner, H.S. and Vallee, R.B., 1989, *Cell,* **59**: 421–432.
8. Kosaka, T. and Ikeda, K., 1983, *J. Neurobiol.* **14**: 207–225.
9. Chen, M.S., Obar, R.A., Schroeder, C.C., Austin, T.W., Poodry, C.A., Wadsworth, S.C., and Vallee, R.B., 1991, *Nature,* **351**: 583–586.
10. Van der Bliek, A.M., and Meyerowitz, E.M., 1991, *Nature,* **351**: 553–563.
11. Robinson, P.J., Sontag, J.-M., Liu, J.P., Fykse, E.M., Slaughter, C., McMahon, H., and Südhof, T.C., 1993, *Nature,* **365**: 163–166.
12. Krueger, B.K., Forn, J., and Greengard, P., 1977, *J. Biol. Chem.* **252**: 2764–2773.
13. Robinson, P.J., and Dunkley, P.R., 1983, *J. Neurochem.* **41**: 909–918.
14. Liu, J.P., Sim, A.T.R., and Robinson, P.J., 1994, *Science,* **265**: 970–973.
15. Robinson, P.J., 1992, *J. Biol. Chem.* **267**: 21637–21644.
16. Robinson, P.J., Liu, J.P., Powell, K., Fykse, E.M., and Südhof, T.C., 1994, *TINS,* **17**: 348–353.
17. Sontag, J.M., Fykse, E.M., Ushkaryov, Y., Liu, J.P., Robinson, P.J., and Südhof, T.C., 1994, *J. Biol. Chem.* **269**: 4547–4554.
18. Herskovits, J.S., Burgess, C.C., Obar, R.A., and Vallee, R.B., 1993, *J. Cell. Biol.* **122**: 565–578.
19. Hinshaw, J.E. and Schmid, S.L., 1995, *Nature,* **374**: 190–192.
20. Takel, K., McPherson, P.S., Schmid, S., and De Camilli, P., 1995, *Nature,* **374**: 186–190.
21. Shirataki, H., Kaibuchi, K., Yamaguchi, T., Wada, K., Horiuchi, H., and Takai, Y., 1992, *J. Biol. Chem.* **267**: 10946–10949.
22. Li, C., Takei, K., Geppert, M., Daniell, L., Steinius, K., Chapman, E.R., Jahn, R., De Camilli, P., and Südhof, T.C., 1994, *Neuron,* **13**: 885–898.
23. Geppert, M., Bolshakov, V.Y., Siegelbaum,S.A., Takei, K., De Camilli, P., Hammer, R.E., and Südhof, T.C., 1994, *Nature,* **369**: 493–497.
24. Chung, S.-H., Takai, Y., and Holz. R.W., 1995, *J. Biol. Chem.* **270**: 16714–16718.
25. Fischer von Mollard, G., Südhof, T.C., and Jahn, R., 1991, *Nature,* **349**: 79–81.
26. Stahl, B., Chou, J.H., Li, C., Südhof, T.C., and Jahn, R., 1996, *EMBO J.* **15**: 1799–1809.
27. Fykse, E.M., Li, C., and Südhof, T.C., 1995, *J. Neurosci.* **15**: 2385–2395.
28. Numata, S., Shirataki, H., Hagi, S., Yamamoto, T., and Takai, Y., 1994, *Biochem. Biophys. Res. Commun.* **203**: 1927–1934.
29. Kato, M., Sasaki, T., Imazumi, K., Takahashi, K., Araki, K., Shirataki, H., Matsuura, Y., Ishida, A., Fujisawa, H., and Takai, Y., 1994, *Biochem Biophys. Res. Commun.* **205**: 1776–1784.
30. Miyazaki, M., Kaibuchi, K., Shirataki, H., Kohno, H., Ueyama, T., Nishikawa, J., Takai, Y., 1995, *Mol. Brain Res.* **28**: 29–36.
31. Fykse, E.M., 1996, *J. Neurochem.* S85C.

SYNAPTIC VESICLE PROTEINS ARE RAPIDLY TRANSPORTED IN THE OPTIC NERVE

Jia-Yi Li and Annica Dahlström

Department of Anatomy and Cell Biology
University of Göteborg
Göteborg, Sweden

1. INTRODUCTION

Neurotransmitter release is facilitated by different synaptic/plasma membrane proteins. According to the manner of association of these proteins, they are subdivided into two major groups: transmembrane proteins and surface adsorbed (peripheral) proteins (1). Each protein has been proposed to be involved in specific steps of synaptic vesicle trafficking — docking, fusion, release and membrane retrieval. The majority of synaptic vesicle proteins in mature neurons are localized in nerve terminals. The targeting of synaptic vesicle components to nerve terminals is mediated by fast axonal transport. Because the machinery of protein synthesis is restricted to the cell body and proximal dendrites, it is of interest to know how each synaptic vesicle protein moves from the cell body to nerve terminals, and to what extent they are recycled in the terminal.

Our previous studies have investigated the intraneuronal dynamics and axonal transport of various synaptic vesicle proteins in the PNS (2–6). The results indicated that most transmembrane proteins were rapidly transported in bidirectional patterns, thus, considerable amounts were recycled in nerve terminals. However, most surface adsorbed proteins were transported predominantly with anterograde axonal transport, while very small amounts were recycled. Since previous investigations have demonstrated some differences in axonal transport between the PNS and the CNS, we have now extended our investigation to study the axonal transport of synaptic vesicle proteins in the CNS, using the optic nerve as a model.

The visual system is a part of the CNS, comprising the retina, the optic nerve and tracts, the lateral geniculate body, the superior colliculus and the visual cortex. The axons of retinal ganglion cells form the optic nerve and transmit signals to the lateral geniculate nucleus, the superior colliculus, etc. Because of the comparative simplicity of the optic nerve, entirely comprised of axons from the retinal ganglion cells and in addition easily accessible, the optic nerve provides a unique opportunity to study axonal transport in the CNS (7–10). Available evidence indicate that in the optic nerve several synaptic vesicle associated proteins, e.g. synapsin I, clathrin and Rab3a, are transported distally in the slow axonal transport phase. However, we have demonstrated that these proteins are also involved in fast axonal transport in peripheral neurons (2–6, 11, 12).

The techniques of stop-flow (crush) and cytofluorimetric scanning (CFS) have been successfully employed by our laboratory for axonal transport studies in the peripheral nervous system (12–14). One important advantage of the CFS-technique is that several different antigens can be studied in a single nerve specimen eliminating biological variations. In addition, the CSF technique simultaneously allows quantitative as well as morphological analyses in the same tissue.

2. MATERIALS AND METHODS

Adult SD male rats (200–250g) were used. Under sodium pentobarbital anesthesia (i.p. 30 mg/kg b.w.), the optic nerves were approached from a lateral aspect and crushed with watchmaker's tweezers for 5 seconds at a position 3–5 mm posterior to the eye ball. One to 8 h post-operatively, the rats were transcardially perfused with 4% paraformaldehyde in 0.1 M PBS (pH 7.4), the optic nerves with the eye balls were dissected and post-fixed in the same fixative for 4 h and rinsed in 10% sucrose/PBS overnight. Longitudinal sections were cut (10 µm in thickness) in a cryostat and immuno-labelling was carried out using the indirect immunofluorescene method as described previously (3).

The following antibodies were used:

1. Rabbit anti-synaptophysin (G96), raised against affinity-purified synaptophysin, (15), dilution 1:2,000.
2. Mouse anti-synaptotagmin I, (Cl 41.1), raised against a recombinant fragment corresponding to the cytoplasmic domain of synaptotagmin I (16), dilution 1:800.
3. Mouse monoclonal anti Rab3a, (hybridoma line Cl 42.2), produced against recombinant Rab3a (17), dilution 1:2,000.
4. Rabbit anti-rabphilin-3A, raised against gluthathion-S-transferase rabphilin-3A (rat rabphilin-3A residues 67–223), (18), dilution 1:800.
5. Rabbit anti-clathrin light chain (neuron-specific splicing variant) (19) dilution 1:800.

3. RESULTS AND DISCUSSION

The accumulated amounts of various synaptic vesicle proteins, 1 to 8 hrs after crush-operation, were registered by CFS. One hour after crushing, the transmembrane proteins synaptophysin and synaptotagmin I were clearly detected in the axons proximal as well as distal to the crush, representing the fast axonal transport of these proteins. The amounts increased markedly with time (Fig. 1 A–B). In contrast, the surface adsorbed proteins Rab3a and rabphilin-3A were predominantly present in axon proximal to the crush, while the amounts distal to the crush, representing retrograde axonal transport, were very small (Fig. 1 C–D). The ratio between distal and proximal accumulations, measured by CFS, was 70–90% for the transmembrane proteins, while the ratio was much smaller for the surface adsorbed proteins, about 10% (Fig. 1 F). This indicates that a large proportion of transmembrane proteins, in the optic nerve as in the sciatic nerve (2, 3, 6), are recycled

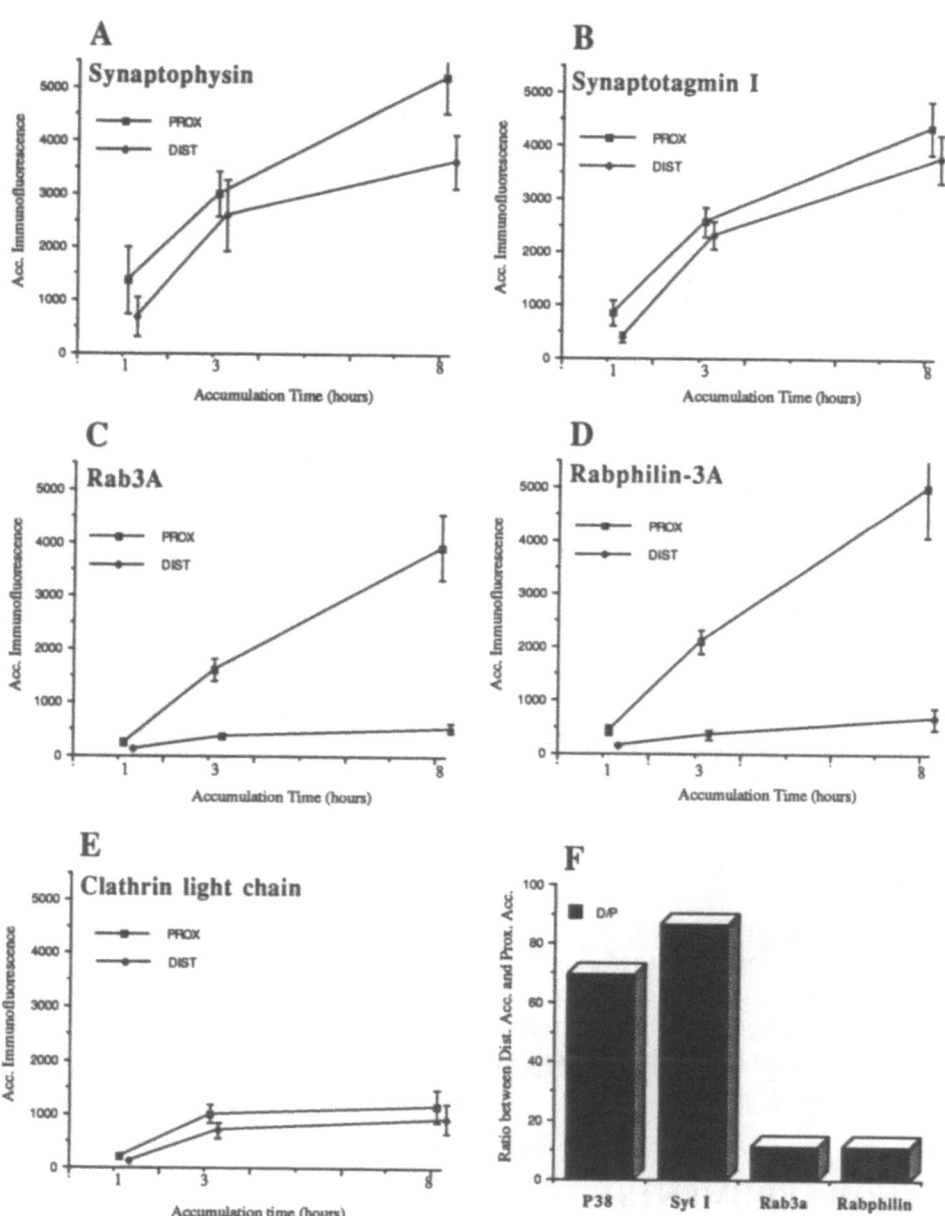

Figure 1. Accumulation of various synaptic vesicle proteins 1 to 8 hours after crush-operation and registered by cytofluorimetric scanning (CFS) (A-E). In F are shown the ratios between distal and proximal accumulations of these proteins.

with retrogradely transported organelles, while surface adsorbed proteins are recycled to a much less degree, indicating that most of these proteins are dissociated from the organelles and degraded or metabolized in the nerve terminals at the end of the lifespan in the terminals.

Interestingly, one surface adsorbed protein, clathrin light chain, demonstrated very little of fast transport in the optic nerve, although this protein was, in the peripheral nerves, clearly transported in the fast phase (4). During the first hour, no distinct accumulation of clathrin light chain could be detected, in contrast to other synaptic vesicle proteins. Even 8 hrs after operation, the accumulated amounts of clathrin light chain were still minute (Fig. 1 E). Previous investigations, using biochemical experimental techniques, showed that synapsin I, Rab3a and clathrin were moving with slow axonal transport (20–23). However, in the sciatic nerve, these proteins accumulated very rapidly, indicating that a considerable proportion of these proteins are indeed transported anterogradely with the fast transport machinery (2, 4, 11). In comparison with the appearance in the sciatic nerve, Rab3a was also transported in a slower rate in the optic nerve, especially during the first hour after operation. However, large amounts of this protein were detected at the longer post-operative intervals (3 and 8 hrs). It has been hypothesized that the proteins or peptides in fast transport are associated with or stored in membrane bounded organelles, while those soluble in the cytosol, or associated with components of the cytoskeleton, are transported with the slow phase (12, 24). The present study, using the short interval crush-operation and CFS, could only examine and quantify the rapidly transported membrane organelles.This, however, does not rule out the existence of a fraction transported with the slow transport system. Synapsin I, Rab3a, or clathrin most probably exist in two fractions; one soluble and the other adsorbed to the transport organelles. A high proportion of synapsin I, clathrin and Rab3a was detected in the soluble pool (21, 25). It is likely to be the case for all surface adsorbed proteins to appear in two states, one membrane-bound (fast transported) and the other, soluble (transported with the slow phase). Since we did not detect any clear accumulations of clathrin light chain shortly after crushing, this may suggest that in the optic nerve this protein is not, or to a very small extent, affiliated with transport organelles, in contrast to the situation in the sciatic nerve (4).

4. CONCLUSIONS

The present study demonstrates that synaptophysin, synaptotagmin I and Rab3a and rabphilin-3A are rapidly transported anterogradely in the rat optic nerve, but not clathrin light chain. Considerable amount of transmembrane proteins synaptophysin and synaptotagmin I are recycled in the nerve terminals and are retrogradely transported, while the surface adsorbed proteins Rab3a and rabphilin-3A are only transported anterogradely, but are not recycled to more than 10%.

ACKNOWLEDGMENTS

Supported by the Swedish MRC (2207), the Royal Academy of Science and Arts in Göteborg, Gustav V's 80-Årsfond, by the Swedish Society for Medical Research and by Stiftelsen Lars Hiertas Minne. For a generous supply of antibody preparations we are grateful to: Dr. R. Jahn, New Haven, USA and Dr. T. Südhof, Dallas (anti-rabphilin-3A).

REFERENCES

1. Jahn, R., and Südhof, T.C., 1994, *Annu. Rev. Neurosci.* **17**: 219–246.
2. Li, J.-Y., Kling-Peterson, A., and Dahlström, A.B., 1992, *J. Neurobiol.* **23**: 1094–1110.
3. Li, J.-Y., Jahn, R., and Dahlström, A.B., 1994, *Neurosci* **63**: 837–850.

4. Li, J.-Y., Jahn, R. and Dahlström, A.B., 1995, *Europ. J. Cell Biol.* **67**: 297–307.
5. Li, J.-Y., 1996, *Synapse* **23**: 79–88.
6. Li, J.-Y., Edelmann, L., Jahn, R., and Dahlström, A.B., 1996, *J. Neurosci* **16**: 137–147.
7. Baitinger, C. and Willard, M., 1987, *J. Neurosci* **7**: 3723–3735.
8. Elluru, R.G., Bloom, G.S. and Brady, S.T., 1995, *Mol. Biol. Cell* **6**: 21–40.
9. Karlsson, J.-O., 1984, *Prog. in Ret. Res.* **3**: 105–116.
10. Petrucci, T.C., Macioce, P., and Paggi, P., 1991, *J. Neurosci.* **11**: 2938–2946.
11. Bööj, S., Goldstein, M., Fischer-Colbrie, R. and Dahlström, A., 1989, *Neuroscience* **30**: 479–501.
12. Dahlström, A., and Li, J.-Y., 1994, *Neurochem. Res.* **19**: 1413–1419.
13. Dahlström, A., Kling-Peterson, A., Bööj, S., Lundmark, K., and Larsson, P.-A., 1989, *J. Microscopy*, **103**: 308–319.
14. Larsson, P.-A., Bööj, S., Lundmark, K., Goldstein, M. and Dahlström, A., 1987, *Exp. Brain Res. Series* **16**: 282–287.
15. Jahn, R., Schiebler, W., Ouimet, C., and Greengard, P., 1985, *Proc. Natl. Acad. Sci. USA* **82**: 4137–4141.
16. Brose, N., Petrenko, A.G., Südhof, T.C. and Jahn, R., 1992, *Science* **256**: 1021–1025.
17. Matteoli, M., Takei, K., Cameron, R., Hurlbut, P., Johnston, P.A., Südhof, T.C., Jahn, R. and De Camilli, P., 1991, *J. Cell Biol.* **115**: 625–633.
18. Geppert, M., Bolshakov, V.Y., Siegelbaum, S.A., Takei, K., De Camilli, P., Hammer, R.E., and Südhof, T.C., 1994, *Nature* **369**: 493–497.
19. Maycox, P.R., Link, E., Reetz, A., Morris, S.A. and Jahn, R., 1992, *J. Cell Biol.* **118**: 1379–1388.
20. Chadan, S., Filliatreau, G., Moya, K.L., DiGiamberardino, L., and Tavitian, B., 1993, *Abstr. in 23rd annual meeting of Society for Neuronsci.* #34.4.
21. Garner, J.A. and Lasek, R.I., 1981, *J. Cell Biol.* **88**: 172–178.
22. Mori, H., and Kurokawa, M., 1981, *Biomed. Res.* **2**: 677–685.
23. Paggi, P., and Petrucci, T.C., 1993, *Mol. Neurobiol.* **6**: 239–251.
24. Ochs, 1982, *Axoplasmic transport and its relation to other nerve functions*, John Wiley and Sons, New York.
25. Fischer von Mollard, G., Südhof, T.C., and Jahn, R., 1991, *Nature* **349**: 79–82.

OVEREXPRESSION OF B-50/GAP-43 INDUCES FORMATION OF FILOPODIA IN PC12 CELLS

L. H. J. Aarts, H. B. Nielander, A. B. Oestreicher, L. H. Schrama,
W. H. Gispen, and P. Schotman

Rudolf Magnus Institute for Neurosciences
Departments for Physiological Chemistry and Medical Pharmacology
Utrecht University
Stratenum, Utrecht, The Netherlands

1. INTRODUCTION

B-50/GAP-43 (also known as neuromodulin and F1) is a membrane associated, calmodulin binding protein kinase C (PKC) substrate that is expressed primarily in neurons (1). Upon synthesis, the protein is attached to Golgi-derived vesicles, via palmitylation of two N-terminal cysteines, and subsequently transported to the plasma membrane and to growth cones, where it accumulates (2,3,4). The expression of the protein is high during neurite outgrowth and regeneration, suggesting a role for B-50/GAP-43 in neuritogenesis. Evidence for this notion was obtained by manipulating expression levels of B-50/GAP-43 in cultured cells. Overexpression of B-50/GAP-43 in rat pheochromocytoma (PC12) cells and in neuroblastoma cells increased their sensitivity towards growth promoting stimuli (5,6). Treatment of primary neurons or PC12 cells with antisense B-50/GAP-43 oligonucleotides interfered with (7,8) and introduction of anti-B-50/GAP-43 antibodies into neuroblastoma cells reduced neurite outgrowth (9). However, a variant PC12 cell line lacking detectable B-50/GAP-43 expression still extended neurites (10) and suppression of kinesin expression using antisense kinesin heavy chain oligonucleotides still allowed neurite extension allthough B-50/GAP-43 was no longer transported down the axon (11).

We have studied the morphogenetic effects of B-50/GAP-43 in rat PC12 cells. These cells form a well-characterized model to study neuritogenesis (12). Upon treatment with nerve growth factor (NGF) these cells grow out neurites, a process that is accompanied by an enhanced B-50/GAP-43 expression as well as a translocation of B-50/GAP-43 from Golgi structures in the cytosol to the plasma membrane (3,13). We have used a PC12 clone (PC-B2) with very low levels of B-50/GAP-43, but still capable of growing out neurites upon differentiation with NGF (14), to examine the effects of overexpression of B-50/GAP-43 on morphology in spreading PC-B2 cells as well as in NGF-differentiated PC-B2 cells.

Neurochemistry, edited by Teelken and Korf
Plenum Press, New York, 1997

2. METHODS

PC-B2 cells were grown essentially according to Greene et al.(12). Cells were maintained in a humidified atmosphere at 37°C and 7% CO_2 in Costar tissue culture flasks, coated with rat tail collagen (40 ng mm^{-2}) and containing Roswell Park Memorial Institute medium (RPMI) 1640 supplemented with 5% fetal calf serum (GIBCO-BRL, Bethesda, MD), 10% heat-inactivated horse serum (GIBCO-BRL) and penicillin (100 IU ml^{-1})/streptomycin (100 µg ml^{-1})(ICN). Cells were transfected with B-50 cDNA under the control of a strong viral promoter (pCDNA1, Invitrogen, USA) in chemically defined N1 medium supplemented with insulin (5µg/ml, Sigma) using lipofectin (Gibco-BRL). Transfected cells (2–5% of the total cell population) could easily be identified by their intense B-50/GAP-43 immunoreactivity. The non-transfected PC-B2 cells, containing low amounts of B-50/GAP-43 immunoreactivity, served as controls.

3. RESULTS

We observed spontaneous formation of filopodia in B-50/GAP-43 transfected PC-B2 cells, 40 minutes after replating onto collagen coated coverslips (Figure 1, lower cell). Careful examination of 1µm thick confocal sections revealed that these filopodia were located mainly at the cell base attached to the substrate. Double staining with monoclonal B-50 antibodies and with phalloidin showed a characteristic F-actin pattern consisting of prominent radial arrays of peripheral actin filaments in the transfected cells, clearly differing from the nontransfected control cells (Fig. 1). In addition, we observed that spreading was delayed in transfected vs. untransfected PC-B2 cells. These phenomena appeared to be more prominent shortly after replating of cells, allthough well-spread (2 days) PC-B2 cells transfected with B-50/GAP-43 still exhibited a more filopodia-rich morphology then nontransfected cells (14). Our results coincide with the finding that overexpression of B-50/GAP-43

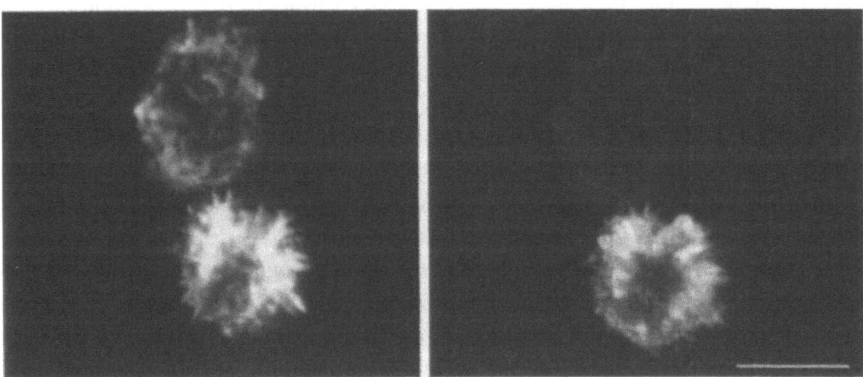

Figure 1. B-50/GAP-43 induces spontaneous formation of filopodia in spreading PC-B2 cells. Two days after transfection, cells were replated onto collagen coated coverslips using trypsin (0.05%)/EDTA (0.1%), fixed 40 min after in 4% paraformaldehyde (PFA) and 0.05% glutaraldehyde for 20 min at 4°C and stored in 1% PFA at 4°C. Upon double fluorescent labeling using B-50 specific monoclonal antibodies (NM4, right) and with phalloidin-FITC (Sigma, left) for F-actin, cells were analyzed by confocal laser scanning microscopy (CLSM).

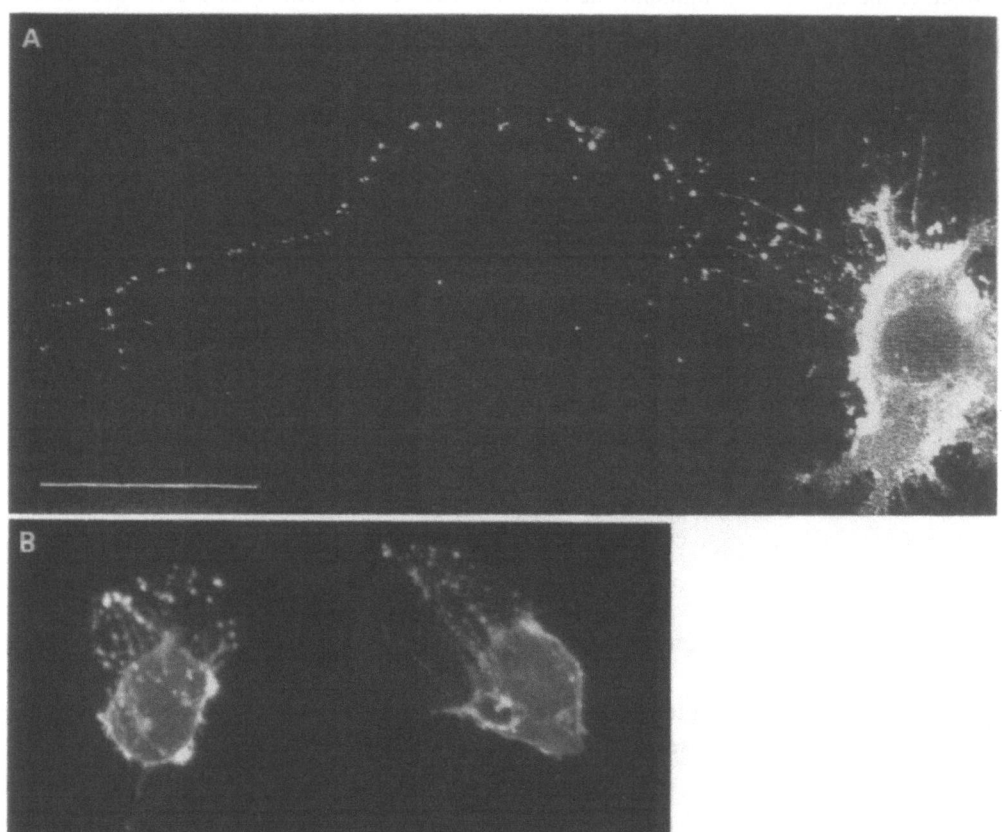

Figure 2. Punctuate B-50/GAP-43 immunoreactivity in NGF-differentiated PC-B2 cells transfected with B-50/GAP-43 cDNA. After transfection, cells were cultured in chemically defined medium (N1) supplemented with ß-NGF (10 µg ml ; Boehringer) and insulin (5 µg ml^{-1}; Sigma) for 2 days before being fixed and immunostained for B-50/GAP-43 (NM4). Cells were analyzed by CLSM.Punctuate B-50/GAP-43 immunoreactivity in NGF-differentiated PC-B2 cells transfected with B-50/GAP-43 cDNA. After transfection, cells were cultured in chemically defined medium (N1) supplemented with ß-NGF (10 µg ml^{-1}; Boehringer) and insulin (5 µg ml^{-1}; Sigma) for 2 days before being fixed and immunostained for B-50/GAP-43 (NM4). Cells were analyzed by CLSM.

affected cell spreading in a variety of nonneuronal cells, an effect that could be modulated by mutation of the PKC-phosphorylation site (15,16).

To compare the effect of B-50/GAP-43 overexpression in undifferentiated vs. differentiated PC12 cells, cells were treated with NGF for 2 days, starting 16 hours after transfection. Allthough no differences could be observed between transfected versus untransfected cells when F-actin patterns were compared (14), we occasionally detected a striking appearance of dotted B-50/GAP-43 immunoreactivity. These dots could be seen both on neurites growing on the collagen substrate (Figure 2a) as well as neurites growing over other (untransfected) cells (Figure 2b), in short as well as in extremely long (up to 100µm) extensions. Punctate B-50/GAP-43 immunoreactivity was observed before in growth cones of dorsal root ganglion neurons and of superior cervical ganglion neurons growing in culture (7,17). We are currently pursuing time-lapse videomicroscopy to see whether the dotted B-50 immunoreactivity we observed are formed as a consequence of active protrusion or indeed more passively as reminiscence of neurite retraction.

4. CONCLUDING REMARKS

The capacity of B-50/GAP-43 to influence neurite outgrowth in cultured cells has been shown to depend strongly on the type of adhesive substrate that is being used, suggesting that the requirement for B-50/GAP-43 in neuritogenesis may be related to adhesiveness (7, 18). Our study suggests that B-50/GAP-43 might augment adhesion by extending filopodia containing strongly adhesive B-50/GAP-43-immunoreactive membrane patches. This is supported by the notion that overexpression of B-50/GAP-43 in RAT1-fibroblasts leads to the formation of a large number of filopodia, accompanied by the formation of specific points containing B-50/GAP-43 as well as vinculin immunoreactivity, a well known component of focal adhesion complexes (Aarts et al., unpublished results). We are currently examining whether the observed B-50/GAP-43 immunoreactive patches in PC-B2 cells also contain vinculin immunoreactivity.

ACKNOWLEDGMENTS

This work was supported by the 'Prinses Beatrix Fonds' and by the Netherlands Organization for Scientific Research (MW-NWO grant 903-42-006).

REFERENCES

1. Gispen, W.H., Nielander, H.B., De Graan, P.N.E., Oestreicher, A.B., Schrama, L.H., and Schotman, P., 1992, *Mol Neurobiol.* **5:** 61–85.
2. Skene, J.H.P., and Virág, I., 1989, *J. Cell Biol.*, **108:** 613–624.
3. Aarts, L.H.J., Van der Linden, J.A.M., Hage, W.J., Van Rozen, A.J., Oestreicher, A.B., Gispen, W.H., and Schotman, P., 1995, *Neuroreport,* **6:** 969–72.
4. Liu, Y., Chapman, E.R., and Storm, D.R., 1991, *Neuron,* **6:** 411–20.
5. Yankner, B.A., Benowitz, L.I., Villa-Komaroff, L., and Neve, R.L., 1990, *Mol. Brain Res.* **7:** 39–44.
6. Morton, A.J., and Buss, T.N., 1992, *Eur. J. Neurosci.* **4:** 910–16.
7. Aigner, L., and Caroni, P., 1993, *J. Cell Biol.* **123:** 417–29.
8. Jap Tjoen San, E.R.A., Schmidt-Michels, M.H., Oestreicher, A.B., Gispen, W.H., and Schotman, P., 1992, *Biochem. Biophys. Res. Commun.* **187:** 839–846.
9. Shea, T.B., Perrone-Bizzozero, N.I., Beermann, M.L., and Benowitz, L.I., 1991, *J. Neurosci.* **11:** 1685–90.
10. Baetge, E.E., and Hammang, J.P., 1991, *Neuron,* **6:** 21–30.
11. Ferreira, A., Niclas, J., Vale, R.D., Banker, G., and Kosik, K.S., 1992, *J. Cell Biol.* **117:** 595–606.
12. Greene, L.A., Sobeih, M.M., and Teng, K.K., 1991, *Culturing nerve cells* (Banker G and Goslin K, eds.). MIT, PP. 207–26.
13. Van Hooff, C.O.M., Holthuis, J.C.M., Oestreicher, A.B., Boonstra, J., De Graan, P.N.E., and Gispen, W.H., 1989, *J. Cell Biol.* **108:** 1115–25.
14. Nielander, H.B., French, P., Oestreicher, A.B., Gispen, W.H., and Schotman, P., 1993, *Neurosci. Lett.* **162:** 46–50.
15. Widmer, F., and Caroni, P., 1993, *J. Cell Biol.* **120:** 503–12.
16. Strittmatter, S.M., Valenzuela, D., and Fishman, M.C., 1994, *J. Cell Sci.* **107:** 195–204.
17. Meiri, K.F., and Gordon-Weeks, P.R., 1990, *J. Neurosci.* **10:** 256–66.
18. Shea, T.B., and Benowitz, L.I., 1995, *J. Neurosci. Res.* **41:** 347–54.
19. Jap Tjoen San, E.R.A., Nielander, H.B., Van Rozen, A.J., Gispen, W.H., and Schotman, P., 1996, *J. Mol. Neurosci.* **6:** 185–200.

Section 40: Brain Proteins

SALT-SOLUBLE MYELIN BASIC PROTEIN IS DEGRADED BY MYELIN-ADSORBED PROTEINASES

U. Haas and H. H. Berlet

Department of Pathochemistry and General Neurochemistry
University of Heidelberg
Im Neuenheimer Feld 220-221
69120 Heidelberg, Germany

1. INTRODUCTION

Myelin basic protein (MBP) is partially released from isolated myelin membranes by high ionic strength (1). Extracts contain traces of endoproteinases, as evident from the limited proteolysis of MBP of extracts on incubation (2,3). The exact origin of these proteinases has remained equivocal (4). We now present evidence in support of apparent extrinsic proteinase activities of myelin being loosely adsorbed onto myelin surfaces. They were found to be soluble in Tris buffer of low ionic strength along with a number of as yet unidentified proteins. The implications of this observation on the in vitro proteolysis of salt-soluble MBP were examined.

2. METHODS

Cerebral bovine myelin was isolated from bovine cerebral white matter by standard methods (5). Adsorbed proteins were released from myelin by washing suspensions of myelin membranes with 5 volumes of 25 mM Tris buffer, pH 7.5. Both untreated and Tris-washed myelin was subsequently treated with 5 volumes of 300 mM Tris-buffered NaCl, pH 7.0, at 4°C for 30 minutes to obtain salt-soluble extrinsic myelin proteins. Associated proteinase activities were assessed by incubating salt-extracts for up to 24 hours at 37°C with buffer solutions from pH 4.0 to 9.0, containing additives commensurate with major cellular proteinases, i.e. pH 4.0 for cathepsin D; pH 5.6 with DTT/Pefabloc/Pepstatin for cathepsin B, pH 7.4 with 5 mM Ca^{2+} and Pefabloc for mCANP, or without Ca^{2+} and Pefabloc for serine proteinase activity, and pH 9.0 with Pefabloc for metalloproteinase activity. The breakdown of MBP in terms of limited proteolysis was assessed by SDS-polyacrylamide gel electrophoresis and quantitative densitometry.

3. RESULTS

The main goal of this study was to distinguish between extrinsic proteins and proteolytic activities that are constitutive to myelin membranes, and those that may become adsorbed onto the myelin surface during the isolation procedure. They were thought to be distinguishable by their differential solubility at low and high ionic strength.

The treatment of myelin membranes with 25 mM Tris buffer of pH 7.5 resulted in the solubilization of small proteins only, i.e. 1.4% of total myelin protein (Table 1). Much more protein was released by high ionic strength equivalent to 300 mM Tris-bufferd NaCl, pH 7.0, The yields, namely as much as 23.2% and 21.2% of total myelin protein, appeared to be grossly independent of whether or not membranes had been treated first with 25 mM Tris buffer.

A representative plain Tris extract of myelin was resolved by SDS-PAGE into about 27 different protein bands (Figure 1). Apart from three major bands with approximate molecular weights of 14 kDa, 23 kDa and 31 kDa, there were mostly traces of other proteins. Altogether molecular weights of proteins were in the range of 97 kDa down to as little as 6 kDa compared to the mobilities of marker proteins. There was a weak band of identical mobility to that of MBP of 18.5 kDa. Whether it is structurally related to MBP remains to be clarified. The same question holds also for the nature and origin of the other protein bands. The latter issue concerns in particular their possible relationship to the myelin sheath despite their being readily soluble at low ionic strength.

The most prominent protein of the two different salt extracts obtained at high ionic strength from untreated and Tris-washed myelin was MBP of 18.5 kDa. Its relative staining intensity was markedly lower in the extract of Tris-treated myelin. In turn, some of the bands of less than 18.5 kDa of the latter were more strongly stained. There even appear to be one or two additional bands in this range. The patterns might suggest a dissociation of peptide aggregates ensuing the removal of adsorbed proteins by Tris. Otherwise there is an additional release of small proteins from myelin. Dissociation of solubilized proteins is unlikely, since the original electrophoretic pattern could not be restored by recombining Tris-soluble material with the salt-extract of Tris-washed myelin (data not shown). An instantaneous proteolytic fragmentation was not experimentally supported by the subsequent studies of limited proteolysis (see Figure 3).

Table 1. Protein content of extracts of isolated myelin membranes. Results are means ± ranges of duplicate determinations performed on two different extracts. The column headings 'Untreated' and 'Tris-washed' refer to the pretreatment of myelin with plain Tris buffer prior to the extraction of myelin with 300 mM Tris-buffered sodium chloride

	Tris-soluble protein	Salt-soluble protein of myelin	
		Untreated	Tris-washed
Total protein, mg/g myelin protein	14 ± 1.2	232 ± 24	212 ± 18
Total protein, mg/g myelin dry weight	3.4 ± 0.1	43.2 ± 5.0	39.6 ± 3.6

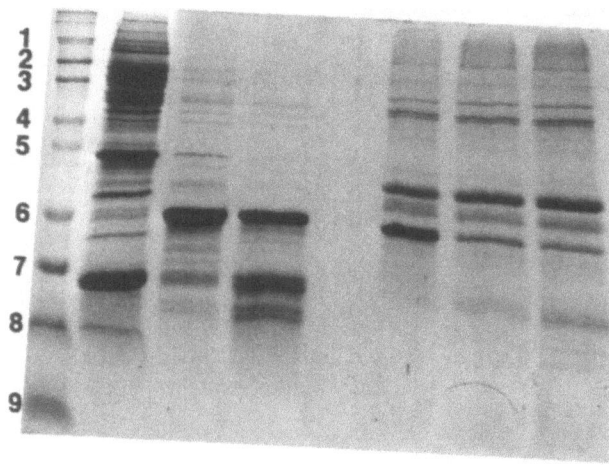

Figure 1. SDS-PAGE of whole and residual myelin, Tris washings of myelin and salt extracts of myelin. Of myelin extracts (**A** to **C**) 10 μg of total protein each were applied onto gels, and of myelin (**D** to **F**) 12 μg of total protein each. Samples were run on homogeneous slab gels (separating gel, 15% TA, stacking gel 6% TA, according to 7). Because of its low protein concentration the Tris washing of myelin was concentrated 20-fold prior to electrophoresis by membrane ultrafiltration. Marker proteins of the following molecular weights were run in lane **M**: 97.4 kDa (1); 66.3 kDa (2); 55.4 kDa (3); 36.5 kDa (4); 31.0 kDa (5); 21.5 kDa (6); 14.4 kDa (7); 6.9 kDa (8) and 3.5 kDa (9).

The electrophoretic patterns of residual myelin proteins of high molecular weight are reflecting the selective removal of Tris-soluble proteins, with only a few proteins remaining associated with myelin nevertheless. The latter are therefore suggested to be more tightly bound to myelin membranes, or even represent 'anchoring' of extrinsic proteins similar to the salt-insoluble fraction of MBP of 18.5 kDa. Their possible intrinsic nature is also conceivable.

Endogenous degradation of salt-soluble myelin proteins. The presence of proteolytic activities in salt extracts in terms of self-degradation of soluble proteins was ascertained by limited proteolysis of soluble MBP and electrophoresis following the incubation of extracts for 24 hours at selected pH values. MBP of *untreated* myelin (Figure 2) was most markedly degraded at pH 4.0 by 75%, most likely by cathepsin D activity. There were lesser changes at pH 5.6 (cathepsin B, 11%), pH 7.4 (serine proteinase, 10%, or CANP, 11%) and pH 9.0 (metalloproteinase, 12%). The percentage values refer to the quantitative densitometry of relative staining intensities by Coomassie-Brillant blue R-250 at 595 nm.

Extracts of *Tris-washed* myelin were treated likewise (Figure 3). Except for the initial dissociation of extracts addressed above there were no further changes during the 24 h-incubation between pH 4.0 and 7.4, as ascertained densitometrically (data not shown). It therefore appears that proteinases pertinent to this pH range were removed by washing myelin with 25 mM Tris buffer. In contrast, there was distinct limited proteolysis at pH 9.0 that was absent from extract of untreated myelin. A spontaneous decomposition of MBP at pH 9.0 was ruled out by incubating samples whose enzymatic activities were inactivated by prior heating (data not shown). Since the alkaline proteolytic activity was

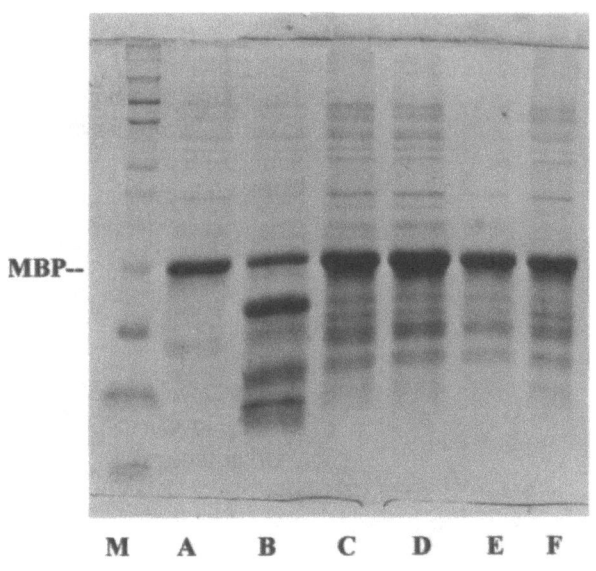

Figure 2. Endogenous proteolysis of salt-soluble proteins of untreated myelin membranes. Incubation assays of 100 μl each (lanes **B** to **F**) and total protein concentrations of 0.5 mg/ml were incubated at 37°C for 24 hours. Aliquots of 10 μg total protein were applied onto the gel.

A: Control, 0 hours of incubation
B to F: 24 hours of incubation
B: pH 4.0
C: pH 5.6
D: pH 7.4 (without Ca^{2+})
E: pH 7.4 (with Ca^{2+})
F: pH 9.0
M: Marker proteins

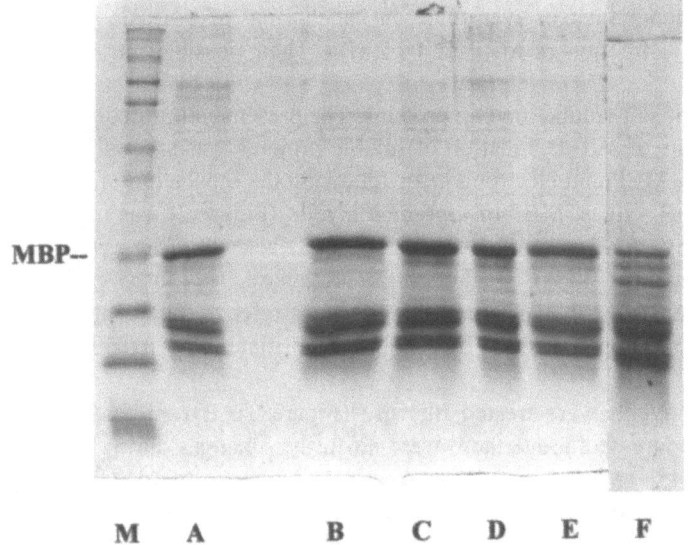

Figure 3. Endogenous proteolysis of salt-soluble proteins of Tris-washed myelin membranes. For experimental details see legend to Figure 2. The percentages of the relative dye binding capacity of 'MBP' were the same in the control (lane **A**) and in lanes **B** through **E**.

A: Control, 0 hours of incubation
B to F: 24 hours of incubation
B: pH 4.0
C: pH 5.6
D: pH 7.4 (without Ca^{2+})
E: pH 7.4 (with Ca^{2+})
F: pH 9.0
M: Marker proteins

observed to be fully soluble in 25 mM Tris buffer with some preparations of myelin, this issue is felt to warrant further close investigation.

4. DISCUSSION

From the data it is concluded that except for the alkaline metalloproteinase of this extract other proteinases that accounted for the limited proteolysis of extrinsic MBP were eliminated by treating myelin membranes with 25 mM Tris buffer. It therefore appears that these activities, though being apparently extrinsic, are only loosely and superficially bound onto myelin membranes. In view of the Tris-soluble cathepsin D activity and the exclusive lysosomal localization of the cellular soluble form, and considering its tendency to bind onto myelin membranes (4), it is questionable whether there are significant extrinsic proteinase activities constitutive to the myelin sheath except for an extrinsic alkaline metalloproteinase. The association of the latter with myelin as found in this study would be in line with its former isolation from myelin resulting in its characterization as an endoproteinase of the myelin sheath (6). In turn, the presence of proteinase activities in Tris extracts of myelin membranes was further corroborated by incubating Tris-soluble myelin proteins with proteinase-free exogenous MBP at the selected pH values from pH 4.0 through pH 9.0 as described above. There was a marked degradation of MBP at all pH values by the Tris extract (unpublished data). Its inhibition by pertinent inhibitors was confirmative of the tentative identification of cellular proteinases like cathepsin D, cathepsin B, neutral proteinase and alkaline metalloproteinase. As a practical consequence of the results of this study the isolation of myelin membranes (e.g. according to ref. 5) should include a final washing of membranes with Tris buffer to remove proteinase activities and other proteins from myelin that might merely become adsorbed during the isolation of myelin. Reassuringly, there was no evidence that it will result in a significant loss of MBP, if any.

ACKNOWLEDGMENT

This study was supported by a grant of the Deutsche Forschungsgemeinschaft.

REFERENCES

1. Berlet, H.H., 1987, *Neurochem. Pathol.*, **7**: 263–274.
2. Berlet, H.H., Ilzenhöfer, H., Schulz, R. and Gass, P., 1987, *Neurochem. Pathol.*, **6**: 195–211.
3. Glynn, P., Chantry, A., Groome, N. and Cuzner, M.L., 1987, *J. Neurochem.*, **48**: 752–759.
4. Berlet, H.H., Ilzenhöfer, H. and Kaefer, M., 1988, *Neurochem. Res.*, **13**: 409–416.
5. Norton, W.T. and Poduslo, S.E., 1973, *J. Neurochem.*, **21**: 749–757.
6. Chantry, A., Gregson, N.A. and Glynn, P., 1989, *J. Biol. Chem.*, **264**: 21603–21607.
7. Laemmli, U.K., 1970, *Nature*, **221**: 680–685.

Section 41: Molecular Neurobiology

SUPPRESSION OF IMMEDIATE EARLY GENE EXPRESSION BY INTRACEREBRALLY APPLIED ANTISENSE OLIGONUCLEOTIDES IMPAIRS MECHANISMS OF LEARNING AND MEMORY

Wolfgang Tischmeyer, Rita Grimm, Klaudia Lohmann, Horst Schicknick, and Eckart D. Gundelfinger

Federal Institute for Neurobiology
Magdeburg, Germany

1. INTRODUCTION

Transcription factors of the *jun* and *fos* families of immediate early genes (IEGs) are supposed to be mediators between short-term neuronal activity and long-term changes in neuronal connectivity during plastic processes of the nervous system (reviewed in ref. 1). Recently, we have observed a differential induction of IEGs such as c-*jun*, *jun* B and c-*fos* in hippocampal and cortical brain structures following training of rats on a footshock-motivated brightness discrimination in a Y-maze (2, 3, 4). To study the functional significance of IEG transcription factors in rat brain during processes of learning and memory formation, experiments assaying the effects of inhibition of c-*jun*, *jun* B or c-*fos* expression by intrahippocampally applied antisense (AS) phosphorothioate oligodeoxynucleotides (S-ODN) on acquisition and retention of the brightness discrimination were performed.

2. MATERIALS AND METHODS

2.1. Distribution of ^{35}S-Labelled S-ODN in Rat Brain

A randomized S-ODN (GTCCCTATACGAAC) was 5'-labelled with ^{35}S (3) and applied twice to the right hippocampus of rats as described below, while the left hippocampus received only the first injection. Two hours after the second injection, brains were removed and sliced. Coronal sections were processed for autoradiography and subsequently Nissl-stained. Autoradiographs and corresponding stained sections were digitized using a CCD-camera and alignment was performed by interactive translation and rotation. Using the program *BrainView* (5, 6), the images were visualized, lines of equal gray-values (equidensites) were extracted from the smoothed autoradiographs and superimposed on the corresponding Nissl-images.

2.2. S-ODNs and Intrahippocampal Injections

The S-ODNs used for behavioral studies were synthesized and purified as described (3) and had the following sequences: AS c-*jun*: TGCAGTCATAGAAC; AS *jun* B: TTTCGTGCACATCC; AS c-*fos*: GAACATCATGGTCGT; control-1 (mismatched AS c-*jun*): TGCACTGATACAAC; control-2 (reversed AS c-*fos*): TGCTGGTACTACAAG. For intracerebral injections, guide cannulae targeted to the dorsolateral hippocampus were bilaterally implanted in male Wistar rats (3). After one week of recovery, 1 µl of a 2 mM S-ODN solution or of physiological saline was bilaterally injected to freely moving rats, followed by a second injection after an interval of 8 h.

2.3. Behavioral Experiments

2.3.1. Open Field Test. Two hours after the last injection, a subset of rats (see Fig. 3) was exposed to an open field arena with an area of 100 cm x 100 cm and squares of 25 cm side length. The number of square crossings during 5 min was counted.

2.3.2. Brightness Discrimination Training. Two hours after the last injection or subsequently to the open field test, rats were trained to acquire a footshock-motivated brightness discrimination in a Y-maze (3). The task was to escape from a mild electrical footshock in the start box into the illuminated alley of the maze. Entering the dark alley was punished by footshock and counted as an error. A training session consisted of 40 trials with an average duration of 1 min. Using the same paradigm, rats were retrained after

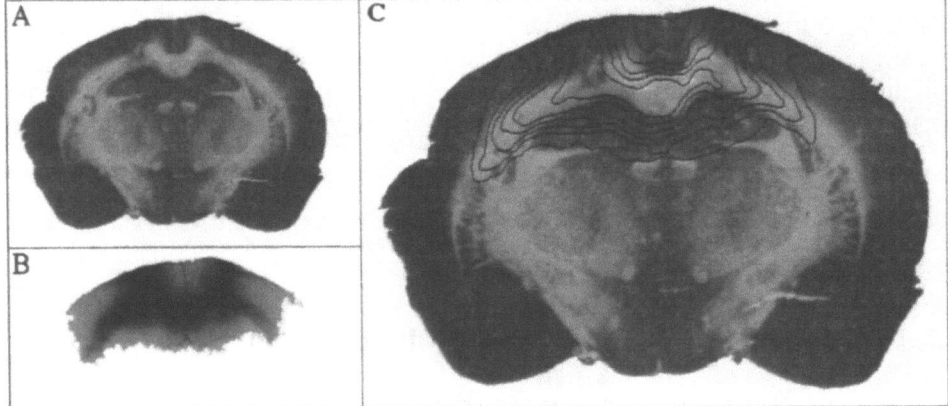

Figure 1. Distribution of ^{35}S-labelled S-ODN in rat brain. Frontal view of (A) a Nissl-stained coronal rat brain section and (B) the corresponding autoradiograph 10 h after bilateral injection of ^{35}S-labelled S-ODN and 2 h after a second injection in the right hippocampus. (C) Equidensites (see Materials and Methods) at 40, 50, 60, 70, 80 and 90 % density calculated from (B) were superimposed on the image shown in (A).

24, 48 and 72 h. The retention performance estimated in the relearning test 24 h after the first training was calculated according to the formula:

savings % = (training errors − relearning errors / training errors) × 100

3. RESULTS

Besides other applications, the program *BrainView* (5, 6) enables the user to visualize the distribution of intracerebrally injected radiolabelled compounds in the precise anatomical context by superimposing equidensites calculated from autoradiographs to images of the corresponding stained brain sections. We used this program to assess the distribution of ^{35}S-labelled S-ODN applied to rat brain via guide cannulae targeted to the dorsolateral hippocampus. As shown in Fig. 1, the distribution of labelled S-ODN was chiefly restricted to brain regions surrounding the injection sites such as the hippocampal formation, adjacent cortical areas and callosal fibres, encouraging us to use this injection protocol for behavioral studies.

Suppression of c-*jun* expression influenced chiefly the acquisition of brightness discrimination as documented by a significantly increased number of errors during the first training session compared to saline-treated controls (Fig. 2). Activity of rats in an open field test was not significantly influenced by any of the S-ODNs examined (Fig. 3), imply-

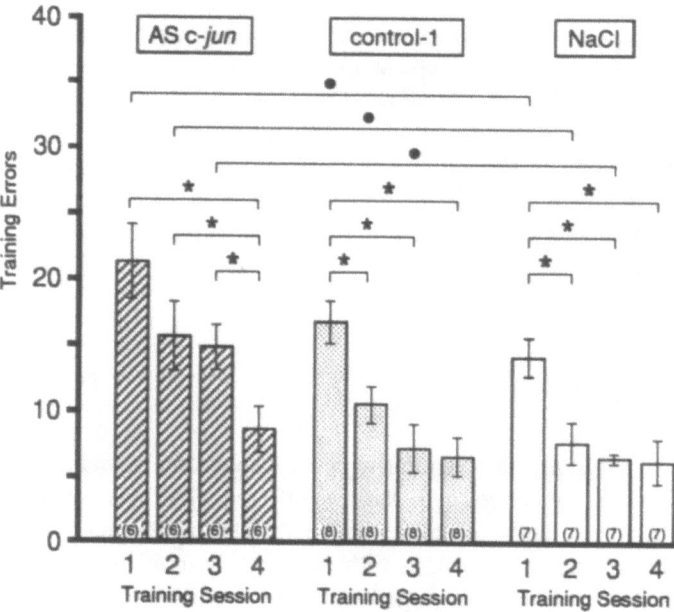

Figure 2. Effect of intrahippocampally applied AS c-*jun* S-ODN on the number of training errors of rats on four consecutive days of brightness discrimination training compared with control-1 S-ODN- and saline-treated controls (adapted from ref. 3). Means ± SEM. Number of animals in brackets. *p<0.05 (Wilcoxon test). •p<0.05 (Mann-Whitney U test).

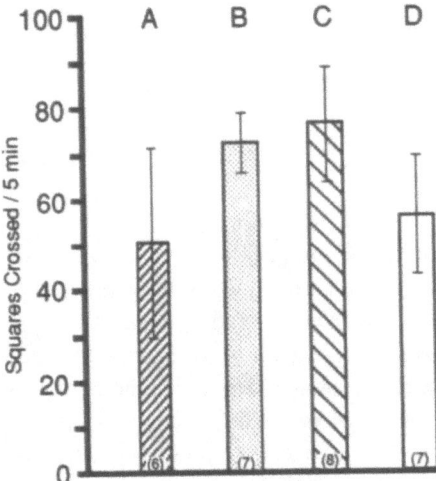

Figure 3. Open field activity of rats following two intrahippocampal injections of (A) AS c-*jun* S-ODN, (B) control-1 S-ODN, (C) AS *jun* B S-ODN or (C) saline. Means ± SEM. Number of animals in brackets.

ing that the impaired learning performance of AS c-*jun* S-ODN-treated rats did not result from an altered state of activity. During subsequent training sessions, AS c-*jun* S-ODN-treated rats behaved almost like naive control animals (Fig. 2), implying that inhibition of c-*jun* expression impaired not only the acquisition of the brightness discrimination but also mechanisms involved in consolidation and/or retrieval. In contrast, inhibition of *jun* B expression did not affect the brightness discrimination score (3), suggesting that either residual expression of *jun* B was sufficient or *jun* B expression is not essential for acquisition and consolidation of the discrimination reaction.

After intrahippocampal injection of AS c-*fos* S-ODN, the number of errors during the first training session resembled the level of control-2 S-ODN- as well as of saline-treated rats, indicating that the acquisition of the discrimination reaction was not impaired and, thus, presenting a reliable basis for comparison of retention. In the relearning test 24 h after the initial training, the percent savings of AS c-*fos* S-ODN-treated animals (28.9 ± 8.0 %; n=15) was significantly reduced compared to the values of control-2 S-ODN-treated (50.3 ± 6.0 %; n=16) and saline-treated (51.2 ± 6.3 %; n=12) rats ($p<0.05$, Mann Whitney U test), suggesting that Fos protein is involved in consolidation of the memory.

4. CONCLUSIONS

Intracerebral antisense intervention with the expression of inducible transcription factors encoded by the *jun* and *fos* gene families differentially impairs the acquisition and retention of a brightness discrimination reaction in rats in a sequence-specific manner. Inhibition of c-*jun* expression interferes with the acquisition of the task, suggesting that Jun protein, expressed at a relatively high basal level in rat brain, is necessary for normal neuronal function. In contrast, inhibition of c-*fos* expression specifically impairs the retention of the discrimination reaction without affecting the acquisition, implying that the induction of Fos protein during learning is of physiological relevance for the formation of a long-term memory trace.

While the target genes regulated by Fos and Jun proteins in the nervous system still remain to be identified, our present findings are indicative of a functional importance of gene regulation by Fos- and Jun-containing transcription factors such as AP-1 for mechanisms involved in learning and memory consolidation.

ACKNOWLEDGMENTS

AS c-*jun* S-ODN and AS *jun* B S-ODN were kindly provided by W. Brysch and K.-H. Schlingensiepen, Göttingen. Supported by the Land Sachsen-Anhalt (MWF 074A1021 and 074A10213), and by the DFG (Innovationskolleg INK 15/A3).

REFERENCES

1. Hughes, P., and Dragunow, M., 1995, *Pharmacol. Rev.* **47**: 133–178.
2. Tischmeyer, W., Kaczmarek, L., Strauss, M., Jork, R., and Matthies, H., 1990, *Behav. Neural Biol.* **54**: 165–171.
3. Tischmeyer, W., Grimm, R., Schicknick, H., Brysch, W. and Schlingensiepen, K.-H., 1994, *NeuroReport* **5**: 1501–1504.
4. Grimm, R., and Tischmeyer, W., 1996, *Behav. Brain Res.*, in press.
5. Lohmann, K., Staak, S., Gundelfinger, E.D., and Hess, A., 1996, *Proceedings of the 24th Göttingen Neurobiology Conference*, Volume 2 (N. Elsner, and H.-U. Schnitzler, eds.), Thieme, Stuttgart-New York, P. 820.
6. Lohmann, K., Staak, S., Gundelfinger, E.D., and Hess, A., in preparation.

189

β-ALANINE BEHAVES AS γ-AMINOBUTYRIC ACID AT *XENOPUS* OOCYTES EXPRESSING γ-AMINOBUTYRIC ACID RECEPTORS

Luis M. Orensanz and Luis C. Barrio

Servicios de Neurobiología e Histología
Departamento de Investigación
Hospital Ramón y Cajal
Madrid, Spain

1. INTRODUCTION

β-Alanine exerts an inhibitory effect in several experimental preparations. The question whether β-alanine acts through its own receptor or through the γ-aminobutyric acid (GABA) or glycine receptor remains unanswered. Parker et al. (1), based on results obtained using the *Xenopus* oocyte translation system, proposed a few years ago that β-alanine may have a specific receptor, given the size of β-alanine responses, relative to those of glycine and GABA, as well as the pharmacological profile of β-alanine responses. However, the most recent results point out that β-alanine actions in neurons of the spinal cord are due to an action at the glycine receptor, either the wild-type (2,3) or the mutated one (2,4). In fact, the conclusion of Parker et al. (1) may be criticized on the basis of our experimental design, in which β-alanine antagonism by strychnine or bicuculline, as well as benzodiazepines and barbiturates potentiation of β-alanine responses, were studied with mRNA fractions eliciting responses to both glycine and GABA. Thus the possibility remains that β-alanine acted as a weak agonist for both glycine and GABA receptors. In the present study, using mRNA preparations that elicited responses to just GABA, the β-alanine responses shared their characteristics with those of the GABA responses, strongly suggesting that β-alanine acts through the GABA receptor.

2. MATERIALS AND METHODS

2.1. Preparation of mRNA and Oocyte Microinjection

The optic lobe of 19-day old chick embryos was used as tissue source. Poly(A)$^+$ RNA was extracted and purified with a Clontech isolation kit. Oocyte preparation was as previously described (5). Dissected oocytes were injected with mRNA (50nl; 1–2μg/μl) and maintained for 4–8 days at 16–18 °C in Leibovit's L-15 medium.

2.2. Electrophysiological Recordings

The oocytes were placed in a bath ($\approx$0.1ml) and perfused with amphibian's Ringer solution (in mM; 116 NaCl, 2 KCl, 1.8 $CaCl_2$, 5 HEPES, pH 7.4). Cells were impaled with two microelectrodes filled with 1 M KCl (0.5–1 MΩ), and were voltage-clamped at –80 mV, unless otherwise indicated. Amino acids and other drugs were bath-applied with an access time ≤0.5 sec.

3. RESULTS AND DISCUSSION

Native (non injected) oocytes of *Xenopus laevis* showed no responses to glycine, GABA, β-alanine or taurine (all of them at 1 mM concentration). Oocytes injected with chick optic lobe mRNA developed responses to GABA or β-alanine, but no responses to glycine or taurine (1 mM concentration).

3.1. Dose-Response Experiments

β-Alanine and GABA elicited inward currents characterized by an initial fast increasing phase lasting 1–3 s until peak current values were reached, followed by a progressive decrease of current with at least two desensitization components, with fast and slow time courses (Fig. 1, sample records). Dose-response curves of peak currents revealed that β-alanine behaved as a full agonist, as compared to GABA (EC_{50}=3.61±0.07 mM and 23.2±0.02 μM, respectively, means±SEM, n=4). Figure 1 also showes similar dose-response curves of steady state current values, measured 25 s after the beginning of agonist application, for both agonists. (Fig. 1, open symbols). This reflects an equal increment of the desensitization degree as concentrations of β-alanine and GABA increase.

3.2. Antagonist Sensitivity

β-Alanine and GABA responses were sensitive to bicuculline (Fig. 2), while they were unaffected by strychnine (results not shown). IC_{50} values for the effect of bicuculline on GABA or β-alanine peak currents were 2.04±0.22 and 2.36±0.29 μM, respectively, while those for bicuculline antagonism of GABA or β-alanine steady state currents were 3.27±0.23 and 3.88±0.75 μM, respectively (in all four cases, means±SEM, n=4). This antagonist profile entirely agrees with that published by Wahl et al. (6) in cultured cerebellar granule cells.

3.3. Potentiation of GABA and β-Alanine Responses

The responses to β-alanine or GABA were potentiated by benzodiazepines and pentobarbital. The potentiation pattern developed by diazepam was similar for β-alanine and GABA (Fig. 3). Chlorazepate increased more the GABA response than the β-alanine response. Pentobarbital increased more the β-alanine response than the GABA response (unshown results). These results agree with previous results from other authors, who have

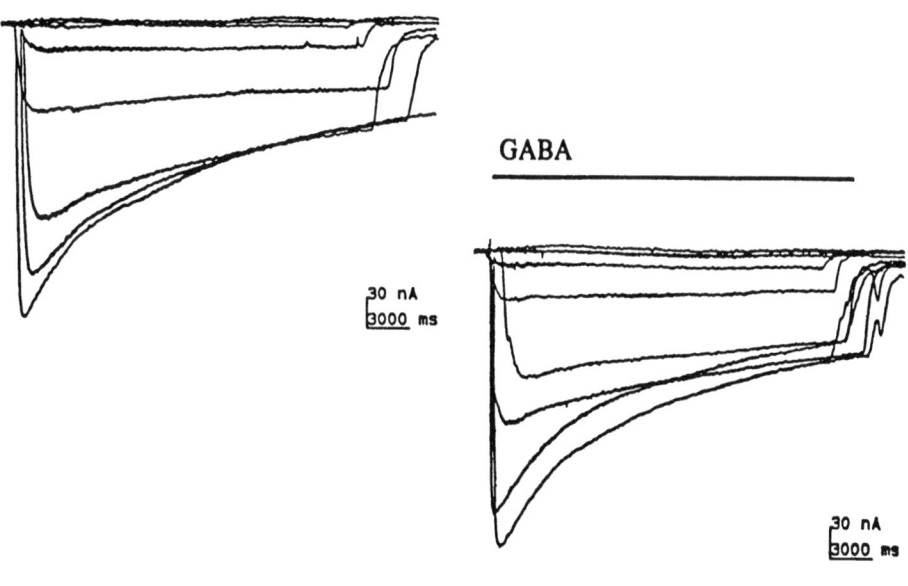
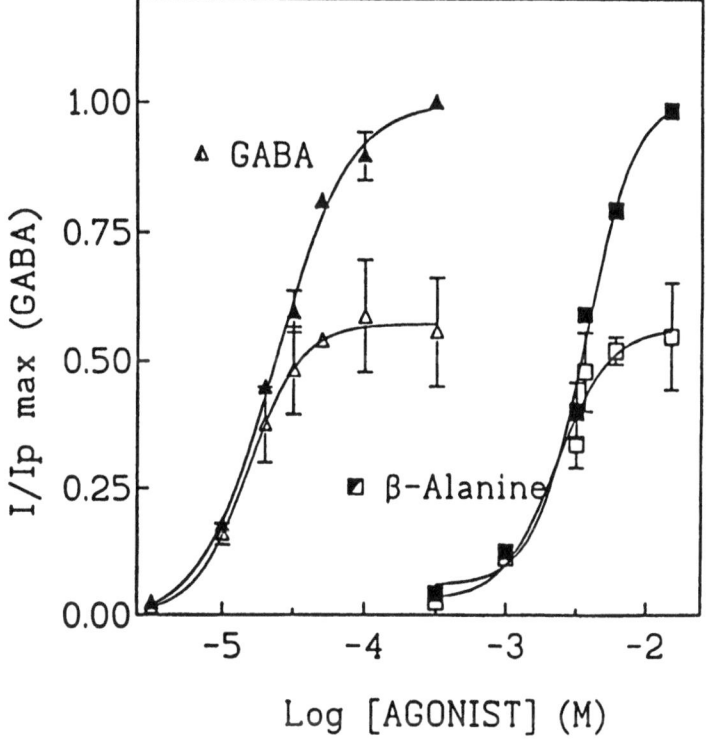

Figure 1. β-alanine and GABA responses in *Xenopus* oocytes expressing chick optic lobe mRNA. Closed symbols: Peak currents (Ip) relative to Ip to 0.3 mM GABA. Open symbols: Steady state currents measured 25 s after the beginning of agonist application (Iss), relative to Ip to 0.3 mM GABA. Sample records represent responses obtained to 1 μM–20 mM β-alanine or to 1 μM–300 μM GABA, with agonist application period indicated by the bar.

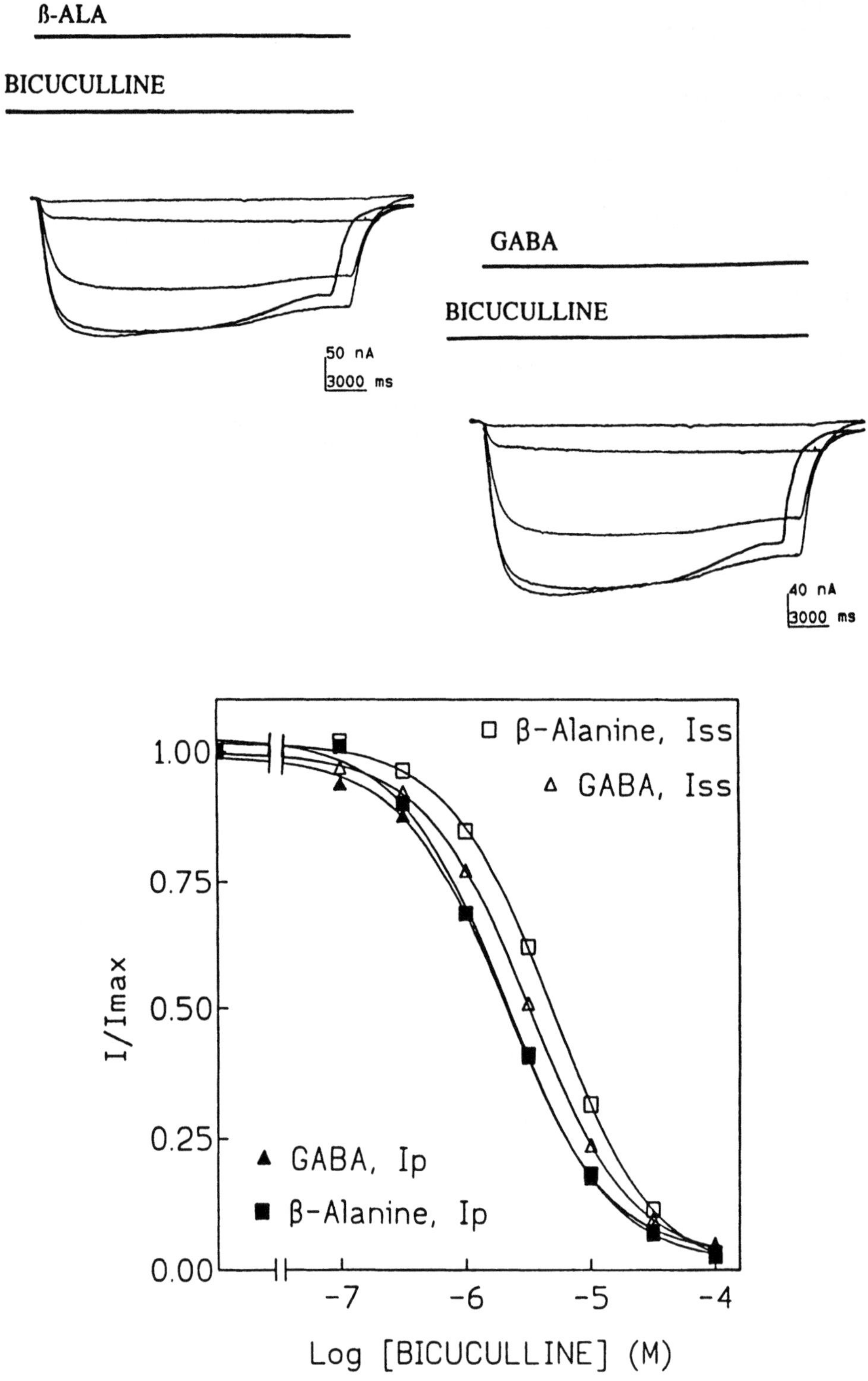

Figure 2. Bicuculline inhibition of agonist responses of *Xenopus* oocytes injected with chick optic lobe mRNA. β-Alanine and GABA concentrations were 3 mM and 30 μM, respectively. Bicuculline application began 1 min before addition of the corresponding agonist. Closed symbols: Ip relative to Ip observed in the absence of antagonist. Open symbols: Iss relative to Iss in the absence of bicuculline: Traces represent agonist response in the presence of 100 nM–100 μM bicuculline.

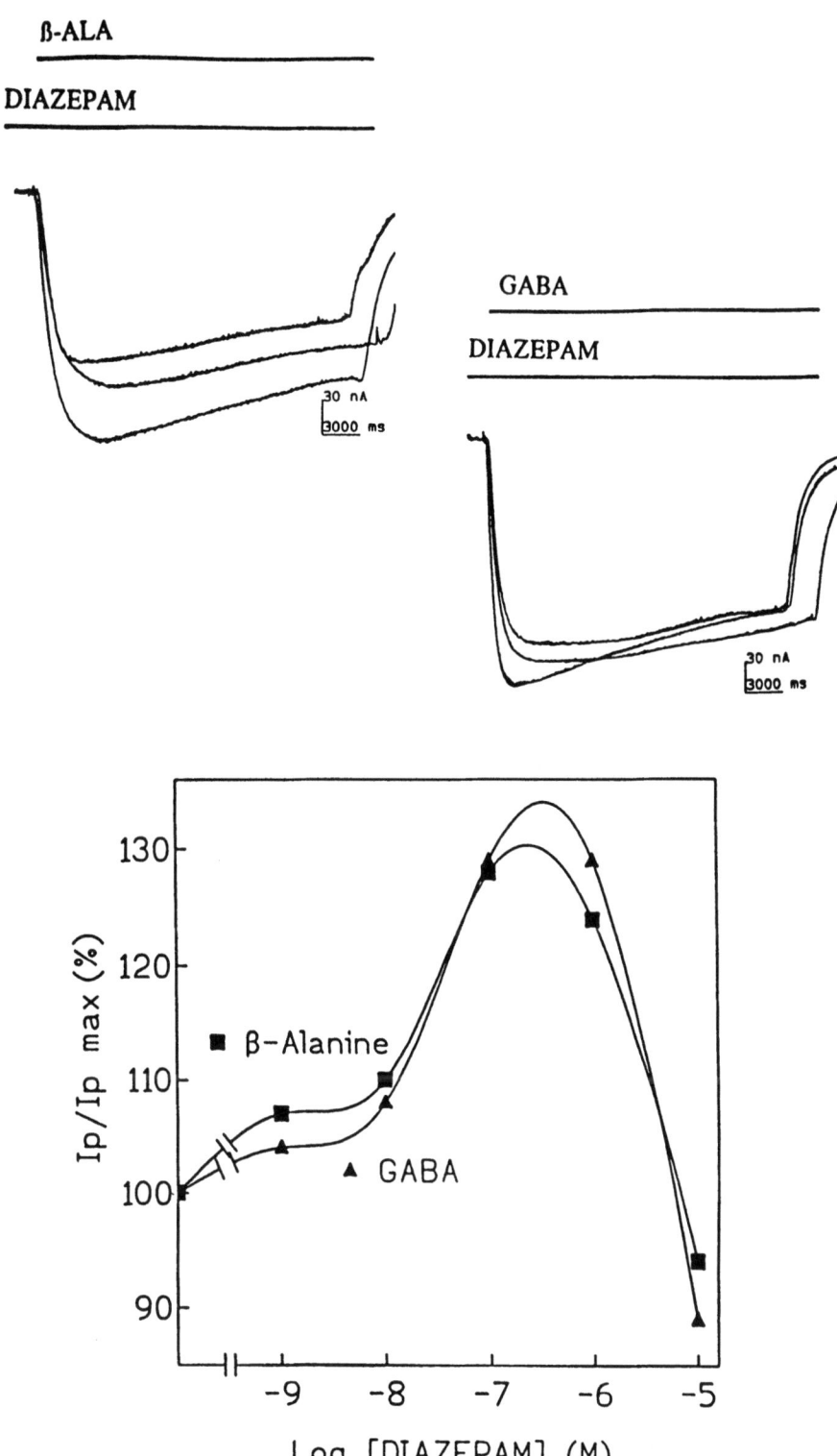

Figure 3. Diazepam potentiation of agonist responses of *Xenopus* oocytes injected with chick optic lobe mRNA. β-Alanine and GABA concentrations were 3 mM and 30 µM, respectively. Diazepam application began 1 min before addition of the corresponding agonist. Traces represent agonist response in the absence or in the presence of 10 or 100 nM diazepam.

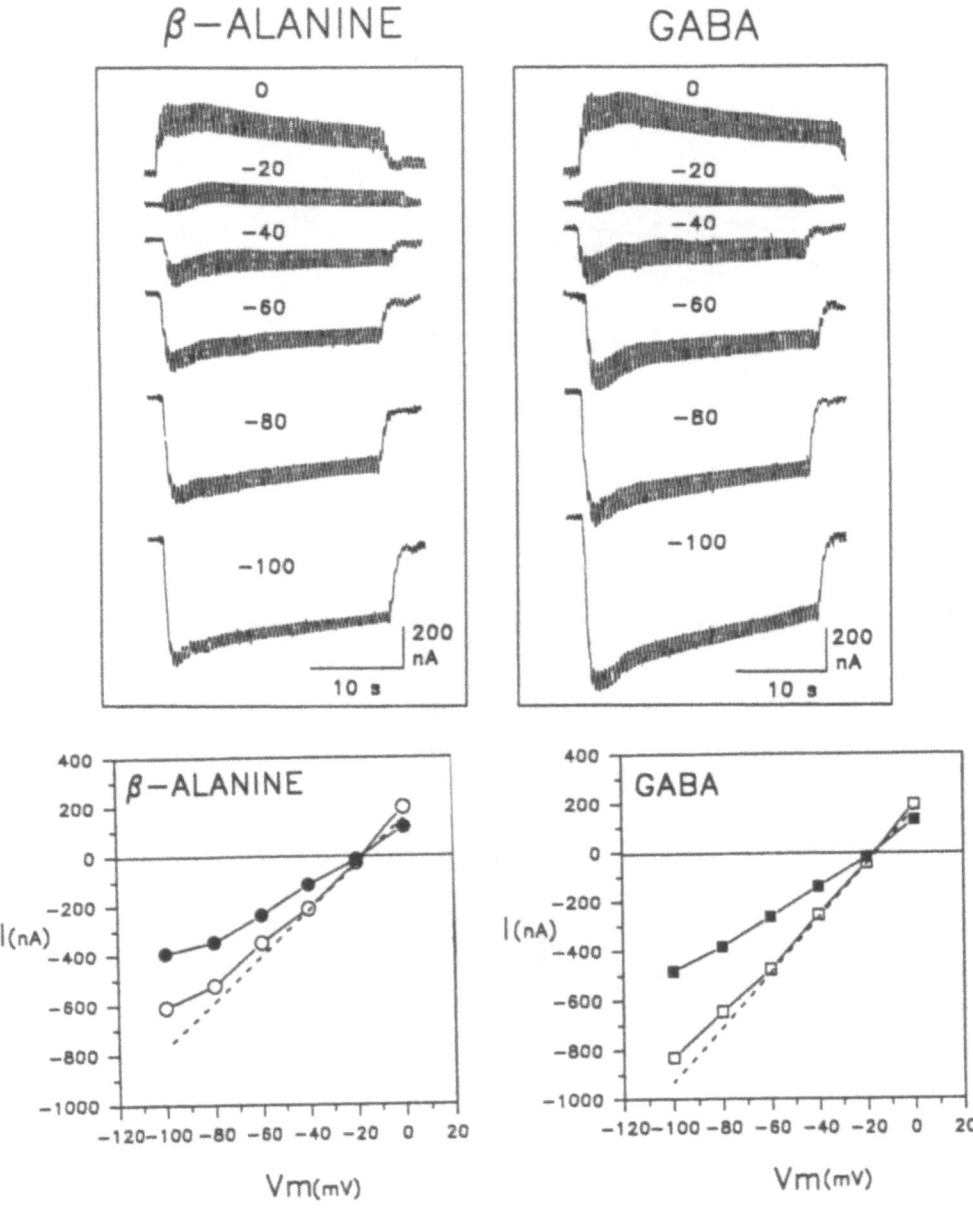

Figure 4. Voltage dependence of β-alanine- and GABA-induced currents in *Xenopus* oocytes injected with chick optic lobe mRNA. Left, sample currents induced by 3 mM β-alanine or 30 μM GABA at several holding potentials (from 0 to −100 mV, increments 20 mV). Changes on conductance was tested applying brief and small voltage pulses (10 mV, 500 ms and 1Hz). Right, Ip (open symbols) and Iss (filled symbols) values of current/voltage curves. Broken lines represent ohmic behavior.

described benzodiazepine and barbiturate potentiation of GABA responses in *Xenopus* oocytes injected with chick optic lobe (1) or rat cortex (7) mRNA.

3.4. Voltage Dependence of β-Alanine and GABA Responses

The currents elicited by 3 mM β-alanine or 30 μM GABA (their IC_{50} values), applied at several holding potentials are shown in Fig. 4. Both agonists evoked outward currents at more positive voltages than −20mV, whereas inward currents were obtained at more negative holding potentials, suggesting that β-alanine- and GABA-induced currents share a similar reversal potential, −20mV, which is the equilibrium potential for Cl^- in *Xenopus* oocytes. The simultaneous measure of conductance, applying small and brief voltage pulses during each agonist application, showed the parallel increment of conductance with the β-alanine- and GABA-induced current until peak value was reached, and its progressive reduction throughout the desensitization phase of responses. Interestingly, the conductance changes varied at different holding potentials since progressively smaller conductance increments obtained for more hyperpolarized potentials for β-alanine and GABA responses (Fig. 4, left). The degree of rectification was slightly more marked in the case of β-alanine, as it is illustrated in the peak and steady state current-voltage curves (Fig. 4, right). The deviation of ohmic behavior, i.e., voltage sensitivity, has been described by Hablitz (8) for GABA response in cultured chick cerebral neurons.

4. CONCLUSIONS

Xenopus oocytes injected with total mRNA isolated from chick embryo optic lobe show β-alanine responses which are entirely similar to GABA responses in their dose-response and pharmacological profiles, as well as in their electrophysiological properties. Taken together, the results in the present work indicate that in oocytes expressing chick optic lobe mRNA, β-alanine acts exclusively through GABA receptors.

REFERENCES

1. Parker, I., Sumikawa K., and Miledi, R., 1988, *Proc. R. Soc. Lond. B* **233**: 201–216.
2. Laube, B., Langosch, D., Betz, H., and Schmieden, V., 1995, *NeuroReport* **6**: 897–900.
3. Wu, F.-S., Gibbs, T.T., and Farb, D.H., 1993, *Eur. J. Pharmacol.* **246**: 239–246.
4. Rajendra, S., Lynch, J.W., Pierce K.D., French, C.R., Barry, P.H., and Schofield, P.R., 1995, *Neuron* **14**: 169–175.
5. Barrio, L.C., T. Suchyna, T. Bargiello, L.X. Xu, R.S. Roginski, M.V.L. Bennett and B.J. Nicholson. 1991. *Proc. Natl. Acad. Sci.* USA **88**: 8410–8414.
6. Wahl, P., Elster, L., and Schousboe, A., 1994, *J. Neurochem.* **62**: 2457–2463.
7. Polenzani, L., Woodward, R.M., and Miledi, R., 1991, *Proc. Natl. Acad. Sci. USA* **88**: 4318–4322.
8. Hablitz, J.J., 1992, *Synapse* **12**: 169–171.

PATCH CLAMP DETECTION OF NEURORECEPTOR MODULATORS IN CAPILLARY ELECTROPHORESIS

Kent Jardemark,[1] Owe Orwar,[2] Ingemar Jacobson,[1]* Alexander Moscho,[2] Harvey A. Fishman,[2] Anders Hamberger,[1] Mats Sandberg,[1] Richard H. Scheller,[3] and Richard N. Zare[2]

[1]Department of Anatomy and Cell Biology
Göteborg University
Göteborg, Sweden
[2]Department of Chemistry
Stanford University
Stanford, California
[3]Department of Molecular and Cellular Physiology
Howard Hughes Institute
Stanford University
Stanford, California

1. INTRODUCTION

Many areas of research in the life sciences have been hampered by the inability to analyse microenvironments for important biologically active species. For example, the identities of many fast acting neurotransmitter substances in the mammalian brain are unknown, even after at least thirty years of active research, and endogenous ligands to numerous orphan receptors (e.g., plasma membrane opiate receptors) already cloned, remain to be discovered. During the past twenty years, development in instrumentation for microanalysis has resulted in techniques capable of sample handling in the low femtoliter range (10^{-15} L). Yet the development of detection devices for bioactive compounds that are compatible with these instrumentations has been extremely slow.

It is fascinating that some of the most selective and sensitive devices for detecting bioactive molecules rely on biological structures such as receptors and enzymes, genetically engineered by nature millions of years ago. Although it took until Clark and Lyons' invention of the enzyme electrode in 1962 (1) for researchers to recognise the tremendous

---

* Present address: Astra Hässle AB, S-43183 Mölndal, Sweden.

potential of biological sensors, since this time biosensors have rapidly become known as sensitive and selective devices for detection of trace quantities of material in complex biological matrices. A functional definition of a biosensor is a detection system that contains a biologically active component that serves to transduce a chemical reaction or molecular interaction. This molecular recognition event confers selectivity and information regarding the physiological role of the analyte. Biological cells are particularly useful as detectors because they combine molecular recognition and biological amplification that results in high-sensitivity detection. Take for instance cell surface metabotropic receptors linked to second messenger cascades (2), and ionotropic receptors that channel the influx or outflux of several thousand ions across the cell membrane (3). In both cases, the agonist-binding step is amplified several-fold, which allows detection of single molecules for some receptor types.

A major problem of biosensors is the inability to discriminate multiple ligands that activate the same receptor type or act as substrates in the same enzymatic reaction. These difficulties can be overcome by coupling the biosensor to a chemical separation that fractionates the sample components prior to detection. Ideally, the separation technique should be rapid, compatible with the sensing device, and able to handle small sample volumes. Capillary electrophoresis (CE) provides an excellent match to each of these requirements (4).

A brief introduction to the principles and applications of CE and receptor-based biosensors is given to provide a context for the proposed analysis system. A rational approach using patch clamp detection in CE for the identification of neuroactive compounds, such as neurotransmitters and drugs that activate ligand-gated ionotropic receptors, is then described.

2. CAPILLARY ELECTROPHORESIS

2.1. Instrumentation

Capillary electrophoresis is a miniaturized separation technique that fractionates chemical species on the basis of differences in their ratios of electrical charge to frictional drag in solution. It is able to separate complex chemical mixtures with high efficiency (up to 10^6 theoretical plates) in typically less than 20 minutes (4). In its simplest and most common embodiment, a CE system consists of a narrow-bore (5 to 75 µm. i.d.) fused-silica capillary (usually 20 to 100 cm in length) filled with an electrolyte solution. The ends of the capillary are placed in electrolyte-containing reservoirs having either a cathode or an anode connected to a high-voltage source. When an electric field is applied across a solution-filled fused silica capillary, a layer of mobile charge that accumulates along the counter-charged fused silica surface induces electroosmosis (bulk solution flow). Under typical operating conditions for CE this sheath of ions is positively charged, and consequently, drags bulk solution from the anode to the cathode. A practical result of electroosmotic flow is that during a separation in free-solution CE, all species — whether possessing positive, neutral, or negative charge — can be made to migrate in the same direction past a single detector.

2.2. Traditional Detection Techniques for Capillary Electrophoresis

A number of different schemes for detection of intrinsic qualities of analytes based on, for example, UV absorption (5), mass spectrometry (6), conductivity (7), nuclear magnetic resonance (8), and electrochemistry (9) have been successfully combined with CE. With the exception of electrochemical methods which can be used, for example, to monitor the release of a single catecholamine-containing synaptic vesicle from a neuronal cell (10), the major disadvantage of these methods is their lack of high sensitivity which is often required for trace analysis of biological samples. Electrochemical methods on the other hand, are limited by that only a few biomolecules are electroactive at potentials useful for analysis of physiologic samples. For biomolecules that do not contain any intrinsic molecular features for their sensitive detection, a promising strategy is to select a highly fluorescent or fluorogenic species (a reagent that becomes fluorescent after conjugation with the analyte) as a probe molecule, and to link the probe to the analyte either before or after separation (11). For many applications, derivatization at low analyte concentrations is not possible. Also, labeling is extremely difficult in water-based solutions for compounds lacking $-NH_2$, -SH, or other reactive groups. Other disadvantages of derivatization include a low selectivity, loss of sample integrity, and extraneous dilution of the original sample. Radioisotope detection is potentially quite sensitive, but usually requires long detection periods that are not easy to achieve using the standard on-line detection employed in CE (12).

2.3. Coupling Biosensors to Capillary Electrophoresis

All the traditional detection techniques mentioned so far provide information that have a physical or chemical meaning, such as molecular mass, molecular structure, relative fluorescence intensity, redox potentials, and so on. They do not yield, however, any information about the biological activity or physiological role of the separated analytes. Such information is of utmost importance in drug testing (e.g. pharmacokinetics), and screening for natural and unnatural receptor ligands from various biological environments, such as brain tissue.

A powerful format for analyte identification that provides biologically relevant information is obtained by coupling a biosensor to a chemical separation. Leal et al. for example used a cockroach antenna as a detector (electroantennographic detection) to screen for insect pheromones following a chiral GC separation (13). Recently, we have developed a liquid-phase analog of this concept that combines capillary electrophoresis with detection by cultured cells or oocytes microinjected with mRNA that encodes specific membrane receptors (both called single-cell biosensors, SCBs) (14–16). Using the capillary electrophoresis/single-cell biosensor system (CE/SCB), components are separated electrophoretically, delivered onto the surface of a cell, and detected at a characteristic migration time through the capillary. For agonists that activate metabotropic receptors, the cell response is measured as increases in the intracellular concentration of Ca^{2+}, by using the Ca^{2+}-chelating dye Fluo-3 AM ester and fluorescence microscopy. In the case of agonists that activate ionotropic receptors, differences in the transmembrane ion-mediated current are detected by means of two-electrode voltage clamp.

2.4. Patch Clamp Detection in Capillary Electrophoresis

Although CE/SCB systems based on measurements of changes in intracellular Ca^{2+} concentrations or transmembrane currents, are powerful devices for analysis of bioactive compounds, the peak responses obtained provide little information. This situation can be improved significantly for compounds that activate ligand-gated ion channels by implementing patch-clamp detection in CE (17). Patch clamp detection provides detailed micro-

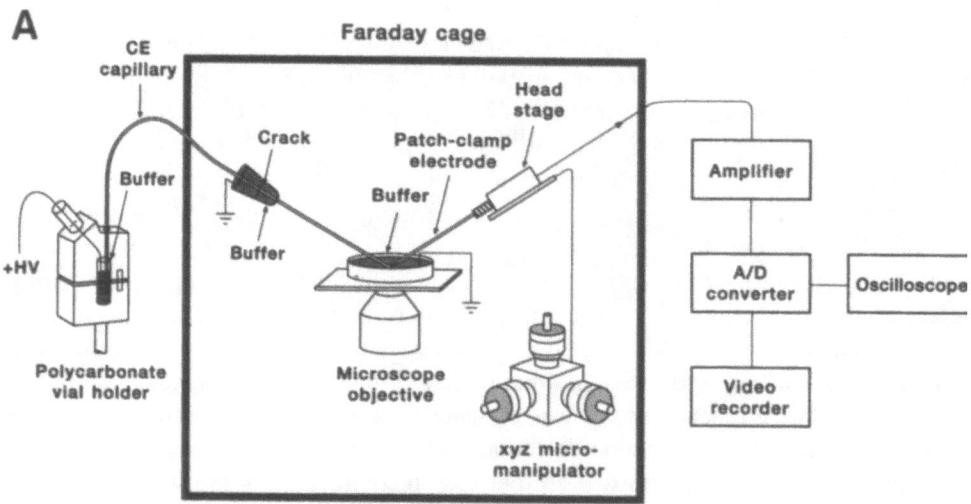

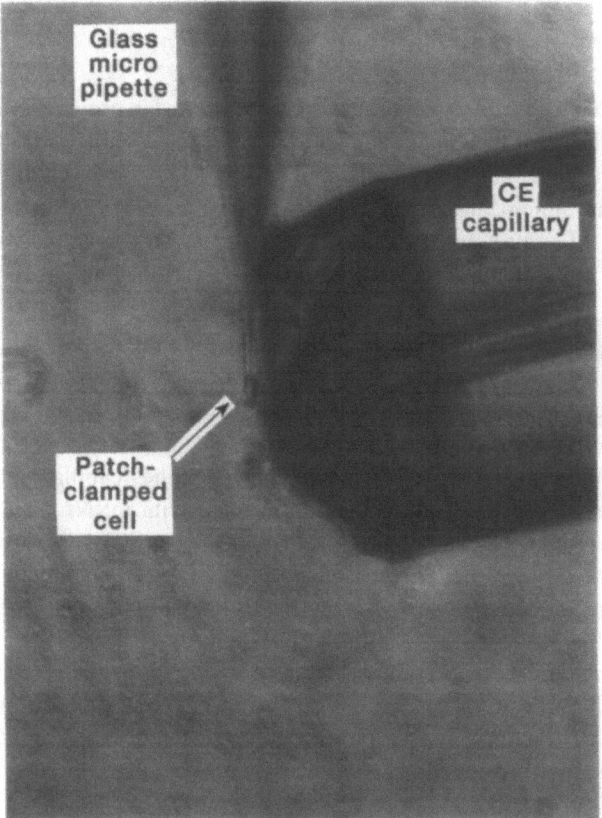

Figure 1. Patch-clamp detection system for capillary electrophoresis (CE). **(A)** The inlet end (sample injection end) of a fused-silica CE separation capillary is connected to a positive high-voltage power supply through a buffer vial housed in a polycarbonate holder equipped with a safety interlock to prevent electric shock. The same buffer used for the cell bath media is used as the electrolyte in the CE capillary and inlet vial, to avoid liquid junction potentials. The tip of the patch clamp electrode is positioned ~ 5 to 25 μm from the capillary outlet by means of a micromanipulator. **(B)** Photomicrograph showing the patch clamp pipette with a whole-cell clamped olfactory interneuron placed at the CE separation column outlet.

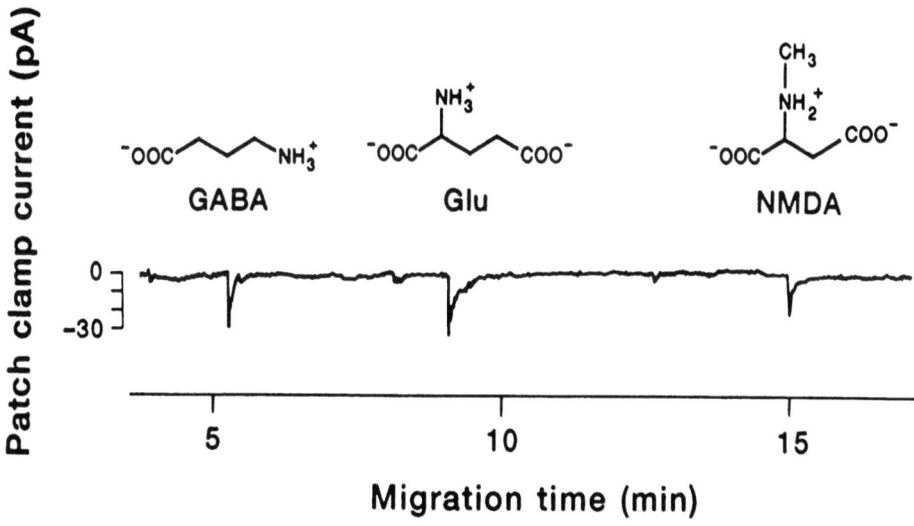

Figure 2. Inward currents recorded from an outside-out patch following separation of GABA, Glu, and NMDA (250 μM of each) by capillary electrophoresis. Current traces were sampled at 2 Hz.

scopic information about the activated ion channels (18). This information includes kinetics, conductance states, open and shut times, and burst lengths, parameters which can be coupled with the electrophoretic mobility of the analyte. Thus, by coupling patch clamp detection to CE, a multidimensional format for agonist identification is provided.

For patch clamp detection in CE, we have used mammalian neurons freshly dissociated from the rat olfactory bulb as biosensors. These cells are known to have GABA (γ-aminobutyric acid) receptors and two types of glutamate receptors: the (R,S)-α-amino-3-hydroxy-5-methyl-4-isoxazole-propionate (AMPA), and N-methyl-D-aspartic acid (NMDA) types. Detection is performed with a standard patch clamp setup (Fig. 1A) with the tip of the patch-clamp electrode placed 5 to 25 μm from the outlet of a 50-μm inner diameter fused silica CE separation capillary (Fig. 1B). The CE capillary is fractured and grounded ~7 cm above the outlet to create an almost field-free region at the position of the patch clamp pipette tip. The residual induced potential is compensated by an offset potential to the patch clamp amplifier system.

In Fig. 2, a mixed standard of GABA, Glu, and NMDA is separated and detected with outside-out patch clamp detection. In addition to migration times, the characteristics of the current responses elicited by the separated agonists further confirm their identity. Spectral analysis of whole-cell current traces of separated components gives power spectra that are fit to the sum of two Lorentzian functions. From these fits, mean single-channel conductance levels (γ/pS) and corner frequencies (f_c/Hz) are obtained (19).

Figure 3 shows a power spectrum and a whole-cell current response elicited by GABA following a CE separation. Furthermore, using the outside-out patch-clamp detection configuration, we can resolve single-channel openings for all ligands separated by CE (17). In essence, this means that it is feasible to detect indirectly a single analyte molecule with this technique.

Potential applications of this technology include screening orphan receptor ligands, neurotransmitters, and excitotoxins in the mammalian brain extracellular fluid and extracts. We are currently investigating the feasibility of performing such analyses.

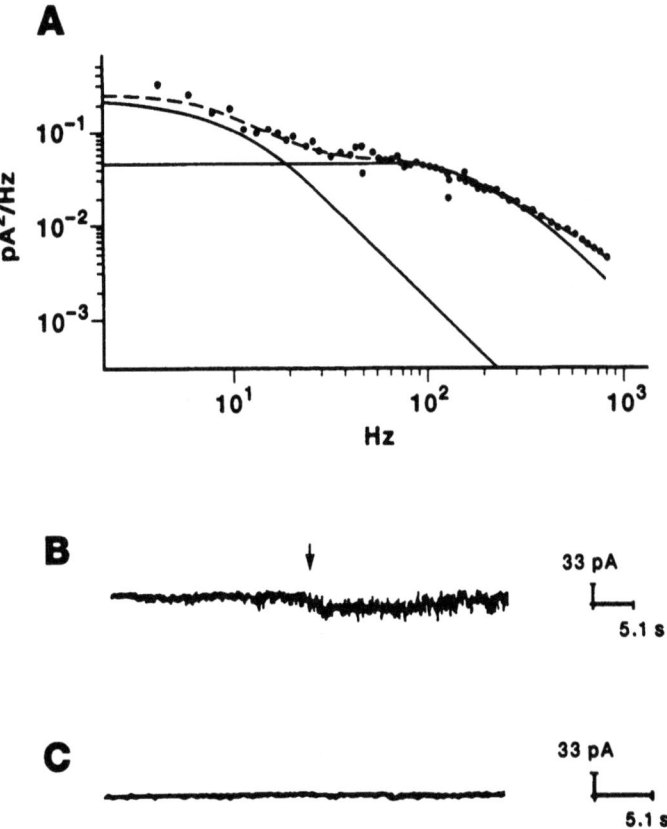

Figure 3. (A) Power spectrum of currents elicited by GABA (250 μM) on $GABA_A$ receptors. Spectral analysis yielded $\gamma = 22.4 \pm 3.1$ pS, $f_{c1} = 10.1 \pm 0.8$ Hz, and $f_{c2} = 214.2 \pm 37.5$ Hz ($n=6$) (B) Example of a whole-cell current elicited by GABA. The arrow shows the onset of the receptor response. (C) Separation of a HEPES saline solution onto the same cell as used in (B), which is used as a control.

In Fig. 4, an electropherogram of a rat brain extract is shown. An olfactory interneuron is used as detector. Even though the peaks have not yet been identified, this separation demonstrates that the technique is capable of analyzing complex biological samples.

By coupling CE to cell-based biosensors, a highly selective and sensitive means for analyzing biologically active components in complex mixtures with a minimum of sample handling is obtained. Unlike traditional techniques for chemical analyses, this system can be engineered for sensitivity to specific biological species, thereby improving the response selectivity and greatly simplifying the challenge of chemical analysis. Patch clamp techniques are especially promising for detection of neurotransmitters because of their (a) high sensitivity, in which unitary ion channel currents elicited by one or a few ligand molecules can be resolved, and (b) high selectivity, in which different receptor–ion channel complexes display characteristic conductance states, open and shut times, and so forth. Also, neuropeptides that normally operate through metabotropic receptors could be detected with this technique if the receptor clones are expressed together with ion channels that couple to the receptor. Because CE offers high separation efficiencies and can be performed in capillary structures capable of handling sample volumes in the low femtoliter

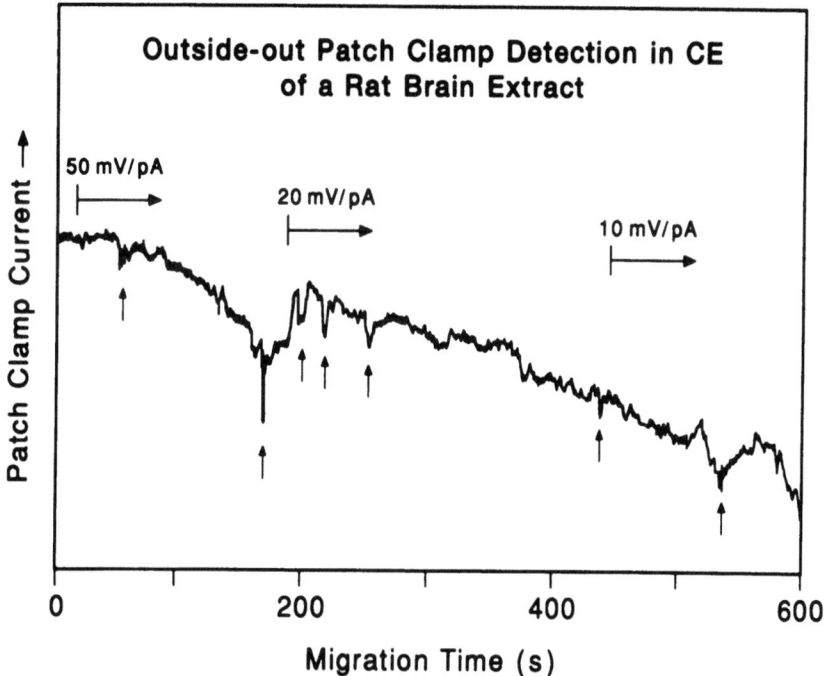

Figure 4. Separation of a rat brain extract detected with an outside-out patch-clamped olfactory interneuron. The arrows indicate peaks with a signal-to-noise ratio greater than two.

range, on-line analysis with CE-patch clamp detection of evoked neurotransmitter release from biological microenvironments such as single cells or discrete nerve terminal areas should be feasible because of conservation of sample integrity.

ACKNOWLEDGMENT

This work was made possible through grants from the National Institute of Mental Health (MH 45423–06) and the National Institute of Drug Abuse (DA 09873–01). O. Orwar is supported by the Swedish Natural Science Research Council (K-PD 10481–303) and A. Moscho is supported by the German Gottlieb Daimler- and Karl Benz-Foundation (2.95.32). Helpful discussions with S. G. Weber, M. E. Meyerhoff, and D. T. Chiu, and technical assistance from H. Zhao, and R. Dadoo are gratefully acknowledged.

REFERENCES

1. Clark, L.C., Jr., Lyons, C., 1962, *Ann. N.Y. Acad. Sci.* **102**: 29–45.
2. Gandhi, C.R., Behal, R.H., Harvey, S.A., Nouchi, T.A., and Olson, M.S., 1992, *Biochem. J.* **287**: 897–904.
3. Hille, B., 1992 *Ionic Channels of Excitable Membranes*, Sinauer, Sunderland, Massachusetts, USA.
4. Li, S.F.Y., 1992, *Capillary Electrophoresis: Principles. Practice and Applications*, Elsevier, Amsterdam, Netherlands.
5. Xue, Y.J., Yeung, E.S., 1994, *Appl. Spect.* **48**: 502–506.
6. Smith, R.D., Wahl, J.H., Goodlett, D.R., Hofstadler, S.A., 1993, *Anal. Chem..*, **65**: 574A.
7. Huang, X., Luckey, J.A., Gordon, M.J., Zare, R.N., 1989, *Anal. Chem..* **61**: 766–770.
8. Olson, D.L., Peck, T.L., Webb, A.G., Magin, R.L., Sweedler, J.V., 1995, *Science*, **270**: 1967–1970.

9. Olefirowicz, T.M., Ewing, A.G., 1990, *Anal. Chem.*, **62**: 1872.
10. Wightman, R.M., Jankowski, J.A., Kennedy, R.T., Kawagoe, K.T., Schroeder, T.J., Leszczyszyn, D.J., Near, J.A., Diliberto, E.J., Jr., Viveros, O.H., 1991, *Proc. Natl. Acad. Sci., USA* **88**: 10754–10758.
11. Cheng, Y.-F., Dovichi, N.J., 1988, *Science*, **242**: 562–565.
12. Tracht, S., Toma, V., Sweedler, J.V., 1994, *Anal. Chem.*, **66**: 2382.
13. Leal, W.S., Shi, X., Liang, D., Schal, C., Meinwald, J., 1995 *Proc. Natl. Acad. Sci., USA* **92**: 1033–1037.
14. Shear, J.B., Fishman, H.A., Allbritton, N.L., Garigan, D., Zare, R.N., and Scheller, R.H., 1995, *Science* **267**: 74–77.
15. Fishman, H.A., Orwar, O., Allbritton, N.L., Modi, B.P., Shear, J.B., Scheller, R.H., Zare, R.N., 1996, *Anal.Chem.* **68**: 1181–1186.
16. Fishman, H.A., Orwar, O., Scheller, R.H., and Zare, R.N., 1995, *Proc. Natl. Acad. Sci. U.S.A.* **92**: 7877–7881.
17. Orwar, O., Jardemark, K., Jacobson, I., Moscho, A., Fishman, H.A., Scheller, R.H., and Zare, R.N., 1996, *Science* **272**: 1779–1782.
18. Neher, E., and Sakmann, B., 1976, *Nature* **260**: 799–80124.
19. Anderson, C.R., Stevens, C.F., 1973, *J. Physiol. (London)* **235**: 655.

Section 43: PET Imaging of Receptors; Experimental Models and Clinical Approaches

NEURORECEPTOR MAPPING *IN VIVO*

The Experimental Approach

Adriaan A. Lammertsma, Susan P. Hume, and Ralph Myers

PET Methodology Group
Cyclotron Unit, MRC Clinical Sciences Centre
Royal Postgraduate Medical School
Hammersmith Hospital
London, United Kingdom

1. POSITRON EMISSION TOMOGRAPHY

Positron emission tomography (PET) is a technique which provides the possibility to non-invasively imaging and measurement of regional physiological, biochemical and pharmacological processes in man *in vivo*. The method was introduced in the late seventies (1,2) and was initially used mainly for flow and metabolism studies in brain, heart and tumours (3). Since the early eighties it has been used increasingly in receptor studies and, more recently, in assessing drug therapies (4).

The increasing use of PET is based on two unique properties, the physical decay characteristics of positron emitters and the types of tracers which can be labelled with these positron emitters. The decay characteristics provide physical, the tracers themselves biological accuracy (3), as discussed below.

A positron is a particle which, except for a positive charge, is identical to an electron. This particle is unstable and, when emitted, will travel at most a few mm in tissue before combining with an electron, resulting in two gamma (annihilation) photons of 511 keV travelling in almost opposite directions. In PET, use is made of this mode of decay by accepting only those events in which two opposing detectors have recorded a pulse simultaneously (in coincidence detection). This electronic "collimation" obviates the need for the lead collimators required in SPET (single photon emission tomography), thereby increasing detection sensitivity. Because of the co-linearity of the two annihilation photons, it is known along which line an event has taken place and, as a result, it is possible to accurately correct for attenuation of the emitted photons within the intervening tissues (3).

From a symposium organized by Albert Gjedde (Aarhus, Denmark) and Adriaan A. Lammertsma (London, United Kingdom). (Supported by Pfizer International Pharmaceuticals Group.)

Neurochemistry, edited by Teelken and Korf
Plenum Press, New York, 1997

The longest lived isotopes of the biological elements carbon, nitrogen and oxygen, which can be used for *in vivo* studies, are all positron emitters: carbon-11, nitrogen-13 and oxygen-15 with half lifes of 20, 10 and 2 minutes, respectively. The short half lifes of these radionuclides dictate the need for an on-site cyclotron and labelling facilities, the main reason why the growth of PET is still limited. On the other hand, carbon-11 in particular, allows labelling of an indefinite number of molecules (tracers) with chemical properties that are identical to those of the non-radioactive molecules (tracees). In other words, no assumptions have to be made about the relationship between the labelled tracer and the tracee: they are (chemically) identical.

Unfortunately, no suitable positron or single photon emitter of hydrogen has been identified. In PET, fluorine-18 is sometimes used as a substitute, but here one has to take into account possible chemical differences between tracer and tracee.

2. RATIONALE FOR PET STUDIES IN SMALL ANIMALS

PET is used extensively in the study of neuroreceptors in man. Based on the properties mentioned above, it is possible to measure receptor binding accurately in both normal subjects and in patients. The success of a PET study, however, depends strongly on the quality of the (radiolabelledĺ) tracer.

The short half life of most positron emitters dictates the need for rapid labelling procedures. In addition, it is necessary that the tracer can be made with high specific activity, to guarantee that a minimal amount of non-radioactive (cold) ligand is injected, so that the tracer injection has no effect on receptor occupancy (i.e. acts like a true tracer).

Apart from these radiochemical considerations, the tracer also needs to fulfil a number of "biological" requirements, taking into account that it is being used for an *in vivo* study. Firstly, to obtain a satisfactory signal (i.e. high signal to noise), the tracer needs to cross the blood-brain barrier. Ideally, metabolism of the tracer should not result in any radiolabelled metabolites. If this does occur, it is important that none of these radiolabelled metabolites cross the blood-brain barrier, since the PET scanner has no chemical resolution. Labelled metabolites can be measured in plasma, but not in brain and would, therefore, result in a contamination of the tissue signal which can not be corrected for. Finally, the level of non-specific binding *in vivo* should be as low as possible.

It is perhaps not surprising that, given these requirements, the number of neuroreceptor studies is still limited by the number of suitable radioligands. For many receptors, still no satisfactory radioligand for *in vivo* use is available.

Therefore, the search for new and better ligands continues. An important step in the evaluation of a potential new ligand is its assessment in laboratory animals. This not only involves testing for selectivity, specificity and tissue metabolites, but also characterization of its kinetic behaviour. These studies require *in vitro* and *ex vivo* methods. *Ex vivo* measurements of tissue kinetics, however, are both animal and labour intensive. In addition, the resulting time-radioactivity curve might suffer from inter-animal variation. The potential to use PET itself to measure tissue time-radioactivity curves in a single animal was realised some years ago (5) and, in our Institute, it is now a standard procedure in the assessment of putative ligands (6,7).

It should be noted that, although PET studies of small animals are a useful addition to the screening procedure of putative ligands, there are also other reasons why this methodology could have much to offer (8). Probably the most important one is the possibility to perform repeat scans in the same animal. This allows for studying both the effects of progression of disease and those of therapeutic intervention (9,10). This is especially important in situations where, for ethical reasons, therapy can not be withheld from patients. PET studies in small animals can also be important in interpreting results obtained in patients. The number of PET studies in patients is usually limited and if there is a change in signal (binding) in a certain patient population, it is easier to perform multiple studies in an animal model of that disease in order to investigate the reason for the change, e.g. change in B_{max}, K_d or endogenous agonist concentrations (9). Finally, the combination of animal PET and postmortem sampling might be useful in the development and validation of tracer kinetic models for the interpretation of PET data, especially if a labelled metabolite enters the tissue. The reference tissue model (11,12) and its recent improvement (13) are examples of models originally developed for animal PET studies, which were then implemented for human studies.

3. DEDICATED SMALL ANIMAL PET SCANNERS

It will be clear that PET scanning of small animals can be performed with a clinical PET scanner (5,9,11). With such a scanner it is even possible to scan several animals simultaneously, and the sensitivity and spatial resolution are such that it is possible to quantify differences in [^{11}C]raclopride binding in unilateral striatal 6-hydroxydopamine lesions.

There are, however, a number of disadvantages in using clinical PET scanners for animal studies. Firstly, at least in our Institute, access to the clinical PET cameras is very limited. Secondly, the availability of a small animal scanner allows studies to be planned in such a way that the same radioligand is used simultaneously for a clinical and an animal study, increasing the cost-effectiveness of its production. This is important since, although less activity is required for the animal study, the requirement for high specific activity is the same, so no shortcuts can be made in the radiochemical production. Finally, sensitivity and spatial resolution of a clinical PET scanner are not optimal for small animal studies. Optimization of these physical parameters is important. It is obvious that, given the size of the rat brain, the spatial resolution of the scanner should be as high as possible. For a given size of detector, both spatial resolution and sensitivity can be improved by placing the detectors in a smaller ring. Although, for animal studies, there are no limitations related to the radioactive dose, sensitivity is still important. This is due to the fact that the ligand has a finite specific activity. Therefore, the injectate is limited by the amount of "cold" ligand that can be injected without actually interfering with the receptor system under study.

Several groups have developed or have under development PET scanners specifically designed for small animal studies (14–18). In our Institute, we have progressed from a prototype dual probe system (19) to a full ring, high resolution, small diameter PET scanner (CTI PET Systems, Knoxville, TN, USA), the physical characteristics of which have been described elsewhere (20). In brief, this tomograph consists of 16 bismuth germanate detector blocks arranged in a ring with diameter of 11.5 cm. The axial field of view is 5 cm and the slice thickness 4.3 mm full width at half maximum (FWHM). The transaxial spatial resolution is 2.3 mm FWHM at the centre of the field of view (FOV), decreasing to 4 mm FWHM at a distance of 1 to 2 cm from the centre.

4. APPLICATION TO THE DOPAMINERGIC SYSTEM

To test the performance of the above mentioned dedicated small animal scanner in practice, studies were performed in adult male Sprague-Dawley rats. In particular, the dopaminergic system was assessed using [^{11}C]raclopride (D_2 receptors), [^{11}C]SCH 23390 (D_1 receptors), [^{11}C]RTI-121 (dopamine transporter) and [^{18}F]fluorodopa. Here only a summary of the results with some practical comments will be given. A full description of the results has been given elsewhere (21).

It is important to realize that the slice thickness of the scanner is large compared to the size of the rat striatum. This means that accurate positioning is extremely important for reproducible results. This was achieved by designing a perspex bed with associated head holder, comprising of ear bars, tooth bar and gaseous anesthetic access (22). Using this bed it was possible to accurately position the striata in the centre of the FOV. Using a stereotaxic reference (23) it was then possible to locate other structures in the brain, which was essential for region of interest (ROI) definition.

Kinetic analysis of the data requires good temporal resolution. For [^{11}C]raclopride and [^{11}C]SCH 23390 a 60 minutes scan was performed, consisting of 21 time frames with a duration ranging from 5 seconds at the beginning to 5 minutes at the end of the scan. For [^{11}C]RTI-121 with its slower kinetics, the scanning protocol was prolonged with a further 3 frames of 10 minutes duration. [^{18}F]fluorodopa kinetics were analyzed with a graphical method (24) and, therefore, a slower protocol of 14 frames over 2 hours was used.

Using the ANALYZE image analysis software package (25), all images were interpolated to give cubic voxels with 0.47 mm dimensions. ROI were defined for left and right striatum (96 voxels each) and for cerebellum (138 voxels). Early images following i.v. tracer injection primarily showed a flow distribution with similar uptake in cerebellum and striatum for all ligands. Thereafter, for the carbon-11 labelled ligands, there was much faster washout from the cerebellum with late images showing bilateral hot spots in the striatum. The improvement in resolution achieved using the small animal scanner compared with a clinical PET scanner was demonstrated by the fact that left and right striatum were clearly separated in the image in contrast to our previous study. It should be noted, however, that despite this expected improvement in resolution, the ratio of striatum to cerebellum counts was still much lower than that obtained from post mortem sampling studies (e.g. for [^{11}C]raclopride the ratio reduced from 9 to 2.5 and for [^{11}C]SCH 23390 from 8 to 4), illustrating the presence of significant partial volume effects. Losses in recovery of tissue counts occur if the size of the structure being sampled is less than twice the spatial resolution of the scanner (26). For [^{18}F]fluorodopa, there were no clear striatal hot spots in the late images, illustrating that the original signal (around two for the striatum to cerebellum count ratio) was too small to be detected in the presence of these partial volume effects.

The importance of human neuroreceptor PET studies does not lie in being able to follow the kinetics of a ligand, but in extracting values of binding parameters from these curves. The same is true for animal studies. An additional problem with studies in rats, however, is the difficulty in obtaining metabolite corrected plasma curves for individual rats. Although it is now possible to measure plasma curves with high temporal resolution (27), measurement of individual plasma metabolites is almost impossible. As an alternative, a reference tissue model has been developed (11), which was later extended and validated for human PET scans (12). This model allows the use of the cerebellum curve as an indirect input function, assuming that the level of non-specific binding in cerebellum and striatum is the same. It does, however, not assume that the time course of free ligand in both tissues is the same, nor does it assume that delivery (flow) to both tissues is the same. The parameter of interest is binding potential (BP), defined as the ratio between rate constants to and from the specifically bound compartment, which is obtained using standard non-linear regression techniques.

Using the reference tissue model, BP values were estimated for [^{11}C]raclopride, [^{11}C]SCH 23390 and [^{11}C]RTI-121. The coefficient of variation (COV) for all three ligands was about 10% with no significant differences between left and right striatum. After predosing with cold ligand, BP reduced significantly by a factor ranging from 2 (raclopride) to 10 (RTI-121). In other words, the COV was small compared to the absolute reduction in BP, opening the possibility to perform studies in animal models of disease. This was confirmed in animals with unilateral pre- or post-synaptic lesions of the nigrostriatal dopaminergic pathway, where a significant reduction in the ipsilateral striatum was observed. Although there were some small effects in the contralateral striatum, which were different for the three ligands, only for [^{11}C]raclopride did this reach significance. This variation can be explained by different cerebral distributions and partial volume effects (21).

5. CONCLUSIONS AND FUTURE DEVELOPMENTS

The small animal PET scanner can be a useful addition to the methodology for screening putative ligands. In fact, the [^{11}C]RTI-121 studies mentioned above were carried out as part of its preclinical evaluation. Another ligand where the scanner has been used in this way is the $5HT_{1A}$ antagonist WAY100635 (7).

More importantly, the scanner can be used in monitoring changes in pre- and postsynaptic dopaminergic function in animal models of disease. In addition, efficacy of various treatment regimes can be assessed objectively.

The successful application of the scanner in studying the kinetics of above mentioned dopaminergic ligands does not imply that the system can be used for ligands which have a more homogeneous distribution in the brain. The lack of an observable signal with [^{18}F]fluorodopa might serve as a warning. At present, due to partial volume effects, the system is only useful if there is a localized signal of sufficient intensity. This, however, does not preclude use of the scanner for more homogeneous distributions, if the purpose of the study is to find a more global effect.

It should also be realized that the studies described here only represent the current status in a rapidly moving field (28). Progress is still being made in improving the performance characteristics of small animal scanners, for example by using other detector material and smaller crystals (29). Even for the prototype scanner mentioned above, significant modifications are in progress. So far, the scanner has been used in 2D mode without applying any scatter correction. Recently, the performance of the scanner in 3D mode has been characterized (30), demonstrating an improvement in sensitivity by a factor of 6 at the centre of the FOV. This should result in better statistics (less noise) and hence more accurate estimates of BP. Implementation of a scatter correction method would improve the sensitivity for true signals even further.

Recently, the robustness of the kinetic analyses has also improved (13). Finally, measurements are in progress to characterize partial volume and spill-over effects in "biological" phantoms with the view of implementing possible corrections. The result of all these improvements is a more accurate and precise estimation of BP. This, in turn, means that smaller, more subtle, changes can be detected, thereby enhancing the pharmacological sensitivity of the method.

REFERENCES

1. Phelps, M.E., Hoffman, E.J., Mullani, N.A., and Ter-Pogossian, M.M., 1975, *J. Nucl. Med.* **16**: 210–224.
2. Ter-Pogossian, M.M., Phelps, M.E., Hoffman, E.J., and Mullani, N.A., 1975, *Radiology* **114**: 89–98.
3. Lammertsma, A.A., and Frackowiak, R.S.J., 1985, *Crit. Rev. Biomed. Eng.* **13**: 125–169.

4. Comar, D. (ed.), 1995, *PET for Drug Development and Evaluation*, Kluwer Academic Publishers, Dordrecht.
5. Ingvar, M., Eriksson, L., Rogers, G.A., Stone-Elander, S., and Widén, L., 1991, *J. Cereb. Blood Flow Metab.* **11**: 926–931.
6. Hume, S.P., Luthra, S.K., Brown, D.J., Opacka-Juffry, J., Osman, S., Ashworth, S., Myers, R., Brady, F., Carroll, F.I., Kuhar, M.J., and Brooks, D.J., 1996, *Nucl. Med. Biol.* **23**: 377–384.
7. Pike, V.W., McCarron, J.A., Hume, S.P., Ashworth, S., Opacka-Juffry, J., Osman, S., Lammertsma, A.A., Poole, K.G., Fletcher, A., White, A.C., and Cliffe, I.A., 1995, *Med. Chem. Res.* **5**: 208–227.
8. Lammertsma, A.A., Hume, S.P., Myers, R., Ashworth, S., Bloomfield, P.M., Rajeswaran, S., Spinks, T., and Jones, T., 1995, *J. Nucl. Med.* **36**: 2391–2392.
9. Hume, S.P., Opacka-Juffry, J., Myers, R., Ahier, R.G., Ashworth, S., Brooks, D.J., and Lammertsma, A.A., 1995, *Synapse* **21**: 45–53.
10. Torres, E.M., Fricker, R.A., Hume S.P., Myers, R., Opacka-Juffry, J., Ashworth, S., Brooks, D.J., and Dunnett, S.B., 1995, *NeuroReport* **6**: 2017–2021.
11. Hume, S.P., Myers, R., Bloomfield, P.M., Opacka-Juffry, J., Cremer, J.E., Ahier, R.G., Luthra, S.K., Brooks, D.J., and Lammertsma, A.A., 1992, *Synapse* **12**: 47–54.
12. Lammertsma, A.A., Bench, C.J., Hume, S.P., Osman, S., Gunn, K., Brooks, D.J., and Frackowiak, R.S.J., 1996, *J. Cereb. Blood Flow Metab.* **16**: 42–52.
13. Lammertsma, A.A., and Hume, S.P., 1996, *NeuroImage* (submitted).
14. Watanabe, M., Uchida, H., Okada, H., Shimizu, K., Satoh, N., Yoshikawa, E., Ohmura, T., Yamashita, T., and Tanaka, E., 1992, *IEEE Trans. Med. Imag.* **11**: 577–580.
15. Tavernier, S., Bruyndonckx, P., Shuping, Z., 1992, *Phys. Med. Biol.* **37**: 635.
16. Marriott, C.J., Cadorette, J.E., Lecomte, R., Scasnar, V., Rousseau, J., and Van Lier, J.E., 1994, *J. Nucl. Med.* **35**: 1390–1397.
17. Hutchins, G.D., Simon, A.J., Winkle, W., and Carlson, K., 1996, *J. Nucl. Med.* **37**: 86P.
18. Cherry, S.R., Shao, Y., Silverman, R.W., Siegel, S., Meadors, K., Mumcuoglu, E., Young, J., Jones, W.F., Moyers, C., Andreaco, M., Paulus, M., Binkley, D., Nutt, R., and Phelps, M.E., 1996, *J. Nucl. Med.* **37**: 86P.
19. Rajeswaran, S., Hume, S.P., Cremer, J.E., Young, J., Bailey, D.L., Ashburner, J., Luthra, S.K., Jones, A.K.P., and Jones, T., 1991, *J. Neurosci. Meth.* **40**: 223–232.
20. Bloomfield, P.M., Rajeswaran, S., Spinks, T.J., Hume, S.P., Myers, R., Ashworth, S., Clifford, K.M., Jones, W.F., Byars, L.G., Young, J., Andreaco, M., Williams, C.W., Lammertsma, A.A., and Jones, T., 1995, *Phys. Med. Biol.* **40**: 1105–1126.
21. Hume, S.P., Lammertsma, A.A., Myers, R., Rajeswaran, S., Bloomfield, P.M., Ashworth, S., Fricker, R.A., Torres, E.M., Watson, I., and Jones, T., 1996, *J. Neurosci. Meth.* (in press).
22. Myers, R., Hume, S.P., Ashworth, S., Lammertsma, A.A., Bloomfield, P.M., Rajeswaran, S., and Jones, T, 1996, *Quantification of Brain Function using PET* (R. Myers, V. Cunningham, D. Bailey, and T. Jones, eds.), Academic Press, San Diego, PP. 12–15.
23. Paxinos, G., and Watson, W., 1986, *The Rat Brain in Stereotaxic Coordinates*, 2nd ed., Academic Press, NSW, Australia.
24. Patlak, C.S., Blasberg, R.G., and Fenstermacher, J.D., 1983, *J. Cereb. Blood Flow Metab.* **3**: 1–7.
25. Robb, R.A., and Hanson, D.P., 1991, *Australas. Phys. Eng. Sci. Med.* **14**: 9–30.
26. Hoffman, E.D., Huang, S.-C., and Phelps, M.E., 1979, *J. Comput. Assist. Tomogr.* **3**: 299–308.
27. Ashworth, S., Ranicar, A., Bloomfield, P.M., Jones, T., and Lammertsma, A.A., 1996, *Quantification of Brain Function using PET* (R. Myers, V. Cunningham, D. Bailey, and T. Jones, eds.), Academic Press, San Diego, PP. 62–66.
28. Jones, T., 1996, *Eur. J. Nucl. Med.* **23**: 807–813.
29. Melcher, C.L., and Schweitzer, J.S., 1992, *Nucl. Instr. Methods Phys. Res.* **14**: 212–214.
30. Bloomfield, P.M., Myers, R., Hume, S.P., Spinks, T.J., Lammertsma, A.A., and Jones, T., 1996, *Phys. Med. Biol.* (submitted).

Section 43: PET Imaging of Receptors; Experimental Models and Clinical Approaches

PET IMAGING OF NEUROTRANSMISSION SYSTEMS IN NEUROLOGY

The Example of the Nigrostriatal Dopaminergic Pathway

Eric Salmon

Cyclotron Research Centre
University of Liege
B30 Sart Tilman, 4000 Liege, Belgium

1. INTRODUCTION

Positron Emission Tomographic (PET) studies of neurotransmission systems provide different types of clinical, physiological or pharmacological informations in neurology. It appears impossible to make an overview of the tremendous amount of data regularly published on new ligands and their distribution in various neurological diseases. In this article, we will concentrate on the most widely investigated nigrostriatal dopaminergic system and on parkinsonian disorders. However, the discussion may easily be transposed to other neurotransmission systems and diseases. Applications of the technique will be illustrated into four main fields of clinical interest by selected PET data.

2. THE MARKERS OF CELL LOSS

A decrease of radiotracer uptake may frequently reflect neuronal loss. So, (^{18}F)fluorodopa is considered as a marker of the integrity of nigrostriatal dopaminergic terminals (1). Decrease of striatal (^{18}F)fluorodopa accumulation in Parkinson's disease is related to nigral dopaminergic neuronal count (2), and to severity of bradykinetorigid symptoms (3). (^{18}F)fluorodopa striatal uptake is also a marker of the extend of nigral lesions in other parkinsonian syndromes, such as progressive supranuclear palsy, and MPTP or post-encephalitis parkinsonism for example (4,5,6). Kinetic analysis of (^{18}F)fluorodopa has been worked out, taking metabolites formation into account (7,8), but simplest methods of analysis, such as ratio method or multiple-time graphical analysis, allow good discrimination between Parkin-

From a symposium organized by Albert Gjedde (Aarhus, Denmark) and Adriaan A. Lammertsma (London, United Kingdom). (Supported by Pfizer International Pharmaceuticals Group).

sonian patients and controls (9,10). (^{18}F)fluorodopa PET could allow neurologists to detect preclinical Parkinson's disease (11), and this would be of major importance for trials of neuroprotective drugs in the disease.

Ligands of presynaptic reuptake sites are also used to assess integrity of nigrostriatal dopaminergic terminals in parkinsonian syndromes (1,3). The first tracer was (^{11}C)nomifensine (12,13), but a series of cocaine analogues have now been developed for both PET and single photon emission computed tomography (SPECT), which have better selectivity and greater specific/nonspecific binding (14,15). Interestingly, a decrease of binding to reuptake sites was observed both in striatum and in the brainstem of Parkinsonian patients (15). The main drawback for studies of reuptake sites is the lack of data on their regulation with disease progression and chronic medication.

There are a lot of PET (and SPECT) ligands for dopaminergic D1 and D2 receptors. Those D1 and D2 receptors are mainly located on striatal neurons with different properties, for example, on striatonigral and striatopallidal neurons, respectively (16). A decrease of striatal D1 and D2 receptors was clearly demonstrated in striato-nigral degeneration, a degenerative parkinsonian disorder not responding to dopatherapy (4,17). The progression of D2 receptors decrease in vivo confirmed the pattern of evolution suggested from neuropathological data in striatonigral degeneration (18). In larger populations, density of striatal D2 receptors was used to differentiate idiopathic Parkinson's disease from other parkinsonian disorders (19). Early studies on progressive supranuclear palsy (PSP), another parkinsonian degenerative disease, also suggested that a decrease of D2 receptors played a role in patient's lack of response to dopatherapy (20). However, receptor studies must clearly be (1) interpreted as a first level of the entire intracellular mechanism of neurotransmission, and (2) integrated into a global view of subcortico-cortical functional pathways. Effectively, further studies showed that D2 receptors decrease in PSP was relatively mild (4), and that impaired metabolism in striato-thalamo-cortical motor network is probably critical to explain the clinical syndrome (21).

In summary, PET imaging of neurotransmission systems may reflect neural pathway integrity, and PET data may first provide in vivo information on brain disease neuropathology.

3. PHYSIOLOGICAL INFORMATIONS

The functional status of neurotransmission system do not only reflect the integrity of pre- and post-synaptic sides, but it may also provide physiological informations. Kinetic analysis of (^{18}F)fluorodopa accumulation suggestedincreased catabolism of dopamine in Parkinson's disease, but this hypothesis was not studied with other markers of the dopaminergic pathway, such as inhibitors of monoamine oxidase, for example (22). A relative increase of tracer binding to D2 dopaminergic receptors was repeatedly demonstrated in drug naïve Parkinsonian patients, and was interpreted as D2 upregulation following decrease of presynaptic nigrostriatal activity (23). However, there is actually no clear explanation for the lack of D1 receptor change in Parkinsonian patients showing increased D2 receptors density (24). D2 receptors density in chronically treated patients with Parkinson's disease is either normal, in sustained responders, or slightly decreased, probably in fluctuating responders (4, 25). Tardive dyskinesia complicating long term neuroleptic treatment is associated with normal striatal D2 receptor density, but the severity of orofacial movement is related to receptor density (26).

4. DEMONSTRATION OF DRUG ACTION

There was a renewed interest in Neurology for monoamine oxidase-B inhibitors (IMAO-B) when those drugs were shown to prevent MPTP neurotoxicity to substantia nigra pars compacta (27). Treatment of de novo Parkinsonians with an IMAO-B (deprenyl) delayed the need for L-dopatherapy, and this result was taken as argument for a neuroprotective action of the drug (28). (^{11}C)-L-deprenyl may now be used as a marker of MAO-B activity, to study the degree and reversibility of MAO-B inhibition by other compounds (29).

5. FOLLOW UP OF THERAPEUTIC STRATEGIES

(^{18}F)fluorodopa is used in most studies of fetal mesencephalic tissue transplantation into the striatum of patients with Parkinson's disease, to assess functional integrity within a surviving graft (30). The tracer accumulates into grafts, but evolution of its metabolism requires further investigations. On the other hand, longitudinal studies using (^{18}F)fluorodopa provide an assessment of the progression of nigro-striatal lesions in Parkinson's disease (31). Since assessment of neuroprotection cannot be accurately performed clinically if a drug has a symptomatic effect, neuroprotection could eventually be demonstrated by a reduction of progression of the nigrostriatal biochemical impairment.

6. CONCLUSION

PET provides accurate informations on integrity and functional status of neurotransmission systems, but there is still a need for new tracers to study relevant steps of biochemical pathways. Neurotransmission systems may be submitted to physiological or pharmacological challenges, but the results must be integrated into a more general functional scheme, taking into account data obtained with sensorimotor or cognitive activation paradigms.

REFERENCES

1. Brooks, D.J., Salmon, E., Mathias, C., Quinn, N., Leenders, K.L., Banister, R., Marsden, C.D., and Frackowiak, R.S., 1990, *Brain* 113: 1539–1552.
2. Snow, B.J., Tooyama, I., McGeer, E.G., Yamada, T., Calne, D.B., Takahashi, H., and Kimura, H., 1993, *Ann Neurol* 34: 324–330.
3. Leenders, K.L., Salmon, E., Tyrrel, P., Perani, D., Brooks, D.J., Sager, H., Jones, T., Marsden, C.D., and Frackowiak, R.S.J., 1990, *Arch Neurol* 47: 1290–1298.
4. Brooks, D.J., Ibanez, V., Sawle, G.V., Playford, E.D., Quinn, N., Mathias, C.J., Lees, A.J., Marsden, C.D., Bannister, R., and Frackowiak, R.S.J., 1992, *Ann. Neurol.* 31: 184–192.
5. Calne, D.B., Langston, J.W., Martin, W.R.W., Stoessl, A.J., Ruth, T.J., Adam, M.J., Pate, B.D., and Schulzer, M., 1985, *Nature* 317: 246–248.
6. Picard, F., Hirsch, E., Salmon, E., Marescaux, C., Collard, M., 1996, *Rev Neurol* 152: 267–271.
7. Kuwabara, H., Cumming, P., Reith, J., Léger, G., Diksic, M., Evans, A.C., and Gjedde, A., 1993, *J. Cereb. Blood Flow Metab.* 13: 43–56.
8. Wahl, L., Nahmias, C., 1993, *J Nucl Med* 37: 432–437.
9. Hoshi, H., Kuwabara, H., Léger, G., Cumming, P., Guttman, M., Gjedde, A., 1993, *J. Cereb. Blood Flow Metab* 13: 57–69.
10. Vingerhoets, F., Schultzer, M., Ruth, T., Holden, J.E., Snow, B. 1996, *J Nucl Med* 37: 421–426.
11. Brooks, D.J., 1991, *Neurology* 41 (suppl 2): 24–27.

12. Salmon, E., Brooks, D.J., Leenders, K.L., Turton, D.R., Hume, S.P., Cremer, J.E., Jones, T., and Frackowiak, S.J., 1990, *J Cereb Blood Flow Method* **10**: 307–316.
13. Salmon, E., Brooks, D.J., 1992, *J Neurol Neurosurg Psychiat.* **55**: 167.
14. Shaya, E.K., Scheffel, U., Dannals, R.F., Ricaurte, G.A., Carroll, F.I., Wagner, H.N., Kuhar, M.J., and Wong, D.F., 1992, *Synapse* **10**: 169–172.
15. Frost, J.J., Rosier, A.J., Reich, S.G., Smith, J.S., Ehlers, M.D., Snijder, S.H., Ravert, H.T., and Dannals, R.F., 1993, *Ann Neurol* **34**: 423–431.
16. Gerfen, C.R., 1992, *Ann. Rev. Neurosci* **15**: 285–320.
17. Shinotoh, H., Inoue, O., Hirayama, K., Aotsuka, A., Asahina, M., Suhara, T., Yamazaki, T., and Takeno, Y., 1993, *J Neurol Neurosurg Psychiat* **56**: 467–472.
18. Salmon, E., Sadzot, B., Maquet, P., Lemaire, Plenevaux, Damhaut and Franck, G.H.P., 1994, *J. Neurol.*, **241** (Suppl. 1): S36.
19. Schwarz, J., Tatsch, K., Arnold, G., Gasser, T., Trenkwalder, C., Kirsch, C.M., and Oertel, W.H., 1992, *Neurology* **42**: 556–561.
20. Barone, P., Braun, A.R., and Chase, T.N., 1986, *Clin. Neuropharmacol.* **9** suppl. 4: 128–130.
21. Salmon, E., Van der Linden, M., Franck, G., 1996, *J Neurol.* **243** (suppl 2): S58.
22. Kuwabara, H., Cumming, P., Léger, G., Gjedde, A., 1992, *J. Nucl. Med.* **33**: 916.
23. Sawle, G.V., Playford, E.D., Brooks, D.J., Quinn, N., and Frackowiak, R.S., 1993, *Brain* **116**: 853–867.
24. Rinne, J.O., Laihinen, A., Någren, K., Bergman, J., Solin, O., Haaparanta, M., Ruotsalainen, U., and Rinne, U.K., 1990, *J Neurosci Res.* **27**: 494–499.
25. Rinne, U.K., Lönnberg, P., Koskinen, V., 1981, *J. Neural Transm* **51**: 97–106.
26. Blin, J, Baron, J.C., Cambon, H., Bonnet, A.M., Dubois, B., Loc'h, C., Mazière, B., and Agid, Y., 1989, *J. Neurol. Neurosurg. Psychiat.* **52**: 1248–1252.
27. Langston, J.W., 1985, *TINS* **2**: 79–83.
28. Tetrud, J.W., Langston, J.W., 1989, *Science* **245**: 519–522.
29. Fowler, J.S., Volkow, N.D., Logan, J., 1993, *Neurology* **43**: 1984–1992.
30. Sawle, G.V., Bloomfield, P.M., Björklund, A., Brooks, D.J., Brundin, P.L., Leenders, K.L., Lindvall, O., Marsden, C.D., Rehncrona, S., Widner, H., and Frackowiak, R.S.J., 1992, *Ann. Neurol* **31**: 166–173.
31. Vingerhoets, F.J.G., Snow, B.J., Lee, C.S., Schulzer, M., Mak, E., Calne, D.B., 1994, *Ann Neurol* **36**: 759–764.

Section 43: PET Imaging of Receptors; Experimental Models and Clinical Approaches

MEASUREMENT OF THE CEREBRAL UPTAKE AND METABOLISM OF L-6-[^{18}F] FLUORO-3,4-DIHYDROXY-PHENYLALANINE IN NEWBORN PIGLETS

P. Brust, R. Bauer,[1] R. Bergmann,[2] B. Walter,[1] J. Steinbach,[2] F. Füchtner,[2] E. Will,[2] H. Linemann,[3] M. Obert,[2] U. Zwiener,[1] and B. Johannssen[2]

[1]Institute of Pathophysiology
Friedrich-Schiller University
Jena, Germany
[2]Research Centre Rossendorf
Dresden, Germany
[3]Clinic for Nuclear Medicine
University of Dresden
Dresden, Germany

1. INTRODUCTION

Dopamine is a major striatal neurotransmitter. Therefore, it has been given special attention in early brain disorders. Dopamine in immature brain plays an important role in the mechanisms of irreversible disturbances in neuronal metabolism due to moderate till severe brain hypoxia (1). Age-related increases in the density of dopamine D_1 and dopamine D_2 receptors during human brain development has been described (2,3). Because the in vivo dopamine synthesis appears to be regulated by D_2 receptors (4) one may expect that changes of the rate of dopamine synthesis occur during brain development. Knowledge of these processes may be of importance for the understanding of the genesis of Parkinsons disease and other brain disorders.

However, available information on the catecholamine metabolism in the immature brain is very fragmentary. L-6-[^{18}F] fluoro-3,4-dihydroxyphenylalanine (^{18}F-DOPA) has been used as a tracer to estimate the activity of the dopamine synthetizing enzyme DOPA-decarboxylase (DDC; EC 4.1.1.28) with positron emission tomography (PET) in human brain (5,6). The clinical significance of this method has been demonstrated (7,8).

From a symposium organized by Albert Gjedde (Aarhus, Denmark) and Adriaan A. Lammertsma (London, United Kingdom). (Supported by Pfizer International Pharmaceuticals Group.)

Neurochemistry, edited by Teelken and Korf
Plenum Press, New York, 1997

Data allowing insights in the regulation of dopamine synthesis during ontogenesis are difficult to obtain in humans. In order to find a useful model, we have performed PET experiments with newborn piglets using ^{18}F-DOPA as tracer.

2. MATERIALS AND METHODS

Six male newborn piglets in age from 2 to 6 days and weighing between 1,8 and 3,2 kg were anesthetized with 0.5% isoflurane in a gas mixture of 70% nitrous oxide and 30% oxygen via endotracheal tube using a volume-controlled ventilator. Blood gases, blood glucose and lactate, blood pressure, EEG and ECG were monitored before and during the PET study. No statistically significant differences were found (data not shown).

Cerebral blood flow (CBF) was measured using the reference sample method with coloured microspheres and was calculated from the following formula:

$$CBF = (A_t \times Q_r)/(A_r \times W_t) \times 100$$

where A_t, Q_r, A_r, W_t are tissue absorption, reference blood withdrawal rate, reference blood absorption and tissue weight.

Cerebral blood volume (CBV) was measured in separate animals of the same age after i.v. injection of 20–30 MBq ^{99m}Tc-O$_4$ in saline under the same conditions as described. Euthanization by i.v. injection of saturated KCl occured 5 min after injection. CBV was calculated from the following formula:

$$CBV = (A_t \times W_b)/(A_b \times W_t) \times 100$$

where A_t, A_b, W_t, W_b are tissue activity, blood activity, tissue weight and blood weight.

PET studies were performed using the positron emission tomograph POSITOME IIIp (Montreal Neurological Institute). The 2-ring tomograph acquires three slices with an axial slice thickness (FWHM) of 15 mm, a spatial resolution (transaxial) of 11 mm and an axial field of view of 5 cm. Attenuation correction was performed using transmission scans with two rotating ^{68}Ge point sources. ^{18}F-DOPA was produced according to the destannylation method (9) by direct fluorination of the tin precursor with [^{18}F]F$_2$ simplifying the procedure (to be published). The process takes 40 min, the overall decay corrected yield is in the range of 16 to 20%.

A total of 30–50 MBq ^{18}F-DOPA in 5–10 ml saline was injected into the lower caval vein over 1 min with an automated infusion pump. Emission scanning began simultaneously with the start of ^{18}F-DOPA injection. Dynamic scan data were acquired in list mode between 0 and 120 min with 35 frames between 30 and 600 s each. The time course of plasma ^{18}F radioactivity was determined by arterial blood sampling from the abdominal aorta. Twenty five 15 sec samples and fifteen 1 min samples were taken by a precision peristaltic pump. Nine discrete samples were taken during the remaining time period. The samples were stored on ice and centrifuged to obtain plasma. Plasma activity in 100 μl was measured in a well counter (COBRA II) cross-calibrated with the tomograph. In additional nine blood samples taken at 2, 4, 8, 12, 16, 25, 50, 90 and 120 min p.i. plasma metabolites of ^{18}F-DOPA were measured using HPLC with radiochemical detection.

The HPLC-system consisted of a Hewlett-Packard 1050 pump with autosampler (0.5 ml sample loop) and a flow-through radioactivity detector (Flow One Beta A 100, Caberra Packard) with high γ-γ coincidence detection efficiency. A C18 reverse-phase column (125 x 4 mm, LiChrospher 100, 5 μm) and a guard column (4 x 4 mm, LiChrospher 100, 5 μm) were used for separation of metabolite peaks with the following eluents: A: H_2O (0.1 % acetic acid), B: Acetonitril:H_2O = 80:20 (0.1 % acetic acid) at a flow rate of 1 ml/min and a temperature of 30 °C (start: A:B = 100:0, end (15 min): A:B = 60:40).

3. RESULTS

The values of CBV measured in separate animals are shown in Fig. 1. The data varied between 1.18 ml/100g (white matter) and 3.41 ml/100g (pons/medulla oblongata). They were given as a constant parameter in the fitting routine which was used to estimate the transfer coefficients of ^{18}F-DOPA.

The CBF was measured before and simultaneously with the PET study (Fig. 1). No significant differences were found between these two measurements. The values during the PET study varied between 40 ml/100g min (white matter) and 84 ml/100g min (mesencephalon). The respective data were used to calculate the regional permeability-surface *(PS)* product of the blood-brain barrier for ^{18}F-DOPA (Tab. 1).

The distribution of ^{18}F-DOPA metabolites in plasma was measured by HPLC (Fig. 2). From the plasma time-activity curves of ^{18}F-DOPA and OMe-^{18}F-DOPA the rate constants of peripheral synthesis (k_0^D =0.0062 ± 0.0003 min^{-1}) and elimination (k_{-1}^M =0.0059 ± 0.0020 min^{-1}) of OMe-^{18}F-DOPA were calculated. The plasma input function for the estimation of ^{18}F-DOPA transfer coefficients was obtained by correction of the plasma time-activity curves of ^{18}F with the fraction of metabolites calculated from the HPLC data.

The distribution of radioactivity occurs between the intravascular space, the extravascular precursor pool and the metabolite compartment and is described by nine transfer coefficients (5,6). This includes transfer coefficients for OMe-^{18}F-DOPA since this compound is able to cross the blood-brain barrier. This complicates the estimation of separate rate constants for ^{18}F-DOPA. Our estimation of the parameters for blood-brain and brain-blood transfer, K_1 and k_2 (Tab. 1), includes both ^{18}F-DOPA and OMe-^{18}F-DOPA which is expected to result in an overestimation of the ^{18}F-DOPA precursor pool and an underestimation of the decarboxylation rate of ^{18}F-DOPA k_3.

Transfer coefficients of ^{18}F-DOPA were obtained from brain time-activity curves of six individual experiments (Fig. 3) and the respective plasma input functions The values of K_1, k_2 and k_3 estimated in the striatum are 19%, 45% and 46% higher than in the frontal

Table 1. Kinetic transfer coefficients of ^{18}F accumulation in the brain, partition volume (V_e) and *PS*-product of blood-brain transfer in newborn piglets

	Frontal Cortex	Striatum
K_1 (ml g^{-1} min^{-1})	0.057 ± 0.017	0.068 ± 0.021
k_2 (min^{-1})	0.056 ± 0.022	0.081 ± 0.033
k_3 (min^{-1})	0.026 ± 0.016	0.038 ± 0.017
k_4 (min^{-1})	0.038 ± 0.019	0.035 ± 0.007
V_e (ml ml^{-1})	1.1 ± 0.2	0.9 ± 0.1
PS (ml g^{-1} min^{-1})	0.060 ± 0.019	0.072 ± 0.024

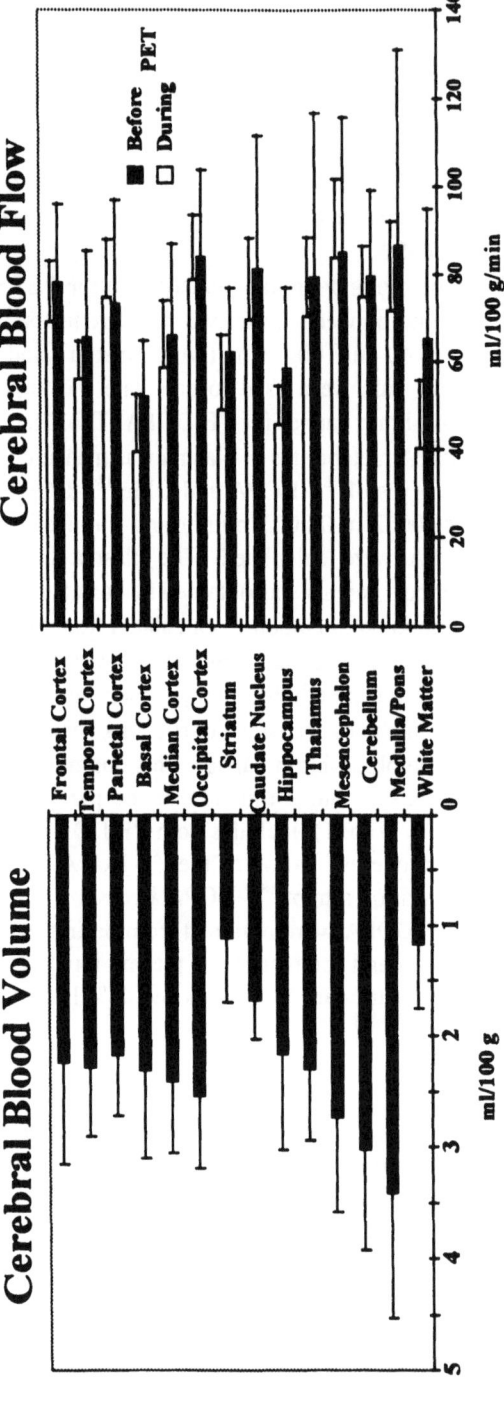

Figure 1. Local cerebral blood volume and cerebral blood flow (CBF) in newborn piglets (means ± S.D.). CBF was measured before and during the PET study. No significant differences were found.

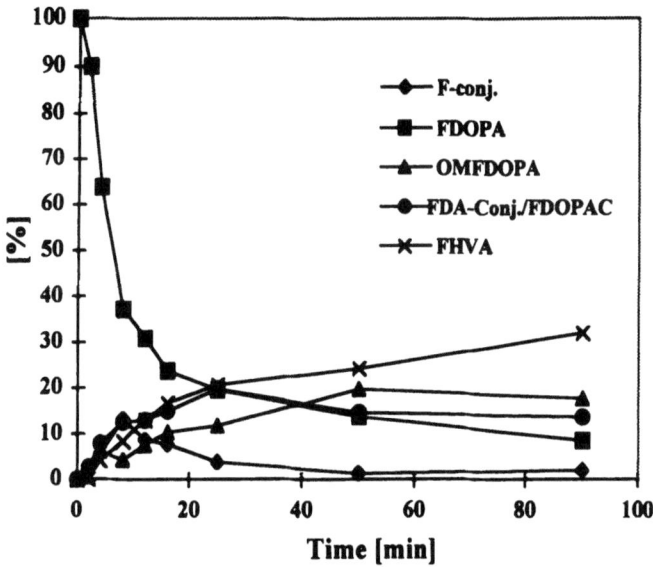

Figure 2. Relative amounts of ^{18}F-DOPA metabolites in porcine plasma as function of time based on HPLC-determined plasma fractions of six piglets (means ± S.D.).

cortex. The rate constant k_4, which accounts for the clearance of labelled metabolites from tissue, is similar in frontal cortex and striatum.

4. DISCUSSION

We have measured the cerebral uptake and metabolism of ^{18}F-DOPA in newborn piglets in order to find a useful model to study the regulation of the dopamine synthesis during ontogenesis. There is a lack of data related to this aspect of dopamine synthesis. On the other hand, conflicting data regarding an influence of normal aging on the presynaptic nigrostriatal dopaminergic function exist (10).

^{18}F-DOPA is a potential substrate for L-DOPA decarboxylase (DDC) the ultimate enzyme in dopamine synthesis in terminals of the dopaminergic nigrostriatal pathway (11,12). From in vitro data it is expected that DDC activity is in great excess with respect to the synthesis of dopamine and tyrosine hydroxylase is the rate limiting enzyme because it is nearly saturated with tyrosine (for references see 12). Despite there is evidence that the activity of DDC is modulated by neuronal activity (13,14) which may be related to the rate-limiting step in the synthesis of 2-phenylethylamine, a putative modulator of dopamine transmission.

Until now there were no data available about the ^{18}F-DOPA metabolism in pigs. Using HPLC we have identified in plasma the major metabolites of ^{18}F-DOPA. In rats and humans (11,15,16) ^{18}F-DOPA undergoes a rapid metabolism by O-methylation (k_{0D}) which mainly reflects hepatic Catechol-O-methyl transferase (COMT) activity. In our study a methylation rate was estimated which is 40% and 90% less than those observed in humans (5,16) and rats (11), respectively. The rate constant of elimination of OMe-^{18}F-DOPA (k_{-1M}), which presumably is related to renal elimination, is about 50% smaller than in rats but comparable to humans.

Our values for K_1 and k_2 are about 50%–80% higher than those in rats and humans (6,17) which may be due to the inclusion of OMe-^{18}F-DOPA in our estimation. It is

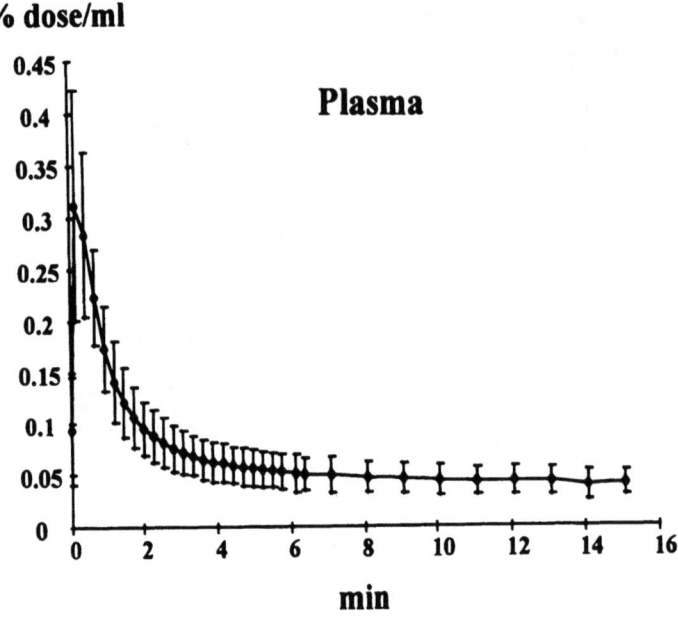

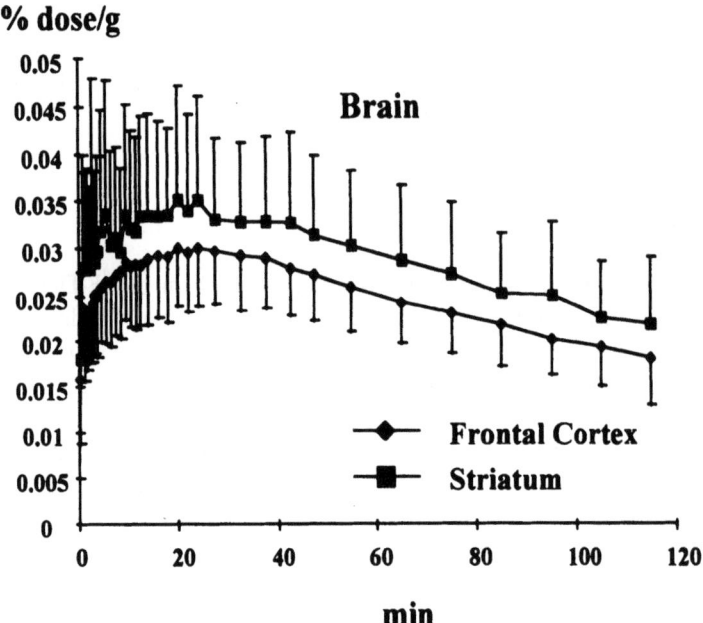

Figure 3. Normalized time-activity curves (n=6) of ^{18}F in plasma and brain of newborn piglets.

reported that the blood-brain transfer (K_1) and brain-blood transfer (k_2) of OMe-^{18}F-DOPA is about twice of ^{18}F-DOPA (17). The average value of the decarboxylation rate constant (k_3) in striatum was 0.036 min^{-1} which is close to values measured in the striatum of adult humans (5,6,18). We expect our value to be underestimated by the reason mentioned above and because of the partial volume effect which is related to the rather low spatial resolution of the tomograph. This may also explain the low ratio of k_3 between striatum and frontal cortex. We have found about 50% higher rate of decarboxylation in striatum than in frontal cortex. In humans a five-fold difference was found (5,18).

We conclude that PET measurements with ^{18}F-DOPA may serve as a useful tool to estimate the activity of DDC in the immature porcine brain. Some of the limitations of the present study will be overcome by future use of a tomograph with a higher spatial resolution and by inclusion of separate input functions of ^{18}F-DOPA and OMe-^{18}F-DOPA.

ACKNOWLEDGMENTS

This work was supported in part by research grants from the Saxon State Ministry of Science and Arts (Grant 7541.82-FZR/309) and from the Thuringian State Ministry of Science, Research and Arts (Grant 3/95–13).The excellent technical assistance of Mrs. R. Scholz, Mrs. R. Herrlich and Mrs. U. Jaeger is gratefully acknowledged.

REFERENCES

1. Huang, C.C., Lajevardi, N.S., Tammela, O., Pastuszko, A., Delivoria-Papadopoulos, M., and Wilson, D.F, 1994, *Neurochem. Res.* 19: 649–655.
2. Srivastava, L.K., and Mishra, R.K., 1994, *Dopamine Receptors and Transporters* (H.B. Niznik, ed.) Marcel Dekker Inc., New York, PP. 437–457.
3. Unis, A.S., 1994, *Biol. Psychiat.* 35: 562–569.
4. Richard, M.G., and Bennett, J.P., 1994, *Exp. Neurol.* 129: 57–63.
5. Gjedde, A., Reith, J., Dyve, S., Leger, G., Guttman, M., Diksic, M., Evans, A., and Kuwabara, H., 1991, *Proc. Natl. Acad. Sci.* 88: 2721–2725.
6. Huang, S.-C., Yu, D.-C., Barrio, J.R., Grafton, S., Melega, W.P., Hoffman, J.M., Satyamurthy, N., Mazziotta, J.C., and Phelps, M.E., 1991, *J. Cereb. Blood Flow Metab.* 11: 898–913.
7. Kuwabara, H., Cumming, P., Yasuhara, Y., Leger, G.C., Guttman, M., Diksic, M., Evans, A.C., and Gjedde, A., 1995, *J. Nucl. Med.* 36: 1226–1231.
8. Ishikawa, T., Dhawan, V.,Chaly, T., Margouleff, C., Robeson, W., Dahl, J.R., Mandel, F., Spetsieris, P., and Eidelberg, D., 1996, *J. Nucl. Med.* 37: 216–222.
9. Namavari, M., Bishop, A., Satyamurthy, N., Bida, G., and Barrio, J.R., 1992, *Appl. Radiat. Isot.* 43: 989–996.
10. Eidelberg, D., Takikawa, S., Dhawan, V., Chaly, T., Robeson, W., Dahl, R., Margouleff, D., Moeller, J.R., Patlak, C.S., and Fahn, S., 1993, *J. Cereb. Blood Flow Metab.* 13: 881–888.
11. Cumming, P., Kuwabara, H., and Gjedde, A., 1994, *J. Neurochem.* 63: 1675–1682.
12. Cumming, P., Kuwabara, H., Ase, A., and Gjedde, A., 1995, *J. Neurochem.* 65: 1381–1390.
13. Rosetti, Z., Krajnc, D., Neff, N.H., and Hadjiconstantinou, M., 1989, *J. Neurochem.* 52: 647–652.
14. Zhu, M.Y., Juorio, A.V., Paterson, I.A., and Boulton, A.A., 1992, *J. Neurochem.* 58: 636–641.
15. Melega, W.P., Luxen, A., Perlmutter, M.M., Nissenson, C.H.K., Phelps, M.E., and Barrio, J.R., 1990, *Biochem. Pharmacol.* 39: 1853–1860.
16. Cumming, P., Léger, G.C., Kuwabara, H., and Gjedde, A., 1993, *J. Cerebr. Blood Flow Metab.* 13: 668–675.
17. Reith, J., Dyve, S., Kuwabara, H., Guttman, M., Diksic, M., and Gjedde, A., 1990, *J. Cerebr. Blood Flow Metab.* 10: 707–719.
18. Dhawan, V., Ishikawa, T., Patlak, C., Chaly, T., Robeson, W., Belakhlef, A., Margouleff, C., Mandel, F., and Eidelberg, D., 1996, *J. Nucl. Med.* 37: 209–216.

Section 44: Multidisciplinary Application of Microdialysis

DOPAMINE RELEASE IN *N. accumbens* IN A CONDITIONED REWARD PARADIGM

K. P. Datla, R. G. Ahier, A. M. J. Young, J. A. Gray, and M. H. Joseph

MRC Behavioural Neurochemistry Group
Department of Psychology
Institute of Psychiatry
London, United Kingdom

1. INTRODUCTION

Monitoring the response of extracellular neurotransmitter levels in a specific brain region(s) to drug treatments has become a standard technique for neuropharmacologists. In a freely moving animal, in vivo microdialysis has the advantage of permitting monitoring of both behaviour and neurotransmitter changes at the same time, and it has become an important tool for directly correlating changes in behaviour with neurotransmitter release (1).

Mesolimbic dopaminergic neurons originating from the ventral tegmental area have been implicated in the mediation of reward and motivational behaviour (2, 3, 4, 5). Nucleus accumbens (NAC) dopamine (DA) is widely believed to be involved in maintaining the rewarding properties of many stimuli.

Studies from a number of laboratories, including our own, have shown that many types of unconditioned stimuli can release DA, and that classical conditioning of neutral stimuli to them results in conditioned DA release. In particular we have shown that during conditioning of a neutral stimulus (light or tone) to an aversive stimulus (foot-shock), the increases of DA were significantly higher than when foot-shock was given alone. Subsequently, presentation of the conditioned stimulus alone increased extracellular DA (6).

Levels of DA have been shown to increase when primary rewarding stimuli [eg., food (7), water (8) and receptive sexual partner (9)] are presented, and although behavioural responses are readily conditioned to such rewards, no study has examined whether DA levels also increase when a stimulus conditioned to reward is subsequently presented alone. The present study aimed to assess whether a conditioned stimulus can also increase DA. To this end we have measured extracellular DA during conditioning to a novel stimulus (tone) in rats trained to lick 10% sucrose solution when light was presented, as well as during testing (when tone and light but not sucrose were presented). As a measure of

---

From a symposium organized by Ben H.C. Westerink (Groningen, The Netherlands) and Michael H. Joseph (London, United Kingdom).

behavioural response to reward we have counted the number of magazine entries during each session.

2. METHODS

Male Sprague-Dawley rats (250–275 g) were allowed to acclimatise for one week before surgery. A stainless steel guide cannula (20G, 10 mm) aimed at nucleus accumbens [AP +1.5 mm; ML −1.5 mm from bregma and DV −5.0 mm from dura; (10)] was implanted under Immobilon (etorphine hydrochloride and methotrimeprazine; 30 µl, i.m.) anaesthesia.

Training started one week after surgery when the animals were fully recovered. Each day for 10 days rats were trained for 30 min in a behavioural box (Campden Instruments Ltd) to obtain 50 µl of 10% sucrose solution for 5 sec, delivered through a dipper at VI-120 schedule (i.e., on an average every 2 min; 15 presentations per session). Each sucrose delivery was preceded by a 1 sec illumination of the magazine light (see Figure 1).

Access to the magazine was via a perspex flap, displacement of which was recorded automatically (magazine entries).

Inappropriate magazine entries (i.e. entries during inter-trial interval) were penalised by a 3 sec delay in the presentation of the next trial. By the end of training animals enter during 90–100% of the presentations of trials. From the beginning of training animals were on controlled diet, allowing the animal to maintain 80% of their initial bodyweight. During training the animals also habituate to the box.

On the 10th day after routine training rats were implanted with a dialysis probe (Cuprophan membrane; 300 µm diameter; 40kD m.w. cut-off; 2 mm active sac) under light anaesthesia (Halothane 2% with N_2O & O_2 50% V/V each). On the following day freshly prepared, filtered (2 µm) and degassed artificial cerebrospinal fluid [composition (mM) NaCl 125; KCl 3.3; $MgSO_4$ 2.4; KH_2PO_4 1.25; $CaCl_2$ 1.85] was pumped through the probe at a flow rate of 2.5 µl/min. After a 90 min equilibration period dialysate samples were collected every 10 min into brown autosampler vials with 2 µl of 1M H_3PO_4. Samples were analyzed by HPLC-ECD (see below).

Conditioning (3 samples): after 1 hr (6 basal samples) animals received a 30 min session during which a light stimulus (for 1 sec) was followed by 5 sec presentation of 10% sucrose, as in training, except that they also received a novel stimulus tone (for 1 sec; 2.5 kHz; 15–20 dB) concomitant with the light stimulus (Figure 1).

Test (3 samples): two hrs later (12 samples) they received the same tone/light stimulus as in conditioning but not followed by sucrose. Three more samples were collected after test. The number of magazine entries during stimulus presentation at conditioning

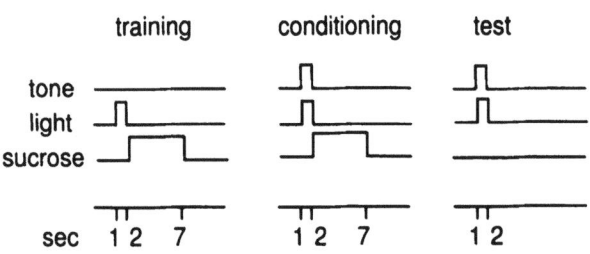

Figure 1. Schematic representation of sequence and duration of presentation of light, tone and sucrose.

and at test was taken as a behavioural index of reward. The activity (movement) of the rat in each session was counted by passive infra-red detector.

The HPLC-ECD system: mobile phase [composition: KH_2PO_4 0.1 M; octyl sodium sulphonate 1 mM; Na-EDTA 0.1 mM and 10% methanol (V/V), pH 2.75 adjusted with orthophosphoric acid] was filtered (2 µm), degassed online and pumped at 0.9 ml/min. Samples were injected into the rheodyne valve (20 µl loop) automatically connected through an Altex Ultrasphere column (3 µm ODS; 4.6 mm x 7.5 cm; Beckman) to an ESA Coulochem detector incorporating a high sensitivity dual electrode analytical cell (Model 5011). Electrode 1 was set at –0.2 V and electrode 2 was set at +0.34 V with respect to the palladium reference electrode and was used to detect DA, 3,4-dihydroxyphenylacetic acid (DOPAC), homovanillic acid (HVA) and 5- hydroxyindoleacetic acid (5-HIAA). The sample rack and column were maintained at 5.8°C and 35°C respectively. Chromatographic data were collected and integrated by computer based Gynkotek software (HPLC Technology).

The data for each animal were expressed as % of the mean basal value from the first six samples. Student's t-test was used to measure level of significance of DA changes from basal.

3. RESULTS

During **training** animals learned quickly to enter the magazine to obtain the sucrose when the magazine light came on. By the end of training more than 90% of magazine entries were made during stimulus presentation. As the number of sucrose presentations were on average 15 per session, animals showed no signs of decreased interest by the end of training.

During **conditioning** when tone was presented with light for the first time followed by sucrose, the number of magazine entries were fewer in the first 10 min than that seen in training session, but increased to the training level during the session. DA levels were significantly higher than basal levels during this 30 min period, with a suggestion of a biphasic response. The return of DA levels to basal was not immediate (Figure 2). Activity of the animals was also higher in the first 10 min but decreased gradually.

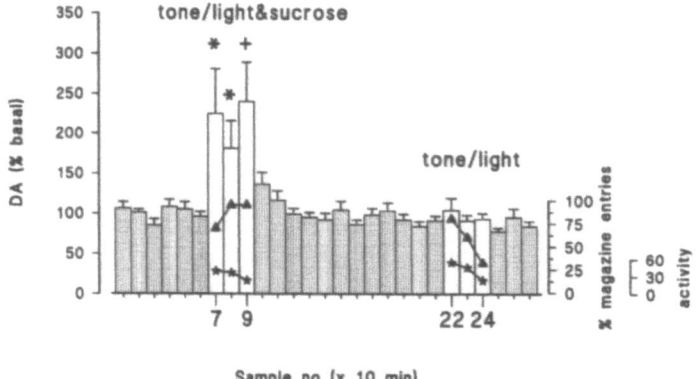

Figure 2. DA release during conditioning and test sessions. Columns are mean ± SEM, n = 11. Filled column represent basal samples. Blank columns represent conditioning and test sessions. Filled triangles and stars represent % successful magazine entries and movement activities respectively. Basal levels (mean of first 6 samples) of DA were 1.56 ± 0.12 pmol/ml. (n = 11). * P<0.05 + P<0.02 vs basal (100%) level based on z score and 't' distribution.

During the first 10 min of the **test** session the number of magazine entries was near maximal but decreased rapidly during the next two 10 min periods, consistent with extinction. DA concentrations however did not change during the test session (Figure 2).

The levels of DA overall did not correlate with either the number of magazine entries or activity ($r = 0.1$ to 0.2). There were no significant changes in the levels of DOPAC, HVA and 5-HIAA (data not shown).

4. DISCUSSION

The formation of neutral-aversive and neutral-neutral associations are accompanied by DA release over and above that seen in response to the component stimuli (6,11). The relatively modest increases in dialysate DA level seen here might suggest that this does not occur with neutral-appetitive associations. The alternative possibility that no association was formed is unlikely, as the animals responded behaviourally to the conditioned stimuli at test.

This response was to the combination of the established conditioned stimulus (light) and the new conditioned stimulus (tone). One remaining possibility is that increased DA release during conditioning, and subsequent conditioned release, are due to new associations but not to established associations, but that no new association was formed between the new stimulus and reward at conditioning because it was "blocked" by the established stimulus (see 12 for an account of this phenomenon).

At present we have no evidence for increased DA release (beyond that expected for reward) during appetitive conditioning or during appetitive conditioned stimuli. To explore this further it will be necessary to require the use of different experimental paradigms. Results so far indicate that shifting the tone to precede the light (to prevent blocking) does result in greater increase in DA during conditioning and a subsequent response to the conditioned stimuli (13). Further work will be required to separate the effects of new and established conditioned stimuli.

ACKNOWLEDGMENT

We thank the Medical Research Council (U.K.) for financial support.

REFERENCES

1. Young, A.M.J., 1993, *Reviews in the Neurosciences* **4**: 373–395.
2. Fibiger, H.C., and Phillips, A.G., 1986, *Handbook of Physiology*, Vol 4, APS, Bethesda, PP. 647–675.
3. Sachs, B.D., and Meisel, R.L., 1988, *The physiology of Reproduction* (E. Knobil, and J. Neill, eds.), Raven, New York, PP. 1393–1485.
4. Le Moal, M., and Simon, H., 1991, *Physiol. Rev.* **71**: 155–234.
5. Blackburn, J.R., Pfaus, J.G., and Phillips, A.G., 1992, *Prog. Neurobiol.* **39**: 247–279.
6. Young, A.M.J., Joseph, M.H., and Gray, J.A., 1993, *Neuroscience* **54**: 5–9.
7. Wilson, C., Nomikos, G.G., and Fibiger, H.C., 1995, *J. Neurosci.* **15**: 5169–5178.
8. Young, A.M.J., Joseph, M.H., and Gray, J.A., 1992, *Neuroscience* **48**: 871–876.
9. Wang, C.T., Huang, R.L., Tai, M.Y., Tsai, Y.F., and Peng, M.T., 1995, *Neurosci. Lett.* **200**: 29–32.
10. Paxinos, G., and Watson, C., 1982, *The Rat brain in stereotaxic coordinates*, Academic Press, New York.
11. Young, A.M.J., Ahier, R.G., Upton, R.L., Gray, J.A., and Joseph, M.H., 1995, *Brain Res. Assoc. Abs.* **12**: 58.
12. Mackintosh, N.J., 1994, *Companion Encyclopedia of Psychology* Vol 1 (A.M. Colman, ed.), Routledge, London, PP. 379–396.
13. Datla, K.P., Ahier, R.G., Young, A.M.J., Gray, J.A., and Joseph, M.H., 1996, *Monitoring molecules in Neuroscience* (M. Mas, et. al., eds., in press).

NEURONAL ACTIVATION VISUALIZED BY FOS-EXPRESSION AFTER INTRACEREBRAL MICRODIALYSIS OF DRUGS

M. Feenstra, M. Bubser, E. Erdtsieck-Ernste, A. van der Wal, M. Botterblom, and H. van Uum

Netherlands Institute for Brain Research
Graduate School Neurosciences Amsterdam
Amsterdam, The Netherlands

1. FUNCTIONAL ANATOMY

Anatomical studies of the central nervous system can show the presence of neuronal pathways and circuits and provide comprehensive three-dimensional descriptions of the connnectivity between neuronal populations, but do not provide information regarding activity or function. On the other hand, functional studies, using either electrophysiological or neurochemical techniques, show cellular responses with restricted anatomical information. The gap between these approaches is now filled with techniques that make use of markers of cellular activity. The expression of immediate early genes (IEGs) in the central nervous system can be visualized on the cellular level using immunohistochemical (IHC) staining of brain sections. IEGs code for nuclear proteins that are transcription factors for certain other genes (1,2). They may be activated by various extracellular stimuli (neurotransmitters, hormones, growth factors), acting through common second messengers, such as Ca^{2+} (1,2,3). Many in vitro and in vivo studies showed that depolarization by excitatory amino acids such as glutamate or by other excitatory neurotransmitters, by stimulation of specific neuronal pathways or by epilepsy-induced synchronized activation results in increased expression of IEGs (2,3). Of this class of genes, *c-fos* has been studied most often, and visualization of (stimulated) expression of its protein product, Fos, by IHC is used in functional anatomical studies, where activational patterns are mapped after stimulation of a particular anatomical pathway (4). Being a nuclear protein, Fos visualizes cell bodies, whereas accumulation of deoxyglucose visualizes activated nerve terminals (5).

However, one should realize that there are differences between the thresholds of neurons for the stimulation needed to induce *c-fos* expression and there is no definite relation between depolarization and Fos induction. Moreover, most antibodies used for the detection of Fos protein by IHC also recognize Fras (fos-related antigens) and the visualized Fos-like immunoreactivity (FLIR) represents a mixture of different IEG products with their own typical temporal and regional patterns of expression (2,6). Finally, IEGs of the leucine-zipper group, to which c-fos belongs, activate gene expression only after dimer formation. Fos has to form a dimer with a protein product of the *jun* family of IEGs and increased FLIR does not automatically mean that there will be consequences for the functioning of the neuron.

Most neurons have a low level of FLIR when the animal is resting or sleeping. This presents apparently ideal conditions to map activational patterns. Stimulation of afferents effected by noxious or non-noxious sensory stimulation resulted in FLIR in spinal cord and several brain areas (7,8). Electrical stimulation of brain areas in anesthetized or awake animals likewise result in specific patterns of FLIR (9,10). Chemical stimulation using microinjection techniques was introduced as it does not activate fibers of passage and offers the possibility to activate or inhibit specific receptors (11).

However, arousal and stress, such as exposure to a novel environment, handling or peripheral or intracerebral injections, increase FLIR in several brain areas (12–16). Repeated stress leads to a decrease in reactivity, but this may differ for various brain regions and will not completely prevent these responses (14,17). Acute penetration of the dural membrane and damage to brain tissue induces widespread c-fos expression (18,19), not only in neurons but also in glia (18). Anesthesia may alter FLIR responses (e.g.20; see 4). Acute mechanical effects in the brain and arousing and stressful manipulations should, therefore, be avoided in functional neuroanatomical studies.

Microdialysis is well suited to locally perfuse brain areas with pharmacologically active substances and to evaluate the effects on cellular activation by visualizing the expression of IEGs. The actual experiment may be performed several hours to several days after probe placement and connection to the perfusion system, so that effects of anesthesia, tissue damage and stress can be controlled for. Rate and duration of perfusion and composition of the fluid can be varied without handling and touching the animal and no fluid is introduced.

Stone et al (21) infused noradrenaline (NA) in a microdialysis cannula placed in the medial prefrontal cortex (mPFC) of rats and observed a marked increase in FLIR in a radius of 1–2 mm around the cannula. Implantation and perfusion with Ringer alone doubled the number of Fos-positive neurons compared to a non-implanted control, NA infusion increased the number tenfold and co-infusion of the β-adrenoceptor antagonist antagonized this effect. Another series of studies, however, showed that local tissue damage induced by the cannula may interfere with the normal patterns of c-fos-expression in brain tissue and with the effects of peripheral administered drugs (22,23). FLIR in the rat striatum was increased around a microdialysis cannula by peripheral administration of a dopamine (DA) reuptake-blocker or a MAO-inhibitor. The number of Fos-positive neurons was high in the direct vicinity of the cannula, irrespective of it being used for dialysis flow. However, in rats without cannulae striatal FLIR was not or far less increased.

2. STIMULATION OF PFC AFFERENTS

The thalamus provides a topographically specific afferent innervation of the neocortex and striatum. The origin of the afferents reaching the PFC is the mediodorsal nucleus of the thalamus (MDT), but the paraventricular (PVT) and other midline/

intralaminar nuclei also send projections to the PFC (24). These projections are excitatory and probably use glutamate as transmitter. Stimulation of these nuclei in anesthetized animals excites cortical neurons and can be inhibited by AMPA-receptor-selective glutamate antagonists (25,26). Moreover, similar effects are obtained by stimulation of another PFC afferent, i.e. from the ventral subiculum (27), and the effects of these stimulations may be inhibited by concurrent stimulation of the ventral tegmental area (VTA), the origin of the DA projection to the PFC (28,29). To further study the actions and interactions of these afferent pathways, which may be important in the regulation of PFC activity and thus of cognitive processes and emotional behaviour, we applied chemical stimulation by drugs perfused through microdialysis cannulae. For stimulation of the thalamocortical pathway we used disinhibition, as suggested by Jones et al (30), who locally injected $GABA_A$- receptor-antagonists in anesthetized rats. We implanted a microdialysis cannula in the MDT and started the perfusion with Ringer solution 1-2 days later. After 4-20 h equilibration, the $GABA_A$-antagonist bicuculline (0.03 or 0.1 mM), was perfused for 20 min. The rat was decapitated for in situ hybridization of c-fos mRNA or perfused with fixative for IHC detection of Fos protein (31). Cannula implantation with or without Ringer perfusion did not induce c-fos mRNA or FLIR. Only in the PVT FLIR was present depending on the arousal state of the rat (32). Rats were generally sleeping but awoke and showed stationary and locomotor activity during the perfusion with bicuculline (31). Thalamic disinhibition increased FLIR around the probe in the MDT, but even more so in the PVT and other midline/intralaminar nuclei. Transsynaptic activation of neurons was observed in medial and lateral PFC, rostral pole and shell of the nucleus accumbens, mediodorsal striatum, claustrum, reticular thalamic nucleus, central and basolateral nuclei of the amygdala. Most of these areas are known to be targets of MDT/PVT efferent fibers. The activation was mostly ipsilateral, but contralateral expression was observed when the stimulation extended to include the complete PVT. When the probe was placed in other thalamic nuclei, transsynaptic c-fos-expression was not observed in PFC or accumbens but in the targets of that thalamic nucleus. The transient nature of the activation was apparent from the fact that mRNA was present 1 h after bicuculline, but was undetectable 4 h after the disinhibition. Quantitative image analysis showed that cortical FLIR was most pronounced in the ventromedial parts of the PFC (ventral prelimbic and infralimbic cortices), was not very different in basal and superficial layers and that staining was dose-dependently related to the bicuculline concentration perfused in the MDT (Fig.1)(33).

With a second microdialysis cannula positioned in the mPFC we monitored extracellular concentrations of DA and glutamate. Both were increased to over 200% of control after MDT disinhibition, but the increase in glutamate was more rapid and of a shorter duration (34,35). The obvious explanation is that disinhibition of the thalamic neurons leads to increased glutamate release in mPFC, which resulted by an action on DA terminals or via multisynaptic pathways in increased DA release. An interesting possibility to show the neuronal pathway involved in these effects is to introduce a retrograde tracer into the mPFC cannula, disconnect the perfusion, wash out with Ringer one day later and perfuse the rat with fixative one week later (36). As Fig.2 shows, the tracer massively enters the mPFC tissue (Fig.2A) and is retrogradely transported to the cell bodies of terminals present in the mPFC, e.g. the VTA (Fig.2C) and the thalamic tissue around the cannula in the MDT (Fig.2B).

Recent results show that lesions of the VTA increase the intensity of FLIR in the mPFC, confirming that DA has an inhibitory effect on the thalamic stimulation (33, Bubser et al, submitted). Furthermore, preliminary results were obtained using stimulation of

Fos–positive nuclei/mm^2

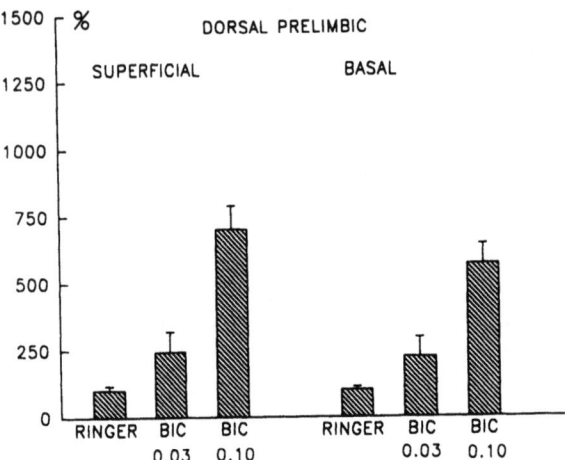

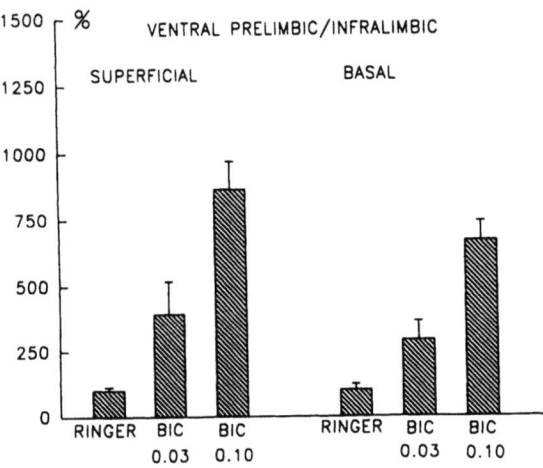

Figure 1. The number (± SEM) of Fos-positive nuclei in the mPFC of rats that were perfused in the MDT with 0.03 or 0.1 mM bicuculline for 20 min (32). After Ringer perfusion of the MDT the number was 15–20/mm^2 (100%).

the subiculum-PFC pathway. Here the glutamate agonist NMDA was found to be more suitable than bicuculline. Perfusion with 0.1 mM NMDA increased FLIR in the PFC similar to the thalamic disinhibition. However, this resulted also in repeated wet dog shakes (WDS), an epileptic phenomenon related to overactivation of limbic systems. Although this was the only sign of epilepsy observed, we found strongly increased FLIR in the dentate gyrus of the hippocampal complex in both hemisphers (see 4). When WDS were absent FLIR was only slightly increased in the ventromedial PFC compared to Ringer perfusion.

These results clearly show the potential of the chemical stimulation technique using microdialysis perfusion of specific brain areas. The effects of stimulation of a neuronal pathway can be evaluated using behaviour in freely moving rats and FLIR throughout the brain. Further anatomical information may possibly be obtained by using doublestaining methods: the cellular targets of the afferent stimulations may be identified as GABAergic interneurons or glutamatergic pyramidal neurons.

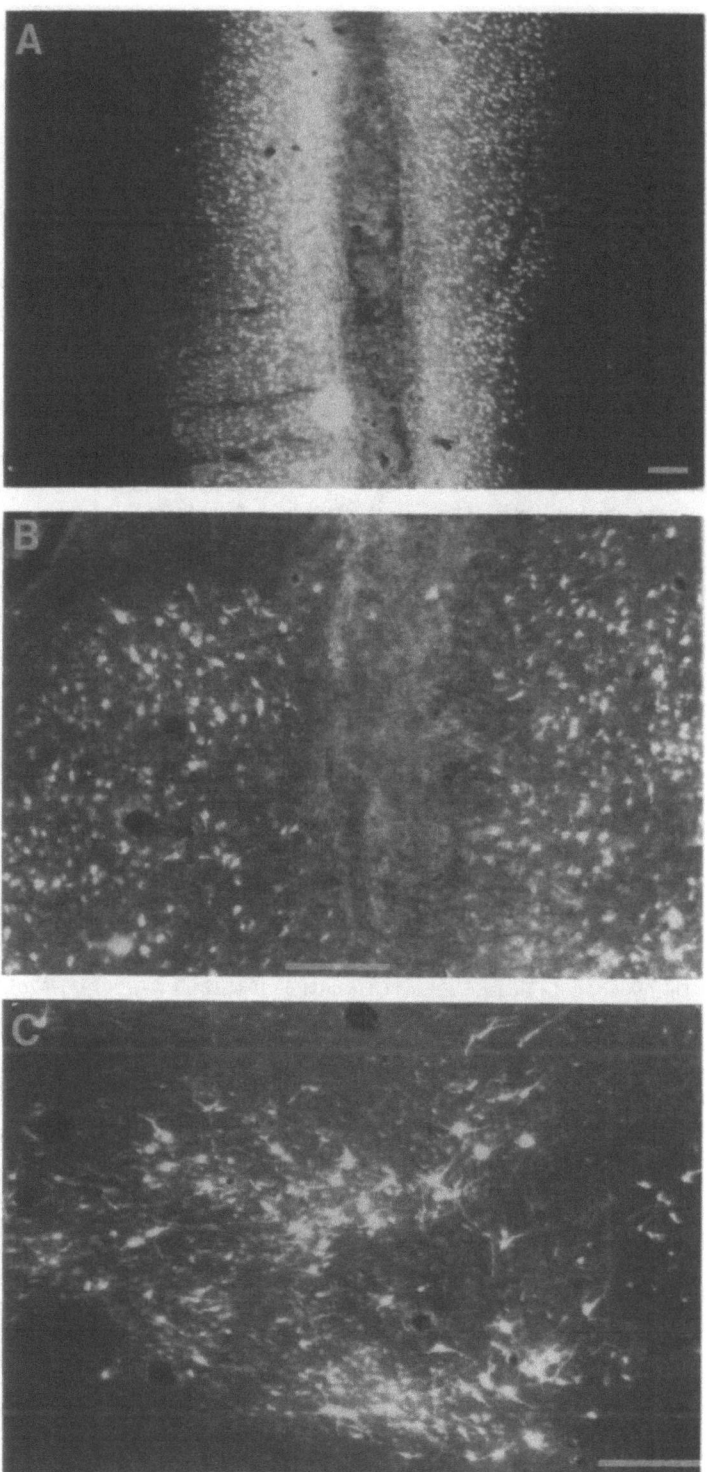

Figure 2. Rat brain sections stained for fluorogold. A. The microdialysis cannula in the right mPFC, where fluorogold was introduced. B. Retrogradely labeled neurons in the right MDT, in a section just before the MDT microdialysis cannula. C. Retrogradely labeled neurons in the right VTA. Scale bars = 500 μm.

REFERENCES

1. Sheng, M., and Greenberg, M.E., 1990, *Neuron* **4**: 477–485.
2. Morgan, J.I., and Curran, T., 1991, *Ann. Rev. Neurosci.* **14**: 421–451
3. Hughes, P., and Dragunow, M., 1995, *Pharmacol. Rev.* **47**: 133–178.
4. Dragunow, M., and Faull, R., 1989, *J. Neurosci. Meth.* **29**: 261–265.
5. Sokoloff, L., 1993, *Devel. Neurosci.* **15**: 194–206.
6. Dilts, R.P., Helton, T.E., and McGinty, J.F., 1993, *Synapse* **13**: 251–263.
7. Hunt, S.P., Pini, A., and Evan, G., 1987, *Nature* **328**: 632–634.
8. Bullitt, E., 1990, *J. Comp. Neurol.* **296**: 517–530.
9. Sagar, S.M., Sharp, F.R., and Curran, T., 1988, *Science* **240**: 1328–1331.
10. Wan, X.S.T., Liang, F., Moret, V., Wiesendanger, M., and Rouiller, E.M., 1992, *Neurosci.* **49**: 749–761.
11. Page, K.J., and Everitt, B.J., *Exp. Brain Res.* **93**: 399–411.
12. Grassi-Zucconi, G., Giuditta, A., Mandile, P., Chen, S., Vescia, S., and Bentivoglio, M., 1994, *J. Physiol.* (Paris) **88**: 91–93.
13. Handa, R.J., Nunley, K.M., and Bollnow, M.R., 1993, *NeuroReport* **4**: 1079–1082.
14. Papa, M., Pellicano, M.P., Welzl, H., and Sadile, A.G., 1993, *Brain Res. Bull.* **32**: 509–515.
15. Asanuma, M., Ogawa, N., Hirata, H., Chou, H., Tanaka, K., and Mori, A., 1992, *J. Neural Transm.* **90**: 163–169.
16. Imaki, T., Shibasaki, T., Hotta, M., and Demura, H., 1993, *Brain Res.* **616**: 114–125.
17. Melia, K.R., Ryabinin, A.E., Schroeder, R., Bloom, F.E., and Wilson, M.C., 1994, *J. Neurosci.* **14**: 5929–5938.
18. Dragunow, M., and Robertson, H.A., 1988, *Brain Res.* **455**: 295–299.
19. Stone, E.A., Zhang, Y., and Carr, K.D., 1995, *Brain Res. Bull.* **36**: 77–80.
20. Krukoff, T.L., Morton, T.L., Harris, K.H., and Jhamandas, J.H., 1992, *J. Neurosci.* **12**: 3582–3590.
21. Stone, E.A., Zhang, Y., John, S.M, and Bing, G., 1991, *Neurosci. Lett.* **133**: 33–35.
22. Morelli, M., Carboni, E., Cozzolino, A., Tanda, G.L., Pinna, A., and Di Ciara, G., 1992, *J. Neurochem.* **59**: 1158–1160.
23. Di Chiara, G., Carboni, E., Morelli, M., Cozzolino, A., Tanda, G.L., Pinna, A., Russi, G., and Consolo, S., 1993, *Neurosci.* **55**: 451–456.
24. Groenewegen, H.J., and Berendse, H.W., 1994, *Motor and cognitive functions of the prefrontal cortex.* (A.-M.Thierry, J.Glowinski, P.Goldman-Rakic, and Y.Christen, eds.), Springer-Verlag, Berlin, pp. 51–77.
25. Gigg, J., Tan, A.M., and Finch, D.M., 1992, *Cereb. Cortex* **2**: 477–484.
26. Pirot, S., Jay, T.M., Glowinski, J., and Thierry, A.-M., 1995, *Europ. J. Neurosci.* **6**: 1225–1234.
27. Jay, T.M., Thierry, A.-M., Wiklund, L., and Glowinski, J., 1992, *Europ. J. Neurosci.* **4**: 1285–1295.
28. Jay, T.M., Glowinski, J., and Thierry, A.-M., 1995, *NeuroReport* **6**: 1845–1848.
29. Ferron, A., Thierry, A.-M., Le Douarin, C., and Glowinski, J., 1984, *Brain Res.* **302**: 257–265.
30. Jones, M.W., Kilpatrick, I.C., and Phillipson, O.T., 1988, *Exp. Brain Res.* **69**: 623–634.
31. Erdtsieck-Ernste, E.B.H.W., Feenstra, M.G.P., Botterblom, M.H.A. van Uum, H.F.M., Sluiter, A.A., and Heinsbroek, R.P.W., 1995, *Neurosci.* **66**: 115–131.
32. Deutch, A.Y., Öngür, D., and Duman, R.S., 1995, *Neurosci.* **66**: 337–346.
33. Bubser, M., Feenstra, M.G.P., Erdtsieck-Ernste, E.B.H.W., Botterblom, M.H.A., and van Uum, H.F.M., 1995, *Soc. Neurosci. Abstr.* **21**: 933.
34. Feenstra, M.G.P., van der Weij, W., Hamstra, J.J., Botterblom, M.H.A., and Buijs, R.M., 1993, *Soc. Neurosci. Abstr.* **19**: 1382.
35. Feenstra, M., Bubser, M., Erdtsieck-Ernste, E., and Botterblom, M., 1995, *Europ. J. Neurosci.* **7** (Suppl.8): 102.
36. Skilling, S.R., Mullett, M.A., Beitz, A.J., and Larson, A.A., 1992, *J. Neurosci. Meth.* **41**: 85–90.

ADDRESSES OF PRESENTING AUTHORS, ORGANIZERS, AND (CO)-EDITORS

Bart Aarts
Utrecht University
Dept. Physiol. Chem.
Stratenum
Universiteitsweg 100
3585 CG Utrecht
The Netherlands
tel. +3130-533634
fax. +3130-539035

Micael Agadjanov
Yerevan State Med. Univ.
Dept. Biochemistry
Koriun Str. 2
Yerevan 375025
Armenia
tel. +3742-582412
fax. +3742-151812

Jan Albrecht
Med. Research Centre
Polish Acad. Sci
Dept. of Neurotoxiocol.
5 Pawinskiego Strr.
Warsaw 02-106
Poland
tel. +4822-608-6490/668-5323
fax. +4822-668-5532
email: ja@ibbrain.ibb.waw.pl

Håkan Aldskogius
Univ. of Uppsala
Dept. of Anatomy
Biomed. Center
P. O. Box 571
Uppsala S-75123
Sweden
tel. +46-18-17411
fax. +46-18-551120
email: hakan.aldskogius@anatomi.uu.se

N. G. Aleksidze
Tbilisi St. University
Abano 23
Tbilisi, 280005
Republic of Georgia
fax. +88-32-233423

Ruth Hogue Angeletti
Develop. & Mol. Biology
Albert Einstein Coll. Med.
1300, Morris Park Ave
Bronx, New York 10461
USA
tel. +1-718-430-3475
fax. +1-718-430-8939/8567
email:angelett@aecom.yu.edu

Frank Angenstein
Inst. Neurobiology
Lab. Cell. Signalling
Brenneckerstr. 6
Magdenburg 39108
Germany
tel. +49-3916263215
fax. +49-391616160
email:angenstein@ifu-magdeburg.de

Arduino Arduini
Dept. of Sciences
University "G. D. Annunzio"
Viale Pindaro, 42
Pescara 65127
Italy
tel. +39-691-39-3843
fax. +39-69139-3988

A. V. Arutjunyan
Ott Inst. of Obst. & Gynecology
Mendeleev Line 3
St. Petersburg 199034
Russia
tel. +7-812-218-98-91
fax. +7-812-218-23-61

Ferenc Auth
Chem. Works Gedeon Richter Ltd.
Pharmacol. Res. Cetr.
Gyomroi U. 19021
H1475 Budapest 10
Hungary
tel. +361-261-2199/1813
fax. +361-260-5000
email:h4277aut@ella.hu

N. F. Avrova
Russian Acad. of Sciences
Inst. of Evol. Physiol. Biochem.
M. Thorez Avenue 44
St. Petersburg 194223
Russia
tel. +7-812-552-7925
fax. +7-812-552-3012

Mimoun Azzouz
Dept. Immunology
Immunopharm. & Pathology
P O Box 24
74, Route du Rhin
67401 Illkirch
France
tel. +33-88-676926
fax. +33-88-660190
email:azzouz@pharma.u-strasbg.fr

Herman Bachelard
M. R. Centre Dept. of Physics
Univ. of Nottingham
Nottingham N67 2RD
United Kingdom
tel. +44-115-9514752
fax. +44-115-9515166
email:ppzwkp@ppn1:nott.ac.uk

P. R. Bär
Utrecht University
Dept. Neurology
P. O. Box 85500
Utrecht 3508 GA
The Netherlands
tel. +3130-2507973
fax. +3130-2542100

Jolanta Baranska
Nencki Inst. of Exp. Biology
Dept. of Molec & Cell. Neurobiology
3 Pasteur Str.
Warsaw Pl-02-093
Poland
tel. +48-22-6598571
fax. +48-22-225342
email:baranska@nencki.gov.pl

Valentina G. Bashkatova
Russ. Ac. of Med. Scs./Res. Inst.
Lab. Neurochem. Pharm.
Baltiyskaya 8
Moscow 125315
Russia
tel. +7-095-155-4797
fax. +7-095-151-1261
email:ksrayif@glasnet.ru

Addresses

Etienne Baulieu
Inserm U. 33
80, Av. du General Leclerc
94276 Le Kremlin-Bicetre Cedex
France
tel. +33-1-4959-1882
fax. +33-1-45211940

Bogdan Beleslin
Faculty of Medicine
Dept. of Patho Physiology
Dr. Subotica I/II
Belgrade 11000
Yugoslavia
tel. +381-11-648-282
fax. +381-11-684-053

Vladimir Berezin
Panum Institute
Protein Lab.
3C Blagdamsvej
Bldg. 6. 2
Copenhagen 2200 N
Denmark
tel. +45-3532-7335
fax. +45-3536-0116

J. J. Bergman
Res. Inst. Neurosciences
Free University
De Boelelaan 1087
1081 HV Amsterdam
The Netherlands
tel. +3120-4447122
fax. +3120-4447123

Peter Bergqvist
Lund Univ.
Inst. for Clin. Pharmacy
Lund Univ. Hospital
Lund S-22185
Sweden
tel. +46-46-173353
fax. +46-46-211-1987

H. H. Berlet
Inst. Pathochem. & Gen. Neurochem.
Dept. of Medicine
Im Neuen Heimer Feld 220/221
Heidelberg 69120
Germany
tel. +49-6221-562-595
fax. +49-6221-564-228

Bruno Berra
Inst. Gen. Psysiol. & Biochem
School of Pharmacy
Via Trentacoste 2
Milan 20134
Italy
tel. +39-221-40814
fax. +39-221-50-464
email:berra@imiucca.unimi.it

Hanna Bielarczyk
Med. Univ. of Gdansk
Dept. Clin. Biochem.
Debinki 7
Gdansk 80-211
Poland
tel. +48-58-478-222-ext. 1575
fax. +48-58-449653

Kaj Blennow
Dept. of Clinical Neuroscience
University of Göteborg
Göteborg
Sweden
tel. +46-31-86-1791
fax. +46-31-86-2426

Elisabeth Bock
Univ. of Copenhagen
Protein Laboratory
Panum Inst.
3 Blegdamsvej, Bld. 6. 2
Copenhagen DK-2200 N
Denmark
tel. +45-35327-334
fax. +45-35360-116

Leidi Bosch
Solvay Duphar B. V.
Cns Farmakologie
C. J. Van Houtenlaan 36
Weesp 1381 CP
The Netherlands
tel. +31294-477561
fax. +31294-410069

Michael Briley
Centre de Rech. Pierre Fabre
Direction
17 Avenue Jean Moulin
Castres 81100
France
tel. +33-63-714231
fax. +33-63-714209

Stefan Bröer
Univ. of Tübingen
Phys. Chem. Institute
Hoppe-Seyler-Str. 4
Tübingen 72076
Germany
tel. +49-7071-29-73331
fax. +49-7071-295360
email:stefan.broeer@uni-tuebingen.de

Annadora J. Bruce
Sanders Brown Center on Aging
Rm 211 / Univ. of Kentucky
800 S. Limestone
Lexington, Kentucky 50536-0230
USA
tel. +1-606-257-6040
fax. +1-606-323-2866

Peter Brust
Forschungszentrum Rossendorf
P O Box 51 01 10
Dresden 01314
Germany
tel. +49-3512-603452
fax. +49-3512-603452
email:brust@fz-rossendorf.de

Jean Lud Cadet
Neurosciences Branch
Mol. Neuro Chemistry Section
4940 Eastern Avenue
Baltimore, Maryland 21224
USA
tel. +1-410-550-2953
fax. +1-410-550-2745/1648

M. Carman-Krzan
Dpet. Pharmacol.
Med. Faculty
Korytkova 2
Ljubljana
1000 Slovenia
tel. +386-61-441-374
fax. +386-61-446-082

Olga V. Chumakova
Inst. Biochemistry
Dept. Morphol. & Neurochem
50 Blk
Grodno 230017
Belarus
tel. +375-0152-337833
fax. +375- 0152-967322
email:tlg@ptb.belpak.grodno.by

Filippo Conti
Dept. of Chemistry
Univ. "La Sapienza"
P. le Aldo Moro, 5
00185 Rome
Italy
tel/fax +39-6-445-5278
email:conti@axrma.uniroma1.it

Gerald Curzon
Inst. of Neurology
Dept. Neurochemistry
1 Wakefield Street
London WC1N 1PJ
United Kingdom
tel. +44-171-278-1328
fax. +44-171-278-6572

Annica B. Dahlström
Univ. of Göteborg
Anatomy Cell Biol.
Medicinargatan 3-5
Göteborg 41390
Sweden
tel. +46-31-7733366
fax. +46-31-829690

Svetlana Dambinova
Russian Acad. of Sciences
Inst. of Human Brain
Acad. Pavlov Street 12A
St. Petersburg 197376
Russia
tel. +7-812-234-9423
fax. +7-812-234-3247
email:dambinova@rsn.spb.su

Pia Davidson
Univ. of Göteborg
Inst. of Clin. Neuroscience
Dept. of Psych. & Neurochem.
Mölndals Sjukhus
Mölndal 431 80, Sweden
tel. +46-31-862-377
fax. +46-31-862-421

Annamaria de Luca
Dept. Pharmacobiologico
Fac. di Farmacia
Universita di Bari
Via Orabona 4
Bari 70125
Italy
tel. +39-80-5442-802
fax +39-80-5442-050/770

Laura Della Corte
Dept. Pharmacology
Viale G. B. Morgagni 65
50134 Firenze
Italy
tel. +39-55-4237411/412
fax. +39-55-4361613

Jan de Vente
Univ. of Limburg
Psych. & Neuropsychol
P. O. Box 616
Maastricht 6220 MD
The Netherlands
tel. +3143-3881168
fax. +3143-3671096

John de Vries
Lab. of Neurobiochemistry
Univ. Hospital
P. O. Box 30.001
9700 Rb Groningen
The Netherlands
tel. +31. 50. 361. 2647

Hugo D'haenen
Acad. Hospital Free Univ. of Brussels
Psychiatric Dept.
Laarbeeklaan 101
Brussels 1090
Belgium
tel. +32-2-477-6405
fax. +32-5230-6440

Wilfried Diener
Eberhard-Karls Univ.
Med. Klinik & Poliklinik
Abt. Innere Med. II
Otfried Müller Strasse
72076 Tübingen
Germany
tel. +49-7071-2984479
fax. +49-7071-2982760

Dusan Dobrota
Dept. of Biochemistry
Jessenius Fac. of Medicine
Comenius University
Malá Hora 4
03601 Martin
Slovakia
tel. +42-842-31565
fax. +42-842-36770/36332
email:dobrota@doktor.jfmed.uniba.sk

Vladimir Dolezal
Inst. of Physiology Cas
Czech Acad. Sciences
Videnska 1083
Prague 142 20
Czech Republic
tel. +422-475-2573
fax. +422-471-9517
email:dolezal@biomed.cas.cz

Krystyna Doma½Ska-Janik
Med. Res. Centre
Dept. of Neurochemistry
5, Pawinskiego Str
Warsaw 02-106
Poland
tel. +48-22-608-6528
fax. +48-22-668-5423

Bas Douma
Univ. of Groningen
Dept. Animal Physiol
Kerklaan 30
Haren 9751 NN
The Netherlands
fax. +3150-3635205

Elizabeth Dovedova
Russian Acad. Med. Sciences
Brain Res. Inst.
Lab. of Cytochemistry
P. Obukha, 5
103064, Moscow
Russia
tel. +7-095-917-9048
fax. +7-095-916-0595

B. Drukarch
Dept. Neurologie
Vrije Universiteit
V. D. Boechorststraat 7
1081 BT Amsterdam
The Netherlands
tel. +31-20-4448103
fax. +31-20-4448100

Yves Dunant
Dept. de Pharmacologie
Centre Medical Univ.
1211 Geneve 4
Switzerland
tel. +41-22-702-5432
fax. +41-22-702-5452
email:dunant@cmu.unige.cb

John Dunlop
Wyeth-Ayerst. Research
Cns Disorders
Cn-8000
Princeton, New Jersey 08543
USA
tel. +1-908-274-4193
fax. +1-908-274-4020

Stephen Dunnett
Univ. of Cambridge
Centr. F. Brain Repair
Foruie Site, Robinson Way
Cambridge CB2 2PK
United Kingdom
tel. +44-1223-33-1160
fax. +44-1223-33-1174
email:sd19@cus.cam.ac.uk

Karine Dzhandzhugazyan
Panum Institute
Protein Lab.
3c Blagdamsvej
Bldg. 6. 2
Copenhagen 2200 N
Denmark
tel. +45-3532-7343
fax. +45-3536-0116

Piet Eikelenboom
Res. Inst. Neurosciences
Free Univ. Valeriuskliniek
Valeriusplein 9
1075 BG Amsterdam
The Netherlands
tel. +31-20-5736686
fax. +31-20-5736687
email:piet.eikelenboom@pca.znw.nl

Addresses

Joel Eyer
Inserm
Dept. Neurobiology
Inserm U298-Chru
Angers 49033
France
tel. +33-41-35-4726
fax. +33-41-73-1630

Matthijs Feenstra
Neth. Inst. for Brain Res.
Meibergdreef 33
Amsterdam Z. O. 1105 AZ
The Netherlands
tel. +31-20-5665500
fax. +31-20-6961006

Marianne Fillenz
Univ. of Oxford
Dept. Physiology
Parks Road
Oxford OX1 3PT
United Kingdom
tel. +44-1865-275908
fax. +44-1865-272469

Abraham Fisher
Israel Inst. for Biol. Research
Dept. of Med. Chemist
P. O. Box 19
Ness-Ziona 74100
Israel
tel. +972-8-381-306
fax. +972-8-401-094

Ulrich Flögel
Univ. Bremen
Inst. Org. Chemie
NW. 2,Fb 2, P. O. Box 330440
Bremen D-28334
Germany
tel. +49-421-2182-951
fax. +49-421-2184-264

Paul T. Francis
Dementia Research Lab.
Dept. Biochemistry
Umds-Guy's Campus
St. Thomas Street
London SE1 9RT
United Kingdom
tel. +44-171-955-2611
fax. +44-171-955-2600/4191

Pam Fredman
Dept. of Clinical Neurosci.
Section of Psych. & Neurochem.
Mölndal Hospital
43180 Mölndal
Sweden
tel. +46-31-862415/2419
fax. +46-31-862426

E. M. Fykse
Norwegian Defence Res. Inst.
Div. for Environm. Toxicology
P O Box 25
N-2007 Kjeller
Norway
tel. +47-6380-7845
fax. +47-6380-7811

József Gaál
Mega Pharma
Baross U 53
1174 Budapest
Hungary
tel/fax +361-258-4078
email:13248gaa@helka.iif.hu

Peter Gebicke-Haerter
Univ. of Freiburg
Dept. of Psychiatry
Haupstr. 5
Freiburg D-79104
Germany
tel. +49-761-270-6835
fax. +49-761-270-6619
email:geha@psylab.ukl.uni.freiburg.de

Lydiya Gershtein
Russian Acad. of Med. Sciences
Brain Res. Institute
Lab. of Cytochemistry
P. Obukha 5
103064 Moscow
Russia
tel. +7-095-917-9048
fax. +7-095-916-0595

Antonio Giuditta
Gen. & Environm. Physiology
Via Mezzocannone 8
Napoli 80134
Italy
tel. +39-81-552-6027
fax. +39-81-552-6194
email: giuditta@unina.it

Anna Maria Giuffrida Stella
Biochemistry
Viale Andrea Doria N. 6
Catania 95125
Italy
tel. +39-95-336990
fax. +39-95-339886

Albert Gjedde
Aarhus General Hospital
Positr. Emis. Tomography
44 Norrebrogade
Aarhus DK-8000
Denmark
tel. +45-8949-3030
fax. +45-8949-3020
email: albert@medfysik.aau.dk

Steven Goldberg
Preclin. Pharmacol. Lab.
Nida Intramur. Res. Progr
4940 Eastern Ave. Bldg. C
P O Box 5180
Baltimore, Maryland 21224
USA
tel. +1-410-550-1522
fax. +1-410-550-1648/1563

Gianfrancesco Goracci
Inst. Biochim. Chim. Med.
Via Del Giochetto
06122 Perugia
Italy
tel. +39-75-5853420
fax. +39-75-5853428

Tatsuya Haga
Univ. of Tokyo
Dept. Biochemistry
Inst. for Brain Res.
7-3-1-Hongo
Tokoy 113
Japan
tel. +81-356-89-7331
fax. +81-338-14-8154

Anders Hamberger
Dept. of Anatomy & Cell Biology
Univ. of Göteborg
Medicinaregatan 5
41390 Göteborg
Sweden
tel. +46-31-853368
fax. +46-31-412805

Bernd Hamprecht
Univ. of Tübingen
Phys. Chem. Inst.
Hoppe-Seyler-Str. 4
Tübingen 72076
Germany
tel. +49-7071-292452

Lise Sofie Haug
Univ. of Oslo
Neurochemical Lab.
P. O. Box 1115
Blindern
Oslo N-0317
Norway
tel. +47-2285-1097
fax. +47-2285-1436

Nariyuki Hayashi
Nihon University
Dept. Emercency Med.
30 Oyaguchi Kami-Machi
Itabashi-Ku
Tokyo 173
Japan
tel. +81-3-3972-8111
fax. +81-3-5995-1069

Heike Heuer
Max-Planck-Inst.
Exp. Endokrinol.
Feodor Lynen Str. 7
P. O. Box 610309
Hannover 30603
Germany
tel. +49-511-5359200
fax. +49-511-5359203
email:100656,722@compuserve.com

Herbert Hildebrandt
Univ. Hohenheim
Inst. Zoology
Garbenstr. 30
Stuttgart 70593
Germany
tel. +49-711-459-2255
fax. +49-711-459-3450
email:hildebra@uni-hohenheim.de

Inge Huitinga
Dept. Pharmacology
Faculty of Medicine
Vrije Universiteit
Van Der Boechorststr. 7
1081 HV Amsterdam
The Netherlands
tel. +31-20-444-8090/8107
fax. +31-20-444-8100

Michael Hüll
Dept. of Psychiatry
Univ. of Freiburg
Hauptstr. 5
Freiburg 79104
Germany
fax. +49-761-270-6619

Maurice Israel
Lab. de Neurobiologie
Cellulaire du CNRS
91190 Gif-sur-Yvette
France
tel. +33-1-6982-3650/3662
fax. +33-1-6982-9466

Natalya Izrarina
Pavlov Inst. Psychol.
Funct. Biochemistry
Nab. Makarova 6
St. Petersburg 199034
Russia
tel. +7-812-1100-642
fax. +7-812-1100-642

Agnieszka Jankowska
Med. Univ. of Gdansk
Dept. Clin. Biochem.
Debinki 7
Gdansk 80-211
Poland
tel. +48-58-478222#1575
fax. +48-58-449653

Kent Jardemark
Univ. of Göteborg
Anat. & Cell Biology
Medicina Regatan 3-5
Göteborg 413. 90
Sweden
tel. +46-31-773-3370
fax. +46-31-773-3330

Jaak Järv
Tartu University
Inst. Chem. Physics
2 Jakobi Str.
Tartu Ee 2400
Estonia
tel. +372-7465-246
fax. +372-7465-247
email:jj@chem.ut.ee

Michael H. Joseph
Inst. of Psychiatry
Dept. Psychology
De Crespigny Park
Denmark Hill
London SE5 8AF
United Kingdom
tel. +44-171-919-3414
fax. +44-171-708-3497

Nebojsa Jovic
Clinic Maxillofacial Surgery
Military Med. Academy
1100 Belgrade
Yugoslavia
tel. +381-11-4885-395
fax. +381-11-4885-395

Romic G. Kamalyan
Inst. of Biochem. Nat. Acad. Sci.
Reg. Enzyme Act.
P. Sevak Str. 5/1
Yerevan 375044
Armenia
tel. +374-39-07048
fax. +374-39-07047

Baruch I. Kanner
Dept. of Biochemistry
Hebrew University
Hadassah Med. School
Jerusalem 91120
Israel
tel. +972-2-758506

Andrzej Kapuócinski
Med. Res. Centre
Polish Acad. of Sciences
Dept. Neuropathology
Ul. Dworkowa 3
Warsau 00784
Poland
tel. +4822-495410
fax. +4822-496973

Jan-Olof Karlsson
Univ. of Göteborg
Inst. Anat. & Cellbiol
Medicinaregatan 3-5
Göteborg 41390
Sweden
fax. +46-31-773-3359
email:janolof.karlsson@anatcell.gu.se

Kiyoshi Kataoka
Ehime Univ. School of Med.
Dept. of Physiology
Shigenobu
Onsen-Gun, Ehime 791-02
Japan
tel. +81-89-960-5240
fax. +81-89-960-5242
email:kataoka@m.chimc-u.ac.jp

Koichiro Kawashima
Kyoritsu College of Pharmacy
Dept. Pharmacology
1-5-30 Shibakoen, Minato-Ku
Tokyo 105
Japan
tel. +81-3-5400-2674
fax. +81-3-5400-2698

Elisabeth Kienzl
Univ. of Technology
Inst. of Appl. Botany & Techn.
 Microscopy
Gertrudemarkt 9
Vienna A-1060
Austria
tel. +43-1-58801-5068/4770
fax. +43-1-587-4835
email:l.puching@fbch.tuwien.ac.at

Vladislav Kiselyov
Panum Institute
Protein Lab.
3c Blagdamsvej
Bldg. 6.2
Copenhagen 2200 N
Denmark
tel. +45-353-27317
fax. +45-353-60116

Addresses

Jochen Klein
Univ. of Mainz
Dept. Pharmacol.
Obere Zahlbacher Str. 67
Mainz D-55101
Germany
tel. +49-6131-174393
fax. +49-6131-176611

Siert Knollema
Dept. Biol. Psychiatry
University Hospital
P. O. Box 30. 001
9700 RB Groningen
The Netherlands
fax. +31-50-3696727

Edward Koenig
Neurobiology Lab.
State Univ. of New York
313 Cary Hall
Buffalo, New York 14214
USA
tel. +1-716-829-3572
fax. +1-716-829-2344

Hermann Koepsell
Inst. of Anatomy
University Würzburg
Koellikerstr. 6
D-97070 Würzburg
Germany
tel. +49-931-312700
fax. +49-931-572338

Jaap Korf
Dept. of Biol. Psychiatry
University Hospital
9713 EZ Groningen
The Netherlands
tel. +31. 50. 361. 2100
fax. +31. 50. 363. 6905
emailj.korf@med.rug.nl

B. Kwiatkowska-Patzer
Med. Research Center
Polish Acad. of Sciences
5 Pawinskiego Str
Warsaw 02-106
Poland
tel. +48-22-668-5388
fax. +48-22-668-5532

Nathalie Lambeng
Inserm U. 289
Hop. de la Salpetriere
Bâtim. Nouv. Pharmacie
47, Blvd. de l'Hopital
Paris 75651
France
tel. +331-42-16-2202
fax. +331-44-24-3658
email:lambeng@cct.jussieu.fr

Adriaan A. Lammertsma
Free University Hosp.
P.O. Box 7057
1007 MB Amsterdam
The Netherlands

Risto Lammintausta
Orion Corporation
Orion Pharma Res. & Development
P. O. Box 425
20101 Turku
Finland
tel. +358-212-72-7528
fax. +358-212-72-7432

A. Lapidot
Weizmann Inst. of Science
Dept. of Organic Chemistry
Rehovot 76100
Israel
tel. +972-8934-3413
fax. +972-8934-4142

Robert O. Law
Univ. of Leicester
Cell Phys. & Pharmacol.
P. O. Box 138
Leicester LE1 9HN
United Kingdom
tel. +44-116-252-3076
fax. +44-116-252-5045

Jerzy W. Łazarewicz
Med. Res. Centre
Dept. of Neurochemistry
5, Pawinskiego Str
Warsaw 02-106
Poland
tel. +4822-608-6528
fax. +4822-668-5423

Ken Lee
Gifu Univ. School of Med.
Dept. Physiology
40 Tsukasa-Machi
Gifu 500
Japan
tel. +81-58-265-1241
fax. +81-58-265-9004

Jan Lehotský
Comenius University
Jesenius Med. Faculty
Dept. Medical Biochemistry
Mulá Hora 4
Sk-03601 Martin
Slovakia
tel. +42-842-31565
fax. +42-842-36770

Anne S. Lesage
Janssen Res. Foundation
Dept. Biochem. Pharmac
Turnhoutseweg 30
Beerse B-2340
Belgium
fax. +32-14-605-380

Giulio Levi
Neurobiol. Section
Lab. of Pathophysiology
Inst. Superiore di Sanita
Viale Regina Elena, 299
Roma 00161
Italy
tel. +39-649-902488
fax. +39-649-902071

Moti Liskovitch
Weizmann Inst. of Sciences
Dept. Hormone Res.
P O Box 26
Rehovot 76100
Israel
tel. +972-8-483-182
fax. +972-8-466-966

Konrad Löffelholz
J. Gutenberg Univ. Mainz
Pharmakol. Inst.
Obere Zahlbacherstr. 67
D-55101 Mainz
Germany
fax. +49-6131-176611

John P. Lowry
Univ. of Oxford
Dept. Physiology
Parks Road
Oxford OX1 3PT
United Kingdom
tel. +44-1865-275907
fax. +44-1865-272469

Paul Luiten
Univ. of Groningen
Dept. Animal Physiology
Kerklaan 30
9751 NN Haren
The Netherlands
tel. +31-50-3632359/2340
fax. +31-50-3635205
email:luitenpg@biol.rug.nl

Nadezda Lukáčová
Inst. of Neurobiology, SAS
Dept. Biochemistry
Soltesovej 4
Kosice 040 01
Slovak Republic
tel. +42-95-765-064
fax. +42-95-765-074

Alla Lysenko
Neurocybernetics Res. Inst.
Phys. & Biochemistry
Stachki Ave. 194/1
Rostov-on-Don 344090
Russia
tel. +7-8632-24-46-93
fax. +7-8632-22-04-24/25-47-23

Adriana Maggi
Inst. Pharmacol. Sci. /Center MPL
University of Milan
Via Balzaretti 9
Milan 20133
Italy
tel. +39-2-204-88375
fax. +39-2-204-88249
email:maggia@isfunix.farma.unimi.it

Pierre Magistretti
Inst. of Physiology
7, Rue du Bugnon
Lausanne 1005
Switzerland
tel. +41-21-692-5542
fax. +41-21-692-5595

Kálmán Magyar
Semmelweis Univ. of Medicine
Dept. of Pharmacodynamics
Nagyvárad Ter 4
Budapest 1089
Hungary
tel. +361-210-4411
fax. +361-210-4411
email:magykal@net.sote.hu

Andrzej Malecki
Silesian Acad. of Medicine
Dept. Pharmacology
Medyków 18
Katowice 40-752
Poland
tel. +48-32-1523835
fax. +48-32-1523835

M. Marangolo
Trinity College
Biochem. Dept.
Dublin–2
Ireland
tel. +353-1-608-1802
fax. +353-1-677-2400
email:mmrngolo@tcd.ie

Gerard J. M. Martens
Univ. of Nijmegen
Dept. Animal Physiol.
Toernooiveld
6525 ED Nijmegen
The Netherlands
tel. +31-24-3652601
fax. +31-24-3652714

Rainer Martin
Univ. Ulm
Sekt. Electr. Microscopy
Albert-Einstein-Allee 11
Ulm D-89069
Germany
tel. +49-731-502-3441
fax. +49-731-502-3383

Angelo R. Massaro
Policlinico Gemelli
Dept. Neurology
Univ. Sacro Cuore
Roma I-00168
Italy
tel. +39-330-423240
fax. +39-6-338-5411

E. R. Mattarredona
Fac. de Farmacia
Dept. Bioquimica
C/Prof. García González
41012 Sevilla
Spain
tel. +34-5-4511580
fax +34-5-4233765

Amanda Mcrae-Dequerce
Univ. of Göteborg
Anatomy & Cell Biol. Dept.
Medicinargatan 3-5
Göteborg 41390
Sweden
tel. +46-31-7733354
fax. +46-31-829690
email:amanda.mcrae@anatcell.gu.se

Karine Melkonian
State Medical Univ.
Dept. Pharmacology
2, Korjun St.
Yerevan 375025
Armenia
tel. +3742-560860/521711
fax. +3742-151812

M. M. Melkonian
State Med. University
Dept. Chemistry
2, Korjun Str.
Yerevan 375025
Armenia
tel. +3742-580343
fax. +8895-529803

Alexander Mendzeritsky
Neurocybernetics Res. Inst.
Phys. & Biochemistry
Stachki Ave. 194/1
Rostov-on-Don 344090
Russia
tel. +7-8632-24-46-93
fax. +7-8632-22-04-24/25-47-23

André Ménez
Dept. d'Ingenierie Et
D'Etudes des Proteines
Cea-Saclay
91191 Gif-sur-Yvette
France
tel. +33-1-6908-9961
fax. +33-1-6908-9137

Daniel M. Michaelson
Tel Aviv University
Dept. of Neurobiochemistry
Ramat Aviv
Tel Aviv 69978
Israel
tel. +972-3-6409624
fax. +972-3-6407643

Mirjana Mijanovic
Dept. of Pharmacology
Med. Faculty
Čekalusa 90
Sarajevo
Bosnia Herzegovina
tel. +387-71-441-895
fax. +387-71-663-625

Jan Minderhoud
Dept. of Neurology
University Hospital
9713 BZ Groningen
The Netherlands
tel. +31. 50. 361. 2430
fax. +31. 50. 361. 1707

M. Teresa Miras-Portugal
Universidad Complutense de Madrid
Dept. de Bioquimica
Facultad Veterinaria
Madrid 28040
Spain
tel. +34-1-394-38-94
fax. +34-1-394-39-09

Analtoly Mokrushin
Pavlov Inst. of Physiology
Higher Nervous Act.
Nab. Makarova 6
St. Petersburg 199034
Russia
tel. +7-812-707-28-06
fax. +7-812-218-05-01
email:chi@physiology.spb.su

J. Molgo
Lab. de Neurobiologie
Cell. Et Moleculaire
C.N.R.S.
1, Ave de La Terrasse
91198 Gif-sur-Yvette
France
tel. +33-1-69-823-030
fax. +33-1-69-829-466
email:molgo@hermes.cnrs-gif.fr

A. Morgan
Physiological Laboratory
University of Liverpool
P O Box 147
Liverpool L69 3BX
United Kingdom
tel. +44-151-794-5333
fax. +44-151-794-5337

Katarzyna Nalecz
Nencki Inst. of Exp. Biology
Dept. Muscle Biochem
Pasteur Str. 3
Warszawa Pl-02-093
Poland
tel. +48-22-659-8571
fax. +48-22-225342

Natalia Nalivaeva
Inst. of Evolutionary Physiology
& Biocehmistry
M. Thorez Av. 44
St. Petersburg
Russia
tel. +7-812-552-7925
fax. +7-812-552-3012
email:nnn@ief.spb.su

Hiroaki Naritomi
Nat. Cardiovascular Center
Cerebral Circ. Lab
5-7-1 Fujishiro-Dai
Suita, Osaka 565
Japan
tel. +81-6-833-5012/872-0010
fax. +81-6-872-7485

V. I. Nazarenko
9, Leontovichstr.
Kiev 252030
Ukraine
tel. +380-44-224-50-10
fax. +380-44-229-63-65

Jerzy Z. Nowak
Polish Academy of Sciences
Dept. of Biog. Amines
Tylna 3, P O Box 225
Lodz 90-950
Poland
tel. +48-42-817006
fax. +48-42-815283

Csaba Nyakas
Haynal Univ. Health Sci.
Centr. Research
Szabolcs U. 35
P. O. Box 112
Budapest H-1389
Hungary
tel. +261-1290222
fax. +261-1208246

Barbara Oderfeld-Nowak
Nencki Inst. Exp. Biol.
Dept. Neurophysiolog
3, Pasteur Str.
Warsaw 02093
Poland
tel. +48-22-659-871
fax. +48-2-222-5342

S. Öhman
Elfin Lab.
P O Box 133 / Evastigen 9
S 59070 Ljungsbro
Sweden
tel. +46-13-219020
fax. +46-13-219021

Simon S. Oja
Univ. of Tampere
Med. School
P. O. Box 607
Tampere FIN-33101
Finland
tel. +358-31-215-6694
fax. +358-31-215-6170

Bart-Jan Oosterink
A. Z. G.
Dept. Neurology
Hanzeplein 1
9700 RB Groningen
The Netherlands
tel. +3150-3624525
fax. +3150-3636905

Luis M. Orensanz
Hosp. Ramon y Cajal
Dept. Investigacion
Carretera de Colmenar Km. 9
Madrid 28034
Spain
tel. +34-1-336-8684
fax. +34-1-336-9016

Neville Osborne
Univ. of Oxford
Nuffield Lab. of Ophtalmology
Walton Street
Oxford OX2 6AW
United Kingdom
tel. +44-1865-248-996
fax. +44-1865-794-508
email:nevilleosborne@eye.ox.ac.uk

Mariusz Papp
Polish Acad. Sciences
Inst. of Pharmacology
12 Smetna Street
Krakow 31-343
Poland
tel. +48-12-374550
fax. +48-12-374500

Marion Paturneau-Jouas
Inserm 134
Hôp. Salpetrière
47, Blvd. de l'Hôpital
F 75651 Paris
France
tel. +33-1-42162157
fax. +33-1-45848008
email:cidjay@worldnel.fr

Giancarlo Pepeu
Univ. of Florence
Dept. Pharmacology
Viale Morgagni 65
Florence 50134
Italy
tel. +39-55-4237-418
fax. +39-55-4361-613

Anna Petroni
Inst. Pharm. Sciences
Via Balzaretti 9
20133 Milan
Italy
tel. +39-2-20488307
fax. +39-2-29404961

Nina Popova
Inst. of Cytology & Genetics
Behavioral Phenogen.
Pr. Lavrentyeva 10
Novosibirsk 630090
Russia
tel. +7-3832-354-753
fax. +7-3832-356-558
email:npopova@ieg.nsk.su

Soren Prag
Panum Institute
Protein Lab.
3C Blagdamsvej
Bldg. 6. 2
Copenhagen 2200 N
Denmark
tel. +45-3532-7322
fax. +45-3536-0116

Alexander Pylaev
Russian State Med. University
Dept. Morphology
Ostrovityanova Str. 1
Moscow 117869
Russia
tel. +7-095-434-0037
fax. +7-095-434-4787
fax. +7-095-131-4863
email:im@mx.iki.rssi.ru

Fenna Radhakishun
Organon International
Medical R & D Unit
P. O. Box 20
5340 BH Oss
The Netherlands
tel. +31-412-661754
fax. +31-412-662555

Hinrich Rahmann
Univ. Hohenheim
Inst. Zoology
Garbenstr. 30
Stuttgart 70593
Germany
tel. +49-711-459-2255
fax. +49-711-459-3450

Rona R. Ramsay
Univ. of St. Andrews
School of Biomedical Sciences
Div. Cell & Mol. Biology
Irvine Building
St. Andrews KY16 9AL
United Kingdom/Scotland
tel. +44-1334-463411
fax. +44-1334-463400
email:rrr@st-and.ac.uk

Ciaran Regan
Univ. College
Dept. Pharmacology
Belfield
Dublin 4
Ireland
tel. +353-1-706-1619
fax. +353-1-269-2749
email:cregan@macollamh.ucd.ie

H. Reiber
Lab. of Neurochemistry
University Göttingen
Robert-Koch-Str. 40
Göttingen D-37075
Germany
tel. +49-551-39-6619
fax. +49-551-39-2028

Russel J. Reiter
Dept. Cell. & Structural Biology
Univ. Texas
Health and Science Center
7703 Floyd Curl Dr
San Antonio, Texas 78284
USA
tel. +1-210-567-3854
fax. +1-210-567-6948

Christiane Richter-Landsberg
Univ. of Oldenburg
Dept. of Neurobiol.
P. O. Box 2503
Oldenburg D-26111
Germany
tel. +49-441-798-3422
fax. +49-441-798-3423

George Robillard
Biochemical Lab. Groningen
Biomol. Sciences & Biotechn.
Nijenborgh 4
9747 AG Groningen
The Netherlands
tel. +31-50-363-4321
fax. +31-50-363-4165

Hans Rollema
Pfizer Inc.
Dept. of Neurosci.
Eastern Point Rd.
Groton, Connecticut 06340
USA
tel. +1-860-441-6374
fax. +1-860-441-1773
email:hans.rollema@groton.pfizer.com

Slavwomir Rump
Mil. Inst. of Hygiene & Epidem.
Pharmacol. & Toxicol.
Kozielska 4
Warsaw 01-163
Poland
fax. +48-22-104391

Eric Salmon
Univ. of Liege
Cyclotron Res. Ctr
B30 Sart Tilman
Liege 4000
Belgium
tel. +32-41-66-36-87
fax. +32-41-66-29-46

V. Sanchez-Margalet
Dept. Biochem. & Molec. Biology
Med. School Univ. Sevilla
Av. Sanchez Pizjvan 4
Sevilla 41009
Spain
tel. +34-5-4557-356
fax. +34-5-4557-481

Reinhard Schliebs
Paul Flechsig Inst. Brain Res.
Dept. Neurochemistry
Jahn Allee 59
Leipzig 04109
Germany
tel. +49-341-9725-766
fax. +49-341-211-4492

Rupert Schmidt
Zentr. Biotechn. Betriebseinheit
Im Strahlenzentrum
Justus-Liebig Univ.
Leihgesterner Weg 217
D-35392 Giessen
Germany
tel. +49-641-99-16500
fax. +49-641-99-15009
email:rupert.schmidt@strz.uni-giessen.de

Arne Schousboe
Dept. of Biol./Neurobiol. Unit
Royal Danish School of Pharm.
2, Universitetsparken
2100 Copenhagen
Denmark
tel. +45-3537-6777 ext. 330
fax. +45-3537-5744

Peter Schubert
Max-Planck-Inst. Psych.
Dept. Neuromorphol.
Am Klopfersmitt 18A
Martinsried 82152
Germany
tel. +49-89-8578-3689
fax. +49-89-8578-3939
email:schubert@neuro.mpg.de

Constanze Seidenbecher
Otto Von Guericke University
Inst. Med. Psychology
Leipzigerstr. 44
Magdeburg 39120
Germany
tel. +49-391-626-3223
fax. +49-391-626-3229

Philippe Siaud
Inserm U. 297
Neuroendocrinol. Exp.
Uer de Médecine Nord.
Blvd. P. Dramard
13916 Marseille Cedex 20
France
tel. +33-91-654311
fax. +33-91-698712

Jerzy Silberring
Karolinska Inst.
Drug Dependence Res.
Stockholm 171 76
Sweden
tel. +46-8-729-4698
fax. +46-8-341-939

Alexander Sinichkin
State University
Dept. Biochemistry
Gorkogostr. 202 Kv. 141
Rostov-on-Don 344022
Russia
tel. +7-8632-537320
fax. +7-8632-649877

D. A. Sival
Dept. of Neurology
Academic Hospital
Hanzeplein 1
9700 RB Groningen
The Netherlands
tel. +31-50-3612464
fax. +31-50-3611707

Malgorzata Skup
Nencki Inst. of Exp. Biology
Dept. Neurophysiol.
M. Pasteura 3
Warsaw 02-093
Poland
tel. +482-659-85-71 ext 360
fax. +4822-225342
email:mskup@nencki.gov.pl

Karl-Heinz Smalla
Inst. Neurobiology
Dept. Neurochem/Mol. B
Brenneckerstr. 6
Magdeburg 39108
Germany
tel. +49-391-626-3215
fax. +49-391-626-3229
email:smalla@ifn-magdeburg.de

G. T. Snoek
Centre for Biomembranes and
Lipid Enzymology
Dept Biochemistry of Lipids
Padualaan 8
3584 CH Utrecht
The Netherlands
tel. +31-30-253-3445/253-4668
fax. +31-30-252-2478

Hermona Soreq
Hebrew University
Dept. Biol. Chem.
Givat Ram
Jerusalem 91904
Israel
tel. +972-2-658-5109
fax. +972-2-6520258

Micha E. Spira
Hebrew Univ. of Jerusalem
Dept. Neurobiology
Life Sciences Inst.
Givat Ram
Jerusalem 91904
Israel
tel. +972-2-6585091
fax. +972-2-637033
email:spira@cc.huji.ac.il

Jeffrey Sprouse
Pfizer Inc.
Dept. Neurosci.
558 Eastern Point Rd.
Groton, Connecticut 06340
USA
tel. +1-860-441-6373
fax. +1-860-441-1017
email:sprouj@pfizer.com

R. Srapionian
Inst. Biochemistry
Nat. Acad. Sci
5/1 Serag Str.
375014 Yerevan
Armenia
tel. +3742-281840
fax. +3742-281951

P. Stinissen
Multiple Sclerosis Res.
Dr. L. Willems Instituut
Universitaire Campus
Drepenbeek
Belgium
tel. +32-11-269211
fax. +32-11-269209

Hans Stoof
Research Inst. Neurosciences
Dept. of Neurology
Free University
De Boelelaan 1087
1081 HV Amsterdam
The Netherlands
tel. +31-20-444-8360
fax. +31-20-444-8100

Paul J. L. M. Strijbos
Cruciform Project Neurosci. Res.
Inst. Neurology
Univ. College London
1, Wakefield Street
WC1N 1PJ London
United Kingdom
tel. +44-171-837-3611 ext. 4954
fax. +44-171-837-1347
email:p.strijbos@ucl.ac.uk

Joanna B. Strosznajder
Polish Ac. of Sciences
MRC, Dept. of Cellular Signalling
Pawinskiego 5
Warszawa 02-106
Poland
tel. +48-22-608-6414
fax. +48-22-668-5223/5423

Bauke Stuiver
Biol. Centre
Dept. Animal Fysiol.
University of Groningen
Kerklaan 30 / P. O. Box 14
9750 AA Haren
The Netherlands
tel. +3150-3632363
fax. +3150-3635205

Joanna Sypecka
Med. Res. Centre
Dept. of Neurochemistry
5, Pawinskiego Str
Warsaw 02-106
Poland
tel. +48-22-608-6528
fax. +48-22-668-5423

Andrea Csilla Szabo
Biol. Res. Center
Inst. of Biophysics
Temesvazi Krt 62
P. O. Box 521
Szeged H-6701
Hungary
tel. +36-62-432232/433465
fax. +36-62-433-133

Andrzej Szutowicz
Med. Univ. of Gdansk
Dept. Clin. Biochem.
Debinki 7
Gdansk 80-211
Poland
tel. +48-58-478222 #1575
fax. +48-58-449653
email:aszut@amedec.amg.gda.pl

Takao Taki
Tokyo Med. and Dental Univ.
Dept. Biochemistry
1-5-45, Yushima, Bunkyo-Ku
Tokyo 113
Japan
tel. +81-3-5803-5165
fax. +81-3-5803-0129

Albert Teelken
Lab. Center
Section of Neurochemistry
University Hospital
9713 GZ Groningen
The Netherlands
tel. +31-50-361-2647
fax. +31-50-361-1707

Addresses

Gert J. ter Horst
University Hospital
Dept. Biol. Psychiatry
Hanzeplein 1
P. O. Box 30. 001
9700 RB Groningen
The Netherlands
tel. +31-50-361-2105/1533
fax. +31-50-361-1699
email:g.j.terhorst@med.rug.nl

Guido Tettamanti
Dept. Medical Chemistry
& Biochem. Univ. of Milano
Via Saldini 50
Milano 20133
Italy
tel. +39-2-7064-5247
fax. +39-2-236-3584

Georg G. Thakhava
Sarajishrili Inst. of Neurology
Dept. Angioneurology
2a, Guda Mataristr.
Tbilisi 380092
Republic of Georgia
tel. +88-32-371936
fax. +88-32-233423

Keith Tipton
Trinity College
Dept. Biochemistry
Dublin 2
Ireland
tel. +353-1-608-1608
fax. +353-1-677-2400
email:ktipton@tcd.ie

F. Titgemeyer
Dept. Biochemistry
Universiteit Groningen
Nijenborgh 4
9747 AG Groningen
The Netherlands
tel. +31-50-363-4170
fax. +31-50-363-4165
email:f.titgemeyer@chem.rug.nl

W. Tischmeyer
Federal Inst. Neurobiology
P O Box 1860
D-39008 Magdeburg
Germany
tel. +49-391-6263222
fax. +49-391-6263229
email:tischmeyer@ifn-magdeburg.de

Maria Tomaszewicz
Med. Univ. of Gdansk
Dept. Clin. Biochem.
Debinki 7
Gdansk 80-211
Poland
tel. +48-58478222 #1575
fax. +48-58-412915/449653

Jack C. de la Torre
Univ. of New Mexico
Dept. Neursurgery
2211 Lomas, N.E.
Albuquerque, New Mexico 87131
USA
tel. +1-505-277-0043/5752
fax. +1-505-277-3891
email:jdelator@medusa.unm.edu

Nina Tsakadze
Sarajishvili Inst. of Neurol.
Dept. Angioneurology
2a Gudamakari St.
Tblisi 380092
Rep. of Georgia
tel. +995-8832-618114
fax. +995-32-001027/001153

Stanislav Tuçek
Inst. of Physiology
Dept. Neurochemistry
Videnska 1083
Prague 14220
Czechia
tel. +42-02-4752620
fax. +42-02-4752488

Hayrettin Tumani
Georg-August Univ.
Neurochem. Lab.
Robert-Koch-Str. 40
Göttingen D-37075
Germany
tel. +49-551-396619
fax. +49-551-392028

Tony Turner
Univ. of Leeds
Dept. Biochem. & Mol. Biology
Leeds LS2 9JT
United Kingdom
tel. +44-113-233-3131
fax. +44-113-242-3187
email:a.j.turner@leeds.ac.uk

Anatoly Uzdensky
Inst. of Neurocybernetics
Rostov University
194/1 Stachky Ave.
Rostov-on-Don 344090
Russia
tel. +7-8632-280577
fax. +7-8632-280367
email:uzd@krinc.rnd.runnet.ru

Anne-Marie Van Dam
Free Univ. A'dam
Res. Inst. Neuroscience
Dept. Pharmacology
Van der Boechorststraat 7
1081 BT Amsterdam
The Netherlands
tel. +3120-444-8095
fax. +3120-444-8100

B. W. Van Oosten
Dept. of Neurology
Free University Hospital
P O Box 7057
1007 MB Amsterdam
The Netherlands
tel. +31-20-444-4444 ext. 455
fax. +31-20-444-0197

Clementina M. Van Rijn
University of Nijmegen
Nici/Comp. & Physiol. Psychology
P. O. Box 9104
6500 HE Nijmegen
The Netherlands
tel. +31-24-361-5612
fax. +31-24-361-6066
email:rijn@nici.kun.nl

Vince Varga
Dept. Physiol./Tampere Brain
Res. Center/Univ. of Tampere
Med. School
P. O. Box 607
Tampere FIN-33101
Finland
tel. +358-31-2156639
fax. +358-31-2156170

Roland Vogel
Univ. of Tübingen
Physiol. Chem. Inst.
Hoppe-Seyler-Str. 4
Tübingen 72076
Germany
tel. +49-7071-293042/2972452
fax. +49-7071-295360

Petra Vöhringer
Univ. Hohenheim
Inst. Zoology
Garbenstr. 30
Stuttgart 70593
Germany
tel. +49-711-459-2255
fax. +49-711-459-3450

Walter Volknandt
Biozentrum, Zool. Inst.
Ak Neurochemie
Marie-Curie-Str. 9 / N210
Frankfurt am Main D-60439
Germany
tel. +49-69-798-29603
fax. +49-69-798-29606

Cinzia Volonté
Inst. of Neurobiology C.N.R.
Viale Marx 43-15
Rome 00137
Italy
tel. +396-868-95963
fax. +396-3243675
email:cinzia@biocell.irmkant.rm.cnr.it

J. S. Wassenaar
BCN–Med. Faculty
Dir. Neuro- & Psychobiology
Bloemsingel 1
9713 BZ Groningen
The Netherlands
tel. +31-50- 363-2779
fax. +31-50-5272527

Hiroshi Watanabe
Toyama Med. Pharmaceut. Univ.
Res. Inst. Orient. Med
2630 Sugitani
Toyama 930-01
Japan
tel. +81-764-34-2281
fax. +81-764-34-5056

Hans Welzl
Eth Behavioral Biology Lab.
Dept. of Toxicology
Schorenstr. 16
Schwerzenbach 8603
Switzerland
tel. +44-1-8257328
fax. +44-1-8250476
email:welzl@toxi.biol.ethz.ch

Dr. Ben H. C. Westerink
Univ. of Groningen
Dept. of Pharmacy
T.A Deusinglaan 1
9713 AW Groningen
The Netherlands
tel. +31-50-363-3307
fax. +31-50-363-6908

Heinrich Wiesinger
Physiological-Chem. Inst.
Univ. of Tübingen
Hoppe Seyler Str. 4
Tübingen D-72076
Germany
tel. +49-7071-2973338
fax. +49-7071-295360

David C. Williams
Univ. of Dublin
Biochemistry Dept.
Trinity College
Dublin 2
Ireland
tel. +353-1-772941 ext. 1802
fax. +353-1-772400

Dr. Paul Willner
Dept. Psychology
Univ. College of Swansea
Swansea 8A2 8PP
United Kingdom
tel. +44-792-295844
fax. +44-792-295679

David Robert Wing
Glyco Biology Inst.
Dept. Biochemistry
South Parks Rd.
Oxford OX1 3QH
United Kingdom
tel. +44-1865-275756/275762
fax. +44-1865-275216

Hans Winkler
Inst. Für Pharmakologie
Univ. Innsbrück
Peter-Mayr Str. 1A
A-6020 Innsbruck
Austria
tel. +43-512-507-3700
fax. +43-512-507-2868

Susan Wonnacott
Univ. of Bath
Dept. Biochemistry
Bath BA2 7AY
United Kingdom
tel. +44-1225-826391
fax. +44-1225-826779
email:s.wonnacott@bath.ac.uk

Martyn Wood
Smithkline Beecham
Psych. Res.
Third Avenue
Harlow, Essex CM19 5AW
United Kingdom
tel. +44-1279-622247
fax. +44-1279-622230

Elena Ivanova Yerlykina
Medical Academy
Dept. of Biochemistr
Brinsky Str. 4/I-6
Nizhy Novgorod 603163
Russia
fax. +7-831-239-0943

Bernard Zalc
Lab. de Neurochimie
Inserm U-134
Hôpital de la Salpetrière
47, Blvd. de l'Hôpital
75651 Paris Cedex 13
France
tel. 33-1-4586-2012
fax. 33-1-4583-7660

Teresa Zalewska
Med. Res. Centre
Dept. of Neurochemistry
5, Pawinskigo Str
Warsaw 02-106
Poland
tel. +48-22-608-6528
fax. +48-22-668-5423

V. A. Zammit
Hannah Research Inst.
Ayr KA6 5HL
Scotland/U.K.
tel. +44-1292-476013 #254
fax. +44-1292-678797
email:zammitv@main.hri.sari.ac.uk

Mzia G. Zhavania
Dept. of Brain Structure Res.
Inst. of Physiology
Georgian Acad. of Sciences
14, Gotuastr.
380060 Tbilisi
Republic of Georgia
fax: +995-3200-1153/1027

Herbert Zimmermann
Biozentrum J. W. Goethe Univ.
Zoologisches Inst.
Ak Neurochem.
Marie-Curie-Str. 9
D-60439 Frankfurt-am-Main
Germany
tel. +49-69-798-29602/1
fax. +49-69-798-29606
email:h.zimmermann
    @zoology.uni-frankfurt.d400.de

AUTHOR INDEX

Aarts, L.H.J., 1107
Ábrahám, C.S., 479
Agadjanov, M.I., 1019
Agid, Y., 259
Agrati, P., 523
Ahier, R.G., 1157
Akhalkatsi, R.G., 909
Aldskogius, H., 549
Aleksidze, N.G., 909
Alvarez, J., 643
Angenstein, F., 1071
Anglade, G., 63
Anglade, P., 259
Annett, L.E., 249
Apelqvist, G., 201
Arduini, A., 1053
Arendash, G.W., 33
Artola, A., 863
Arutjunyan, A.V., 457, 529
Audet, R.M., 201
Aureli, T., 1065
Auth, F., 215
Avgustinovich, D.F., 209
Avrova, N.F., 973
Azuhata, T., 97
Azzouz, M., 485

Bachelard, H., 1
Baker, C.R., 773
Balshüsemann, U., 925
Baranowska, B., 137
Barahska, J., 1011
Barcikowska-Litwin, M., 137
Bardahchjan, E.A., 291
Barrio, L.C., 1123
Bartolini, P., 955
Bashkatova, V., 977
Bastiaens, P.I.H., 999
Bauer, J., 27, 105, 967
Bauer, K., 691
Bauer, R., 1149

Becker, C.G., 863
Begeer, J.H., 475
Beleslin, B.B., 983
Belik,Ya.V., 9
Benbassat, D., 647
Benech, J.C., 643
Bengtsson, F., 201
Berent, U., 1059
Berényi, S., 215
Berezin, V., 295, 857, 891, 897
Berg, P.A., 281
Berger, M., 27, 535
Bergman, J.J., 655
Bergmann, R., 1149
Bergqvist, P.B.F., 201
Berlet, H.H., 1111
Berners, M.O.M., 561
Berra, B., 913
Bianchi, L., 949, 955
Bielarczyk, H., 737, 821, 993
Bigl, V., 829
Bittar, P.G., 555
Blasevich, M., 1005
Blennow, K., 39, 415
Blomgren, K., 415
Bock, E., 295, 451, 707, 857, 891, 897
Boer, A.-K., 773
Bolam, J.P., 949, 955
Boldyrev, A.A., 177
Bolotashvili, T., 909
Borg, J., 485
Bosch, A.I., 59
Botterblom, M., 1161
Boutelle, M.G., 561
Brammer, N., 1089
Braun, N., 701
Briley, M., 193
Bröer, S., 619
Brown, A.M., 69
Bruce, A.J., 165

Brugg, B., 259
Brust, P., 1149
Bubser, M., 1161
Burgoyne, R.D., 719
Burov, Y.V., 345
Butterworth, R.F., 201
Buwalda, B., 147

Cadet, J.L., 323
Calvani, M., 1053, 1065
Cano, J., 309
Capuani, G., 1065
Čarman-Kržan, M., 501
Carnevale, A., 451
Cazevieille, C., 611
Charyeva, I.G., 853
Chekulaeva, U., 919
Chen, Y., 21
Chen, Y.-J., 929
Chumakova, O.V., 837
Clark, R.A.C., 929
Colivicchi, A., 955
Colombo, I., 913
Conte Camerino, D., 461
Conti, F., 1065
Conti, R., 1053
Cooper, T., 315
Crandall, B.M., 33
Crispino, M., 637, 643
Csutorás, Cs., 215
Curzon, G., 187
Czarny, M., 1011

D'Amico, T., 583
Dahlström, A., 1101
Dambinova, S.A., 265, 281
Danneberg, U., 281
Datla, K.P., 1157
Dauplais, M., 783
Davidsson, P., 39
De Groot, C.J.A., 967

De Jongste, M.J.L., 141
De la Torre, J.C., 173
Deli, M.A., 479
Della Corte, L., 949, 955, 959
De Luca, A., 461
Demestre, M., 561
De Vries, J., 125
De Vries, K.J., 999
Di Cocco, M.E., 1065
Diener, W., 281
Dijk, S.N., 153
Dijkstra, C.D., 105, 967
Dirksen, R., 275
Dixon, H.B.F., 959
Dobrota, D., 177 183
Dohovics, R., 733
Dolezal, V., 843, 987
Domañska-Janik, K., 233, 369, 407
Dormann, A., 647
Dottori, S., 1053
Douma, B.R.K., 401, 507
Dovedova, E.L., 287
Drukarch, B., 243
Dubrovskaya, N., 919
Dunlop, J., 1089
Dunnett, S.B., 249
Dvoláková, L., 389
Dwek, R.A., 929
Dzhandzhugazyan, K.N., 707

Edvardsen, K., 891
Eikelenboom, P., 15
Ennis, C., 1089
Erdtsieck-Ernste, E., 1161
Eriksson, N.P., 549
Esfandiari, A., 751
Evans, N., 1089
Eyer, J., 227

Feenstra, M., 1161
Fellows, L.K., 561
Fiebich, B.L., 27
Fillenz, M., 561, 577
Fischer, S., 625
Fishman, H.A., 1131
Flögel, U., 571
Fox, G.B., 877
France-Lanord, V., 259
Francescangeli, E., 1029
Francis, P.T., 153
Füchtner, F., 1149
Fujii, T., 813
Fujimoto, K., 813
Furukawa, K., 165
Fyske, E.M., 1095

Gaal, J., 315
Gager, T.L., 69

Galli, C., 1005
Galoyan, A., 595
Garnier, M., 523
Gaynullin, M.R., 757
Gebicke-Haerter, P.J., 535
Geraerts, W.P.M., 655
Gerardy-Schahn, R., 885
Gershtein, L.M., 161
Gispen, W.H., 1107
Giuditta, A., 637, 643
Glinkina, V.V., 853
Glushchenko, T.S., 797
Goberna, R., 589
Goldberg, S.R., 327
Goracci, G., 1029
Gordon-Krajcer, W., 361
Gottfries, C.G., 415
Gray, J.A., 1157
Grimm, R., 1117
Gromova, L.G., 265
Gundelfinger, E.D., 901, 1117
Gurney, M., 491
Gyulkhandanyan, A.V., 1077

Haas, U., 1111
Haga, K., 807
Haga, T., 807
Hagberg, H., 361
Hakopian, V.P., 1083
Hamberger, A., 1131
Hamprecht, B., 765
Handa, S., 933
Hashim, A., 315
Haug, L.S. 389
Hayashi, N., 97
Hayashi, M.K., 807
Heider, M., 829
Heilbronn, A., 701
Heinonen, E., 331
Henn, C., 743
Hermann, A., 733
Hermann, C., 675
Hertting, G., 843, 987
Hery, F., 63
Heuer, H., 691
Hilberer-Ehret, S., 773
Hildebrandt, H., 885
Hjorth, S., 201
Hogue Angeletti, R., 583
Holland, V.L., 69
Hoppenstedt, W., 239
Hoveyan, G.A., 761
Hovsepian, L.M., 761
Huitinga, I., 105
Hüll, M., 27
Hume, S.P., 1139

Izvarina, N.L., 797

Jacobson, I., 1131
Jakowic, I., 55
Janáky, R., 733
Jankowska, A., 737, 847, 993
Janson, I., 415
Jardemark, K., 1131
Järv, J., 791
Jellinger, K., 33
Jo, N., 97
Johannsen, B., 1149
Joó, F., 479
Joseph, M.H., 1157
Jovanovic, Z., 983
Jovic, N., 469
Jovic, R., 469
Jozanc, Lj., 963
Juric, D.M., 501

Kalaóny, T., 605
Kamalyan, R.G., 1077
Kameyama, K., 807
Kanayan, A.S., 1083
Kaplan, B.B., 643
Kaplán, P., 375, 383
Kapuściński, A., 729
Karageuzyan, K.G., 761
Karapetyan, T.D., 1077
Karlsson, I., 415
Karlsson, J.-O., 415
Kárpáti, E., 215
Kasahara, T., 813
Kascheyeva, T.K., 457
Kast, K., 281
Kataoka, K., 83
Kawa, A., 295
Kawashima, K., 813
Kendall, A.L., 249
Kevorgian, G.A., 1083
Khvatova, E.M., 757
Khyazeva, L.A., 853
Kienzl, E., 33
Kinosita, K., 97
Kinugawa, H., 91
Kiselyov, V.V., 891
Kisielevski, Y., 737, 847
Kiss, B., 215
Klein, J., 743
Klein, R., 281
Knollema, S., 73 79
Koczyk, D., 497
Koenig, E., 667
Kohen, R., 21
Konstantinova, N.N., 457
Korenevsky, A.V., 529
Korf, J., 401
Korinthenberg, R., 281
Körtje, K.-H., 395
Kosina, L.S., 457
Kovács, J., 479

Author Index

Krizbai, I., 479
Krug, M., 905
Krupka-Matuszczyk, I., 1035
Kucia, K., 1035
Kwiatkowska-Patzer, B., 137

La Spada, P., 1005
Lajtha, A., 315
Lambeng, N., 259
Lammertsma, A.A., 1139
Lammintausta, R., 331
Lapidot, A., 631
Laszlovszky, I., 215
Law, R.O., 943
Łazarewicz, J.W., 361
Leclerc, N., 491
Lee, K., 987
Lehotský, J., 375, 383
Leibfritz, D., 571
Lepeohin, V.A., 291
Levi, G., 541
Li, J.-Y., 1101
Lieb, K., 27
Linemann, H., 1149
Lipkowski, A.W., 137
Liptaj, T., 177 183
Litvin, S., 631
Liu, L., 549
Lligoña-Trulla, L., 1053
Lobanov, N.A., 1059
Löffelholz, K., 743
Lohmann, K., 1117
Lomnitski, L., 21
Long, S.K., 59
Lòw, M., 215
Lowry, J.P., 577
Luiten, P.G.M., 147, 351, 401, 507
Lukáčová, N., 465
Lysenko, A., 339
Lysenko, A.V., 419
Lyzlova, L.V., 457

Maar, T.E., 891
Machado, A., 309
Maggi, A., 523
Magistretti, P.J., 555
Magyar, K., 303
Maienschein, V., 701
Mäki-Ikola, O., 331
Makleit, S., 215
Malecki, A., 1035
Marangolo, M., 959
Marchini, C., 83
Marcilhac, A., 63
Marquette, C., 967
Marsala, J., 465
Marsala, M., 465
Martin, R., 643, 661, 667

Massaro, A.R., 451
Matarredona, E.R., 309
Mateo, J., 695
Matsionis, A., 339
Matsionis, A.E., 419
Matthies, Jr., H., 905
Mattson, M.P., 165
Melkonian, K.V., 1083
Melkonian, M.M., 761
Mendzeritsky, A., 339
Ménez, A., 783
Merlo, D., 357
Meunier, F.A., 713
Mézesová, V., 375
Michaelson, D.M., 21
Michel, P.P., 259
Mickaleva, I.I., 757
Mijanovic, M., 963
Mikaelyan, E.R., 1077
Mikhailov, A.V., 457
Mikoyan, V., 977
Minghetti, L., 541
Miras-Portugal, T., 695
Misawa, H., 813
Mlynárik, V., 177 183
Mohrmann, M., 281
Mokrushin, A.A., 519
Molgó, J., 713
Moret, C., 193
Morgan, A., 719
Moscho, A., 1131
Mroczkowska, J., 1059
Murphy, K.J., 877
Myers, R., 1139

Nagel, J.G., 141
Nakamura, F., 807
Nałęcz, K.A., 1059
Nałęcz, M.J., 1059
Nalivaeva, N., 919
Naritomi, H., 91
Narkevich, V., 977
Nazarenko, A.V., 9
Nelson, N., 173
Németh, L., 479
Nerini, M., 949
Nicolini, A., 541
Nielander, H.B., 1107
Nikulina, E.M., 209
Nilova, N.S., 797
Nowak, J.Z., 605
Nyakas, C., 147, 351

O'Connell, A.W., 877
Obert, M., 1149
Oderfeld-Nowak, B., 497
Oestreicher, A.B., 1107
Oganesjan, N., 847
Ogata, T., 83

Öhman, S., 433
Oja, S.S., 733, 939
Oosterink, B.J., 351, 401
Oparina, T.I., 529
Oren, R., 647
Orensanz, L.M., 1123
Orwar, O., 1131
Osborne, N.N., 611
Ostvold, A.C., 389

Pahor, V., 501
Papini, N., 1005
Papp, M., 197
Paronian, Z., 595
Pass, I., 1089
Patrone, C., 523
Paturneau-Jouas, M., 751
Pavlenko, A.V., 457
Pavlova, N.G., 457
Pedersen, N., 857, 897
Pellerin, L., 555
Perrone Capano, C., 637
Persson, J.K.E., 549
Peter, J.B., 443
Peterson, A., 227
Petroni, A., 1005
Pfeuffer, J., 571
Pierno, S., 461
Plekhanov, A.Y., 519
Plesneva, S., 919
Poindron, P., 485
Polazzi, E., 541
Polman, C.H., 121
Popova, N.K., 209
Porcellati, S., 1029
Povilaitite, P.E., 419
Prag, S., 897
Prokopenko, V.M., 529
Puccetti, C., 1065
Puchinger, L., 33
Puka-Sundvall, M., 361
Pylaev, A.S., 853
Pylaeva, S.A., 853

Račay, P., 375, 383
Raeymaekers, L., 375, 383
Rahmann, H., 395, 885, 925
Ramsay, R.R., 1039
Rapelli, S., 913
Raus, J., 113
Rayevsky, K., 977
Regan, C.M., 877
Regland, B., 415
Reiber, H., 423
Reiter, R.J., 599
Ricciardi-Castagnoli, P., 625
Ricciolini, R., 1053, 1065
Richter, K., 901
Richter-Landsberg, C., 239

Roβner, S., 829
Robakidze, T.N., 345
Roberg, B., 389
Robillard, G.T., 773
Rosiak, J., 605
Rösner, H., 885
Rosser, A.E., 249
Rotllán, P., 695
Ruberg, M., 259
Rump, S., 55
Rusina, E.I., 457
Russell, J., 583
Ruuls, S.R., 105, 967
Rybkowski, W., 361

Sabala, P., 1011
Safa, R., 611
Saido, T.C., 407
Salinska, E., 361
Salmon, E., 1145
Sánchez-Margalet, V., 589
Sandberg, M., 1131
Santagati, S., 523
Santiago, M., 309
Santos-Alvarez, J., 589
Saransaari, P., 733, 939
Sargsian, A., 761
Sawada, T., 91
Scalibastri, M., 1053
Schachner, M., 863
Scheller, R.H., 1131
Schicknick, H., 1117
Schliebs, R., 829
Schloss, P., 773
Schmidlin, A., 625
Schmidt, R., 869
Schobert, A., 843
Schomburg, L., 691
Schotman, P., 1107
Schousboe, A., 891
Schrama, L.H., 1107
Schubert, P., 83
Sciarroni, A.F., 1053
Seidenbecher, C., 901
Sellin, L.C., 713
Serfozo, Z., 733
Sergutina, A.V., 161
Sershen, H., 315
Shakarisvili, R., 1025
Shelayeva, E.V., 457
Shen, G.Q., 443
Shibuya, T., 97
Shimizu, T., 91
Shohami, E., 21
Siaud, P., 63
Sinichkin, A.A., 291
Sival, D.A., 475
Skrinskaya, J.A., 209
Skup, M., 497, 513

Smalla, K.H., 905
Smit, A.B., 655
Snoek, G.T., 999
Sokolova, T.V., 973
Soliakov, L., 801
Soroka, V., 891
Sottocornola, E., 913
Spira, M.E., 647
Sprouse, J., 47
Srapionian, R., 595
Staak, S., 905, 1071
Steinbach, J., 1149
Stepanov, M.G., 529
Stienstra, C.M., 351, 401
Stinissen, P., 113
Stoof, J.C., 243
Stuiver, B.T., 351, 401
Stvolinsky, S.L., 177
Sutherland, R.J., 173
Suzuki, T., 813
Svendsen, C.N., 249
Svensson, M., 549
Syed, N.I., 655
Sypecka, J., 233
Szabó, C.A., 479
Szutowicz, A., 737, 821, 847, 993

Tabagari, S., 1025
Tajima, S., 813
Taki, T., 933
Ter Horst, G.J., 73, 79, 141
Thakhava, G., 1025
Thomas, D.R., 69
Tilders, F.J.H., 967
Tipton, K., 299
Tipton, K.F., 959
Tischmeyer, W., 1117
Titgemeyer, F.M., 773
Tkáè, I., 177 183
Tokarev, A.V., 797
Tomaszewicz, M., 737, 821, 847, 993
Tonali, P., 451
Topuria, M., 1025
Torgner, I., 389
Torzewska, D., 513
Trzeciak, H.I., 1035
Tsiang, H., 967
Tsuga, H., 807
Tumani, H., 443
Turner, A.J., 683
Turwitt, S., 691
Tyurin, V.A., 973
Tyurina, Y.Y., 973

Ugarte, M., 611
Uskova, N.I., 419
Utagawa, A., 97
Uzdensky, A.B., 345

Valjevac, K., 963
Van Calker, D., 535
Van Dam, A.-M., 967
Van der Wal, A., 1161
Van der Werf, Y.D., 141
Van der Zee, E.A., 507
Vandevyver, C., 113
Van de Witte, S.V., 73
Vanin, A., 977
Van Kesteren, E.R., 655
Van Minnen, J., 655
Van Oosten, B.W., 121
Van Rijn, C.M., 275
Van Uum, H., 1161
Vardanyan, A.G., 1077
Varga, V., 733
Vartanian, G.S., 1019
Vegeto, E., 523
Vitskova, G., 977
Vogel, R., 765
Vöhringer, P., 395
Volk, B., 27
Volknandt, W., 675
Volonté, C., 357

Walaas, S.I., 389
Wallin, A., 415
Walmod, P., 295
Walmod, P.S., 897
Walter, B., 1149
Warter, J.-M., 485
Wassenaar, J.S., 221
Watts, C., 249
Wawrzenczyk, A., 1059
Welzl, H., 863
Wiersma, A., 79
Wiesinger, H., 625, 765
Wiktorek, M., 1011
Will, E.,1149
Willems-van Bree, E., 275
Williams, D.C., 773, 959
Willker, W., 571
Willner, P., 197
Wing, D.R., 929
Wirtz, K.W.A., 999
Wonnacott, S., 801
Wood, M.D., 69

Yamada, S., 813
Yasar, S., 327
Yerlykina, E.I., 757
Young, A.M.J., 1157
Yuki, N., 933

Zablocka, B., 407
Zakharova, I.O., 973
Zalewska, T., 369, 407

Author Index

Zamfirova, R., 949
Zamze, S.E., 929
Zare, R.N., 1131
Zaremba, M., 497, 513
Zawilska, J.B., 605

Zhang, J., 113
Zhuravin, I., 919
Zhvania, M.G., 491
Ziembowicz, A., 361
Zieminska, E., 361

Zimmermann, H., 701
Zisterer, D., 959
Zouheiry, H., 583
Zwart, J.P.C., 275
Zwiener, U., 1149

SUBJECT INDEX

A10 nervous system, hypothermia treatment, 102
Acetyl coenzyme A (acetyl CoA), 737, 847, 993, 1039
  effect of ethanol, 847
  neuropathies, 821
N-Acetylaspartate, 181
Acetylcarnitine, 822, 1053
  ischemic injury, 1048
N-Acetylcarnosine, 117
Acetylcholine, 695, 993
  carnitine, 1049
  diadenosine polyphosphates (ApnA), 695
  estrogen-induced differentiation, 525
  plasma and blood, 813
  release of neurotransmitters, 801
  thiamine deficiency, 737
Acetylcholine esterase, 148, 830
  anencephaly, 458
  tacrine, 345
Acetylcholine receptors, 648, 794, 801, 813
  calcium channels, 843
  hypertensive state, 853
  phosphorylation, 807
Acetylcholine release, 830
  botulinum neurotoxins, 713
  effect of ethanol, 847
Acetylcholine synthesis, 837
  carnitine, 1062
Acetylcholinergic neurons, effect of nerve growth factor, 831
N-Acetyldehydrosphingosine, 261
N-Acetyl-D-glucosamine, 911
N-Acetylglucosaminidase, 932
N-Acetylhistidine, 177
N-Acetylhomocarnosine, 177
N-Acetyl-D-mannosamine, 878
N-Acetylserine, 177
Acid phosphatase activity, histamine, 479
Aconitase, effect of nitric oxide, 993
ACTH: see Adrenocorticotropic hormone
Actin, synthesis in axons, 673

Actin cytoskeleton, NCAM-B expressing cells, 900
Actinopterygian fish, ependymin mRNA, 874
Activated T lymphocytes, cytokines, 130
Active oxygen forms, 798
Acute phase proteins, 16, 27
Acyl coenzyme A (acyl CoA), carnitine, 1040, 1047
Acylcoenzyme A (acyl CoA) dehydrogenase, 755
Acylcarnitine, 1040
Adenosine, 701, 732
  effect of ethanol, 847
  effect on glial cell activation, 83
  microglial activation, 535
  release of prostaglandin E2, 535
Adenosine diphosphate (ADP), glutamine hydrolyse, 1079
Adenosine kinases, 695
Adenosine 5'-(α,β-methylene)diphosphate, reduction in neurite formation, 703
Adenosine monophosphate (AMP), effects of adenosine, 88
Adenosine 3',5'-monophosphate (cAMP)
  catecholamines, 731
  ceramid-induced apoptosis, 261
  chromaffin cell exocytosis, 722
  effect of gangliosides, 919
  glycogen resynthesis, 556
Adenosine 3',5'-monophosphorothioate, 545
Adenosine receptor (P1), 357
Adenosine receptors, 703
Adenosine triphosphate (ATP), 701, 707, 757, 1084
  diadenosine polyphosphates (ApnA), 695
  effect on platelet activation, 1032
  hexokinase, 757
  phosphatidylserine synthesis, 1011
  relation with NCAM, 707
  thiamine deficiency, 739
Adenosine triphosphatase (ATPase)
  calcium transport, 375
  relation with NCAM, 707
Adenylate, 695

Adenylate cyclase, 88
  cardiac arrest and resuscitation, 729
  effect of gangliosides, 919
Adenylyl-β,γ-imidodiphosphate, 358
Adhesion factors, 16, 27
Adhesion molecule on glia (AMOG), 711
Adrenal chromaffin, 987
  effect of nitric oxide, 987
Adrenal chromaffin cells, exocytosis, 719
Adrenal cortex, stress response, 151
Adrenal grafts, Parkinson's disease, 250
Adrenal medulla, 595
  Parkinson's disease, 250
Adrenal medullary proteins, 595
Adrenalectomy, calcium homeostasis, 405
Adrenaline, chromaffin cells, 720
Adrenaline release, effect of nitric oxide, 990
β-Adrenergic activation, cAMP levels, 545
Adrenocorticotropic hormone (ACTH), 63
Aggrecan, 901
Aging
  behavioral dysfunctions, 149
  dysfunction of the respiratory chain, 751
AIDS dementia, prostanoid levels, 542
Alanine, 632
  γ-aminobutyric acid receptor, 1123
  taurine uptake, 959
β-Alanylhistidine, and brain ischemia, 177
Albumin, 292
  cerebrospinal fluid, 426
  immunoreactivity in brain, 142
Albumin CSF/serum quotient, 423
Albumine transport, histamine, 483
Alcoholism, acetylcholine synthesis, 821
Alkaline metalloproteinase, effect on myelin basic protein, 1115
1-O-Alkyl-2-acetyl-sn-glycerol, 1029
1-O-Alkyl-2-acetyl-sn-glycero-3-phosphocholine, 1029
1-O-Alkyl-2-(longchain)acyl-sn-glycero-3-phosphocholine, 1030
1-O-Alkyl-2-lyso-sn-glycero-3-phosphocholine, 1030
Alloxan monohydrate, 632
Alpha-fetoprotein
  anencephaly, 458
  estradiol requesting, 526
Alpha-tocopherol, sodium, potassium-ATPase activity, 973
Aluminum, effect on ACh-synthesis, 822
Alzheimer's disease (AD), 15, 21, 27, 39, 259, 268
  acetylcholine synthesis, 821
  amiridine and tacrine, 345
  calcium homeostasis, 402, 415
  carnitine, 1049
  cerebrovascular insufficiency, 173
  cortical cholinergic dysfunction, 829
  ependymin, 869
  glutamine synthese, 443
  IP3-receptor, 389
  neurotransmitter-based treatments, 153

Alzheimer's disease (AD) (cont.)
  nitric oxide (NO), 977, 993
  phospholipid metabolites, 33
  potassium channels, 538
Amacrine cells, 605, 613
  serotonin and γ-aminobutyric acid, 612
Amantadine, 102
Aminoacyl-tRNA synthetases, in axons, 638
2-Aminobicyclo[2,2,1]heptane-2-carboxylic acid, aminoacid transport, 619
γ-Aminobutyric acid (GABA), 421, 525, 529, 632
  amacrine cells, 613
  capillary electrophoresis patch clamp det, 1135
  carnitine, 1049
  effect of taurine, 964
  ethanol, 748
  extracellular after optic nerve cut, 395
  hypokinesia, 1083
  inhibitory neurotransmitter, 275
  metabolic trafficking, 1069
  release from globus pallidus, 955
  release from substantia nigra, 949
  transporters, 774
  uptake system, 1042
γ-Aminobutyric acid receptor, 275, 834, 1123
  alanine, 1123
  relation with ACh-receptor, 801
γ-Aminobutyric acid-aspartate transaminase, calpain activity, 421
γ-Aminobutyric acid-glutamate transaminase, calpain activity, 421
1-Aminocyclopropanecarboxylic acid, 197
1-Aminocyclopentane-dicarboxylic acid, 734, 1071
  neuroprotection, 939
Aminoethylarsonic acid, taurine uptake, 961
Aminoethylphosphonic acid, taurine uptake, 961
Aminoguanidine, 970
2-Amino-3-hydroxy-5-methyl-4-isoxazolpropionate receptors, dopamine release, 733
2-Amino-3-hydroxy-5-methyl-4-isoxazolpropionate acid (AMPA), 266, 733
  capillary electrophoresis patch clamp, 1135
6-Aminomethyl-3-methyl-1,2,4-benzothiadiazine-1,1-dioxide, taurine uptake, 961
Aminopeptidase, 162, 163
Aminopropylarsonic acid, taurine uptake, 961
Aminoquinoline, 16
  endplate potentials, 715
Amiridin, senile dementia of Alzheimer's type (SDAT), 345
Amitriptyline, 194
  phospholipase A2, 1035
Ammonia, 201
  phosphate activated glutaminase, 1077
Amnesia, role of ependymin, 873
AMP: see Adenosine monophosphate
AMPA: see Amino-3-hydroxy-5-methyl-4-isoxazolpropionic acid
AMPA/quisqualate receptors, epilepsy, 266

Subject Index

Amphetamine, 198, 304, 315
  dopamine and serotonin, 327
Amphetamines, monoaminergic systems, 323
Amygdala, synaptic structures, 492
β-Amyloid, 156, 345, 389, 822
β-Amyloid associated proteins, 15
β-Amyloid peptides, calcium homeostasis, 402
Amyloid plaques, 28
Amyloid precursor protein (APP), 27, 156, 345
  potassium channels, 536
Amyloid precursor protein (sAPP), secreted form, 165
Amyloid-protein, 27
Amyotrophic lateral sclerosis (ALS), 259
  glutamine synthase, 444
  superoxide dismutase, 485
Anatoxin-a, 801
Anencephaly, creatine kinase isoenzymes, 457
Angiotensin converting enzyme, 425
Anhedonia, 198
Anion transport protein, 417
Annexin II, effect on exocytosis, 721
Anserine, 177
Antagonist F, diacylglycerol and arachidonic acid, 1019
Anterior horn, motor neurons, 485
Anterograde degeneration, microglial reaction, 549
Anthraquinone sulfonic acid derivative, 358
Anti-epileptic pharmaca, 275
Anti-inflammation drugs, 18
Anti-neural antibodies, Guillain-Barré syndrome, 933
Antibody, to gangliosides, 282
α_1-Antichymotrypsin, 27
Antidepressant drugs, 197
Antimycin A, beta-oxidation activity, 752
Antisense phosphorothioate oligodeoxynucleotides, inhibition of c-jun, jun B, or c-fos expression, 1117
Antisense intervention, jun and fos gene families, 1120
α_1-Antitrypsin, 27
Anxiety
  cholestokinin type B receptor, 1089
  dopamine metabolism, 209
  role of serotonin (5HT), 48
Aplysia californica, 864
  N-CAM-related adhesion molecules, 864
Aplysia neurons, 647
Apolipoprotein E, 16
  allele E4, 21
Apolipoproteins, 27
Apomorphine, 215
Apoptosis, 300, 323
  amiridine, 346
  dopaminergic neurons, 259
Apyrase, 698, 700
Arachidonic acid, 37, 126, 1010
  fenaridin and antagonist F, 1019
  neuroprotection, 541
Arachis hypogaea agglutinin, 910
Arachnoid villi, 423

Arginine
  catecholamine release, 988
  transport in neural cells, 625
Aromatic aminoacid decarboxylase, 605
Arsonoalanine, taurine uptake, 961
Aryl sulphatase, 479
Ascorbate, 599
Ascorbic acid, 614
  leech ganglia, 983
Aspartate
  deoxyglucose uptake, 556
  P2 purinoceptor, 358
Astacus leptodactilus, amiridine and tacrine, 345
Astrocyte culture, 502
Astrocytes, 73, 501
  5'-nucleotidase, 702
  aminoacid transport, 619
  ecto-5'-nucleotidase, 702
  glucose uptake, 555
  malic enzyme isoforms, 765
  NGF and NGF-receptor expression, 499
  Parkinson's disease, 253
  regulation by adenosine, 88
  role in neuronal metabolism, 561
  study to assess the effects, 751
Astroglial cells
  NO synthase, 627
  trimethyltin intoxication, 499
Astroglioma cells, human U373, 30
ATP: *see* Adenosine triphosphate
ATP analogs, ecto-ApnA hydrolase, 698
ATP receptor (P2), glutamate release, 357
ATP-binding protein, glutamine synthetase, 443
ATP-citrate lyase, effect of nitric oxide, 997
ATP-utilizing ecto-enzymes, 707
ATPase: *see* Adenosinetriphospatase
Atrial natriuretic peptide family, 683
Atropine, 793
Autoantibodies
  to central nervous system and phospholipid, 281
  to nervous tissue antigens, 265
Autographa californica, 693
Autoimmune diseases
  nitric oxide, 542
  prostanoid levels, 542
Autophagosomes, 341
Autoreactive T lymphocytes, multiple sclerosis, 113
Axon injury, microglial reaction, 549
Axon, gene expression, 637
Axonal segments
  de novo protein synthesis, 649
  long-term survival, 647
Axonal transport, and Rab3a, 1102
Axons, long term survival, 647
Axotomy, microglial reaction, 549
Axotremorine, 794

B-50: *see* Protein B50
Bacterial endotoxin, 541

Baculovirus, serotonin transporter expression, 778
Barbiturates, 275
  alanine responses, 1123
Basal nucleus of Meynert (BNM), effect of N-methyl-D-aspartate, 352
Basilen blue, 358
Bay k 8644, 844
Bcl-2 expression, 323
Benzodiazepines, 275
  alanine responses, 1123
Bergmann glial cells, 5′-nucleotidase, 702
Beta-oxidation activity, mitochondrial electron transport, 752
Betaine, transporters, 774
Bicuculline
  alanine responses, 1123
  fos-expression after microdialysis injection, 1163
  hypokinesia, 1085
Bicuculline-methochloride, 276
Bilateral olfactory bulbectomy, effect on ACTH and corticosterone, 65
Biogenic amines, 469
Biosensors, capillary electrophoresis, 1132
Blood, 1026
  acetylcholine (ACh), 813
Blood-brain barrier (BBB), 291, 550
  aminoacid transport system y^+, 627
  aminoacid transport, 619
  catecholamines, 729
  effects of brain temperature, 91
  effect of reoxygenation, 465
  histamine, 480
  in multiple sclerosis (MS), 454
  selective dysfunction, 141
  transport of glucose, 565
  transport of ketone, 634
Blood-CSF barrier, 423
Borna disease, induction of NO-synthase, 968
Botulinum neurotoxins, acetylcholine release, 713
Bradykinin, 469
  idiopathic neuralgia, 472
Brain damage, effect on reperfusion, 465
Brain disorders, positron emission tomography (PET), 1149
Brain injury, 83
  nitric oxide, 993
Brain paroxysmal activitiy, autoantibodies, 265
Brain spectrin, 408
Brain tissue temperature, 97
Brain, sexual differentiation, 523
Brain-derived neurotropic factor (BDNF), 507
Brain-derived neurotropic factor receptor (trkB), 507
Brain-derived proteins, 427
Brainstem, motor neurons, 485
BrainView program, help in equidensity lining, 1119
Branched-chain amino acids
  carnitine, 1040
  transport in glial cells, 619

Brevican, hyaluronic acid-binding protein, 901
Bromocriptine, 334
Brush borders, ectopeptidase, 683
Bulbectomized rats, secretion of ACTH and corticosterone, 64
α-Bungarotoxin, 844
Bunodosoma granulifera, dendrotoxin, 787
t-Butylbicyclo-orthobenzoate, GABA-receptor, 275

C-reactive-protein, 27
C6 glioma cells, NO synthase, 627
Cadherin superfamily, cell-cell adhesion molecules, 857
Cadmium, dopamine release, 803
Caffeine
  leakage of calcium, 463
  P1 purinoceptor antagonist, 358
Calbindin, 169
Calbindin-D28K, 832
Calcitonin gene related peptide, endplate potentials, 715
Calcium, 401, 720, 926, 1083
  AChR-mediated dopamine release, 802
  action of pancreastatin, 591
  after optic nerve cut, 395
  axonal protein synthesis, 645
  axonal segments, 647
  binding to ependymin, 871
  ecto-ApnA hydrolase, 699
  effect of NO-donors on catecholamine release, 990
  effects of adenosine, 88
  effects on noradrenaline release, 843
  generation of oxygen radicals, 300
  intracellular store, 383
  mdx mouse, 461
  phosphate activated glutaminase, 1077
  phosphatidylserine synthesis, 1011
  SH-SY5Y cells, 793
  stretch receptor neuron firing, 346
  transport ATPases, 708
Calcium binding proteins, role in homeostasis, 402
Calcium channel blockers, neuroprotection, 402
Calcium channels, 351, 802, 843
  endplate potentials, 716
Calcium homeostasis, 415
  chromogranin A, 583
  developing neurons, 147
Calcium, intracellular
  γ-aminobutyric acid, 1083
  neuroprotective mechanisms, 401
Calcium leak channels, mdx mouse, 461
Calcium mobilisation, cholestokinin type B receptor, 1089
Calcium overload, ischemic-reperfusion injury, 375
Calcium release, calcium-induced, 361
Calcium transport, 383
  effect of taurine, 941
  N-methyl-D-aspartate (NMDA) receptors, 361
Calcium transport ATP dependent, role in homeostasis, 402

Subject Index

Calcium transport ATPase, 375
Calcium uptake, P2 purinoceptor, 358
Calcium-activated kinases, ischemia, 369
Calcium-activated phosphatases, ischemia, 369
Calcium-binding proteins, 383
Calcium-channel proteins, effect of calpains, 408
Calcium-dependent protein kinase, 645
Calcium-dependent proteolysis, calpains, 407
Calcium/calmodulin-dependent protein kinase, phosphoylation of synaptic proteins, 1095
Calmodulin, 645
Calmodulin-dependant kinase II (CaM KII), ischemia, 369
Calpain, 407
  NMDA-glutamate receptors, 343
Calpastatin, 407, 415
Calreticulin, 376
  antibodies, 384
cAMP: see Adenosine 3',5'-monophosphate; Cyclic AMP
Campylobacter jejuni, anti-neural antibodies, 933
Canavanine, 626
Cancer, free radicals, 611
Capillaries, sharply narrowed, effect of γ-aminobutyric acid, 1084
Capillary electrophoresis, patch-clamp detection, 1133
Carassius carassius, effect of temperature on gangliosides, 926
Carbachol, 793, 794
Carbohydrate epitope L2/HNK1, cell adhesion molecules, 702
Carbon-11, use in positron emission tomography, 1140
Carbonyl cyanide m-chlorophenylhydrazone, endplate potential, 716
Cardiac arrest, norepinephrine and epinephrine, 729
Cardiac ganglia, M-cholinoreception and beta-adrenoreception, 853
Cardiolipin, 762
Carnitine, 1039, 1053
  cell function, 1047
  taurine uptake, 961
  transport in neuronal mitochondria, 1061
Carnitine acetyltransferase, 1055
  mitochondria, 1048
  thiamine deficiency, 739
Carnitine acyltransferases, 1039, 1047
  energy modulating, 1040
  mitochondria, 1048
Carnitine octanoyl transferase, membrane formation, 1048
Carnitine palmitoyltransferase
  energy modulating, 1040
  lipid biosynthetic pathways, 1053
  membrane formation, 1048
Carnitine translocase, 1040
Carnosine, brain ischemia, 177
Casein kinase II, 679
  phosphorylation of MVP100, 679

Cataract, free radicals, 611
Catechol-O-methyl transferase (COMT), 209
  positron emission tomography, 1153
Cathecholamine release, 987
Catecholamine secreting primary neurons, 251
Cathecholamine transport carrier, 316
Catecholamines
  adrenal medulla, 595
  cardiac arrest and resuscitation, 729
  effect of xenobiotics, 529
  effect on pancreastatin, 589
  estrogen-induced differentiation, 525
Cathepsin B, effect on myelin basic protein, 1115
Cathepsin D, 341
  effect on myelin basic protein, 1115
Caudate nucleus, 162, 288
CD98 surface antigen, aminoacid transport, 622
Cell adhesion molecules, 869
  carbohydrate epitope L2/HNK1, 702
Cell communication, 707
Cell cycle, effect of methamphetamine, 324
Cell movement, neural cell adhesion molecule (NCAM), 897
Cell surface antigen A2B5, 239
Cell-cell adhesion molecules, 857
Cell-cell adhesion, role of NCAM, 894
Cell-cell interactions, 929
  neural cell adhesion molecule (NCAM), 885
Central nervous system, autoantibodies, 281
Central nervous tissue factor, 513
Centruroides sculpturatus, toxin, 784
Ceramides, 169
  in P12 cells, 260
Cerebellar granule cells, ecto-5'-nucleotidase, 702
Cerebellar granule neurons, 357
Cerebellum, neuroprotective effect of taurine, 939
Cerebral amyloid plaques, 27
Cerebral blood flow, 97, 561, 1150
Cerebral blood volume, 1150
Cerebral cortex
  cortical cholinergic dysfunction, 829
  dopamine release, 733
Cerebral ischemia
  magnetic resonance imaging, 184
  mild hypothermia, 91
Cerebrosides, multiple infarct dementia, 1026
Cerebrospinal fluid (CSF), 423, 433, 443
  lipid spectrum and lysosomal enzymes, 1026
Cerebrovascular insufficiency, 173
CG4 cell line, 239
cGMP: see Guanosine monophosphate
CGP 37849, 199
CGP 40116, 199
Charybdotoxin, 784
Chelerythrine, 240, 1072
Chinese hamster ovary cell line, cholestokinin type B receptor, 1089
Chlordiazepoxide, 198
Chloride-channels, 275

N-(Chloroethyl)-N-ethyl-2-bromo-benzylamine (DSP-4), 299, 305
p-Chlorophenylalanine, 189
  pain feeling, 473
m-Chlorophenylpiperzine, 188
Cholecystokinin, 1089
Cholecystokinin type B receptor, 1089
Cholesterol esters, 1026
Cholesterol
  leech ganglia, 983
  multiple infarct dementia, 1026
  effect of temperature, 926
Choline
  taurine uptake, 961
  transport through neuronal membranes, 837
Choline acetyltransferase, 525, 813, 830
  thiamine deficiency, 739
Choline uptake, 830
Cholinephosphoglycerides, 1030
Cholinergic nerve terminals, 713
Cholinergic neurons, nerve growth factor (NGF), 513
Chondroitin sulphate, 901
Choroid plexus, 423
Chromaffin cells, diadenosine polyphosphates (ApnA), 695
Chromaffin granules, chromogranin A, 584
Chromogranin A
  pancreastatin, 589
  properties of peptides, 583
Chromogranins, adrenal medulla, 595
Chronic experimental hepatic encephalopathy, 201
Chronic mild stress, 198
Cichlid fish, extracellular calcium, 395
Citalopram, 194, 774
Citrate, 740
  release from astrocytes, 1066
Citrulline, transport in neurons, 626
Clathrin, axonal transport and Rab3a, 1102
Clomipramine, 194
Cloramphenicol-acetyltransferase activity, 17-beta-oestradiol, 524
Clostridium botulinum, neurotoxins, 713
Clozapine, N-acetyltransferase, 607
Clusterin, 16
Cocaine, 329, 774
Cocaine analogues, use in positron emission tomography, 1146
Coenzyme A: see Acetyl coenzyme A; Acyl coenzyme A
Cognitive functions, n-3 fatty acids, 132
Colony-stimulating factor, 535
Complement, mediated neurotoxocity, 17
Complement proteins, 16
Complement receptor, 3
  microglial cells, 550
Complement receptors, 17
Complement system, nervous system, 550
Complement-regulatory protein, 901
COMT: *see* Catechol-O-methyl transferase

Concanavalin A, 910
Conditioning of a neutral stimulus, effect on dopamine release, 1157
ω-Conotoxin, 844
Convulsants, 275
Convulsions, platelet-activating factor (PAF), 1029
Convulsive states, effect on nitric oxide, 977
Copolymer-1, 115
Cortex synaptosomes, effect of glutamate on Na^+, 973
Cortical neurones, effect of carnitine on acetylcholine synthesis, 1062
Cortical pyramidal neurones, glutaminergic function, 153
Corticospinal tract, 485
Corticosterone, 63
  eicosanoid synthesis, 132
  stress response, 150
Corticotropin-releasing hormone, 63, 1088
Countershaping process, 222
Crayfish, amiridine and tacrine, 345
Creatine kinase, 457
Creatine kinase isoenzymes, brain development disorders, 457
Crucian carps, effect of temperature on gangliosides, 926
CSM14.1 neurons, expression of Bcl-2, 323
Cyano-7-nitroquinoxaline-2, 734
Cyclic nucleotides: *see* Adenosine 3′,5′-monophosphate; Guanosine monophosphate
Cyclic AMP dependent protein kinase A
  phosphorylation of synaptic proteins, 1095
  phosphorylation of MVP100, 679
  protein exo2, 722
Cyclic voltammetry, 24
Cyclo-oxygenase, 127, 541
Cysteine proteases, 407
Cysteine, 734
  carnitine transport, 1060
Cystinuria, mutations in the NBAT/rBAT/D2 protein, 622
Cytidine triphosphate (CTP), effect on platelet-activating factor, 1032
Cytidine monophosphate (CMP), degradation of platelet-act 1030
Cytidine diphosphocholine (CDPcholine), 1029
Cytochalasin D, 295
Cytochrome c oxidase, 753
Cytofluorimetric scanning, axonal transport studies, 1102
Cytokine synthesis, microglial activation, 535
Cytokines, 129, 142
  in Alzheimer's disease, 27
  multiple sclerosis, 121
Cytosine Arabinoside C, microglia, 551
Cytoskeletal proteins, effect of calpains, 408
Cytoskeleton organization, carnitine, 1056
Cytosolic hydrolases, 695

Dantrolene, 363

DARPP-32, 390
Deferoxamine, 614
Degeneration, regulation of neuronal function by astrocytes, 501
Degenerative diseases, nitric oxide, 542
Delta-sleep inducing peptide (DSIP), 287
    calpain activity, 419
    hypokinesia, 339
    hypoxia, 757
    phospholipids, 761
Dementia, neuroprotective agents, 401
Demyelination, role of nitric oxide, 969
Dendrotoxin, 787
Dentate granule cells, 497
2-Deoxy-D-galactose, long-term potentiation, 905
Deoxyglucose
    marker of glucose utilization, 556
    nerve terminals, 1161
Deprenyl, 315
    dopamine, 303, 327
    monoamine oxidase-B inhibitor, 299
    neuroprotective effect, 79
    use in positron emission tomography, 1147
Depression, 197
    5-hydroxytryptamine, 187
    monoamine hypothesis, 47
    noradrenaline, 193
    phospholipase A2, 1035
    role of serotonin (5HT), 48
    serotonin, 193
Desferrioxamine, MPP$^+$-induced dopaminergic toxicity, 309
Desipramine, 305, 774
    effect on ACTH and corticosterone, 64
L-Desmethyldeprenyl, 315
Developing neurons, hypoxia and ischemia, 147
Development
    motor behavior, 475
    regulation of neuronal function by astrocytes, 501
2,6-Di-t-butyl-1,4-benzohydroquinone, 387
Diabetes mellitus
    acetylcholine synthesis, 821
    cholinergic activity, 825
    creatine kinase, 457
    glucose in brain, 632
Diacylglycerol, 591
    fenaridin and antagonist F, 1019
    pancreastatin, 591
Diadenosine polyphosphates (ApnA), 695, 701
Dialysis encephalopathy, acetylcholine synthesis, 821
3,4-Diaminopyridine, endplate potentials, 716
Dibutyryl cyclic AMP, 260
Dictyostelium discoideum, major vault protein, 679
Diethyldithiocarbamate, 978
Dihydroalprenolol, 855
5,6-Dihydrohydroxytryptamine, pain feeling, 472
Dihydro-β-erythroidine, dopamine release, 802
Dihydroxyphenylacetic acid (DOPAC), 209, 216
    microdialysis measurement, 1159

L-Dihydroxyphenylalanine (L-Dopa), 161, 250, 331
Dimethyl sulfoxide (DMSO), 173
N,N-Dimethyltaurine, taurine uptake, 961
6,7-Dinitroquinoxaline-2,3-dione (DNQX), 734
    taurine and GABA, 956
    taurine release, 950
Dioxane, 529
Discopyge ommata, 701
Distal motor latency, sciatic nerve, 487
Dithio-bis-2-nitrobenzoate, 734
Dithiothreitol, 734
Dizocilpine, 197, 734
    5HT1A receptor binding, 55
DL-2-amino-5-phosphonovalerate, 734
DM-20 protein, PLP molecule variant, 234
DMPP, 844
DNA damage, 324
    effect of methamphetamine, 324
DNA polymerase α, 695
DNA, of neostratial cells, 391
DNA-RNA network, 223
DNQX: *see* 6,7-Dinitroquinoxaline-2,3-dione
Docosahexaenoic acid, 126
Domperidone, 336
L-Dopa: *see* L-Dihydroxyphenylalanine
L-Dopa decarboxylase, positron emission tomography, 1153
Dopamine, 209, 215, 605, 774
    antidepressant drugs, 197
    cardiac arrest and resuscitation, 729
    estrogen-induced differentiation, 525
    fos-expression after microdialysis injection, 1162
    hypothermia treatment, 102
    methamphetamine, 323, 327
    microdialysis measurement, 1159
    positron emission tomography (PET), 1149
    selegiline, 331
    transporters, 774
Dopamine cells, transplantation, 249
Dopamine delivery systems, Parkinson's disease, 250
Dopamine metabolism, mental diseases, 161
Dopamine neurotransmission, glutathione, 246
Dopamine receptor agonists, effects of light, 606
Dopamine receptors, 199, 209
    positron emission tomography, 1145
Dopamine release
    effect of rewarding stimuli, 1157
    ionic basis of AChR mediation, 802
    ionotropic glutamate receptors, 733
    nicotinic agonist, 802
Dopamine system
    morphochemical plasticity, 161
    positron emission tomography, 1142
Dopamine-β-hydroxylase, 595
Dopaminergic neurons, 259, 324
    glutathione, 245
Dopaminergic nigrostriatal neurons, 312
Down's syndrome, 417
Drug abuse, neurotransmitter transporters, 773

DSIP: *see* Delta-sleep inducing peptide
DSP-4, 315; *see also* N-(chloroethyl)-N-ethyl-2-bromo-benzylamine
  effect of (−)deprenyl, 305
Duchenne muscular dystrophy, role of calcium, 461
Dynamin, phosphorylation of proteins, 1095
Dyskinesias, selegiline, 332
Dystrophin, role of calcium, 461

EAE: *see* Experimental allergic encephalomyelitis or Experimental autoimmune encephalomyelitis
Ecto-5′-nucleotidase, neuritic differentiation and neuron survival, 701
Ecto-ApnA hydrolase, 696
Ecto-apyrase, 705
Ecto-ATP diphosphohydrolase, ecto-nucleotidases, 700
Ecto-ATPase, neural differentiation, 705
Ectoenzyme, degradation of thyrotropin-releasing hormone, 692
Ectoenzymes, 683
Ectopeptidase, central nervous system, 683
ED1 positive macrophages, NO and NO synthase in EAE, 969
Effect of nitric oxide, 993
Egg-laying hormone, translation of the mRNA, 656
Eicosanoids, 127, fenaridin and antagonist F, 1019
Eicosaptentaenoic acid, 126
Electric organ, vault proteins, 675
Electric ray, ecto-5′-nucleotidase, 701
Electrochemical detection, biogenic amines, 530
Electron microscopy (EM), 466
  brevican localization, 904
  extracellular calcium, 395
  of mRNA in axons, 657
  ribosomes in axons, 667
  transfer of proteins to axons, 651
Electron paramagnetic resonance (EPR), nitric oxide complex, 978
Electron spectroscopic imaging (ESI)
  visualization of ribosomes, 661
  axonal ribosomes, 667
Eliprodil, 197
Encapsulated cell lines, 252
Encephalopathies, nitric oxide, 993
Endogenous opioids, 469
  role in pain feeling, 472
Endoneuraminidase, 864
Endopeptidase 24.11, endothelin converting enzyme, 685
Endoplasmic reticulum, 383
  in axonal segments, 659
  inositol triphosphate and glutaminase, 389
  inositol-triphosphate, 390
  phosphatidylserine synthesis, 1011
  synthesis of phosphatidylinositol, 999
Endoproteinases, myelin basic protein, 1111
Endorphines, eicosanoids, 1019

Endothelial cells, 479
  aminoacid transport, 619
  diadenosine polyphosphates (ApnA), 695
  ecto-5′-nucleotidase, 702
Endothelial lysosomes, histamine, 479
Endothelin converting enzyme mRNA, 686
Endothelin converting enzyme, 685
Endothelin peptide family, endopeptidase 24.11, 685
Endplate potential, 714
Energy metabolism, 739
  catecholamines, 729
Energy substrate, astrocytes, 555
Enkephalinase, endopeptidase 24.11, 684
Enzyme immunoassay
  of glutamine synthese, 444
  of nerve growth factor, 502
Ependymin mRNA, 871
Ependymin, memory formation, 870
Epidermal growth factor, 901
Epilepsy, 287
  autoantibodies, 265
  immunological abnormalities, induced by oxygen, 291
  neuroprotective agents, 401
  neuroprotective effect of taurine, 939
  nitric oxide (NO), 977
  penicillin-induced, 287
Epileptogenesis, effect on nitric acid, 980
Epinephrine, cardiac arrest and resuscitation, 729
EPP: *see* Endplate potential
Erythrocytes, calpastatin activity, 415
Escherichia coli, serotonin transporter expression, 775
Essential fatty acids, 126
Estradiol
  TRH-degrading ectoenzyme, 692
  SK-ER3 cells, 524
Estrogen receptor antagonists, SK-ER3 cells, 524
Estrogen receptor, neural cells, 523
Estrogens, neuroblastoma cell line, 523
Ethanol
  acetylcholine release, 743
  effect on ACh release, 826
  effect on cholinergic mechanism, 847
Ethanolamine O-sulfoether, glutamine hydrolysis, 1077
Ethanolamine plasmalogens, effect of reoxygenation, 465
Ethanolamine, glutamine hydrolysis, 1077
N-Ethylmaleimide-sensitive fusion protein, exocytosis, 723
2-(Ethylthio)-apomorphine, 215
Eugerres plumieri, 396
Evan's blue, labelled albumin, 480
Excitation-contraction coupling, calcium release and reuptake, 461
Excitatory amino acids, nimodipine and 8-OH-DPAT, 351
Excitatory amino-acid, neurodegeneration, 155
Exo1 protein, exocytosis, 722

Subject Index

Exo2 protein, exocytosis, 722
Exocytosis, proteins involved, 719
Experimental allergic encephalomyelitis (EAE), 105, 113
   by prostaglandin E2, 542
   cytokines, 121
   n-3 fatty acids, 131
   induction of NO-synthase, 968
Experimental autoimmune encephalomyelitis (EAE), suppression by spinal cord protein hydrolysate, 137
Experimental epilepsy, 287
Extensor digitorum longus (EDL), calcium release and reuptake, 461
Extracellular glucose, measurement with biosensor, 577
Extracellular matrix, proteoglycans, 901
Extrapyramidal structures, synaptic structures after hypokinesia, 491

F1: *see* Protein B50
F1, neurite outgrowth and regeneration, 1107
Factors M31 and M41, neurohormones, 596
$FADH_2$: *see* Flavin adenine dinucleotide
Fatty acid beta-oxidation, astrocytes, 751
Fatty acid homeostatis, retinoic acid, 1005
Fatty acid residues, 798
Fatty acids, 574
   in Alzheimer's disease, 33
Fenaridin, 1019
Fenfluramine, 188
Fentanyl, diacylglycerol and arachidonic acid, 1019
Ferritin, oxygen convulsions, 292
Ferrous chloride, free radicals in leech ganglia, 983
Fetal motor behavior, during development, 475
F2hc surface antigen, aminoacid transport, 620
Fibroblasts
   Parkinson's disease, 253
   spectrin degeneration, 417
Fibronectin, effect on 5'-nucleotidase, 703
Fish, N-CAM-related adhesion molecules, 864
Fisher syndrome, anti-neural antibodies, 936
Flavin adenine dinucleotide ($FADH_2$), beta-oxidation activity, 751
Fluorine-18, use in positron emission tomography, 1140
*p*Fluoro-amphetamine, 317
Fluoro-2-deoxyglucose, uncoupling from oxygen consumption, 559
*p*Fluorodeprenyl, 316
*p*Fluorodesmethyl-deprenyl, 317
Fluoro-dihydroxyphenylalanine, positron emission tomography, 1149
Fluorodopa, use in positron emission tomography, 1142, 1145
*p*Fluorometamphetamine, 317
5'-Fluorosulphonylbenzoyl adenosine, 708
Fluoxetine, 194, 774
   effect on ACTH and corticosterone, 64
   interaction with selegiline, 336

Fluvoxamine, 194
Focal potential amplitudes, 519
Fodrin, 408, 410
Foetal disturbances, creatine kinase isoenzymes, 457
Footshock-motivated brightness discrimination, role of transcription factors, 1117
Formaldehyde, 529
Forskolin, adenylate cyclase, 921
Fos protein, expression after microdialysis of drugs, 1162
Fos- and Jun-containing transcription factors, 1121
Fos-like immunoreactivity, expression after microdialysis of drugs, 1162
Free fatty acids, multiple infarct dementia, 1026
Free radical hypothesis, Parkinson's disease, 244
Free radical oxidation, 530
Free radicals, 17, 309, 375
   melatonin, 611
   methamphetamine, 323
Frontal cortex, 5-hydroxytryptamine, 189
1,6-Fructose diphosphate (FDP), 173
Fucose, long-term potentiation, 905
2'-Fucosyllactose, long-term potentiation, 905

G protein, pancreastatin, 591
G protein-coupled receptor kinases, phosphorylation of ACh-receptor, 807
G protein-coupled receptor mechanism, chromogranin A, 585
G protein-coupled receptors, 72, 792
G proteins, 807
G proteins, 797, 798, 807
GABA: *see* γ-Aminobutyric acid
GABA-transaminase, 278
Galactocerebroside, 235, 239
Galactosidase, 479
   multiple infarct dementia, 1026
Gamma-glutamyltranspeptidase, 244
Gamma-glutamylcysteine synthetase, 244
Ganglion nodosum, M-cholinoreception and beta-adrenoreception, 853
Ganglioside GM1, 975
Gangliosides
   adenylate cyclase system, 920
   calcium dependent proteolytic activity, 417
   effect of temperature, 925
   meningiomas, 913
   sodium, potassium-ATPase activity, 973
GAP43: *see* Protein B50
Gastric secretion, effect on pancreastatin, 589
Gastrocnemius muscle, 486
Gene Bcl-2 expression, 323
Gene expression
   carnitine, 1056
   in axons, 638
Gene transfer, Parkinson's disease, 253
Gerbil forebrain, calcium-homeostasis, 383
Gerbils, 376
GFAP: *see* Glial fibrillary acidic protein

Glia cells, malic enzyme isoforms, 765
Glial cell progenitors, 239
Glial cells, 83, 389, 619
 aminoacid transport, 619
 effect of adenosine, 83
Glial fibrillary acidic protein (GFAP), 73, 83, 239, 390, 498, 549
 in multiple sclerosis, 451
Glioma (Bu17 and C6), 685
Glioma C6 cells, phosphatidylserine synthesis, 1012
Glioma cells, lipid metabolism, 572
Glioma x neuroblastoma hybrid cell line 108CC15, 626
Gliosis, inositol-triphosphate and glutaminase, 390
Globus pallidus, taurine and GABA release, 955
Glucocorticoids, calcium homeostasis, 405
Glucocorticosteroids, eicosanoid synthesis, 132
Glucose biosensor, brain extracellular concentration, 577
Glucose metabolism
 pancreastatin receptors, 590
 under acute osmotic stress, 573
Glucose oxidase electrode, 564
Glucose, 821
 astrocytes, 555
 energy for neuronal metabolism, 561
 measurement with biosensor, 577
 trafficking in fetal brain, 631
Glucose-6-phosphate, hexokinase, 757
Glucose-N-acetyl transferase, meningiomas, 915
Glucosidase, multiple infarct dementia, 1026
Glutamate, 375, 525
 capillary electrophoresis patch clamp, 1135
 deoxyglucose uptake, 556
 effect of taurine, 964
 flux from glia to neurons, 1066
 fos-expression after microdialysis injection, 1163
 neurodegeneration, 155
 nitric oxide (NO), 977
 release of neurotransmitters from, 801
 sodium, potassium-ATPase activity, 973
 trafficking in fetal brain, 631
Glutamate binding proteins, 267
Glutamate decarboxylase (GAD), 421, 525, 631
Glutamate dehydrogenase (GDH), 390, 1078
Glutamate receptor
 antibodies, 283
 antidepressant drugs, 197
Glutamate receptor antagonists, 556
Glutamate receptors, 88, 153, 421
 calcium transport, 361
 paroxysmal activity, 266
 retina cells, 612
Glutamate release, P2 purinoceptor, 357
Glutamate-glutamine cycle, 631
Glutamatergic function, in Alzheimer's disease, 153
Glutamatergic neurotransmission, dopamine release, 733
Glutaminase, 154, 631, 1077
 neurons, 390

Glutamine
 flux from neurons, 1066
 trafficking in fetal brain, 631
Glutamine synthetase, 443, 631
 cerebrospinal fluid, 430
 enzyme-linked immunoassay, 444
Glutamylcysteine, 734
Glutathione (GSH)
 dopamine release, 733
 dopaminergic neurons, 245
 melatonin, 602
 Parkinson's disease, 243
Glutathione peroxidase, 244, 983
 melatonin, 602
Glutathione S-transferase, 244
 serotonine transporters, 774
Glycerol, 574
Glycine, 733
 transporters, 774
Glycine receptor
 alanine, 1123
 relation with ACh-receptor, 801
Glycoconjugates (38 and 104 KD), nucleo-cytoplasmic relation, 911
Glycogen, 926
 energy reserve of brain, 555
Glycogenphosphorylase, 597
Glycoproteins, long-term potentiation, 905
Glycoproteins gp110 and gp116, 929
Glycosaminoglycan, 901
Glycosphingolipids, effect of temperature, 925
 meningiomas, 913
Glycosylphosphatidylinositol, 683
Glycosyltransferase, 913, 914
GM1 ganglioside, similarity with lipopolysaccharides of *C. jejuni*, 933
Goldfish
 ependymin genes, 870
 retinal ganglion cells, 870
Gonadoliberin, effect of xenobiotics, 529
Good laboratory practice, 436
Gpp[NH]p, adenylate cyclase, 921
GR 127935, $5HT_{1D}$ receptor antagonist, 62
GR 46611, $5HT_{1D}$ receptor agonist, 62
Graded postischemic reoxygenation (GPIR), 465
Growth-associated protein GAP43: see Protein B50
GTP-binding proteins, pancreastatin receptors, 590
Guanosine, 705
Guanosine monophosphate (cGMP)
 APP-mediated neuroprotection, 166
 pancreastatin, 591
Guanosinetriphosphatase (GTPase), 798
Guillain-Barré syndrome, anti-neural antibodies, 933

Haemopis sanguisuga, reactive oxygen species, 984
Haloperidol, 161
Head injury
 closed, 21

Subject Index

Head injury (cont.)
  dimethyl sulfoxide (DMSO) and 1,6-fructose diphospate (FDP), 173
  hypothermia therapy, 97
  mild hypothermia therapy, 91
Heat shock protein 72, 142
Helper T cells, n-3 fatty acids, 131
Hemolytic disease, creatine kinase, 457
Heparan sulphate proteoglycans, 16
Hepatic encephalopathy, 201
  glutamine synthetase, 443
  taurine efflux, 944
Hepatic necrosis, n-3 fatty acids, 133
Heroin, MPTP contamination, 332
Herpes simplex, induction of NO-synthase, 968
Hexokinase, delta-sleep inducing peptide, 757
Hexosaminidase, 932
N-2-Hexyl-N-methylpropargylamine, 307
5HIAA: see 5-Hydroxyindoleacetic acid
Hippocampal CA1 region, long-term potentiation, 905
Hippocampal long-term potentiation, platelet-activating factor, 1029
Hippocampal slices, 1071
  phosphatidylethanol, 744
Hippocampus, 61, 149, 189, 497, 702, 801, 863, 939
  cholinergic and serotonergic innervation, 149
  5-hydroxytryptamine, 61, 189
  neuroprotective effect of taurine, 939
  NGF-expression in astrocytes, 497
  5'-nucleotidase, 702
  release of neurotransmitters, 801
Hippocampus CA3, brevican immunostaining, 903
Histamine, acid phosphatase activity, 479
Histamine receptors, 483
Histology image analysis, of protein kinase C, acetylcholine receptors and trkB, 509
Histone acetylation, carnitine, 1056
Homoanserine, 177
Homocarnosine, 177
Homovanillic acid (HVA), 209, 216
  microdialysis measurement, 1159
Hormones, eicosanoids, 1019
HPC-1: see Syntaxin
5-HT: see Serotonin
Huntington's disease, 259
  nitric oxide, 823, 993
HVA: see Homovanillic acid
12-Hyaluronic, 901
5-Hydroeicosatetraenoic acid (12HETE), 1022
Hydroeicostetraenoic acid (5HETE), 1022
Hydrogen peroxide, 133, 312, 983
  free radicals in leech ganglia, 983
  n-3 fatty acids, 133
Hydroxy radical scavenger, melatonin, 611
12-Hydroxy-5,8,10-heptadecatrienoic acid (HHT), 1022
11-Hydroxy-N-n-propylnorapomorphine, 215
Hydroxyalkenals, 600

Hydroxybutyrate, 825
  diabetic rabbit brains, 633
8-Hydroxydipropylaminotetralin (8-OH-DPAT), 188
  calcium homeostatis, 404
  NMDA-induced neurodegeneration, 352
6-Hydroxydopamine (6-OHDA), 245, 250, 299, 312
5-Hydroxyindole-3-acetic acid (5-HIAA), 203, 216, 601, 733
  microdialysis measurement, 1159
Hydroxyindole-O-methyltransferase, 605
Hydroxyl radicals, 309
  effect on melatonin, 599
5-Hydroxytryptamine (5-HT), 193, 201, 773; see also Serotonin
  depression, 187
Hyperalbuminization, of pituitary tissue at oxygen convulsion, 292
Hyperammonaemia, taurine efflux, 944
Hyperbaric oxygenation, 291
Hyperosmolality, taurine biosynthesis, 944
Hypertensive state, 853
Hypoglossal nerve, microglial reaction, 549
Hypokinesia
  γ-aminobutyric acid, 1083
  delta-sleep inducing peptide (DSIP), 339
  synaptic structures, 491
Hypokinetic stress
  delta-sleep inducing peptide (DSIP), 339
  synaptic structures, 492
Hypomyelination, 234, 235
Hypo-osmolality, release of taurine, 944
Hypothalamic circadian rythm, xenobiotics (dioxane, formaldedehyde), 529
Hypothalamic neurohormones, 595
Hypothalamic-pituitary-adrenal (HPA) axis, 63
  5-hydroxytryptaminefunction, 188
Hypothalamo-pituitary-adrenocortical axis, n-3 fatty acids, 131
Hypothalamus, effect of xenobiotics, 529
Hypothermia therapy
  human head injury, 97
  in human acute embolic stroke, 92
  in human head injury, 91
Hypoxia
  calpain activity, 419, 421
  carnitine palmitoyltransferase, 1055
  catecholamines, 729
  delta-sleep inducing peptide, 757
  developing neurons, 147
  neuronal lactate, 566
  neuroprotective agents, 402
  neuroprotective effect of taurine, 939
  nimodipine and 8-DH-DPAT, 351
  protective pretreatments, 79
  transient, 73

ICI-182, 780
  SK-ER3 cells, 524
Idiopathic neuralgia, 469

IgG
    cerebrospinal fluid, 426, 440
    immunoreactivity in brain, 142
IgG-index, cerebrospinal fluid, 440
IgM, cerebrospinal fluid, 426, 434
Imipramine, 194, 774
    phospholipase A2, 1035
Immediate early genes
    c-jun, jun B, and c-fos, 1117
    expression of fos protein, 1161
Immortalised cell lines, Parkinson's disease, 255
Immune cells, 130
Immune system, n-3 polyunsaturated fatty acids, 125
Immunoblot analysis, proteins of neostratum, 390
Immunocytochemistry, GFAP and NGF, 498
Immunoglobulin (Ig) superfamily, 863, 891
    cell-cell adhesion molecules, 857
Immunoglobulin-like domains, neural cell adhesion
    molecule, 891
Immunohistochemical staining, expression of fos protein, 1161
Immunological reactions, interleukin-1 (IL-1), 501
Immunotoxin 192IgG-saporin, effect on cholinergic
    function, 830
Inermediate filament proteins, 227
Infarction, NGF-release, 513
Infection
    microglial activation, 535
    prostanoid levels, 542
Inflammation, 17
    interleukin-1 (IL-1), 501
    prostanoid levels, 542
    related proteins, 18
Inherited pyruvate dehydrogenase deficiencies, acetylcholine synthesis, 821
Injury
    interleukin-1 (IL-1), 501
    microglial activation, 535
Inonophore A23187, 1012
Inosine, 705
Inositol, 591
Inositol phosphate production, 830
Inositol triphosphate (IP3)
    calcium homeostasis, 402
    neurons, 389
Inositol triphosphate-receptor (IP3-R), 376, 383, 389
Insomnia, selegiline, 333
Insulin B chain, degrading activity, 684
Insulin, estrogen receptors, 525
Insulin-like growth factor 1 (IGF-1), estrogen receptors, 525
Insult, glutamine synthetase, 443
Integrin (CR3), 109
Integrin (VLA-4), 109
Integrin family, cell-cell adhesion molecules, 857
Integrin receptor, 551
Interaction between neurons, 5'-nucleotidase, 703
Intercellular adhesion molecule-1 (ICAM-1), 27, 122
Interfacing process, 222
Interferon, 536

Interferon-β, 115
Interferon-γ, 121
    arginine transport, 625
    balance between prostanoids and NO, 544
    n-3 fatty acids, 131
Interleukin-1 (IL-1), 27, 121
    effect on nerve growth factor, 501
    n-3 fatty acids, 131
Interleukin-3 (IL-3), 535
Interleukin-4 (IL-4), 121
Interleukin-6 (IL-6), 27
Interleukin-10 (IL-10), 121
Intermediate filament protein, vimentin, 239
Interplexiform cells, 605
Ion channels, axonal segments, 649
Ion pumps, in axonal segments, 649
Ionotropic glutamate receptors, 733
IP3: see Inositol triphosphate
IP3-receptor, Alzheimer's disease (AD), 389
Ipsapirone, calcium homeostasis, 404
Iron, 309
Iron chelator, MPP^+-induced dopaminergic toxicity, 309
Ischemia
    calcium-homeostasis, 383
    calmodulin-dependant kinase II, 370
    calpain activation, 407
    carnitine palmitoyltransferase, 1055
    catecholamines, 729
    developing neurons, 147
    effect on phospholipids, 465
    free radicals, 611
    melatonin, 616
    neuroprotective agents, 402
    neuroprotective effect of taurine, 939
    nimodipine and 8-OH-DPAT, 351
    platelet-activating factor (PAF), 1029
    protective pretreatments, 79
    study by MRI methods, 183
    transient, 73
Ischemic injury
    acetylcarnitine, 1048
    sodium, potassium-ATPase activity, 973
Ischemic insult, glutamine synthetase, 443
Ischemic-reperfusion injury, InsP3 receptor and calcium transport, 375
Isethionic acid, taurine uptake, 961
Isoleucine, aminoacid transport, 622
Isoprenaline, 565
Isoproterenol, cAMP levels, 545
Isovalerylcarnitine, calcium dependent proteolytic activity, 417

JL-18
    N-acetyltransferase, 608

Kainic acid
    CAM kinase II, 373
    dopamine release, 733
    melatonin, 600
    taurine and GABA release, 949, 955

Ketanserin, 70
α-Ketoglutarate, 631

L 929 cell line, screening toxicity, 295
L-365,260
  cholestokinin type B receptor, 1089
L-740,093
  cholestokinin type B receptor, 1089
L2/HNK-1 carbohydrate epitope, neuron-cell adhesion molecule 929
Lactate dehydrogenase (LDH), isoenzymes, 557
Lactate, 181
  effect of glutamate release, 567
  energy for neuronal metabolism, 561
  extracellular glucose, 557
  metabolic trafficking, 1070
  trafficking in brain, 632
Laminin/nidogen complex, effect on 5′-nucleotidase, 703
Lazabemide, 80
Learning, role of transcription factors, 1117
Lectin-binding capacity, nuclear glycoproteins, 910
Leech, reactive oxygen species, 983
Leiurus quinquestriatus hebreus, charybdotoxin, 784
Leucine, aminoacid transport, 622
Leukemic T-cell line, acetylcholine receptors, 813
Leukocytes, selectin family, 929
Leukotriene A_4 (LTA_4), 1022
Leukotriene B_4 (LTB_4), 1022
  proinflammatory agent, 129
Leukotriene, 130
Leukotrienes, 127, 130
Light microscopy, 466
Limbic structures
  adenylate cyclase, 919
  synaptic structures after hypokinesia, 491
α-Linolenic acid, 126
Lipid biosynthetic pathways, carnitine system, 1053
Lipid metabolism, osmotic stress, 571
Lipid peroxidation, 797
  6-hydroxymelatonin, 600
  convulsive states, 977
  delta-sleep inducing peptide (DSIP), 339
  leech ganglia, 983
  melatonin, 600, 614
Lipid spectrum, multiple infarct dementia, 1026
Lipogenic acetyl-CoA, carnitine, 1056
Lipopolysaccharide (LPS)
  arginine transport, 625
  cytokine production, 536
  melatonin, 601
  mRNA expression of K^+-channels, 538
  prostanoids, 543
  tumor necrosis factors, 536
Lipopolysaccharides (LPS) of *C. jejuni*, similarity with GM1-ganglioside, 933
Lipoxygenase, 127
Liver membranes, pancreastatin receptors, 590
Locus coeruleus, 243

Long-chain acylcarnitine, membrane repair, 1040
Long-chain acylcarnitine transferases, mitochondrial membrane, 1054
Long-chain fatty acids, carnitine, 1040
Long-term potentiation (LTP), 507, 797
  effect of 2′fucosyllactose, 905
  of brain polypeptides, 519
  polysialic acid associated with NCAM 864
  role of NCAM, 881
Long-term potentiation, phospholipase A2, 1038
Long-term survival, axons and axonal segments, 647
Lumbar ganglia, M-cholinoreception and beta-adrenoreception, 853
Lung resistance-related protein, 675
Luteinizing hormone (LH), hypothermia treatment, 102
Lymnaea stagnalis, translation in isolated axons, 655
Lymphocytes
  acetylcholine receptors, 813
  calpain activity, 416
  n-3 fatty acids, 131
Lymphotoxin, 121
Lysine
  aminoacid transport, 622
  transport in neurons, 626
LysoPAF acetyltransferase, 1030
Lysophosphatidylcholine, multiple infarct dementia, 1026
Lysophospholipid acyltransferase, 1041
Lysosomal enzymes, multiple infarct dementia, 1026
Lysosomal membranes, 341
Lysosomal proteinases, 343
Lysosomes, histamine, 479

Macrophage depletion
  microglia cells, 106
  T cell-induced EAE, 106
Macrophage-specific markers, microglial cells, 549
Macrophages
  cytokines, 130
  ED2+ perivascular, 107
  experimental allergic encephalomyelitis (EAE), 105
Magnesium
  ecto-ApnA hydrolase, 699
  phosphatidylserine synthesis, 1011
  transport ATPases, 708
Magnetic resonance imaging (MRI), 117, 122, 440
  of ischemia, 183
Magnocellular basal nucleus, calcium homeostasis, 402
Magnocellular neurons, brevican immunostaining, 903
Main pelvic ganglion, M-cholinoreception and beta-adrenoreception, 853
Major vault protein (MVP100), torpedo electric organ, 677
Malate, 739
  release from astrocytes, 1066
Malate dehydrogenase, delta-sleep inducing peptide, 757
Malate: NADP oxidoreductase (decarboxylating), 765
Malic enzyme, isoforms, 765

Malondialdehyde, 600
Malonyl coenzyme A (Malonyl CoA), 1050
Malonyl-CoA-sensitive acyltransferase, energy modulating, 1040
Maltose binding protein, 891
Mammalian proteins (Kell and Pex), endothelin converting enzyme, 687
Manganese, ecto-ApnA hydrolase, 699
MAO: see monoamine oxydase
MAP kinase: see Mitogen-activated protein kinase
Mauthner cell axoplasm, presence of RNA, 668
Mauthner cell axon, ribosomes, 667
Mdx mouse, role of calcium, 461
Mealworm, carnitine, 1039
Mecamylamine, dopamine release, 802
Medial eminence of hypothalamus, effect of xenobiotics, 529
Median eminence, axonal ribosomes, 664
Mediodorsal nucleus of thalamus, injection of drugs by microdialysis, 1162
Melatonin
  neural antioxidant, 599
  retinal cells, 611
  synthesis in the retina, 605
Membrane fluidity, n-3 fatty acids, 132
Membrane-associated enzymes, n-3 fatty acids, 132
Membrane-associated proteins, effect of calpains, 408
Memory formation
  neural cell adhesion molecule (NCAM), 878
  role of transcription factors, 1117
Memory retention, effect on pancreastatin, 589
Memory, platelet-activating factor (PAF), 1029
Meningiomas, glycosyltransferase activities, 913
Mental illness, neurotransmitter transporters, 773
MEPP: see Miniature endplate potential
Messenger RNA (mRNA)
  alanine expression, 1129
  axonal protein synthesis, 644
  choline acetyltransferase, 817
  in axons, 667
  L-amino acid transporter, 620
  translation in axons, 655
Messenger RNA expression
  in axons, 638
  of outward rectifying K^+-channels, 537
Mesulergine, 70
Metabolic trafficking
  dopamine, 209
  glucose, 555
  glucose and lactate, 561
  lipids, 571
Metabotropic glutamate receptors, 733
  phosphorylation of brain proteins, 1071
Methamphetamine, 315, 327
  dopamine (DA) levels, 323
Methionine, 649
Methylaminoisobutyric acid, aminoacid transport, 619
Methylamphetamine, 304
  dopamine and serotonin, 327

N-Methyl-D-aspartic acid (NMDA), 733
  capillary electrophoresis patch clamp, 1135
N-Methyl-D,L-aspartic acid (NMDLA), 978
N-Methyl-D-aspartate antagonists, 197
N-Methyl-D-aspartate receptor channels, 401
N-Methyl-D-aspartate receptors, 832, 941
  calcium transport, 34
  dopamine release, 733
  effect of nimodipine, 351
  in learning and memory processes, 153
5-Methyl-10,11-dihydro-5H-dibenzo(a_1d)cyclohepten-5,10-imine: see Dizocilpine
α,β-Methylene adenosine 5'-diphosphate, ecto-nucleotidases, 700
Methylenedioxy-N-n-propylnorapomorphine, 215
N-Methyl-D-glucamine, 802
Methyllycaconitine, dopamine release, 802
8-Methyl-6-(4-methyl-1-piperazinyl)-11H-pyrido[2,3-b][1,4]benzodiazepine, N-acetyltransferase, 608
1-Methyl-4-phenyl-1,2,3,6-tetrahydropyridine (MPTP), 246, 301, 303, 332
  effect of iron chelators, 309
1-Methyl-4-phenyl-2,3-dihydropyridine, effect of iron chelators, 309
1-Methyl-4-phenylpyridine (MPP^+), 299, 303
  effect of iron chelators, 309
Methylpiperidinyl benzilate, 794
Methysergide, 70
MGluR1–8: see Metabotropic glutamate receptors
Mianserin, 70
  phospholipase A2, 1035
Microdialysis
  5HT analyses, 59
  acetylcholine and ethanol, 744
  brain extracellular concentration, 577
  calcium transport, 362
  dopamine metabolism, 1158
  extracellular glucose concentration, 562
  injection of noradrenaline or dopamine, 1162
  serotonin metabolites, 202
Microglial activation
  adenosine, 535
  cytokine synthesis, 535
  potassium, 535
Microglial cells, 73, 453
  macrophage depletion, 106
  nerve injury, 550
  neurotoxicity and neuroprotection, 541
  NO and NO synthase in EAE, 969
  NO synthase, 627
  phagocytose of disintegrated neuroplasm, 551
  prostanoids, 543
  regulation by adenosine, 88
Microsomes, platelet-activating factor (PAF), 1029
Microtubule-associated protein-2, effect of calpains, 408
Mild hypothermia, cerebral ischemia, 91
Milnacipran, 194
Miniature endplate potentials (MEPP), 714

Subject Index

Mitochondria, 376
    malic enzyme, 765
Mitochondrial electron transport, beta-oxidation activity, 752
Mitochondrial NADH dehydrogenase, 312
Mitogen-activated protein (MAP) kinase, signal transduction cascade, 166
MK-801, 55, 199, 299, 404, 833; see also Dizocilpine
MOLT-3, 813
MOLT-3 human leukemic T-cell line, acetylcholine receptors, 813
Monoamine hypothesis, of depression, 47
Monoamine oxidase (MAO), 79, 209, 287, 731, 773
    delta-sleep inducing peptide, 757
    role in neuroprotection, 303
Monoamine oxidase-B (MAO-B), 309, 315, 331
    effect of L-deprenyl, 70
Monoamine oxydase inhibitors, 189, 198
Monoamine oxidase-B inhibitor, (−)-deprenyl (Selegiline), 299
Monoamine oxidase-B inhibitors, use in positron emission tomography, 1147
Monoaminergic systems, amphetamines, 323
Monoamines
    cardiac arrest and resuscitation, 729
    diadenosine polyphosphates (ApnA), 695
    role in pain feeling, 472
Monoclonal antibody (192196), 830
    nerve growth factor (NGF) receptors, 830
N-Monomethyl-L-arginine, 544, 626
Mononuclear leukocytes, acetylcholine receptors, 813
Monophosphoinostide, 762
Morphine, 198, 1019
Morphochemical plasticity, dopamine system, 161
Mossy fiber terminals, 5′-nucleotidase, 702
Motor activity, n-3 fatty acids, 132
Motor behavior, during development, 475
Motor neurons, 485
Motor unit number, 487
Mouse L-cells, screening toxicity, 295
MPTP: see N-Methyl-4-phenyl-1,2,3,6-tetrahydropyridine
mRNA: see Messenger RNA
Multi-infarct dementia, lipid spectrum and lysosomal enzymes 1025
Multiple sclerosis (MS), 105, 113, 137, 268, 433
    cytokines, 121
    n-3 fatty acids, 131
    NO synthase induction, 970
    prostanoid levels, 542
Muscarinic acetylcholine receptors, phosphorylation, 807
    spatial learning, 507
Muscarinic M1 receptors, 157
Muscimol, effect of taurine to GABA-receptor binding, 962
Muscle action potential, gastrocnemius muscle, 486
Myelin
    autoreactive T cells, 114
    endoproteinases, 1111

Myelin basic protein (MBP), 137, 239, 970
    autoantigen in MS, 114
    endoproteinases, 1111
    in multiple sclerosis, 451
Myelin oligodendrocyte glycoprotein (MOG), 137
    autoantigen in MS, 114
Myelin proteolipid, 234
Myelin-associated glycoprotein (MAG), 137, 239
Myelin-associated proteins, 236
Myelogenesis, in pt rabbit brain, 233
Myo-inositol, 70

Na,K-ATPase: see Sodium, potassium (Na$^+$, K$^+$) ATPase
NADH: see Nicotinamide adenine dinucleotide
NADH ubiquinone reductase, 752
NADPH: see Nicotinamide adenine dinucleotide phosphate
NADPH-diaphorase, 832
NADPH-oxidase complex, 17
NBAT/rBAT/D2 protein, aminoacid transport, 620
NCAM: see Neural cell adhesion molecule
Necrosis
    amiridine, 346
    use of Bcl-2 expression, 323
Neocortex, 206
    cholinergic and serotonergic innervation, 149
Neostriatum
    dopamine release, 733
    inositol triphosphate and glutaminase, 389
Neprilysin, nervous system, 684
Nerve endings, 5′-nucleotidase, 702
Nerve growth factor (NGF) receptors, monoclonal antibody (192196), 830
Nerve growth factor (NGF), 260, 501, 507, 703
    astrocytes, 501
    effect on cholinergic function, 831
    expression in astrocytes, 498
    expression in non-neuronal cells, 513
    role of protein B-50/GAP-43, 1107
Nerve growth factor receptor (TrkA)
    cholinergic neurons, 513
    expression in astrocytes, 498
Nerve injury, microglial cells, 550
Nerve terminals, gene expression, 637
Nervous tube defects, creatine kinase isoenzymes, 457
Neural cell adhesion molecule (NCAM), 857, 863
    ATPase activity, 710
    cell movement, 897
    immunoglobulin-like domains, 891
    in multiple sclerosis, 451
    L2/HNK-1 carbohydrate, 929
    memory formation, 878
    polysialic acid (PSA), 885, 929
Neural cell surface, binding of NCAM, 893
Neural cells, arginine transport, 625
Neural crest, estrogen receptor, 524
Neural differentiation, 5′-nucleotidase, 705
Neural tube defects, 295

Neuralgia, 469
Neurite differentiation, ecto-5'-nucleotidase, 705
Neurite extension, 5'-nucleotidase, 703
Neurite outgrowth, protein B50, 1107
Neuro-organic system, 221
Neuroactive steroids, 275
Neuroblastoma cell line, estrogen receptor, 524
Neuroblastoma cells, overexpression of B-50/GAP-43, 1107
Neuroblastoma N115 and OLN-93 cells, 240
Neuroblastoma NB-2a cells, effect of carnitine on acetylchol, 1062
Neuroblastoma SK-N-BE, retinoic acid, 1005
Neuroblastoma SHSY-5Y, 685, 885
Neurocan, proteoglycans, 901
Neurodegeneration, during aging, 147
Neurodegenerative diseases, 538
  dysfunction of the respiratory chain, 751
  neuroprotective agents, 401
  nimodipine and 8-OH-DPAT, 351
Neurodegenerative disorders, 259, 315
Neurofibrillary tangles, 15
Neurofilament proteins
  effect of calpains, 408
  synthesis in axons, 673
Neurofilaments, cytoskeleton, 227
Neurogranin, 43
Neurohypophysis, 291
  axonal ribosomes, 664
Neuroleptics, 198
Neurological severity score (NSS), 23
Neuromodulin: see Protein B50
Neuron damage, dimethyl sulfoxide and 1,6-fructose diphospate, 173
Neuron death, amiridine, 346
Neuron differentiation, retinoic acid, 1005
Neuron survival, ecto-5'-nucleotidase, 705
Neuron-glia interactions, trimethyltin intoxication, 497
Neuron-specific phosphoprotein B-50: see Protein B50
Neuronal damage, 73
Neuronal degeneration
  calpain activation, 407
  inositol-triphosphate and glutaminase, 390
Neuronal energy metabolism, role of glucose and lactate, 561
Neuronal interaction, 5'-nucleotidase, 703
Neuronal plasticity, neuropeptides, 419
Neuronal specific, 719
Neurons
  diadenosine polyphosphates (ApnA), 695
  inositol triphosphate and glutaminase, 389
  malic enzyme isoforms, 765
Neuropeptide Y, adrenal medulla, 595
Neuropeptides, neuronal plasticity, 419
Neurophysins, at oxygen epilepsy, 291
Neuropil threads, 15

Neuroprotection
  (−)-deprenyl, 299
  amyloid precursor protein (sAPP), 165
  cerebral hypothermia treatment, 100
  melatonin, 601
  role of cyclooxygenase-2, 541
  taurine, 939
  tumor necrosis factor, 165
Neurorescue, (−)-deprenyl, 299
Neurotensine, adrenal medulla, 595
Neurotoxicity, 312
  protection by (−)-deprenyl, 300
  role of cyclooxygenase-2, 541
Neurotransmission, platelet-activating factor (PAF), 1029
Neurotransmitter receptors
  adenylate cyclase, 919
  in axonal segments, 649
Neurotransmitter transporters, 773
Neurotransmitter-based treatments, Alzheimer's disease (AD), 153
Neurotransmitters, eicosanoids, 1019
Neurotrauma, taurine efflux, 944
Neurotrophins, 507
Neutral glycolipids, effect of temperature, 926
Neutral proteinase, effect on myelin basic protein, 1115
Neutral proteinases, 343
Nicotinamide adenine dinucleotide, beta-oxidation activity, 751
Nicotinamide adenine dinucleotide phosphate, malic enzyme isoforms, 765
Nicotine receptors, 843
  ethanol, 748
Nicotine, dopamine release, 802
Nigral grafts, Parkinson's disease, 249
Nigrostriatal dopaminergic pathway, positron emission tomography, 1145
Nimodipine, 148
  calcium homeostasis, 402
  NMDA-induced neurodegeneration, 352
Nissl staining, 77
Nitric oxide (NO), 17, 625, 967
  ACh release, 823
  activated microglia, 544
  calcium dependent proteolytic activity, 417
  during epilepsy, 977
  effect on catecholamine release from adrenal c, 987
  hypothermia, 99
  neurotoxicity, 541
Nitric oxide pathway, oligodendrocytes, 626
Nitric oxide synthase (NOS), 967
  melatonin, 602
Nitrite, n-3 fatty acids, 133
N-nitro-L-arginine, 626, 978, 990
Nitrogen-13, 1140
  use in positron emission tomography, 1140
p-Nitrophenylethylamine, MAO-B substrate, 287
p-Nitrophenylphosphate, 480

Subject Index

3-Nitropropionic acid, 752
NMDA channel antagonists, neuroprotection, 402
NMDA glutamate receptor, tacrine, 345
NMDA receptor-mediated transmission, ethanol, 748
NMDA: see N-Methyl-D-aspartic acid
NMDA-glutamate receptors, calpain I, 343
NMR: see Nuclear magnetic resonance spectroscopy
NO synthase (NOS), neurotoxicity, 541
NO synthases (NOS), 625
Nomifensine, use in positron emission tomography, 1146
Non-AD dementia, glutamine synthase, 444
Non-steroidal anti-inflammatory drugs, 29
Noradrenaline, 305, 469, 843
  depression, 193
  fos-expression after microdialysis injection, 1162
  glycogenolysis, 555
  release of neurotransmitters from, 801
  role in pain feeling, 472
Noradrenaline receptor, hypertensive state, 853
Noradrenaline release
  effect of nitric oxid, 988
  factors M31 and M41, 597
Noradrenaline reuptake inhibitors, 194
Noradrenergic neurons, locus coeruleus, 243
Norepinephrine, 601
  antidepressant drugs, 197
  cardiac arrest and resuscitation, 729
  transporters, 774
Normoxia, 566
Northern blot analysis, 556
  acetylcholine receptors, 813
de novo protein synthesis, axonal segments, 649
NSF: see N-Ethylmalcimide-sensitive fusion protein
Nuclear DNA, axonal expression, 640
Nuclear glycoproteins, lectin-binding capacity, 910
Nuclear magnetic resonance spectroscopy, 183, 631
  compartmentation of glucose metabolism, 1065
  glial lipid metabolism, 571
  of carnosine, 178
Nuclear membrane lectin, nucleo-cytoplasmic relationship, 911
Nuclear transcription factor NFκB, 168
Nucleo-cytoplasmic relationship, nuclear membrane lectin, 911
Nucleosides, as neurotransmitters, 357
Nucleotidase, neuritic differentiation and neuron survival, 701
Nucleotides
  adrenal medulla, 595
  as neurotransmitters, 357
Nucleus accumbens, 209
  dopamine release, 1157
Nucleus basalis, calcium homeostasis, 401
Nucleus basalis magnocellularis, NGF and TrkA expression, 513
Nucleus basalis of Meynert, cortical cholinergic dysfunction, 829
Nucleus caudatus, synaptic structures, 492

Nucleus of thalamus, 1162

Oedema, 480, 943
  histamine, 480
  regulation of osmotic pressure, 943
8-OH-DPAT: see 8-Hydroxydipropylaminotetralin
6-OHDA: see 6-Hydroxdopamine
Oleic acid, 126
Olfactory cortex tract, of brain polypeptides, 519
Olfactory tract fibres, of brain polypeptides, 519
Oligoclonal IgG bands, in cerebrospinal fluid, 436
Oligodendrocyte progenitors, 239
Oligodendrocytes, 227
  hypomyelination, 235
  malic enzyme isoforms, 765
  nitric oxide pathway, 626
Oligodendroglial cells, ecto-5'-nucleotidase, 702
OLN-93 cell line, 239
Oncogene transduction, 239
Oncoproteins, trkB, 507
Opiate addiction, 268
Opioid peptides, 529
  adrenal medulla, 595
  eicosanoids, 1019
Opioid receptors, 1019
Optic nerve, axonal transport of proteins, 1101
Optic tectum, extracellular calcium, 395
Oral tolerance, of EAE by spinal cord protein hydrolysate, 137
Oreochromis mossambicus, 395
Organum vasculosum of lamina terminalis, effect of xenobiotics, 529
Ornithine, transport in neurons, 626
Ortho-phosphoric monoester hydrolase, 480
Osmosis, regulation of osmotic pressure, 943
Osmotic stress, 571
Ouabain, 557
Oxidation, carnitine translocase, 1040
Oxidative phosphorylation
  amiridine, 348
  role of phosphate activated glutaminase, 1077
Oxidative stress, 17
Oxoglutarate, 740, 1079
  release from astrocytes, 1066
Oxoglutarate dehydrogenase, 737, 824
Oxygen based radicals, effect of L-deprenyl, 79
Oxygen consumption, uncoupling from blood flow and glucose utilization, 559
Oxygen radicals, 300
Oxygen-15, use in positron emission tomography, 1140
Oxytocin mRNA, in axons, 664

P_1 Purinoceptor, adenosine, 357
P_2 purinoceptor, glutamate release, 358
P_2-purinoceptor subtypes, 698
PAF: see Platelet-activating factor
Pain threshold, n-3 fatty acids, 132
Pallidonigral neurons, taurine and GABA release, 957
Palmitic acid, beta-oxidation, 752

Palmitoylcarnitine transferase, transport in neuronal membranes, 1059
Pancreastatin, action in liver membranes, 589
Pancreastatin receptors, liver membranes, 590
Pancreatic islets, chromogranin A, 584
Pancreatic secretion, effect on pancreastatin, 589
Panic attacks, cholestokinin type B receptor, 1089
Paralytic tremor, PLP gene mutant rabbit, 233
Parathormone release, effect on pancreastatin, 589
Parathyroid gland, chromogranin A, 583
Parathyroid hormone, chromogranin A, 583
Parkinson's disease (PD), 259, 309, 315, 324, 328
   cells for transplantation, 249
   (−)-deprenyl, 308
   free radicals, 611
   glutamine synthase, 444
   glutathione (GSH), 243
   nitric oxide (NO), 823, 977
   PET study of dopamine receptors, 1146
   positron emission tomography (PET), 1149
   selegiline, 331
Parkinsonism, 268
Parlodel, 102
Paroxetine, 194, 774
Paroxysmal activity, autoantibodies, 265
Parvalbumin, 830
Patch-clamp detection, capillary electrophoresis, 1133
PC12 cells, 722
   apoptosis, 260
   ecto-5′-nucleotidase, 702
   exocytosis, 725
   overexpression of B-50/GAP-43, 1107
PC12 phaeochromocytoma, secretion, 252
PCR: see Polymerase chain reaction
Penicillin-induced epilepsy, 287
Penthylenetetrazol, 978
Pentoxifylline, 123
Periaxoplasmic plaques, 667
Peroxidation of lipids, delta-sleep inducing peptide, 761
Peroxisome proliferator-activating receptors, retinoic acid, 1005
Peroxyl radical, effect on melatonin, 599
Pertussis toxin-sensitive G protein, pancreastatin, 591
Pethidine, interaction with selegiline, 336
Phenobarbital, 304
Phenyalanine, aminoacid transport, 622
Phenylethylamine, 304
Phenytoin, n-3 fatty acids, 133
Pheochromocytoma cell line (PC12), 526
Pheochromocytoma, overexpression of B-50/GAP-43, 1107
Phorbol-12-myristate-13-acetate, phosphorylation of brain proteins, 1071
Phosphate activated glutaminase
   isoforms, 1077
   neurons, 390
Phosphate, glutamine hydrolyse, 1079
Phosphathidylcholine, 762

Phosphatidic acid, effect of reoxygenation, 465
Phosphatidylcholine, 1026
   multiple infarct dementia, 1026
Phosphatidylethanolamine, 762
Phosphatidylethanol, phospholipase D, 744
Phosphatidylethanolamine, multiple infarct dementia, 1026
Phosphatidylinositol, 507
   effect of reoxygenation, 465
Phosphatidylinositol transfer protein, location and characterization, 999
Phosphatidylinositol-4,5-bisphosphate, plasma membrane, 999
Phosphatidylserine, 762
   effect of reoxygenation, 465
Phosphatidylserine synthesis, role of calcium and protein kinase C, 1011
Phosphocholinetransferase, platelet-activating factor (PAF), 1029
Phosphodiesterase, 596
Phosphoinositide, 1019
Phospholipase, pancreastatin, 591
Phospholipase A2, tricyclic antidepressants, 1035
Phospholipase C, 1019
Phospholipase D, phosphatidylethanol, 744
Phospholipid antibodies, 281
Phospholipid glycerides, delta-sleep inducing peptide, 761
Phospholipid turnover, carnitine palmitoyltransferase, 1053
Phospholipids, 129
   delta-sleep inducing peptide, 761
   effect of temperature, 926
   in Alzheimer's disease, 33
   transfer between membranes, 999
   tricyclic antidepressants, 1035
   under acute osmotic stress, 573
Phosphoprotein phosphatase, 479
Phosphoramidon, 684
Phosphorylation
   effect of protein kinase C, 1071
   of acetylcholine receptors, 807
Pichia pastoris
   neural cell adhesion molecule (NCAM), 891
   serotonin transporter expression, 778
Picrotoxin, 276
Pineal organ, 611
Piperidine-3-sulphonic acid, taurine uptake, 961
Piracetam, hypokinesia, 1085
Pirenzepine, 148
Pirrolidine-2–4-dicarboxylate, 567
Pisum Sativum agglutinin, 910
Pituitary, effect of xenobiotics, 529
Pituitary hormones, hypothermia treatment, 102
Pituitary tissue, hyperalbuminization, 292
Plasma membrane
   Ca^{2+}-transport ATPase, 376
   phosphatidylinositol, 999
Plasma, acetylcholine (ACh), 813

Platelet-activating factor (PAF), cytidine 5'-monophosphate, 1029
PLP: see Proteolipid protein
Polyacrylamide gel electrophoresis (PAGE)
 brain polypeptides, 519
 CAM kinase II, 370
 glutamine synthase, 444
 MVP100, 676
Polymerase chain reaction (PCR), axonal mRNA, 639
Polymyxin B, 975
Poly(o-phenylenediamine), 577
Polypeptides
 brain slices, 521
 synthesis in axons, 673
Polyphosphoinositides, calcium dependent proteolytic activity, 417
Polysialic acid (PSA), differentiation and neuritic outgrowth, 885
Polysialic acid, 863
 binding to NCAM, 879
Polysialyltransferase, 886
Polysomes
 axonal protein synthesis, 644
 squid giant axon, 663
Polyunsaturated fatty acids, 37
 leech ganglia, 983
 melatonin, 600
 multiple sclerosis, 125
 neuron differentiation, 1005
n-3 Polyunsaturated fatty acids, multiple sclerosis, 125
n-6 Polyunsaturated fatty acids, multiple sclerosis, 125
Porcine pancreas, pancreastatin, 589
Portacaval shunt, 201
Positron emission tomography (PET)
 activation-induced glycolysis, 558
 dopamine receptor mapping, 1139, 1145
 fluoro-3,4-dihydroxyphenylalanine, 1149
Potassium channels
 in microglia, 536
 scorpion toxins, 784
 tacrine, 345
Potassium cyanide, melatonin, 601
Potassium transport, tetra-butylhydroxide, 983
Potassium
 AChR-mediated dopamine release, 802
 microglial activation, 535
 uptake of arginine, 626
Prealbumin, 425
Preoptic area of hypothalamus, effect of xenobiotics, 529
Presenilin genes, 21
Probenecid, effect on lactate transport, 565
Prolactin, 188
 gonadoliberin, 530
Proline, 774
 transporters, 774
Proline directed kinase, phosphoylation of synaptic proteins, 1095

Promedol, 1019
Propranolol, 565
N-n-Propylnorapomorphine, 215
N-Propylpentanoic acid, neural tube defects, 295
Prostacyclin I$_3$, vasodilator, 130
Prostaglandins, 127, 469, 535
 piracetam, 1087
Prostaglandin D$_2$, 470
 interferon-γ, 544
Prostaglandin E$_2$, 470
 cAMP levels, 545
 inflammatory responses, 129
 interferon-γ, 544
 neuroprotective action, 541
Prostaglandin F$_{2\alpha}$, 470
Prostanoids, neuroprotection, 541
Proto-oncogene, estrogens, 524
Protein B50, neurite outgrowth and regeneration, 1107
Protein 14–3–3, exocytosis, 722
Protein acylation, carnitine, 1056
Protein Exo1, exocytosis, 722
Protein Exo2, exocytosis, 722
Protein fraction PF-2, nucleo-cytoplasmic relationship, 911
Protein kinase A, 556
 phosphorylation of ACh receptors, 807
 protein exo2, 722
 serotonine transporters, 774
Protein kinase C, 169, 556, 692, 715, 798, 830, 973, 1033, 1107
 effect of calpains, 408, 410
 palmitoylcarnitine, 1062
 phosphatidylserine synthesis, 1012
 phosphorylation of ACh receptors, 807
 phosphorylation of brain proteins, 1071
 phosphorylation of MVP100, 679
 phosphorylation of synaptic proteins, 1095
 serotonine transporters, 774
 spatial learning paradigm, 507
Protein kinase M, formation by calpains, 411
Protein of the synaptic vesicle membrane, 714
Protein S-100, in multiple sclerosis (MS), 451
Protein synthesis
 axonal segments, 647
 effect of reoxygenation, 465
 in axons, 667
 synaptosomal fractions, 643
de novo Protein synthesis, axonal segments, 649
Protein turnover, 667
Proteins, axonal segments, 647
Proteoglycans, 858, 901
Proteolipid mRNA, 236
Proteolipid protein (PLP), 234, 239
 autoantigen in MS, 114
Proteolysis, myelin basic protein, 1111
Proteolytic activity
 calcium dependent proteolytic activity, 417
 lymphocytes, 417

Proton magnetic resonance spectroscopy (1H-MRS)
   activation-induced glycolysis, 558
   of carnosine, 177
Prozac, 774
Purines, 357
Purinoreceptors, 707
Purkinje cells, brevican immunostaining, 903
Pyramidal cells, 497
Pyridine-3-sulphonic acid, taurine uptake, 961
Pyrithiamine, 737
Pyruvate, 739, 821
   acetylcarnitine, 1048
   metabolic trafficking, 1070
Pyruvate carboxylase, 631
   metabolic trafficking, 1067
Pyruvate dehydrogenase, 631, 1044
   acetylcholine release, 737
   inherited deficiencies, 821, 825
   metabolic trafficking, 1067
Pyruvate utilization, 993
   effect of ethanol, 847

Quaking mouse, 211
Quality assurance, 433
Quinolinic acid, inositol-triphosphate and glutaminase, 390
Quinpirole, 606
Quinuclidinyl benzilate (QNB), 148, 825, 855
Quisqualate, 266
   phosphorylation of brain proteins, 1071

Rab3a, anterograde transport in optic nerve, 1104
Rabies, induction of NO-synthase, 968
Rabphilin, phosphorylation of proteins and dynamin, 1095
Rabphilin-3A, 1104
   anterograde transport in optic nerve, 1104
Raclopride
   N-acetyltransferase, 607
   use in positron emission tomography, 1142
Raphe nucleus, 5-hydroxytryptamine, 189
Rasmussen-encephalitis, 281
Rassmussen disease, 272
Rat glioma BT4Cn, neural cell adhesion molecule (NCAM), 897
Ray electric organ, major vault protein, 679
Reactive blue 2, 358
Reactive oxygen species (ROS), 33, 244
   leech ganglia, 983
   melatonin, 614
   sodium, potassium-ATPase activity, 973
Receptor binding kinetics, interleukin-1, 502
Regeneration
   regulation of neuronal function by astrocytes, 501
   role of ependymin, 872
Reoxygenation, effect on phospholipids, 465
Reperfusion, effect on phospholipids, 465
Reproductive system, effect of xenobiotics, 529
Resuscitation, norepinephrine and epinephrine, 729

Retina, melatonin, 605, 613
Retinoic acid, role in neuron differentiation, 1005
Retzius nerve cell, reactive oxygen species, 983
Reverse transcription-polymerase chain reaction, acetylcholine receptors, 813
Rewarding stimuli, effect on dopamine release, 1157
Rheobase voltage, dystrophic mdx mice, 462
Ribonuclease (RNase), 644
Ribonucleoprotein particles
   in axons, 664
   in nerve terminals, 675
Ribosomal RNA (rRNA), 667
   in axons, 638, 667
Ribosomes, in axonal segments, 659
Ricinus communi agglutinin, 910
Rotenone, beta-oxidation activity, 752
RTI-121, use in positron emission tomography, 1142
Rutilus rutilus, 396
Ryanodine, 363
Ryanodine receptors [RyR], 361
   calcium homeostasis, 402

S-100: *see* Protein S-100
Sambucus nigra agglutinin, 910
Sarco-endoplasmic reticulum Ca^{2+} transport ATPase, 376, 383
Sarcolemma, role of calcium, 461
Sarcoplasmic reticulum, calcium ion release, 461
SB 206553, 70
SCH 23390, 211
   use in positron emission tomography, 1142
Schizophrenia, 268
   glutamine synthase, 444
Schwann cell, 228
Sciatic nerve
   distal motor latency, 487
   microglial reaction, 549
Scorpion toxins, potassium channels, 784
Selectin family
   cell-cell adhesion molecules, 857
   leukocytes, 929
Selegiline, 315
   dopamine, 327
   monoamine oxidase-B inhibitor, 299
   neuroprotective effect, 79
   Parkinson's disease, 331
Senile dementia of Alzheimer's type (SDAT), tacrine and amiridin, 345
Sensomotor cortex, 162, 288
Sensorimotor structures, adenylate cyclase, 919
Sensory-motor deficits, dimethyl sulfoxide (DMSO) and 1,6-fructose, 173
Serine, carnitine transport, 1060
Serine/threonine kinases, 369
Serotonin (5-HT), 148, 187, 201, 469, 773; *see also* 5-Hydroxytryptamine
   amacrine cells, 612
   antidepressant drugs, 197
   depression, 193

Subject Index

Serotonin (5-HT) (cont.)
  effect of xenobiotics, 259
  estrogen-induced differentiation, 525
  MAO-A substrate, 287
  methamphetamine, 327
  rabies virus, 968
  role in pain feeling, 472
Serotonin (5-HT$_{1A}$) agonist, neuroprotection, 402
Serotonin N-acetyltransferase, dopamine receptor blockers, 605
Serotonin receptor, 157
  5-HT$_{1A}$ subtype, 47, 55, 351
  5-HT$_{2C}$ subtype, 69
  5-HT$_{1D}$ subtype, 59
  relation with ACh-receptor, 801
Serotonin reuptake inhibitors (SSRIs), 189, 193, 198
Serotonin transporter, 773
Serum amyloid P, 16
Sexual differentiation, brain, 523
SH-SY5Y cells, 793
SH-SY5Y, retinoic acid induced differentiation, 886
Shaping process, 222
Sialic acid transferases, meningiomas, 917
Sialo-glycosphingolipids, effect of temperature, 926
Sialyltransferases, 886
Signal transduction processes, carnitine, 1056
Signal transduction, calcium, 369
Single photon emission computed tomography (SPECT), dopamine receptors, 1146
Singlet oxygen, effect on melatonin, 599
SK-ER3 cells, 17-beta-oestradiol, 524
SKF-525A, 304
Small proteins, calcium dependent proteolytic activity, 417
SNAP-25: see Synaptosomal-associated protein, 25KDa
SNAPs: see Soluble NSF attachment proteins
SNAREs: see Soluble NSF attachment protein receptors
SOD: see Superoxide dismutase
Sodium, 376, 973, 1060
  AChR-mediated dopamine release, 802
  choline transport, 837
  effect on taurine levels, 945
Sodium azide, 709
  amiridine, 348
  beta-oxidation activity, 752
Sodium, calcium antiporter, role in homeostasis, 402
Sodium channel, polysialic acid (PSA), 885
Sodium channels
  AChR-mediated neurotransmitter release, 803
  polysialic acid, 867
  α-toxins, 783
Sodium cyanide, beta-oxidation activity, 752
Sodium-dependent active transport, 316
Sodium fluorescein, 480
Sodium fluoride, adenylate cyclase, 921
Sodium nitroprusside, 824
  catecholamine release, 988
  generation of nitric oxide, 994

Sodium, potassium-ATPase, 557, 797
  effect of glutamate, 973
Sodium, potassium pump, 222
Solanum tuberosum agglutinin, 910
Soluble NSF attachment protein receptors, exocytosis, 723
Soluble NSF attachment proteins, exocytosis, 723
Somatostatin, adrenal medulla, 595
Soybean agglutinin, 910
Spatial learning, polysialic acid associated with NCAM, 864
Spatial learning paradigm, changes in protein kinase C, acetylcholine receptors and trkB, 508
Spatial orientation task, changes in protein kinase C, acetylcholine receptors and trkB, 508
Spectrin, 408, 417
Sphingolipids
  as second messenger, 260
  multiple infarct dementia, 1027
Sphingomyelin, 762
Sphingomyelin-dependant transduction system, 259
Sphingomyelinase, 168
Sphingosine, 169
Spina bifida aperta, motor behavior, 475
Spinal cord ischemia, MRI study, 184
Spinal cord trauma, neuroprotective agents, 401
Spinal cord, superoxide dismutase, 489
Spinocerebellar atrophy, glutamine synthetase, 443
Spiperone, 211
Spiroperidol, 606
  N-acetyltransferase, 607
Spodoptera frugiperda, 693
SQ22, 536, 545
Squid giant axon
  polysomes, 663
  protein synthesis, 644
Squid photoreceptor neurons, 663
Staurosporine, 1072
Stem cells, Parkinson's disease, 254
Steroid treatment, in multiple sclerosis (MS), 454
Stratum fibrosum et griseum superficiale, γ-aminobutyric acid, 397
Stratum marginale, γ-aminobutyric acid, 397
Stress factor, aminobutyric acid, 1083
Stretch receptor neurons, amiridine and tacrine, 345
Striatonigral neurons, taurine and GABA release, 949, 955
Striatum, dopamine release, 801
Stroke, 83
  mild hypothermia therapy, 92
  nitric oxide (NO), 977
Strychnine, alanine responses, 1123
Subarachnoid space, 423
Substance P, 469
  adrenal medulla, 595
  idiopathic neuralgia, 472

Substantia nigra, 76, 259, 309
   glutathione (GSH), 243
   Parkinson's disease, 249
   taurine and GABA release, 949
Succinate ubiquinone reductase, 752
Succinate, 1079
Sulfamoylbenzo(f)quinoxaline, 734
Sulpiride, N-acetyltransferase, 607
Superoxide anions, n-3 fatty acids, 133
Superoxide dismutase, 169, 304, 323, 983
   amyotrophic lateral sclerosis (ALS), 485
   lipid peroxidation, 489
   sodium, potassium-ATPase activity, 973
Superoxide radicals, methamphetamine, 323
Suppressor T cells, n-3 fatty acids, 131
Suramin, 709
   ecto-ApnA hydrolase, 698
   ecto-nucleotidases, 700
Sympathetic ganglion neurons, acetylcholine, 843
Synapsin, axonal transport and Rab3a, 1102
Synapsins, phosphorylation of synaptic proteins, 1095
Synaptic membrane, 857
Synaptic proteins
   GAP-43, 43
   rab-3a, 43
Synaptic structures, 223
   amygdala, 492
   nucleus caudatus, 492
Synaptic terminals, diadenosine polyphosphates (ApnA), 695
Synaptic vesicle cycling, 1095
Synaptobrevin, exocytosis, 723
Synaptophysin, 43
   anterograde transport in optic nerve, 1104
Synaptosomal fractions, protein synthesis, 643
Synaptosomal membranes, neutral proteinases, 343
Synaptosomal-associated protein, 25 kDa
   acetylcholine release, 714
   exocytosis, 723
Synaptosomes
   choline transport, 838
   torpedo electric organ, 676
Synaptotagmin, 43
Synaptotagmin I, anterograde transport in optic nerve, 1104
Syntaxin, 714
   exocytosis, 723

TBARS: see Thiobarbituric acid-reactive substances
T cell-induced EAE, macrophage depletion, 106
Tacrine, senile dementia of Alzheimer's type (SDAT), 345
Tacrine-like compounds, endplate potentials, 715
Taltrimide, taurine agonist, 942
Tartrate, effect on acid phosphatase, 481
Taurine, 299, 529, 959
   effect glutamate and GABA, 963
   neuroprotection, 939
   regulation of osmotic pressure, 943
   taurine and GABA release, 949, 955

Teleosts, retinotectal projection, 870
Temperature of the brain, regulation mechanisms, 97
Temperature, effect on gangliosides, 925
Tenascin-R, N-glycosylation sites, 929
Tenebrio molitor, carnitine, 1039
Tenidap, 30
Tetra-butylhydroxide, potassium transport, 983
Tetrapeptide tuftsin (Thr-Lys-Pro-Arg), 287
Tetrodoxin, 803, 844
   taurine release, 950
Thalamic nuclei, NGF and trkA expression, 513
Thapsigargin, 384, 1012
Thermoregulation, n-3 fatty acids, 132
Thiamine deficiency
   acetylcholine release, 737
   acetylcholine synthesis, 821
   effects on ACh-synthesis, 824
Thiamine pyrophosphate, 738
Thiamine pyrophosphate-dependent enzymes, acetylcholine release, 737
Thiamine triphosphates, 741
Thiobarbituric acid-reactive substances (TBARS), 33, 614, 979
Thiopental, 278
Thiosemicarbazide, 978
5'-O-3-Thiotriphosphate, ecto-nucleotidases, 700
Threohydroxyaspartate, 556, 566
Thromboxane A_2, vasoconstrictor and thrombosis, 129
Thromboxane A_3, 130
Thromboxane B_2, interferon-γ, 544
Thyroid hormones, TRH-degrading ectoenzyme, 692
Thyrotropin-releasing hormone (TRH), degrading ectoenzyme, 691
TNF: see Tumor necrosis factor
Tocopherol, 599, 761
   liver enzymes, 333
Torpedo electric organ, ecto-nucleotidases, 696
Torpedo marmorata, vault proteins, 675
α Toxins, sodium channels, 783
β Toxins, activation of sodium channels, 784
Trace protein assay, 424
Trafficking in fetal brain, cerebral glucose, glutamate and glutamine, 631
Transcription factor NFκB, 260
Transforming growth factor beta (TGF-beta), 116, 121, 970
   effect on NO production in, 970
Transgenic mice, mature NFH-lacZ, 229
Transgenic G93A mice, muscle action potential, 486
Transplantation, dopamine cells, 249
Transport of acyl moieties, carnitine, 1047
Transport system y+, amino acids, 626
Transthyretin, 425
Transverse tubules, calcium ion release, 461
Trauma, nitric oxide (NO), 977
TRH: see Thyrotropin-releasing hormone
Tricarboxylic acid (TCA) cycle
   glucose, glutamate, and glutamine trafficking, 631
   metabolic trafficking, 1065

Tricyclic antidepressants, 193, 774
  phospholipase A2, 1037
Triethyltin, n-3 fatty acids, 133
Trigeminal nerve, 469
Triglyceride fatty acid turnover, carnitine palmitoyltransfe 1053
Triglycerides, multiple infarct dementia, 1027
Trimetaphan, dopamine release, 802
Trimethaphosphatase, 479
4-N-Trimethylammonium-3-hydroxybutyric acid: see Carnitine
N,N,N-Trimethyltaurine, taurine uptake, 961
Trimethyltin, NGF-expression in astrocytes, 497
Triphosphate, pancreastatin, 591
Transfer RNA, in axons, 638
Tryptophan hydroxylase, 605
Tryptophan, 201, 605
  in depression, 187
Tubulin, 673
  acetylation, 1056
  synthesis in axons, 673
Tuftsin, 288
Tumor necrosis factor (TNF), 27, 130, 165
Tumor necrosis factor α (TNF-α), 121, 259
  selective BBB-damage, 142
Tumor necrosis factor β (TNF-β), 121
Tumor necrosis factors, 536
Tumor-promoting phorbol ester, phosphorylation of brain proteins, 1071
Tyrosine hydroxylase, estrogen-induced differentiation, 525
Tyrosine protein kinase, phosphorylation of MVP100, 680

Ubiquinol cytochrome c reductase, 752
Unitas multiplex, 221
Uridine monophosphate (UMP), effect on platelet-activating factor, 1032
Uridine triphosphate, 701
UTP: see Uridine triphosphate
UV-A light, effect on N-acetyltransferase, 606

Valproic acid, neural tube defects, 295
VAMP/synaptobrevin, acetylcholine release, 714
Vanadate, 387
Vascular cell adhesion molecule-1 (VCAM-1), 122, 858
Vascular dementia, lipid spectrum and lysosomal enzymes, 1025
Vasoactive intestinal polypeptide
  adrenal medulla, 595
  glycogenolysis, 555

Vasogenic brain oedema, histamine, 480
Vasopressin mRNA, in axons, 664
Vault protein, homolog of lung resistance-related protein, 675
Vault, ribonucleoprotein particle, 675
Venlafaxine, 194
Ventral tegmental area, injection of drug by microdialysis, 1163
Veratridine, 803
Versican, proteoglycans, 901
Vigabatrine, 278
Vimentin, 239
Vinculin, NCAM, 899
VIP: see Vasoactive intestinal polypeptide
Viral infections, induction of NO-synthase, 968
Vitamin A metabolite, retinoic acid, 1005
Vitamin E, 489
Voltage-dependant Ca^{2+} channels, effect of nimodipine, 351

Wallerian degeneration, microglial reaction, 549
WAY100635, use in positron emission tomography, 1143
Western blot
  calcium-binding proteins, 384
  InsP3 receptor, 376
  glutamine synthase, 444
  MVP100, 676
  proteins of neostriatum, 390
Wheat germ agglutinin, 910
Wistar rats, 738
Wolfgram protein (CNP), 239

Xenobiotics
  hypothalamic circadian rythm, 529
  Parkinson's disease, 246
Xenografts, Parkinson's disease, 254
Xenopus laevis oocytes, neutral amino acid transport, 620
Xenopus oocytes, alanine expression, 1129

Yeast, 778
  neural cell adhesion molecule (NCAM), 891
YM022, cholestokinin type B receptor, 1089
YOYO-1, nucleic acid binding, 668

Zinc metalloenzymes, 683
Zinc metallopeptidase, 692
Zinc, binding to ependymin, 871
Zinc-proteases, 714

MIX
Papier aus verantwortungsvollen Quellen
Paper from responsible sources
FSC® C105338

If you have any concerns about our products,
you can contact us on
ProductSafety@springernature.com

In case Publisher is established outside the EU,
the EU authorized representative is:
**Springer Nature Customer Service Center GmbH
Europaplatz 3, 69115 Heidelberg, Germany**

Printed by Libri Plureos GmbH
in Hamburg, Germany